AF598224

NORTH EAST WALES INSTITUTE

3 5196 00126433 7

FOOD COLLOIDS

other AVI books

Food Science

BASIC FOOD CHEMISTRY Cloth and Soft Cover *Lee*
BEVERAGES: CARBONATED AND NONCARBONATED *Woodroof and Phillips*
CARBOHYDRATES AND THEIR ROLES *Schultz, Cain and Wrolstad*
ECONOMICS OF NEW FOOD PRODUCT DEVELOPMENT *Desrosier and Desrosier*
ELEMENTARY FOOD SCIENCE Cloth and Soft Cover *Nickerson and Ronsivalli*
EGG SCIENCE AND TECHNOLOGY *Stadelman and Cotterill*
ENCYCLOPEDIA OF FOOD ENGINEERING *Hall, Farrall and Rippen*
ENCYCLOPEDIA OF FOOD TECHNOLOGY *Johnson and Peterson*
DIETARY NUTRIENT GUIDE Soft Cover *Pennington*
FABRICATED FOODS *Inglett*
FISHERY BY-PRODUCTS TECHNOLOGY *Brody*
FLUID MILK INDUSTRY, 3RD EDITION *Henderson*
FOOD CHEMISTRY *Aurand and Woods*
FOOD CHEMISTRY Soft Cover *Meyer*
FOOD COLORIMETRY: THEORY AND APPLICATIONS *Francis and Clydesdale*
FOOD ENZYMES *Schultz*
FOOD PROCESS ENGINEERING Cloth and Soft Cover *Heldman*
FOOD PRODUCTS FORMULARY, VOL. 1 *Komarik, Tressler and Long*
VOL. 2 *Tressler and Sultan* VOL. 3 *Tressler and Woodroof*
FOOD SCIENCE, 2ND EDITION *Potter*
FOOD TEXTURE *Matz*
FREEZING PRESERVATION OF FOODS, 4TH EDITION, VOLS. 1, 2, 3, AND 4 *Tressler, Van Arsdel and Copley*
FUNDAMENTALS OF DAIRY CHEMISTRY, 2ND EDITION *Webb, Johnson and Alford*
FUNDAMENTALS OF FOOD ENGINEERING, 2ND EDITION *Charm*
MICROBIOLOGY OF FOOD FERMENTATIONS *Pederson*
MICROWAVE HEATING, 2ND EDITION *Copson*
PRACTICAL FOOD MICROBIOLOGY AND TECHNOLOGY, 2ND EDITION *Weiser, Mountney and Gould*
PRACTICAL MEAT CUTTING AND MERCHANDISING, VOLS. 1 AND 2 Soft Cover *Fabbricante and Sultan*
PROTEINS AND THEIR REACTIONS *Schultz and Anglemier*
QUALITY CONTROL IN FOOD SERVICE Cloth and Soft Cover *Thorner and Manning*
SAFETY OF FOODS *Graham*
SNACK FOOD TECHNOLOGY *Matz*
SUGAR CHEMISTRY *Shallenberger and Birch*
TECHNOLOGY OF WINE MAKING, 3RD EDITION *Amerine, Berg and Cruess*

FOOD COLLOIDS

Edited by HORACE D. GRAHAM, Ph.D.
Professor of Chemistry
Department of Chemistry
University of Puerto Rico
Mayaguez, Puerto Rico

N.E.W.I. DEESIDE
LIBRARY
CLASS No. 541·18:664
ACCESS No. K39667

THE AVI PUBLISHING COMPANY, INC.
WESTPORT, CONNECTICUT

LIBRARY
ATHROFA GOGLEDD DDWYRAIN CYMRU
THE NORTH EAST WALES INSTITUTE
of higher education CONNAH'S QUAY
DEESIDE, CLWYD CH5 4BR

© *Copyright 1977 by*
THE AVI PUBLISHING COMPANY, INC.
Westport, Connecticut

All rights reserved. No part of this work covered by the copyright hereon may be reproduced or used in any form or by any means—graphic, electronic, or mechanical, including photocopying, recording, taping, or information storage and retrieval systems—without written permission of the publisher.

Library of Congress Catalog Card Number: 76–12722
ISBN-0-87055-201-5

Printed in the United States of America

Contributors

IAN W. COTTRELL, Ph.D., Assistant Technical Director, Kelco Company, San Diego, California

A. JEROME GANZ, Research Food Scientist, Food and Fragrance Development Department, Food Technology Center, Hercules Incorporated, Middletown, New York

HORACE D. GRAHAM, Ph.D., Professor of Chemistry, Department of Chemistry, University of Puerto Rico, Mayaguez, Puerto Rico

ERWIN HECKMAN, Group Leader, Special Products Laboratory, Celanese Corporation, Stein-Hall Division, Long Island City, New York

KENNETH S. KANG, Ph.D., Section Head, Biochemical Development, Kelco Company, San Diego, California

PETER KOVACS, Section Head, Food Development and Regulatory Compliance, Kelco Company, San Diego, California

CHIFA F. LIN, Ph.D., Senior Research Chemist, Food Ingredient Section, Stauffer Chemical Company, Eastern Research Center, Dobbs Ferry, New York

WILLIAM A. MEER, Ph.D., Director of Research, Meer Corporation, North Bergen, New Jersey

ARTHUR L. MOIRANO, Technical Director, Marine Colloids, Inc., Springfield, New Jersey

DENNY B. NELSON, Ph.D., Senior Research Chemist, Products Research and Development Division, Sunkist Growers, Inc., Ontario, California

ARTHUR F. NOVAK, Ph.D., Professor of Food Science and Department Head, Department of Food Science, Louisiana State University, Baton Rouge, Louisiana

HERBERT W. OCKERMAN, Ph.D., Professor of Animal Science, Department of Animal Science, The Ohio State University, Columbus, Ohio

RAMU M. RAO, Ph.D., Professor of Food Science, Department of Food Science, Louisiana State University, Baton Rouge, Louisiana

C. J. B. SMIT, Ph.D., Professor of Food Science, Department Head and Division Chairman, Department of Food Science, University of Georgia, Athens, Georgia

DURWARD A. SMITH, Research Associate, Department of Food Science, Louisiana State University, Baton Rouge, Louisiana

FRANK SOSULSKI, Ph.D., Professor of Crop Science, Department of Crop Science, University of Saskatchewan, Saskatoon, Saskatchewan, Canada

WILLIAM J. STADELMAN, Ph.D., Professor of Animal Sciences, Department of Animal Sciences, Purdue University, West Lafayette, Indiana

ROBERT McL. WHITNEY, Ph.D., Professor of Food Chemistry, Department of Food Science, University of Illinois, Urbana, Illinois

RALDON R. WILES, Senior Chemist, Pectin Development Section, Lemon Products Division, Sunkist Growers, Inc., Corona, California

Preface

All natural products used as foods contain materials which exhibit colloidal properties. Processed and prepared foods contain even more of such materials since various food additives are colloidal in nature. In addition to their essentiality as components of a well-balanced, nutritious diet, several proteins, for example, wheat gluten, confer highly desirable qualities on products into which they are incorporated.

The proteins of milk interact with other food components under varying conditions to form complexes of a desirable or undesirable nature. Likewise, various and diverse carbohydrates are important naturally occurring food colloids. Their use for the emulsification and stabilization of a myriad of food preparations is well known and, through their colloidal behavior, they exert their beneficial roles.

Several semi-synthetic colloidal substances are also allowed for use in food preparations. The cellulose ethers are the primary examples and their use competes strongly with that of naturally occurring hydrocolloids such as carrageenan, tragacanth, etc.

Recently, microbial polysaccharides have received much attention as food colloids and one, xanthan gum (Keltrol), has been approved for use in various food preparations.

Although several books have been published describing the plant hydrocolloids, a comprehensive treatment which includes the animal and plant proteins also is considered desirable. In addition, previous treatments have not emphasized the colloidal aspects. This volume on food colloids attempts to present an overview of those substances which occur naturally in, or are added to, foods and which when added to water exhibit the classical phenomena of colloids, namely dispersibility and the Tyndall effect.

The text is intended to serve fourth year students in Chemistry, Biology and Food Science and Food Technology. Graduate students in these disciplines should find the text useful, also. In addition, it should serve as reference for professional workers in these areas.

HORACE D. GRAHAM

February 1976

Acknowledgements

Sincere thanks are due to each of the contributors to this volume for his time, patience and profound cooperation. To the various publishers who have granted permission for the use of data, graphs and other materials, I wish to extend my gratitude. To all others who contributed in typing, proofreading etc. and to Dr. Donald K. Tressler, President of the Avi Publishing Company, for his understanding and patience, I am very grateful.

Contents

CHAPTER 1

Robert McL. Whitney

Chemistry of Colloid Substances: General Principles

INTRODUCTION

Nearly all food products contain substances in the colloidal state and thus colloidal phenomena are of considerable importance to food scientists because they are intimately related to the changes that occur in food products during processing.

The existence of the colloidal state was first recognized by Thomas Graham (1850), the father of colloid chemistry, when he differentiated between crystalloids and colloids by their rates of diffusion: crystalloids such as salts and sugars diffusing rapidly and colloids such as proteins and polysaccharides diffusing slowly. He attributed the difference to their particle size. For many years it was assumed that colloids were distinct substances until Ostwald (1907) was able to demonstrate that colloids were biphasic systems in which the colloidality was a function of the degree of subdivision of one substance in another rather than the chemical nature of the material concerned. He illustrated this principle by obtaining a colloidal dispersion of NaCl, a crystalloid, in benzene. Von Weimarn (1908) further confirmed this concept of the colloidal state by demonstrating that crystalloids such as $BaSO_4$ could exist either in true solution, colloidal suspension, amorphous or crystalline precipitates dependent upon the conditions such as concentration, temperature, rate of nucleation and rate of crystallization. Therefore, while for some substances the molecular weight is sufficiently large for their dispersions to always possess colloidal properties, it is now generally believed that all substances can exist in the colloidal state provided the conditions are suitable. Therefore, we can define the colloidal state as a dispersion of one substance in another of such a degree of subdivision as to possess properties different from those of the substance either en masse or in true solution.

Among the properties that are characteristic of the colloidal state are the slow diffusion of the dispersed phase, the large surface to volume ratio, Brownian movement, the ability to scatter light or Tyndall phenomenon, high interfacial energy, and high adsorptive capacity. These properties are a function of the particle size and therefore delineate the colloidal range. While the limits of this range vary with the substance under investigation and the property of interest, we generally considered this to be from 1 mμ, which is somewhat greater than the diameter of simple molecules and ions in solution, to 0.5 μ, which is near the lower limit of the resolving power of an ordinary microscope. Numerous exceptions to these limits exist as, for example, the existence of microscopically visible casein particles in HTST-sterilized condensed milk.

The various colloidal systems can be classified according to the physical state of the two phases present: the dispersed phase, the substance which is subdivided, and the dispersing medium, the substance in which the dispersed phase is subdivided. Eight types of systems are recognized (Table 1.1).

Of particular interest in this text and therefore in this chapter, however, is the chemistry of those substances that exist in the colloidal state due to their being macromolecules in their native state and therefore our discussion of colloidal phenomena will be largely limited to those related to the proteins and hydrocolloids.

CONFIGURATION OF COLLOIDAL SUBSTANCES

The phenomena associated with the colloidal behavior of macromolecules and aggregates of molecules are, in many cases, intimately associated with their primary structure and the configuration that the macromolecule assumes as dictated by this structure, its environment and the treatment to which it has been subjected. Therefore, before discussing their colloidal behavior, it is necessary to elucidate their structure and the factors involved in their configuration.

Proteins

All monomeric protein molecules consist of a specific combination of amino acids arranged in a genetically determined order and linked together head to tail in a chain by the covalent peptide bonds which

TABLE 1.1

CLASSIFICATION OF COLLOIDAL SYSTEM

Dispersed Phase	Dispersing Medium	Name	Example
Liquid	gas	aerosol, fog	aerosol sprays
Solid	gas	aerosol, smoke	spray dried milk
Gas	liquid	lyosol, foam	whipped cream
Liquid	liquid	lyosol, emulsion	mayonnaise
Solid	liquid	lyosol, sol	whey
Gas	solid	solid froth	bread
Liquid	solid	liquid inclusion	water in butter
Solid	solid	solid sols	candies

arise from the elimination of the elements of water from the carboxyl group of one amino acid and the α-amino group of the next (Fig. 1.1). While these polypeptide chains may consist of hundreds of amino acids, they are not polymers of random chain length but for each protein the chain length is fixed. Due to presence of the amino acid cystine in some proteins, they may consist of more than one polypeptide chain held together by a disulfide bridge. The proteins can be divided into two major classes on the basis of their composition: simple and conjugated. Simple proteins are those which on hydrolysis yield only amino acids, while conjugated proteins yield not only amino acids but also other organic or inorganic components called prosthetic groups. The conjugated proteins are classified ac-

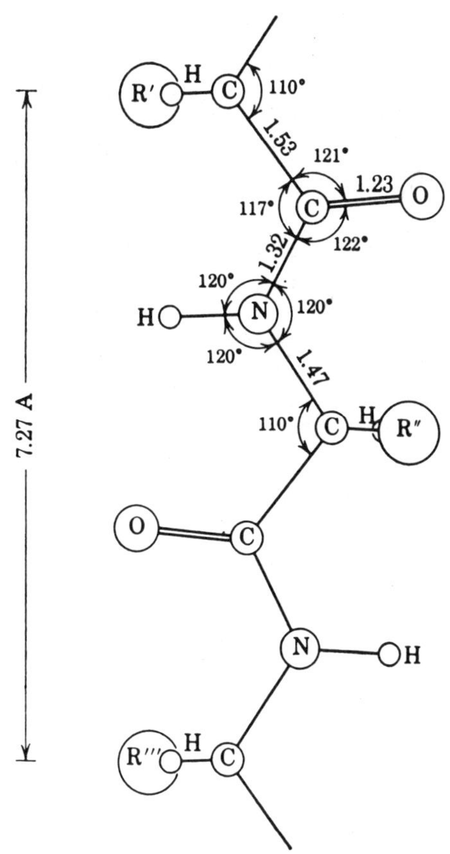

From Pauling et al. (1951)

FIG. 1.1. EXTENDED PEPTIDE CHAIN INDICATING BOND DISTANCES AND ANGLES

cording to the nature of the prosthetic group such as liproproteins which contain lipids, phosphoproteins containing phosphate, metaloproteins containing metals, glycoproteins containing carbohydrates and nucleoproteins containing nucleic acids.

The 20 amino acids commonly encountered in proteins are listed in Table 1.2. However, a number of other amino acids have been isolated from the hydrolysates of a few special proteins such as 4-hydroxyproline and 5-hydroxylysine from collagen and desmosine and isodesmosine in elastin. The amino acids except for glycine are optically active and, as present in proteins, possess an absolute configuration related to L-glyceraldehyde and therefore are considered L-α-amino acids regardless of the actual optical rotatory power.

TABLE 1.2

THE COMMON NATURALLY OCCURRING AMINO ACIDS IN PROTEIN

Amino Acid	Symbol	Structural Formula
Alanine	Ala	$CH_3-CH(NH_2)-COOH$
Arginine	Arg	$H_2N-C(=NH)-NH-(CH_2)_3-CH(NH_2)-COOH$
Asparagine	Asn	$H_2N-C(=O)-CH_2-CH(NH_2)-COOH$
Aspartic acid	Asp	$HO-C(=O)-CH_2-CH(NH_2)-COOH$
Asn + Asp	Asx	
Cysteine	½ Cys	$HS-CH_2-CH(NH_2)-COOH$

TABLE 1.2 (*Continued*)

Amino Acid	Symbol	Structural Formula
Cystine	Cys	$HO-C(=O)-CH(NH_2)-CH_2-S-S-CH_2-CH(NH_2)-COOH$
Glutamine	Gln	$H_2N-C(=O)-(CH_2)_2-CH(NH_2)-COOH$
Glutamic acid	Glu	$HO-C(=O)-(CH_2)_2-CH(NH_2)-COOH$
Gln + Glu	Glx	
Glycine	Gly	$H-CH(NH_2)-COOH$
Histidine	His	$HC{=}C(-N{=}CH-NH-)-CH_2-CH(NH_2)-COOH$ (imidazole ring: $HC{=}C$, N, N–H, C–H)
Isoleucine	Ile	$C_2H_5-CH(CH_3)-CH(NH_2)-COOH$
Leucine	Leu	$CH_3-CH(CH_3)-CH_2-CH(NH_2)-COOH$
Lysine	Lys	$H_2N-(CH_2)_4-CH(NH_2)-COOH$

TABLE 1.2 (*Continued*)

Amino Acid	Symbol	Structural Formula
Methionine	Met	$CH_3{-}S{-}(CH_2)_2{-}CH(NH_2){-}COOH$
Phenylalanine	Phe	$C_6H_5{-}CH_2{-}CH(NH_2){-}COOH$
Proline	Pro	$H_2C{-}CH_2{-}CH(COOH){-}NH{-}CH_2$ (ring: $H_2C{-}CH_2$, $H_2C{-}N(H)$, $C(H){-}COOH$)
Serine	Ser	$HO{-}CH_2{-}CH(NH_2){-}COOH$
Threonine	Thr	$CH_3{-}CH(OH){-}CH(NH_2){-}COOH$
Tryptophan	Trp	indole ring: $C{-}CH_2{-}CH(NH_2){-}COOH$, $C{=}CH$, $N{-}H$
Tyrosine	Tyr	$HO{-}C_6H_4{-}CH_2{-}CH(NH_2){-}COOH$
Valine	Val	$CH_3{-}CH(CH_3){-}CH(NH_2){-}COOH$

Since proteins consist of a large number of amino acids, the number of possible conformations to be expected would be extremely large. However, for most proteins there exists one configuration that has the least energy and therefore is the most stable under normal conditions of temperature, ionic strength and pH and in the absence of denaturing agents. This is called the native conformation. From X-ray diffraction studies Pauling *et al.* (1951) deduced the structure and dimensions of the peptide bond in backbone of the protein (Fig. 1.1). Their most important observation was that the C—N bond which is shorter than most C—N single bonds possesses some double bond character and thus some rigidity. Thus the four atoms of the peptide bond and the two α-carbons lie in a single plane. Therefore, the protein molecule is more rigid than would be expected since every third bond in the peptide chain is rigid and not free to rotate.

The forces involved in determining the conformation that a given protein will assume in a specific environment are individually weak noncovalent forces such as ionic interactions, hydrogen bonding, hydrophobic interactions and steric effects. Ionic interactions result from the presence of charged groups on the protein such as the positively charged (cationic) groups: the ϵ-amino group on lysine residues, the imidazolium group of a histidine residue and the guanadinium group of arginine; and the negatively charged (anionic) groups: the γ and δ carboxyl groups of aspartic and glutamic acid residues, the sulfhydryl group of cysteine and the phenolic hydroxyl group of tyrosine. These forces can be both attractive and repulsive since two like charges will tend to repel each other while two unlike charges will tend to form ion pairs. The dipole nature of an O—H or N—H bond results in what is called hydrogen bonding between these groups and the electronegative atoms in a molecule such as oxygen, nitrogen and fluorine. Thus, if an O—H or N—H group falls in the vicinity of an electronegative atom, the partially positively charged hydrogen will be attracted by the pair of unshared electrons on the electronegative atom forming a bond the length of which will depend upon the structural geometry and electron distribution in the molecules involved. For example in ice, tetrahedral bonding occurs with each water molecule bound to four others by a hydrogen bond 1.77Å in length. The hydrogen bond also possesses a high degree of directionality which is the result of arrangement of the bonding orbitals of the hydrogen and electronegative atoms. While a single hydrogen bond is very weak, the energy required to separate two structures held together by a number of hydrogen bonds is much greater than the sum of the bond energies of the individual hydrogen bonds. Thus such configurations can exist even in water dispersion even though the water itself is capable of bonding to these groups. Due to the presence of nonpolar R-groups on proteins such as those present in alanine, valine, leucine, isoleucine, proline and phenylalanine, the proteins possess regions on the peptide chain which are hydrophobic in nature. To insert these nonpolar groups into water would require energy since the surrounding water molecules would be forced to form a more regular arrangement than in pure water or, in other words, one with less entropy or randomness. The more stable position would be that in which these hydrophobic groups would be brought close together providing a nonpolar environment and thus allowing the water molecules to assume a state of maximum randomness or entropy. Such an orientation is called hydrophobic interaction and results in considerable stability for the configuration that allows it to occur. In contrast with hydrogen bonding, hydrophobic interaction has little directionality. The size and shape of the R-group of the amino acids also influence the possible configurations that the protein can assume, for example, the presence of a proline residue with its five membered amino acid ring structure lends additional rigidity to the molecule.

Of the natural proteins the first to be investigated by X-ray diffraction were the fibrous proteins. Three different types of primary spacings or repeating units were observed along the fiber axis: 5.0–5.5Å as in α-keratin, 3.5Å as in silk fibroin and 2.86Å as in collagen. Pauling *et al.* (1951) concluded that the 5.0–5.5Å spacing could be explained by a right-hand helical structure called the α-helix with about 3.7 amino acid residues per turn (Fig. 1.2). This system is maximally hydrogen bonded since the planes of the peptide bonds lie in the cylindrical surface of the helix allowing each hydrogen attached to the nitrogen atom of the peptide bond to form a hydrogen bond with the oxygen of the carbonyl group of the fourth amino acid residue behind it. In such a structure the repeating unit, a single turn of the helix, would measure 5.4Å in the direction of the axis. The R-groups of the amino acids would extend outward from the helix. The 3.5Å spacing can be explained by the β-configuration or pleated sheet (Fig. 1.3) in which the peptide bonds lie in the planes of the sheet and the R-groups project above and below the sheet. Stability is secured by maximal interchain hydrogen bonds between all the N—H of the peptide bond of one chain to the C=O of the peptide bond of the adjacent chain. In some protein the polypeptide chains are parallel and in other adjacent chains run antiparallel or in the opposite direction. The third major type of fibrous protein with a primary spacing of 2.86Å, as encountered in collagen, can be accounted for by a triple or three-stranded helix consisting of three polypeptide chains twisted together.

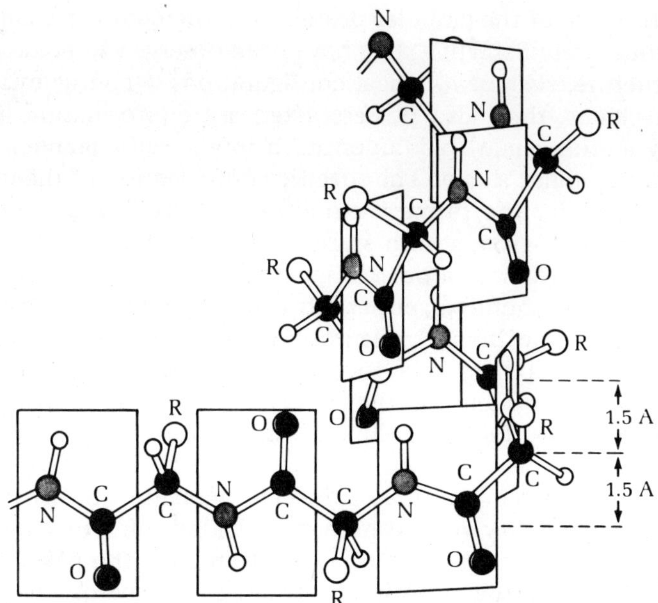

From Leninger (1970)

FIG. 1.2. THE α-HELIX FORMATION OF THE RIGHT-HANDED α-HELIX. THE PLANES OF THE RIGID PEPTIDE BONDS ARE PARALLEL TO THE LONG AXIS OF THE HELIX FACILITATING HYDROGEN BOND FORMATION

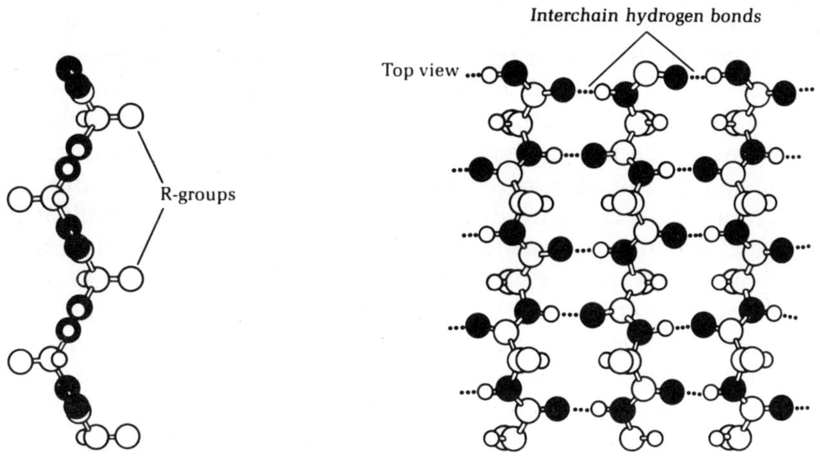

From Leninger (1970)

FIG. 1.3. EDGE AND TOP VIEW OF THE β-CONFORMATION OF POLYPEPTIDE CHAINS WITH THE ANTIPARALLEL ARRANGEMENT OF ADJACENT CHAINS NOTE THE MAXIMAL HYDROGEN BONDING TO FORM A PLEATED SHEET

The structure of the globular proteins is more complex than that of the fibrous proteins. While they have been observed to possess secondary structure, such as α-helical configurations, β-configurations and random chains, they also possess a tertiary conformation in which the polypeptide chain folds upon itself in a specific manner to yield an overall globular shape. The specific conformation of the molecule is determined by the primary structure and the forces encountered by the various R-groups both steric effects and bonding forces. The native configuration of a protein in an aqueous system is thus the result of hydrogen bonding, either intrachain as in the α-helical regions or interchain as in the β-configuration regions, and the tendency to bend at certain points due to steric or charge effects, or hydrophobic interaction such that the molecule assumes a position of minimum energy. Disulfide cross linkages also add to the conformational stability.

Changes in the temperature or the environment of the protein may decrease the stability of the native configuration. For example, in many globular proteins, increasing or decreasing the pH of the system will increase the negative or positive charges on the protein, respectively, due to the ionization of the anionic and cationic groups present. Thus the presence of a high density of positive or negative charge in the given region of the protein will result in repulsive forces which overcome the native configuration and cause the molecule to unfold. This rupture of secondary and tertiary structure is called denaturation. Certain other substances such as urea or guanidine-HCl produce the same effect. Depending upon the protein involved removal of the denaturing agent or adjustment of the pH may again restore the native configuration. Proteins have a characteristic temperature of maximum stability usually at refrigerator temperature. As the temperature increases, most proteins become unstable and unfold at temperatures above 40–50°C. The higher the temperature or the longer the time of exposure, the more random the possible configuration of the unfolded chain. Thus upon cooling rapidly after extended heating, the protein usually does not return to the native configuration. However, if slowly cooled at room temperature to allow sufficient time for the molecules to assume the configuration of minimum energy, it will often return to its native conformation.

In addition to the secondary and tertiary structures present in all proteins, some proteins called oligomeric proteins exist in the native state as complexes of two or more polypeptide chains. The polypeptide chains involved may be the same or different. The same forces that are responsible for the secondary and tertiary structure of a protein are also active in the formation of this quarternary structure, and as in the case of tertiary structure, there exists a configuration of minimum energy for the complex in the native state. Changes in temperature and pH as well as the presence of denaturing agents such as urea can cause the complex to dissociate into its component polypeptide chains.

Polysaccharides

In contrast with proteins, the primary structure of which consists of as many as 20 different amino acids arranged in a characteristic sequence of fixed length in polypeptide chains of 50 or more amino acid residues, the polysaccharides consist of polymers of varied length of either a single monosaccharide or its derivative, such as D-glucose in starch (homopolysaccharide), or of a very few monosaccharides or their derivatives, such as chondroitin which contains D-glucuronic acid and *N*-actetyl-D-galactosamine, (heteropolysaccharides).

Of the basic monosaccharides, building block of the polymers, the most prevalent is D-glucose, but polysaccharides of D-mannose, D-fructose, D-galactose, D-xylose, L-arabinose and L-fucose are common. Derivatives of monosaccharides, such as D-glucosamine, D-galactosamine, D-glucuronic acid, D-galacturonic acid and its methyl ester, D-mannuronic acid, D-galactose sulphate, 3, 6 anhydro-D- and L-galactoses and their sulphates have been found in natural polysaccharides. These monosaccharides are either pentoses with a 5-carbon chain or hexoses with 6-carbons or their derivatives (Table 1.3). The monosaccharides exist in the polysaccharides in the form of 6-membered rings, pyranoses, as the D-glucose in starch, or a 5-membered ring, furanoses, as the D-fructose in inulin (Fig. 1.4). The Haworth pro-

TABLE 1.3

THE COMMON MONOSACCHARIDES FOUND IN POLYSACCHARIDES

Pentoses	
D-Xylose	L-Arabinose
CHO	CHO
|	|
HCOH	HCOH
|	|
HOCH	HOCH
|	|
HCOH	HOCH
|	|
CH_2OH	CH_2OH

TABLE 1.3 (*Continued*)

Hexoses

D-Glucose	D-Mannose	D-Galactose	D-Fructose	L-Fucose
CHO	CHO	CHO	CH_2OH	CHO
HCOH	HOCH	HCOH	C=O	HOCH
HOCH	HOCH	HOCH	HOCH	HCOH
HCOH	HCOH	HOCH	HCOH	HCOH
HCOH	HCOH	HCOH	HCOH	HOCH
CH_2OH	CH_2OH	CH_2OH	CH_2OH	CH_3

Uronic Acids

D-Glucuronic Acid	D-Mannuronic Acid	D-Glacturonic Acid
CHO	CHO	CHO
HCOH	HOCH	HCOH
HOCH	HOCH	HOCH
HCOH	HCOH	HOCH
HCOH	HCOH	HCOH
COOH	COOH	COOH

Other Hexose Derivatives

D-Glucosamine	D-Galactosamine	D-Galactose-6 Sulphate	3,6-Anhydro-D-galactose
CHO	CHO	CHO	CHO
$HCNH_2$	$HCNH_2$	HCOH	HCOH
HOCH	HOCH	HOCH	CH (—O—, bridge to C-6)
HCOH	HOCH	HOCH	HOCH
HCOH	HCOH	HCOH	HCOH
CH_2OH	CH_2OH	$CH_2OSO_3^-$	CH_2 (—O—, bridge to C-3)

jection formulas shown in this figure are somewhat misleading as they suggest a planar ring structure. Actually the pryanose ring may exist in several configurations: 2-chair, 6-boat and 6-skew forms. The chair form indicated is the most stable and therefore predominates in aqueous solution of hexoses while in the pentoses there is less tendency for this form to be the most stable and in some such as D-arabinose the other chair form predominates (Durette and Horton

α-D-GLUCOPYRANOSE

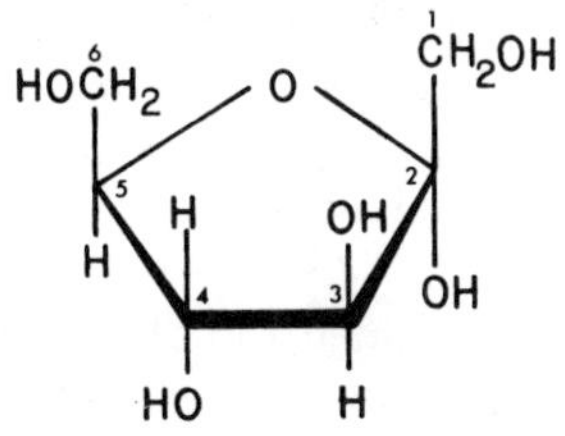

β-D-GLUCOPYRANOSE

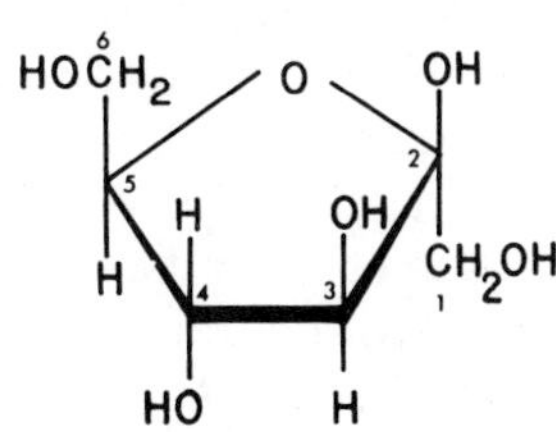

α-D-FRUCTOFURANOSE

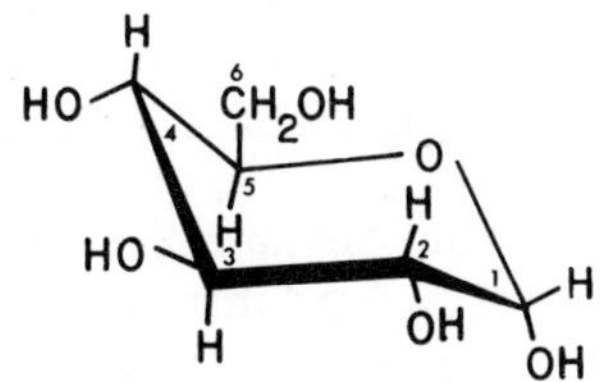

β-D-FRUCTOFURANOSE

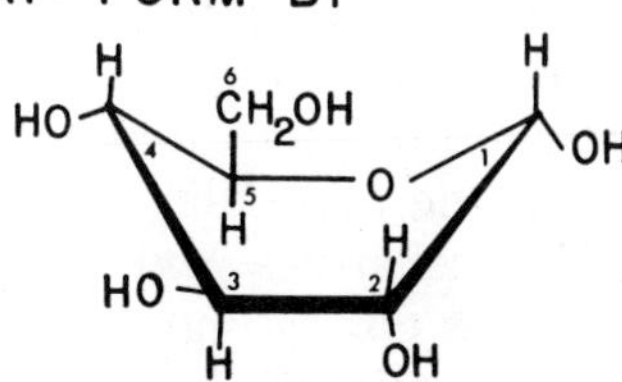

CHAIN FORM CI

BOAT FORM BI

FIG. 1.4. HAWORTH PROJECTION FORMULAE AND CONFORMATIONAL STRUCTURES OF HEXOSES

1971). However, the actual shape assumed in a polysaccharide will depend upon the structure of the particular monosaccharide involved and its environment. The formation of the ring results in the creation of another asymmetric carbon atom, the 1 carbon of the aldoses and the 2 carbon in the ketoses. The two optical isomers, anomers, resulting from this ring closure are designated by the Greek letters α and β (Fig. 1.4).

In the formation of a polysaccharide chain of aldoses, the monosaccharide units are linked together by a glycosidic bond between the 1 carbon of one unit to either the 3, 4 or 6 carbon of the adjacent unit depending upon the polysaccharide. The glycosidic bond is formed by the elimination of the elements of water from the hydroxyl groups attached to the two carbon atoms involved in the linkage. In the case of the polysaccharides of ketoses, the 2 carbon replaces the 1 carbon in the linkage. Since the configuration of the 1 carbon in the aldoses and the 2 carbon in ketoses can exist in either the α or β form, the linkage may be α or β depending upon the anomer involved. For example, in amylose the linkage is $\alpha(1\rightarrow4)$ between the D-glucopyranose units, while in cellulose, it is $\beta(1\rightarrow4)$.

In addition to straight chain polysaccharides, branched chains are commonly found in nature as in amylopectin where the branch is attached to an $\alpha(1\rightarrow4)$-D-glucopyranose chain by an $\alpha(1\rightarrow6)$-D-glucopyranosidic linkage. Depending upon the polysaccharide, the lengths of the chain and the extent of the branching will vary considerably. In amylose the chains are polydispersed and vary in molecular weight from a few thousand to 500,000. In amylopectin the branches are about 12 glucose residues long and occur on the average at every twelfth residue.

The configurations of the polysaccharide depend upon the monosaccharide units involved and the environment. The secondary forces involved consist of hydrogen bonding, electrostatic forces, if charged groups exist on the polysaccharide, and steric effects. In the natural state, the linear homopolysaccharides usually exist in extended chains linked together by hydrogen bonds as in the case of amylose in starch and cellulose. The conformation of the branched chain polysaccharides and heteropolysaccharides are quite varied depending upon the particular molecular structure. In aqueous dispersions, the polysaccharides exist either as random-chain linear dispersions, micelles consisting of helical coils bound by intra- and intermolecular hydrogen bonds, as in amylose, or in linear bundles, or cross linked gels as in the case of pectins. The ability to form gels is related to the number of different monomers in the polysaccharide. If it consists of only one type of monomer, the molecules are able to form attachments at any place along the molecule or theoretically along the whole molecule. Thus the result is a complete coagulation. While in the case of heteropolysaccharides, such as agar, the lack of uniformity in the monosaccharide units limits the regions of attachment and results in stable crosslinked gels.

ELECTRICAL PROPERTIES OF COLLOIDS

Source of Charge on Dispersed Phase of a Colloid

Under most conditions, the dispersed phase of an aqueous sol will possess a charge. This charge results from either the ionization of an ionizable group on the particle or macromolecule or the adsorption of charged ions from solution due to the presence of dipoles in the macromolecule. Thus such macromolecules as amylose, which lacks ionizable groups, can bind ions from solution as is demonstrated by the amylose-iodine complex in which both I_2 and I^- are involved (Morawetz 1975).

The common ionizable groups in food colloids consist of the sulphate on polysaccharides such as carrageenan, the carboxyl group of both proteins and some polysaccharides such as algin, the sulphydryl group in some proteins such as β-lactoglobulin, the tyrosine hydroxyl group in proteins, the phosphate group in phosphoproteins, the amino group of proteins and those polysaccharides containing hexosamine, and the guanidinium and histidine groups in proteins. The sulphate group behaves as a strong electrolyte and remains ionized throughout the pH range except in the presence of ions capable of binding to it. The other groups are weak electrolytes and their ionization is dependent upon the pH, ionic strength of the environment and temperature. The acidic or anionic groups dissociate in the following manner as illustrated by the carboxyl group:

$$\mathrm{RCOOH} \overset{K}{\rightleftharpoons} \mathrm{RCOO^-} + \mathrm{H^+}$$

while the cationic groups dissociate as illustrated by the amino group:

$$\mathrm{RNH_3^+} \overset{K}{\rightleftharpoons} \mathrm{RNH_2} + \mathrm{H^+}$$

The ionization constant for such dissociations,

$$K = \frac{(\mathrm{RCOO^-})(\mathrm{H^+})}{(\mathrm{RCOOH})} \text{ or } K = \frac{(\mathrm{RNH_2})(\mathrm{H^+})}{(\mathrm{RNH_3^+})} \tag{1.1}$$

can be related to pH by the Henderson-Hasselbach equations

$$\mathrm{p}H = \mathrm{p}K + \log\frac{(\mathrm{RCOO^-})}{(\mathrm{RCOOH})} \text{ or } \mathrm{p}H = \mathrm{p}K + \log\frac{(\mathrm{RNH_2})}{(\mathrm{RNH_3^+})} \tag{1.2}$$

where pK, analogous to pH, is the negative logarithm of the ionization constant and the quantities in parentheses represent the activities of the indicated species. The characteristic pK values for the ionizable groups on proteins are listed in Table 1.4.

Amphoterism

If a macromolecule is capable of possessing both negative and positive charges, it is said to be amphoteric. The pH at which a macro-

TABLE 1.4

INTRINSIC pK VALUES OF THE IONIZING GROUPS ON PROTEINS

Group	pK (25°C)	
	Model Compounds[1]	Proteins in Dilute Salt[2]
Carboxyl (α)	3.8	3.6
Carboxyl (aspartic acid)	4.0	(4.6)
Carboxyl (glutamic acid)	4.4	(4.6)
Imidozolium (histidine)	6.3	6.6
Amino (α)	7.5	?
Sulphydryl (cysteine)	9.5	?
Phenolic hydroxide (tyrosine)	9.6	9.8
Amino (ϵ, lysine)	10.4	10.0
Guanidinium (arginine)	12.0	12.0

Source: Nozaki and Tanford (1967).
[1] Corrected for electrostatic and other effects to give pK representative of the unperturbed group.
[2] Obtained by application of equation 1.4; values vary from protein to protein.

molecule possesses a net charge of zero is called its isoelectric point. In the case of a macromolecule in which the source of charge is due to ionization only and no other ions are bound to the macromolecule, we can define the isoionic point as the p*H* at which the number of anionic groups that have lost a proton equal the number of charged cationic groups.

The amphoteric properties of macromolecules can best be illustrated by the titration curves of proteins. Consider the titration of a protein sol which has *m* groups per molecule capable of ionizing to yield a proton, each with a dissociation constant which is dependent not only on the particular group involved but also upon the number of groups ionized on the molecule and on the ionic interaction between it and the other groups in its neighborhood. Assuming that the ionization of the *m* groups is sufficiently weak so that the concentration of added base is equal to the concentration of protons released by the protein and that the activity coefficients of the various protein-ion species present are independent of p*H*, it can be shown (Edsall 1943) that the total moles of protons released, or base required, per mole of protein at a given pH is

$$\bar{h} = \frac{\partial \log}{\partial \mathrm{p}H} \left[\sum_{h=0}^{m} \frac{1}{(H^+)^h} \sum_{\lambda=1}^{n} \frac{1}{y_{h\lambda}} \prod^{h} k_{h\lambda} \right] \quad (1.3)$$

In this equation h is the number of protons released (from 0 to m) per molecule by each of the various ion species present. The Greek letter, λ, represents the number of a particular arrangement of charges on the protein ion specie that has lost h protons and varies from 1 to n, the number of possible arrangements. The symbol $y_{h\lambda}$ is the activity coeeficient of the protein specie that has lost h protons in the λth arrangement.

$$\prod^{h} k_{h\lambda}$$

represents the product of the dissociation constants for the λth arrangement that have lost from 0 to h protons. This expression is true regardless of the number and type of groups present and whether or not interaction exists between the groups and is subject only to the two assumptions employed in its derivation.

Computations based upon the above expression are difficult and tedious. It is much simpler to use the following equation (a good approximation) for each type of group present:

$$\text{p}H = \text{p}k - 0.868\, wZ - \log \frac{1-\alpha}{\alpha} \tag{1.4}$$

where pk is the negative logarithm of the intrinsic dissociation constant of the particular type of group such as the carboxyl group; w the interaction coefficient; Z is the net charge on the protein, and α is the fraction of the groups that are ionized. Once the isoionic point is determined, the number of cationic groups on the molecule can be determined from the maximum acid binding capacity of the protein. Then the titration curve usually can be divided in three regions that can be roughly identified as (1) carboxyl groups, (2) imidazole plus α-amino groups and (3) phenolic hydroxyl, ϵ-amino and sulphydryl groups. In this manner the number of groups involved can be roughly estimated. Finally, equation 1.4 can be applied to the various types of groups present and the titration curve resolved. The detailed calculations for this procedure are described by Tanford (1962).

Association or Binding of Small Ions

The association or binding of small ions by macromolecules can take place either by binding to a charged site on the molecule and thereby neutralizing that charge on the molecule or it can be held by the presence of dipoles in the macromolecule. The study of small ion binding is analogous to the binding of the proton discussed under amphoterism. In a manner similar to titration theory, the expression for the moles of small ion, A, bound per mole of macromolecule can be shown to be

$$A_b = \frac{\partial \ln}{\partial \ln (A)} \left[\sum_{i=0}^{m} (A)^i \sum_{\lambda=1}^{n} \frac{1}{y_{MA_{i\lambda}}} \prod^{i} k_{i\lambda} \right] \tag{1.5}$$

where (A) represents the activity of the small ion and i is the number of small ions bound per macromolecule (from 0 to m, the number of binding sites). The symbol λ represents the number of the arrangement of the bound ions on the macromolecule that has bound i small ions and varies from 1 to n, the number of possible arrangements. The letter $y_{MA_{i\lambda}}$ is the activity coefficient of the macro-ion which has bound i small ions in the λth arrangement. $\overset{i}{\Pi} k_{i\lambda}$ represents the product of the association constants for the λth arrangement that has bound from 0 to i small ions.

If each binding site on the macromolecule has the same affinity for the small ion and no interaction exists between the sites, the above equation simplifies to

$$A_b = \frac{mk\,(A)}{1 + k\,(A)} \tag{1.6}$$

If there are several different types, ω, of binding sites, but no interaction the association phenomenon can be described by

$$A_b = \sum_{\gamma = 1}^{\omega} \frac{m_\gamma k_\gamma\,(A)}{1 + k_\gamma\,(A)} \tag{1.7}$$

where γ is number of the type of binding site from 1 to ω.

The association phenomenon in the presence of interaction between the binding sites is more complex. Scatchard *et al.* (1950), using some approximations which are accurate for m greater than 4, deduced the following relationship for an isoionic macromolecule with only one type of binding site

$$A_b = \frac{mk\,[A]\mathrm{e}^{-2wA_b}}{1 + k\,[A]\mathrm{e}^{-2wA_b}} \tag{1.8}$$

where w represents the interaction coefficient.

A wide variety of experimental procedures have been proposed to measure the bound and free small ion in colloidal sols including those based upon: the solubility product of a sparingly soluble salt of the small ion (Norbo 1939); the dissociation constant of a weakly ionized salt of the small ion (Sundararajan and Whitney 1975); potentiometry (Scatchard *et al.* 1950); dialysis equilibrium (Vink 1963); sedimentation analysis (Zittle *et al.* 1958); electron spin resonance (Malmstrom *et al.* 1958); and titration curves (Lal 1959).

Electrokinetic Phenomena

The presence of charges on the particles or macromolecules of the dispersed phase of a colloidal sol will result in an electrical double layer at the interface, as suggested by Helmholtz, formed by an ion atmosphere with a preponderance of charges of opposite sign to that of the macromolecule. Four different phenomena are associated with the presence of the electrical double layer in colloids. In the presence of an external electrical field the particles of the dispersed phase move relative to the dispersing medium in a direction determined by the sign of the charge and the direction of the applied field. This is called electrophoresis. Also the dispersing medium, due to the charges in the ion atmosphere of the particle, move in the direction determined by the sign of the preponderant charges in the ion atmosphere. This phenomena can best be seen in gels or membranes where the dispersed phase is fixed. This is called electroosmosis. If instead of applying an electrical field, the dispersing medium is forced through a gel, capillary or membrane made up of the dispersed phase by the application of pressure, a potential is developed across the dispersed phase which is called the streaming potential. Finally, if the dispersed phase sediments through the dispersing medium either due to gravity or the application of centrifugal force, a potential develops which is called the sedimentation potential.

Electrophoresis

Of all the above phenomena, electrophoresis is the most commonly used and therefore we shall consider it in more detail. Let us consider a system of charged conducting spheres of effective radius, r, and specific conductance, χ_p, dispersed in a dispersing medium of specific conductance, χ_m. Upon the application of an external electrical field, the charged particles will accelerate due to the force of the electric field until the force of friction created by its movement through the dispersing medium is equal to that of the electrical field. Then it will continue to move at a constant velocity. We generally express this movement as the electrophoretic mobility which is the ratio of the velocity to the intensity of the electric field. Assuming that the applied field and the electrical field of the particle with its ion atmosphere are independent and additive, Henry (1931) derived the following expression for the electrophoretic mobility in this system:

$$\mu = \frac{D}{\pi\eta}\left[\psi_0 + \sigma r^3\left(3r^2 \int_\infty^r \frac{1}{\rho^5}\frac{\partial\psi}{\partial\rho}\,d\rho - 2\int_\infty^r \frac{1}{\rho^3}\frac{\partial\psi}{\partial\rho}\,d\rho\right)\right] \quad (1.9)$$

where D and η are the dielectric constant and viscosity coefficient of the dispersing medium, respectively, and Ψ_0 is the effective potential at the surface of the particle. σ is equal to $(\chi_m - \chi_p)/(2\chi_m - \chi_p)$ where χ_m and χ_p are the specific conductances of the medium and the particle, respectively, and Ψ is the potential at a distance, ρ, from the center of the particle.

If the ionic strength of the dispersing medium is very small, the electrical double layer will be very thick and the potential, Ψ, will be proportional to the reciprocal of the radial distance from the center of the particle. Introducing these conditions into Henry's equation yields the equation

$$\mu = \frac{D\psi_0}{6\pi\eta} \tag{1.10}$$

In contrast, when the ionic strength is large, the double layer will be very thin and

$$\mu = \frac{D\psi_0}{6\pi\eta}\,[1 + \sigma] \tag{1.11a}$$

If the particle is dielectric in nature, which frequently is the case, $\sigma = 1/2$ and the above equation reduces to

$$\mu = \frac{D\psi_0}{4\pi\eta} \tag{1.11b}$$

For the intermediate condition,

$$\mu = \frac{D\psi_0}{4\pi\eta} f(\kappa r) \tag{1.12a}$$

where κ is the Debye-Hückel coefficient,

$$\kappa = \frac{4\pi\epsilon^2}{DkT} \sum_i n_i Z_i^2 \tag{1.12b}$$

ϵ is the charge on the electron k, the Boltzman constant, T is the absolute temperature, and n_i is the number of ions per cubic centimeter of the ith specie with a valence of Z_i. The function, $f(\kappa r)$, varies from 1 for high ionic strengths to 2/3 for low ionic strengths.

There are two complicating factors that tend to reduce the mobility below those predicted by Henry's equation. The movement of the dispersing medium in the opposite direction to the particle tends to reduce the apparent velocity of the particle. Also the ion atmosphere of the particle tends to lag behind the particle and to lack the radial symmetry assumed by Henry. These factors can have a serious effect upon the results (Longsworth 1941).

The treatment of the electrophoresis of macromolecules of shapes other than spherical is more complex. This problem has been investigated by Hermanns and Fujita (1955).

Numerous electrophoretic techniques have been used to investigate the electrical behavior of macromolecules, to isolate and characterize various macromolecules, to test the homogeneity of preparations, and to study the effect of treatment and environment upon the macromolecules (Bier 1959, 1967). The free-boundary method is not as effective in separating the various species of macroions present as other methods, but is most useful in exploring the electrical phenomena associated with the macromolecules. Zonal electrophoresis makes use of gels, filter paper, or other supporting media which overcome any convection effects and because of the longer path yields better resolution of the components of a mixture. However, the movement is not only dependent upon the charge of the macromolecule but also upon its size and shape due to the size of the channels in the supporting medium. In fact, use has been made of this effect for the determination of molecular weights of proteins by eliminating the charge effect with sodium dodecyl sulphate. Isoelectric focusing separates the proteins on the basis of their isoelectric points by moving them through a pH gradient gel under an applied electrical field.

HYDRATION OF COLLOIDAL SUBSTANCES

Nearly all colloidal substances in food have an affinity for water and bind it to a greater or lesser extent depending upon the structure and conformation.

Theory of Bound Water

While much remains to be learned about the nature of bound water, it is well known that intermolecular forces exist in liquid water due to hydrogen bonding between the water molecules resulting in transient molecular aggregates with an open structure due to the directional nature of the hydrogen bond. At 0°C each water molecule is hydrogen bonded to 3.4 other water molecules on a statistical average.

When macromolecules possessing nonpolar groups are dispersed in an aqueous system, an ice-like structure forms around the nonpolar groups in the macromolecular surface with the liberation of heat. It is the reverse of this process that results in hydrophobic bonding. Since the water involved in the ice-like structure behaves differently from free water, it may be considered bound water. However, if the partially negatively charged atoms such as oxygen and nitrogen are present in the macromolecular surface hydrogen bonds are formed between the water molecule and these electronegative atoms resulting in a different type of binding to the macromolecule (Morawetz 1975). The amount of water bound by the macromolecule is a function of the temperature, pressure and the environment of the macromolecules in the system.

Methods for the Measurement of Bound Water

To measure the water bound by macromolecules in food, investigators have defined bound water as that portion of the water in the colloid system that is so closely associated with the macromolecules that it has physical properties different from liquid or free water. Thus the amount of bound water determined will depend in part upon the physical property employed. Some of the properties attributed to bound water that have been used in its determination are: an inability to freeze at low temperatures, an inability to act as solvent, a slower rate of evaporation than free water, a lower vapor pressure than free water, an ability to move with the particle, and the ability to yield a nuclear magnetic resonance signal in a strong radio-frequency magnetic field.

Calorimetric methods have been used to measure the heat of fusion and therefore the amount of free water present (Moy *et al.* 1971). The bound water is obtained by difference. In dilatometry the amount of free water can be determined by the change of sample volume during freezing (Jones and Gortner 1932).

McMeekin and Warner (1942) investigated the bound water content of β-lactoglobulin in solution by two different methods: one depended upon the inability of water to act as a solvent and involved the measurement of the concentration of sucrose inside and outside a crystal of the protein while in the other he noted a break in the drying curve when the free water was completely removed.

Defining bound water as the water that is sufficiently tightly held to move with the macromolecule, the bound water can be determined by measuring the volume occupied by the macromolecule and its bound water by dielectric constant dispersion or sedimentation.

Shanbag *et al.* (1970) used wide line nuclear magnetic resonance to determine the amount of bound water in wheat flour, corn starch and egg white. They found that, at high radio frequency power, the free water signal is saturated and thus becomes negligible while the bound water signal remains significant.

STABILITY OF COLLOIDS

The stability of colloid systems is dependent upon the magnitude of the potential barrier between the colloidal particles or macromolecules. This barrier can be electrical and/or mechanical in nature dependent upon the particular system involved, and the magnitude of the barrier will be a function of the composition of the dispersing medium and the structure of the macromolecule. In the absence of a barrier the colloid will coagulate or flocculate due to London-Van der Waals forces or other secondary forces such as electrostatic attraction, hydrogen bonds and hydrophobic bonding.

For hydrophobic systems, the barrier usually originates from the repulsion experienced by particles or macromolecules of like charge. The potential energy of repulsion not only depends on the magnitude of the charge and its distribution but the ion atmosphere surrounding the particles. For spherical particles with a thin double layer of charges due to high ionic strengths, Derjaguin (1939) obtained the following approximate expression for the potential energy of repulsion:

$$V_R = \frac{Dr\psi_0^2}{2} \ln (1 + e^{-\kappa H_0}) \tag{1.13}$$

where H_0 is the distance between two particles measured along the line connecting their centers. For the opposite extreme when the ionic strength is low and the double layer is thick, Verwey and Overbeek (1948) derived the useful and approximate expression:

$$V_R = \frac{Dr^2 \psi_0^2 e^{-\kappa H_0}}{R} \tag{1.14}$$

where R is the distance between the centers of the particles. Assuming only London-Van der Waals attraction between the spherical particles, Hamaker (1937) derived the following expression for the potential energy of attraction:

$$V_A = \frac{-A}{6} \left(\frac{2r^2}{R^2 - 4r^2} + \frac{2r^2}{R^2} + \ln \frac{R^2 - 4r^2}{R^2} \right) \tag{1.15}$$

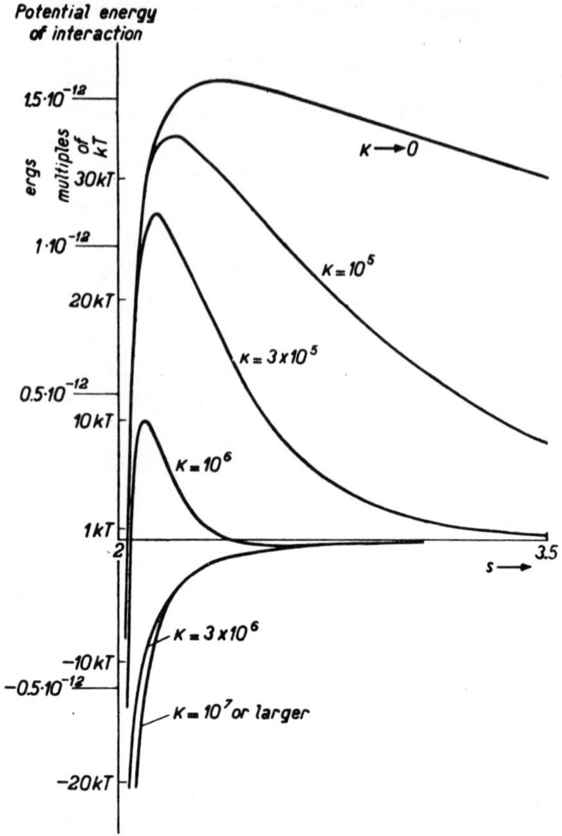

From Verwey and Overbeek (1948)

FIG. 1.5. THE POTENTIAL ENERGY OF INTERACTION AS INFLUENCED BY THE IONIC STRENGTH EXPRESSED IN TERMS OF THE DEBYE-HÜCKEL COEFFICIENT, κ

$r = 10^{-5}$ cm; $A = 10^{-12}$ ergs; $\Psi_o = 25.6$ mv.

where A is a constant of the order of 10^{-12} ergs. The total potential energy can therefore be obtained by adding the appropriate equations for the system of interest. Figure 1.5 illustrates the resulting potential energy of interaction as influenced by the ionic strength expressed in terms of the Debye-Hückel coefficient, κ. From these curves it is possible to derive a criterion for the stability of spherical hydrophobic colloids. Verwey and Overbeek (1948) demonstrated that when the quantity

$$W = 2r \int_{2r}^{\infty} e^{V/kT} \frac{dR}{R^2} \tag{1.16}$$

is greater than 10^5 for dilute colloids or 10^9 for very concentrated colloids, they are stable for all practical purposes. To achieve this, the maximum potential of the curve should be 15 kT or 25 kT, respectively.

For hydrophilic systems, the barrier, in addition to its electrical component possesses a mechanical component due to its hydration. While generally speaking the greater the degree of hydration the more stable the system in the absence of charge, the type and strength of binding is also a factor.

Association of Macromolecules

Depending upon the composition of the dispersing medium and temperature of the colloidal system, certain macromolecules associate to form polymers with themselves or other macromolecules. In these associations the size distribution of polymers formed appears to be determined by the particular configuration of the polymers, the concentration of the monomers, the temperature, pH, ionic strength and composition of the dispersing medium. Consider a stable colloidal system in monomeric form. By altering the pH or ionic strength of the system the electrical component of the potential barrier can be reduced sufficiently to allow for association of the monomers to occur and to form a particular distribution of polymers. The arrangement of the monomer in the polymers and the resulting conformational changes in the monomers that occur may further alter the potential barrier by altering the groups in the surface and therefore prevent further aggregation. For example, β-lactoglobulin A associates in the vicinity of 4.6 to form an octamer with the predominate species the dimer and octamer with only small amounts of tetramer and hexamers (Timasheff and Townend 1964). Apparently the law of mass action is involved in this association, as in the case of small-ion association, and equilibria exist between the monomer and the various polymer species present.

Macromolecular association can also be brought about by changes in temperature. If the macromolecule possesses a sufficient number of hydrophobic groups and increases in temperature will "melt" the icelike structure of water around these groups and allow these groups to associate by hydrophobic bonding to form polymers as in the case of β-casein (Payens and van Markwijk 1963).

In those systems in which the conformation of the resulting polymer is not a limiting factor, the weight average molecular weight of the polymers present will increase linearly with increasing concentration of macromolecules until equilibrium is reached as shown by Schmidt and Payens (1964) in their study of the association of α_{s1}-casein.

Coagulation

When there is no limitation placed upon the association of macromolecules by the configuration of the polymer and its accompanying potential barrier, and equilibrium has not been obtained, the macromolecules will continue to aggregate and coagulation occurs. Von Smoluchowski (1917) demonstrated that in the absence of a potential barrier coagulation is rapid and the phenomena for spherical particles follows the relationship:

$$n = n_0/(1 + 8\pi \mathfrak{D} r n_0 t) \tag{1.17}$$

where n and n_0 are the number of particles per unit volume at times t and 0, respectively, and $\mathfrak{D}$ is the diffusion constant. For a spherical particle, $\mathfrak{D}$ equals $kT/6\pi\eta r$ where k is the Boltzman constant. Substituting this in equation 1.17 we have

$$n = \frac{3\eta n_0}{3\eta + 4kTn_0 t} \tag{1.18}$$

From the discussion of this phenomena by Verwey and Overbeek (1948), and the above equation it is easily demonstrated that for coagulation in the presence of a potential barrier the rate is slowed such that

$$n = \frac{3\eta W n_0}{3\eta W + 4kTn_0 t} \tag{1.19}$$

Thus as the potential barrier increases, W will increase and the rate of coagulation will decrease.

Hydrophilic sols with a high degree of hydration are quite stable and a high concentration of salt is needed to cause them to coagulate. This procedure is called salting out and is believed to be due to a competition for the water between the added salt and the hydrophilic colloid. The amount of salts needed to precipitate the hydrophilic sol is dependent primarily upon the salt employed, but contrary to the hydrophobic system, where the effectiveness of the salt in precipitating the colloid is dependent upon the valence of the counter ion with the higher the valence the greater the flocculating effect, the nature of the anion is more important regardless of the charge on the hydrophilic particles. In the salting out of proteins it has been observed that the logarithm of the concentration of protein is a linear function of the ionic strength of the salt.

$$\log c_2 = a - b\,\Gamma/2 \tag{1.20}$$

where c_2 is the concentration of protein, $\Gamma/2$ is the ionic strength and a and b are empirical constants which are dependent upon the protein; b is also dependent upon the salt employed.

Hydrophilic colloids can be coagulated not only by salts but by liquids which are miscible with the dispersing medium and cannot act as dispersing agents themselves. For example, gelatin can be dispersed in water or glycerol (at sufficiently high temperature) but not in alcohol; hence a gelatin sol can be precipitated by alcohol but not glycerol. The amount of precipitant required for precipitation is dependent upon: (1) its structure, (2) the nature of the colloid, and its concentration, (3) the temperature and the pH of the system. Generally speaking, the more lipophilic the precipitant and the more concentrated the sol, the less precipitant is required. For many systems the precipitability increases with increasing temperature. Numerous exceptions to these observations exist and the phenomenon in many cases is more complicated than these statements suggest.

Gelation

The phenomenon of gelation is a particular type of coagulation and usually requires that the macromolecules involved be highly asymmetric either in the form of random chains or rods. For globular macromolecules to be involved in gelation a prior aggregation into polymers consisting of chains of monomers appears to be necessary. A gel is an elastic coherent colloid which, while it usually assumes the shape of the container in which it is formed, is more or less capable of maintaining that shape after removal from the container. The relative amounts of dispersed phase to dispersing medium in a gel can vary widely. However the amount of dispersed phase present can be extremely low and still maintain the shape of the container. For example agar-agar gels can be formed containing 99.8% water.

In gelation the coagulation occurs not in the form of relatively compact aggregates but by binding at selective sites along the chain of the linear macromolecules or chains forming a network which is capable of entrapping the dispersing medium. However, it must be emphasized that gelation does not always occur upon the coagulation of linear macromolecules but varies with the particular colloid involved and the method employed for coagulation. For example, agar-agar or gelatin which are dispersed in the medium by heat will set to a gel upon cooling, but, if the coagulation is brought about by the addition of alcohol, a precipitation occurs without immobilizing much of the dispersing medium. The gelation phenomenon can be characterized by (1) the time of gelation, (2) the gelation temperature, and by (3) the minimum concentration of dispersed phase required for gelation. During the gelation process the viscosity of the system increases with time, but it is impossible by this means to follow the entire course of gelation although the rate of increase in viscosity can be also used to characterize the system.

The stability of gels is dependent upon the type of bonding that is involved in the formation of the network. If the forces involved are only weak London-Van der Waals forces the gels are unstable and frequently demonstrate thixotropy which is defined as an isothermal sol-gel interchange produced by agitation. The gel system is broken down by shaking to a sol but resets to a gel upon standing at rest. When the bonds involved are hydrogen bonds or bonds of comparable strength, a metastable framework is encountered as in the case of some protein gels which can be transformed to sols by heating. Finally, the presence of primary valence bonds between the individual units in the network result in a stable gel.

Since most of the water in a gel is simply mechanically immobilized, it behaves like free water and can be removed from the gel by evaporation in dry air or vacuum. When water is removed, the gel shrinks and becomes dry and hard. This form is called a xerogel. For many food gels such as gelatin and pectins this process is almost completely reversible and the xerogel swells in the appropriate dispersing medium.

Many gels such as starch and casein gels upon aging undergo syneresis, which is the spontaneous liberation of the dispersing medium, even in moist air and at low temperatures. This is believed to be due to the contraction of the gel resulting from the formation of additional bonds between the mobile elements of the structural framework. Syneresis is generally promoted by increases in temperature and ionic strength of the dispersing medium as would be expected from coagulation theory.

OPTICAL PROPERTIES OF COLLOIDS

The optical properties of colloids are extremely useful in the study of the size, shape and configuration of food macromolecules in dispersion and in following the changes that they undergo with processing, alterations in their environment and storage. Spectroscopy can be used not only to determine the amount of a particular macromolecule present in the system but also to study its conformation. Polarimetry has been applied in studies of the optically active macromolecules. This property of the molecules is not only dependent upon the character of the asymmetric carbons present but also on the configuration and has been invaluable in the study of helix-coil transition phenomena. Radiation which is not absorbed can be scattered by a colloidal system, the Tyndall effect, which is one of the most characteristic properties of a colloid. From this phenomena it is possible to secure information as to the size and shape of the particles of the dispersed phase.

Absorption Spectra

The absorption of electromagnetic radiation by colloids can be studied in the three regions of the spectra: ultraviolet, visible, and infrared. The absorption in the ultraviolet and visible regions is the result of electronic excitations in the sample while that of the infrared is due to transitions in the vibrational and rotational states of the atoms in the systems.

Absorption in the visible range is absent from most of the colloidal food systems except for those containing transition metal ions such as hemoglobin which contains the iron-porphyrin complex associated with the macromolecule. We shall leave the discussion of this rather specialized system to discussion in later chapters involving the pertinent proteins.

Studies of the ultraviolet absorption spectra of macromolecules containing aromatic acids such as tyrosine and tryptophane have contributed to our knowledge of their conformation and extent of ionization. This work, performed largely on proteins and synthetic polypeptides, involves the use of difference spectra in which the spectra of a reference system containing the protein is compared with that of the protein in the system of interest. The differences observed appear to be due to changes in the polarity and polarizability of the group responsible for the absorption. If following an environmental change, the absorption band increases in intensity, the effect is called hyperchromic; if the absorption decreases, the effect is hypochromic. If the absorption maximum moves to longer wave, the shift is called bathochromic while a shift to shorter wave lengths is hypsochromic. This approach has been helpful in suggesting changes in the conformation proteins. For example the difference spectra for bovine serum albumin at pH 2.1 compared with the protein at pH 5.1 (Fig. 1.6) indicates a hypsochromic shift in the tyrosine absorbance with a negative difference maximum at 287 nm and a small peak near 278 nm. This suggests a conformational change in which the tyrosine residues in the protein molecule which were buried in the interior of the molecule in a region of high index of refraction or high polarizability were moved to the surface in a solvent environment of lower polarizability. This conclusion was supported by sedi-

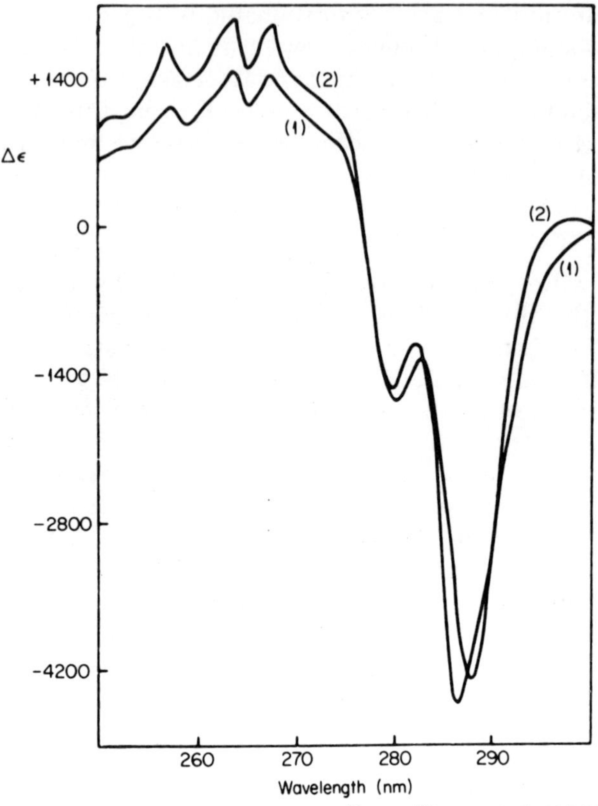

From Glazer et al. (1963)

FIG. 1.6. DIFFERENCE SPECTRA (MOLAR ABSORPTIVITY DIFFERENCE, Δϵ VERSUS WAVE LENGTH IN NM), 2MM CELL AT 25° C

1—Protein at pH 2.1 versus protein at pH 5.1 both 0.15 *M* in NaC1.
2—Protein at pH 5.1 in 7 *M* urea versus protein at pH 5.1, both 0.15 *M* in NaC1.

mentation, viscosity and optical rotation studies (Glaser *et al.* 1963). The approach has also been helpful in pH titration curves in following the ionization of the tyrosine residues by the bathochromic shift in their absorbance from 279 to 295 nm (Crammer and Neuberger 1943).

Studies of the infrared spectra of macromolecules have been useful in suggesting their conformation and the nature of hydrogen bonding (Susi 1969). While the rigorous interpretation of infrared spectra of macromolecules is currently beyond our capabilities, qualitative interpretations based upon the assignment of certain absorption maxima to certain vibrational modes of groups in the macromolecule have been helpful. The study of macromolecules in aqueous dispersions is further complicated by the strong interference of the absorbance of water. However, differential techniques have been developed that to a certain extent overcome this difficulty (Susi *et al.* 1967). Infrared as a technique for conformational study has largely been involved with the configuration of proteins. Of the numerous infrared absorption bands present in the spectra of protein, attention has primarily been directed toward those related to the peptide bond: (1) the amide A and B vibration mode at 3300 cm^{-1} and 3100 cm^{-1}, respectively, primarily related to hydrogen bonded N—H stretching vibration; (2) the amide I mode at $\sim$1650 cm^{-1} primarily related to C=O stretching vibration; and (3) the amide II mode at $\sim$1560 cm^{-1} related in the main to N—H deformation or bending. With the aid of films of fibrous proteins of known configuration the relationship of the wave numbers of the major components of the amide I and II vibration modes to configuration were calculated (Table 1.5). These values were observed to change only slightly in water solution. Further supporting information can be obtained by infrared dichroism. When a solid sample is oriented in plane polarized light the infrared absorbance is dependent upon the orientation. Maximum absorption is observed if the transition moment is parallel to the electric vector of the light. For helical structure, the N—H and C=O stretching vibrations would be parallel to the axis of the helix and the N—H deformation would be perpendicular. In contrast for a sheet structure, the N—H and C=O stretching vibrations would be perpendicular and the N—H deformation would be parallel to the axis of the polypeptide chain. Infrared dichroism studies have confirmed this orientation.

TABLE 1.5

WAVE NUMBERS FOR AMIDE I AND II VIBRATIONS FOR DIFFERENT PROTEIN CONFORMATIONS

	cm^{-1}		
Conformation	Strongest Amide I Component	Strongest Amide II Component	Weak Amide I Component
Unordered	1658	1520	—
Antiparallel pleated sheet	1632	1530	1685
Parallel pleated sheet	1632	1530	—
Parallel chain polar sheet	1648	1550	—
α-helix	1650	1546	
Triple helix	1648	1558	

Source: Krimm (1962).

Fluorescence

When chemical compounds containing certain chromophoric groups are irradiated with light at a wave length corresponding to their absorption maximum they will fluoresce, that is, emit light of a wave length different than that of the absorbed light usually of longer wave length. This phenomenon is due to the transition of the electrons in the chromophore from the ground to the excited state upon the absorption of light energy and the subsequent emission of that energy after a short period of time, of the order of 10^{-8} seconds, when the electron returns to a lower energy level. The emitted fluorescence has a greater wave length or lower energy than the absorbed light because initially the electron falls first to an intermediate energy state before returning to the ground state by thermal deactivation.

Phenylalanine, tyrosine and tryptophane in neutral aqueous solution yield fluorescent maxima of 282 nm, 303 nm, and 348 nm when irradiated at 280 nm. Proteins containing these amino acids also exhibit fluorescence. However, after examining a large number of proteins, Teale and Weber (1959) were able to divide them into two classes: (1) those containing phenylalanine and tyrosine but no tryptophane which yield a fluorescence maximum at 303 nm only; and (2) those containing all three aromatic amino acids which yield only the peak due to tryptophane. The reason for this phenomenon has not been agreed upon completely. However, the phenomenon does demonstrate the interaction of amino acid residues within the protein molecule and has been used to demonstrate relatively minor structural differences.

Fluorescence polarization is an even more powerful tool for structural studies. When a macromolecule is fluorescent or is made fluorescent by the binding of a simple fluorescent molecule such as the dye, 1-dimethyl-aminonaphthalene-5-sulfonyl chloride, a study of the polarization of the fluorescence yields information on its rotational Brownian movement and thus its size and shape. When polarized light is used for excitation only those molecules will absorb appreciable energy that have the direction of their absorption oscillation parallel to the plane of excitation. Since each molecule emits a plane-polarized wave parallel to its emission oscillation, the emitted fluorescence is also polarized. However there is another factor which determines the observed polarization, which is the interval between excitation and emission, 10^{-8} seconds. With macromolecules the relaxation time is quite long and therefore the rotatory Brownian movement will produce little disorientation during this time interval and partially polarized fluorescence will be observed. Measurement of fluorescence polarization as a function of temperature and viscosity permits the calculation of the relaxation time due to rotatory Brownian movement and therefore information on the size and shape of the macromolecule. The equation of Perrin shows the relationship between the fluorescence polarization and the molecular volume:

$$\frac{1}{p} = \frac{1}{p_0} + \left(\frac{1}{p_0} - \frac{1}{3}\right)\left(\frac{RT}{V}\right)\frac{\tau}{\eta} \tag{1.21}$$

where p is the polarization; p_0, the maximal p value obtained in a rigid medium; V, the molecular volume; τ, the interval between excitation and emission; η, the viscosity; and R, the gas content.

$$p = \frac{F\| - F\perp}{F\| + F\perp} \tag{1.22}$$

In equation 1.22 $F\|$ is the fluorescence intensity measured with the polarizing and analyzing prisms oriented with their electric vectors parallel and $F\perp$ is the fluorescence intensity with the electric vectors crossed. Fluorescence polarization studies of macromolecules give reliable estimates of molecular size when account is taken of molecular symmetry (Udenfriend 1962).

Optical Rotatory Dispersion and Circular Dichroism

Optical activity has long been of interest in its application to molecular structure and conformation. However, the theory pertaining to this phenomenon is very complex and only a brief overview can be presented here. For those interested in a more thorough treatment numerous excellent treatments are available for example, Moscowitz (1962) and Urry (1968). For a substance to be optically active it must lack planes or centers of symmetry. The asymmetry present may be of two types: (1) that due to the presence of an asymmetric carbon atom, a carbon atom to which four different groups of atoms are attached, or (2) a helical coil, in which the structure as a whole is asymmetric.

Linearly polarized light, in which the oscillation of the wave traces a straight line perpendicular to the direction of propagation, can be considered to be the resultant of two coherent circularly polarized components of opposite sense of rotation. Circularly polarized light can be produced by two orthogonal linearly polarized waves of equal amplitude but differing in phase by a quarter of a wave. If the refractive index of the medium due to the presence of an optically active substance depends upon the sense of rotation of the circularly polarized rays, the difference in their velocities will result in their being out of phase when they emerge from the medium. Addition of the two circular components will then result in a linear polarized beam which has rotated by an angle, α, with respect to the incident beam and

$$\alpha = \frac{\pi L}{\lambda_v} (n_l - n_d) \tag{1.23}$$

where L is the distance traversed in the medium, λ_v is the wave length of the light in vacuum, and n_l and n_d are the indices of refraction of the two circularly polarized components of the light.

In the spectral region in which the medium absorbs the two circular components equally the resulting wave is circularly polarized. However, when the components are unequally absorbed the resulting wave is eliptically polarized. The major axis of the ellipse still forms an angle with the original line of polarization equal to α. However, a new value called the angle of ellipticity, v, can be defined as the ratio of the arc tangent of the minor axis to the major axis of the ellipse. Since the ellipse is very elongated, the value of v can be closely approximated by

$$v = 0.576Lc(k_l - k_r) \tag{1.24}$$

where c is the concentration of the optically active material in gram per deciliter, k_l and k_r the absorption coefficients. The value $(k_l - k_r)$ is the circular dichroism of the optically active substance and is directly related to the ellipticity.

In molecular units, the molecular rotation at wave length λ is defined by

$$[\phi]_\lambda = \frac{M\alpha_\lambda}{Lc} \tag{1.25}$$

and the molecular ellipticity is

$$[\theta]_\lambda = \frac{Mv_\lambda}{Lc} \tag{1.26}$$

The molecular ellipticity is related to the molecular dichroism $(\epsilon_l - \epsilon_r)$ by

$$[\theta]_\lambda = 3305\,(\epsilon_l - \epsilon_r) \tag{1.27}$$

where ϵ_l and ϵ_r are the molecular extinction coefficients.

The combined effect of circular dichroism (CD) and optical rotatory dispersion (ORD) in the vicinity of an absorption band is called the Cotton effect (Fig. 1.7). When the ORD peak is on the long-wave length side, the Cotton effect is positive and, when there is an ORD trough on the long-wave length side, it is negative. These two phenomena can be related by the following equations

$$[\phi]_\lambda = \frac{2}{\pi} \int_0^\infty [\theta]_\lambda \frac{\lambda' d\lambda'}{\lambda^2 - \lambda'^2} \tag{1.28}$$

and

$$[\theta]_\lambda = -\frac{2}{\pi\lambda} \int_0^\infty [\phi]_\lambda \frac{\lambda'^2\, d\lambda'}{\lambda^2 - \lambda'^2} \tag{1.29}$$

Since the optical rotation of a macromolecule in solution is the sum of contributions of the mutual interactions of the various groups and structures present in the molecule, one might expect a very complex

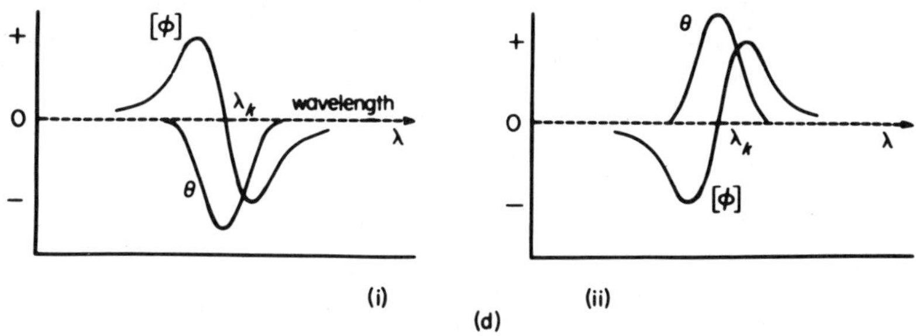

From McKenzie (1970)

FIG. 1.7. COTTON EFFECTS IN THE VICINITY OF AN ABSORPTION BAND: (i) NEGATIVE EFFECT; (ii) POSITIVE EFFECT

$[\phi]$ = molecular optical rotation; θ = molecular ellipticity; λ_k = wavelength of absorption maximum

behavior. In many cases this is not the case. For example, the ORD of randomly coiled proteins in the region well removed from absorption bands, 350–600 nm, can be expressed by the simple Drude equation

$$[m]_\lambda = \frac{n^2+2}{3} \frac{a\lambda_c^2}{\lambda^2 - \lambda_c^2} \tag{1.30}$$

where $[m]_\lambda$ is the mean residue rotation and a and λ_c are empirical constants of the order of −600° and 200 nm respectively.

Moffitt and Yang (1956) observed that the helical polypeptides follow the equation

$$[m]_\lambda = \frac{n^2+2}{3} \left[\frac{a_0\lambda_0}{\lambda^2 - \lambda_0^2} + \frac{b_0\lambda_0^4}{(\lambda^2 - \lambda_0^2)^2}\right] \tag{1.31}$$

where a_0, b_0 and λ_0 are constants. λ_0 is usually selected for the best fit of the data and has a value near 212 nm. While the use of this equation for the interpretation of protein configuration falls short of complete and accurate elucidation, except in the case of pure configuration, it has been used to estimate the distribution of the peptide chain between the various conformations on the basis of the parameters a_0 and b_0. The values of these parameters for the various configurations are given in Table 1.6.

In recent years attention has been given to the use of ORD and CD in the ultraviolet from 185–300 nm where absorption maxima occur. Excellent agreement has been obtained between the ORD and CD curves calculated according to equations 1.28 and 1.29 with known polypeptides (Urry 1968). Some characteristics of these curves are shown in Table 1.7.

TABLE 1.6

OPTICAL ROTATORY DISPERSION MOFFITT-YANG PARAMETERS FOR VARIOUS POLYPEPTIDE CONFORMATIONS

Configuration	a_0	b_0
α-helix	0	−630
β-configuration	+300 to +700	−200 to +400
Coil	−650	0

Source: McKenzie (1970).

TABLE 1.7

SOME OPTICAL ROTATORY DISPERSION AND CIRCULAR DICHROISM EXTREMA FOR VARIOUS PEPTIDE CONFORMATIONS

	α-helix		β-form		Coil	
ORD	λ(nm)	$[m']^1_\lambda$	λ(nm)	$[m']^1_\lambda$	λ(nm)	$[m']^1_\lambda$
Trough	232-233	-15,000	229-230		288	
Peak	198-199	+67,000	205		228	
Trough	182-184		190		205	-15,000
Peak					189	+17,000
CD	λ(nm)	$[\theta']^2_\lambda$	λ(nm)	$[\theta']^2_\lambda$	λ(nm)	$[\theta']^2_\lambda$
Minimum	221-222	-30,000	217-218	-15,000	235-238	-200
Minimum	207-210	-29,000				
Maximum	190-192	+54,000	195-197		217-218	+2000 to +5000
Minimum					196-200	-25,500 to -35,000

Source: McKenzie (1970).
[1] Reduced mean residue rotation.

$$[m']_\lambda = \frac{3}{n^2 + 2}[m]_\lambda$$

[2] Reduced mean residue molecular ellipticity.

$$[\theta']_\lambda = \frac{3}{n^2 + 2}[\theta]_\lambda$$

Light Scattering

The initial fundamental concepts dealing with light scattering were developed by Lord Rayleigh (1881) in his study of gases. The phenomenon is caused by the induction of an oscillating dipole by the oscillating electric field of the incident light. Neglecting multiple scattering and assuming electrically isotropic molecules, he demonstrated that the Rayleigh ratio is

$$R_\theta = \frac{I_\theta r^2}{I_0} = \frac{8\pi^4 (D-1)^2 M}{\lambda_v^4 N\rho}(1+\cos^2\theta) \tag{1.32}$$

where I_0 and I_θ represent the intensity of the incident light and the scattered light measured at an angle θ to the incident beam and a distance r from the unit volume irradiated, respectively. D is the dielectric constant; ρ, the density of the gas of molecular weight M; λ_v, the wave length of light in vacuum; and N, Avogadro's number.

If the scattering molecules are optically anisotropic, the light scattered is not completely polarized. The intensity of the light scattered at right angles is increased over the value predicted by equation 1.32 by the Cabannes' factor $(6 + 6d)/(6 - 7d)$ where d is the depolarization or the ratio of the intensity of the horizontally polarized component of the light to the vertically polarized component.

This approach could also be used with liquids with their partially ordered structure but the procedure would be laborious. Therefore Debye (1944) used a more convenient approach based upon the fluctuation of an intensive thermodynamic property of the system, density in a system of one component or concentration in a multicomponent system. Although macroscopically the system will appear uniform, we may visualize a fluctuation in the local value of the property on a scale small compared to the wave length of light. For such a system, the contribution to the Rayleigh ratio resulting from fluctuations in macromolecular concentration can be calculated from

$$\Delta R_\theta = \frac{\pi^2 (\partial D/\partial c_2)_T^2}{2\lambda_v^4 (\partial \Delta P/\partial c_2)_T} kTc_2 (1+\cos^2\theta) \tag{1.33}$$

where the $(\partial D/\partial c_2)_T$ is the change in dielectric constant with concentration of the macromolecules, c_2, in gm per ml at constant temperature; $(\partial \Delta P/\partial c_2)_T$, the change in osmotic pressure with concentration, and k, the Boltzman's constant. With appropriate substitutions and rearrangements in this equation we have

$$K_\theta c_2/\Delta R_\theta = 1/M_2 + 2A_2c_2 + 3A_3c_2^2 + \ldots \tag{1.34}$$

where

$$K_\theta = [2\pi^2 n_0^2 (\partial n/\partial c_2)_T^2/\lambda_v^4 N](1 + \cos^2\theta) \tag{1.35}$$

n and n_0 = the index of refraction of the system and dispersing medium, respectively, and

A_2 and A_3 = the osmotic virial coefficients.

A plot of $K_\theta c_2/\Delta R_\theta$ against c_2 will yield the reciprocal molecular weight of the macromolecular specie present or, if the system is polydispersed, the weight mean molecular weight, $\bar{M}_w$, as the intercept.

When the dimensions of the scattering particles exceeds 1/20 of the wavelength of light used, the intensity of the scattered light will be reduced by interference between light scattered from different parts of the particle. If we define $\mathscr{P}_\theta$ as the attenuation of the scattered light at the angle, θ, due to destructive interference, equation 1.34 can be modified for a single molecular specie to give

$$\lim_{c_2 \to 0} (K_\theta c_2/\Delta R_\theta) = 1/M_2 \mathscr{P}_\theta \tag{1.36}$$

Calculations of $\mathscr{P}_\theta$ have been carried out for spheres of uniform density (Rayleigh 1911), rigid thin rods (Neugebauer 1942), randomly coiled chains (Debye 1947), and polydispersed randomly coiled chains with a normal distribution of chain lengths (Zimm 1948).

Zimm (1948) has developed a general method for the elimination of this effect and obtaining the correct value for the molecular weight. If data is obtained for a number of macromolecular concentrations at a number of angles, θ, and the data plotted according to Fig. 1.8, by double extrapolation, first to zero concentration and then to zero θ, a value for $K_0c_2/\Delta R_0$ is obtained from which $\bar{M}_w$ can be determined free of both deviations from ideality and destructive interference.

The shape of the macromolecule can be determined from dissymmetry measurements of the scattered light. Dissymmetry is defined as ratio of scattered light intensities measured at two angles symmetrically distributed around 90°, usually 45° and 135°. Thus the

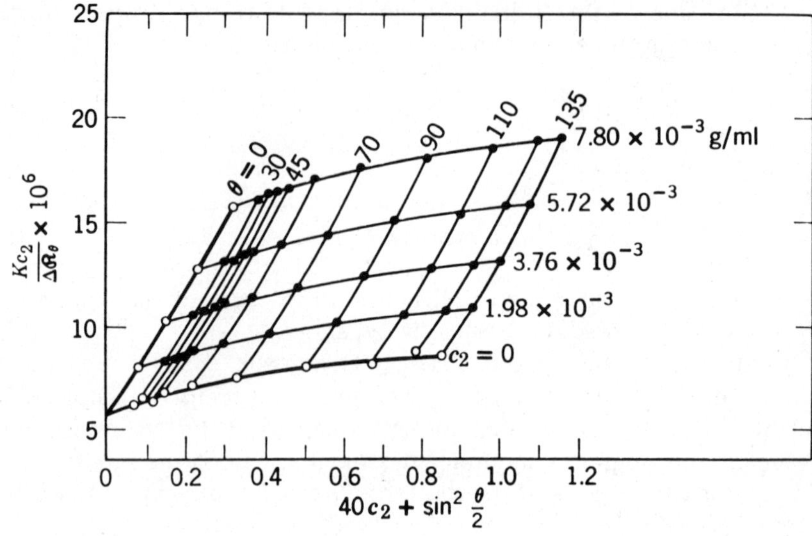

From Zimm (1948)

FIG. 1.8. ZIMM'S PLOT OF LIGHT-SCATTERING DATA, ILLUSTRATED WITH POLYSTYRENE, $\bar{M}_w$ = 170,000, IN DIOXANE

c_2 = concentration of polystyrene; θ = angle of observation; ΔR_θ = Rayleigh ratio contribution of macromolecules at angle θ; K = light scattering constant

dissymmetry, will be equal to $\mathscr{P}_{(45°)}/\mathscr{P}_{(135°)}$. The relationship between dissymmetry and $1/\mathscr{P}_{(90°)}$ is shown in Fig. 1.9. From the molecular weight obtained from a Zimm's plot and the value of $K_{(90°)}c_2/\Delta R_{(90°)}$ at c_2 equal to zero the value of $\mathscr{P}_{(90°)}$ can be evaluated and by inspection of Fig. 1.9 the shape of the macromolecule deduced. Thus light scattering can not only yield information as to the size of the macromolecule but in some cases yield information as to its shape.

DIFFUSION PHENOMENA OF COLLOIDS

The particles of dispersed phase in a colloidal sol are continuously engaged in movement both translational and rotational due to their bombardment by molecules of the dispersing medium. This motion was named Brownian movement after its discoverer who observed it in a suspension of pollen grains.

If the distribution of particles is not uniform throughout the system, their translational diffusion will tend to establish a uniform distribution. If the orientations of the axes of the particles or macromolecules are not arranged so as to be uniformly oriented in all direc-

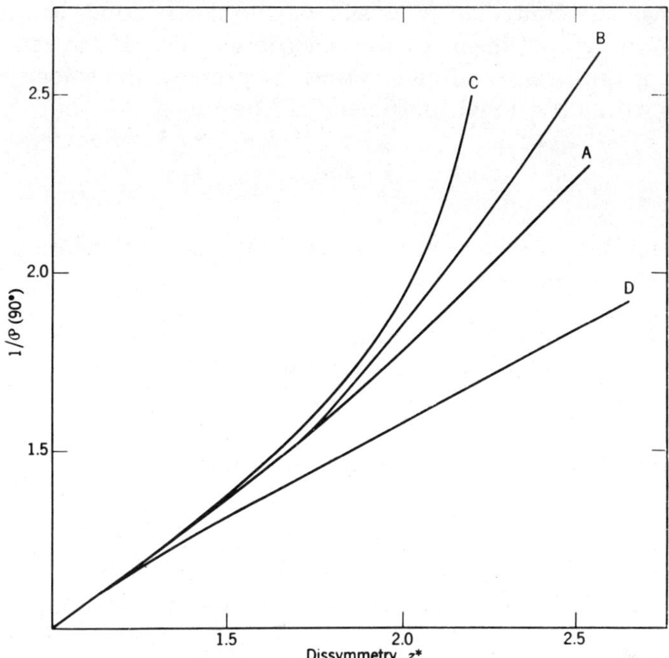

FIG. 1.9. DEPENDENCE OF THE RECIPROCAL OF THE ATTENUATION OF SCATTERED LIGHT AT AN ANGLE OF 90° TO THE INCIDENT BEAM ON THE DISSYMETRY Z FOR DIFFERENT PARTICLE SHAPES

A—Monodispersed coils. B—Coils with normal chain length distribution. C—Rods. D—Spheres.

tions, their rotational diffusion will tend to establish this condition. These phenomena are functions of the size and shape of the macromolecules involved.

Translational Diffusion

The diffusion of particles or macromolecules in a concentration gradient may be thought of as a motion of the particles in response to a driving force provided by the gradient of their chemical potential, $\bar{G}_2$. The force acting on the macromolecule in the x direction is

$$F = -1/N(\partial \bar{G}_2/\partial x) = -kT\left(\frac{\partial \ln c_2}{\partial x}\right)\ [1 + \partial \ln y_2/\partial \ln c_2] \qquad (1.37)$$

where k is the Boltzman constant, c_2 the molar concentration and y_2 the activity coefficient of the macromolecules. If the activity coefficient is uniform at all concentrations present, the velocity of the particles with a frictional coefficient of f becomes

$$v = F/f = -(kT/f)(1/c_2)(\partial c_2/\partial x) \quad (1.38)$$

or the amount of dispersed phase crossing a unit plane perpendicular to the x axis per second is

$$J = \frac{dS}{dt} = vc_2 = -(kT/f)(\partial c_2/\partial x) \quad (1.39)$$

which compares with Fick's first law of diffusion,

$$J = -\mathfrak{D}(\partial c_2/\partial x) \quad (1.40)$$

For diffusion in three dimensional space in vector notation,

$$J = -\mathfrak{D}\,\mathrm{grad}\ c_2 \quad (1.41)$$

If we wish to observe the increase in concentration with time, it can be demonstrated that

$$\frac{\partial c_2}{\partial t} = \mathfrak{D}\,\mathrm{div\ grad}\ c_2 \ \text{(Fick's second law)} \quad (1.42)$$

Fick's laws are only valid in highly dilute solution not only because of the thermodynamic factor involved in their derivation, but also a hydrodynamic factor since we assume that the frictional coefficient is independent of the presence of other similar particles. Therefore

$$\frac{\partial c_2}{\partial t} = \mathrm{div}\,(\mathfrak{D}\,\mathrm{grad}\ c_2) \quad (1.43)$$

If we consider an initial system in which a colloidal sol of concentration c_0 is separated by a sharp boundary from its dispersing medium in the middle of a tube of sufficient length that in the time of observation the concentrations at the ends of the tube do not change, the solution for Fick's second law is

$$c_2 = \frac{c_0}{2}\left[1 - \frac{2}{\sqrt{\pi}}\int_0^q e^{-q^2}\,dq\right] \tag{1.44}$$

where

$$q = \frac{x}{\sqrt{4\mathfrak{D}t}} \tag{1.45}$$

and the concentration gradient at any x and t is

$$\frac{\partial c_2}{\partial x} = \frac{-c_0}{\sqrt{4\pi\mathfrak{D}t}}\, e^{\frac{-x^2}{\sqrt{4\mathfrak{D}t}}} \tag{1.46}$$

From this relationship the apparent diffusion coefficient can be obtained. Extrapolation of a plot of $\mathfrak{D}$ versus c_0 will yield the true diffusion constant at the intercept. Polydispersity or other deviation from the laws of free diffusion can be detected by deviations from the above equations.

From equations 1.39 and 1.40 it can be seen that the diffusion coefficient is related to the frictional coefficient by the relationship

$$\mathfrak{D} = kT/f \tag{1.47}$$

Therefore information can be secured on the shape and degree of hydration of the macromolecule from diffusion studies. The frictional coefficient for a spherical molecule with no hydration is

$$f_0 = 6\pi\eta r = 6\pi\eta\left(\frac{3M\bar{v}_2}{4\pi N}\right)^{1/3} \tag{1.48}$$

where $\bar{v}_2$ is the partial specific volume. If the macromolecule is hydrated and spherical, the degree of hydration can be calculated from

$$\frac{f}{f_0} = \left(1 + \frac{h}{\bar{v}_2\,\rho}\right)^{1/3} \tag{1.49}$$

where h is the hydration or bound water per gram of protein and ρ is the density of water. Numerous treatments of the effect of shape upon the frictional coefficient have been attempted. For example, Debye and Bueche (1948) considered chain molecules to fall between two limiting situations. The macromolecule can be considered as rigid "beads" connected by infinitely thin linkages. In the first case the "beads" are relatively far apart and the effective frictional coefficient of the coil will be proportional to the number of units composing it. In the second case the interaction between the units is so large that the macromolecule may be treated as a rigid "hydrodynamically equivalent sphere" with an effective radius proportional to the root-mean-square radius of gyration $\langle s^2\rangle^{1/2}$ or

$$f_{coil} = 6\pi\eta C \langle s^2 \rangle^{1/2} \tag{1.50}$$

where C is a characteristic constant.

Rotational Diffusion

In considering rotational diffusion of rigid particles, we can reason by analogy to translational diffusion. If we consider that the orientation of particles in a two dimensional plane can be specified by the distribution function, $W(\Psi)$, such that $W(\Psi)d\Psi$ is the number of particles whose axes of reference fall between Ψ and $\Psi + d\Psi$ with a given direction of the coordinate system. If the system is completely disordered, $W(\Psi)$ will be independent of Ψ. However if the particles in the system have been somewhat oriented, the Brownian movement will tend to disorient the system and the orientations will rotate from more occupied to less occupied Ψ's. Then by analogy to Fick's first law of translational diffusion

$$\frac{\partial\psi}{\partial t} = [\mathfrak{D}_r/W(\psi)]\, \partial W(\psi)/\partial\psi \tag{1.51}$$

where $\mathfrak{D}_r$ is the rotational diffusion constant or by Fick's second law

$$\frac{\partial W(\psi)}{\partial t} = \mathfrak{D}_r \frac{\partial^2 W(\psi)}{\partial\psi^2} \tag{1.52}$$

Since the rotational diffusion constant must be inversely proportional to the rotational frictional coefficient,

$$\mathfrak{D}_r = kT/f_r \tag{1.53}$$

where f_r is equal to the force-couple in dyne-centimeter required to impart to the particle a rotational velocity of one radian per second around an axis perpendicular to the axis of reference.

For spheres of radius r, Stokes obtained

$$f_r = 8\pi\eta r^3 \tag{1.54}$$

while for ellipsoids of revolution with a semiaxis of revolution a and an equatorial axis b, Perrin (1934) demonstrated that

$$f_r = 8\pi\eta ab^2/y(p) \tag{1.55}$$

where p is the axial ratio a/b and

$$y(p) = \frac{3}{2}\frac{p^2}{p^4 - 1}\left[1 + \frac{2p^2 - 1}{2p\sqrt{p^2 - 1}} \ln \frac{p + \sqrt{p^2 - 1}}{p - \sqrt{p^2 - 1}}\right] \text{ for } p > 1 \tag{1.56}$$

and

$$y(p) = \frac{3}{2}\frac{p^2}{1 - p^4}\left[1 + \frac{1 - 2p^2}{2p\sqrt{1 - p^2}} \tan^{-1} \frac{\sqrt{1 - p^2}}{p}\right] \text{ for } p < 1 \tag{1.57}$$

Thus as p increases f_r increases and the resistance to rotation increases sharply with elongation of the ellipsoid of revolution. For ellipsoids with $p > 5$, a good approximation is

$$f_r = 8\pi\eta a^3/(3 \ln p + 0.57) \tag{1.58}$$

For the rotational diffusion of flexible polymer chains, as in the case of translational diffusion, the chains may be considered as a "necklace" of rigid "beads." In the case of the free draining coil, the "beads" are relatively far apart and the rotational frictional coefficient can be shown to be

$$f_r = 4\pi\eta r_s n_s \langle s^2 \rangle \tag{1.59}$$

where r_s is the radius of the beads, n_s is the number of beads per chain and $< s^2 >$ is the mean-square radius of gyration. Since the mean square radius of gyration is proportional to the chain length, the rotational frictional coefficient in free draining coils is proportional to the square of the chain length. For the opposite extreme where the interactions between the "bead" are so large that the chain can be considered an inpermeable spherical body, Riseman and Kirkwood (1949) demonstrated that

$$f_r = 8\pi\eta R_r \tag{1.60}$$

where R_r is the radius of a sphere which would encounter the same frictional resistance to rotation as the chain, or

$$R_r = 0.89 \langle s^2 \rangle^{1/2} \tag{1.61}$$

Several methods are available for studying rotational diffusion including fluorescence depolarization, streaming birefringence and electrical birefringence. Fluorescence depolarization has been discussed previously under Optical Properties of Colloids. Streaming birefringence is the birefringence resulting from the anisotropy induced in a dispersion by the partial orientation of asymmetric particles with the aid of a velocity gradient. If the test solution is placed in the annulus between two cylindrical surfaces, one stationary and one rotating, Boeder (1932) has demonstrated that

$$\frac{\partial W(\psi)}{\partial \psi} + \frac{\beta}{\mathfrak{D}_r} W(\psi) \sin^2 \psi = \text{constant} \tag{1.62}$$

where β is the velocity gradient. Numerical integrations of this differential equation for various values of the ratio, $\alpha = \beta/\mathfrak{D}_r$, yield the curves in Fig. 1.10. The angle of the maximum can be determined by use of crossed Nicol prisms since at that angle the polarized light from the polarizer will not be altered by the dispersion and will therefore be cut off by the analyzer. Thus the appropriate ratio α can be selected and $\mathfrak{D}_r$ calculated.

In electrical birefringence, the dispersion is placed in an electrical field which results in a partial orientation of the macromolecules in the field due to their polarizability. This orientation will produce a birefringence. If the field is now cut off, Brownian motion will disorient the particles so that the mean value of cos Ψ will decay according to

$$(\overline{\cos \psi})_t = (\overline{\cos \psi})_0 / e^{-t/\tau} \tag{1.63}$$

which leads to a decay in the birefringence

$$\Delta n_t = \Delta n_0 \, e^{-t/\tau} \tag{1.64}$$

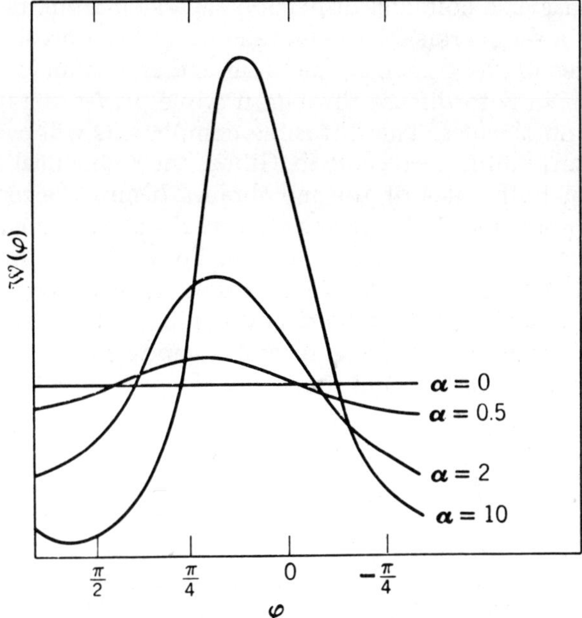

From Boeder (1932)

FIG. 1.10. THE DISTRIBUTION OF ANGULAR ORIENTATIONS (ψ) TO THE FLOW LINES IN STREAMING BIREFRINGENCE AS A FUNCTION OF THE RATIO OF THE VELOCITY GRADIENT TO THE ROTARY DIFFUSION COEFFICIENT ($\alpha = \beta/D_r$)

where Δn_t and Δn_0 are the difference between the refractive indices of light polarized so that its electric vector vibrates parallel to the optic axis and light vibrating perpendicular to this axis at time t and 0, respectively. The relaxation time τ is related to the rotational diffusion constant by

$$\tau = 1/6\mathfrak{D}_r \tag{1.65}$$

This technique was first used to determine rotational diffusion constants by Benoit (1950).

MEMBRANE EQUILIBRIA IN COLLOIDAL SYSTEMS

Membrane equilibria are useful in both securing information as to the size of macromolecules and in the investigation of small ion binding. In addition they have been employed for fractionation and processing procedures.

Let us consider a colloidal dispersion separated from its dispersing medium by a semipermeable membrane of such a porosity to allow the molecules of the dispersing medium and any small ions or molecules present in it to diffuse through it while preventing the passage of the macromolecules. The diffusible components will pass through the membrane until, according to Gibbs, their chemical potentials are equal on both sides of the membrane. It must be remembered that, by component, Gibbs means only a substance that can exist by itself as a molecular not an ionic specie. For example, the chemical potential of NaCl not Na^+ or Cl^-. From thermodynamics, we can express the chemical potential of any component at constant temperature in terms of the pressure and composition of the system. Therefore at membrane equilibrium

$$\int_{P_0}^{P_\alpha} {}_\alpha\bar{V}_i\, dP + \int_0^{{}_\alpha m_2} \left(\frac{\partial \bar{G}_i}{\partial m_2}\right)_{PT} dm_2 + \sum_j \int_0^{{}_\alpha m_j} \left(\frac{\partial \bar{G}_i}{\partial m_j}\right)_{PT} dm_j$$

$$= \int_{P_0}^{P_\beta} {}_\beta\bar{V}_i\, dP + \sum_j \int_0^{{}_\beta m_j} \left(\frac{\partial \bar{G}_i}{\partial m_j}\right)_{PT} dm_j \qquad (1.66)$$

where P_0 is the standard state pressure; ${}_\alpha\bar{V}_i$ and ${}_\beta\bar{V}_i$ are the partial molal volume of the ith component in phase α and β, respectively, where α is the phase containing the macromolecules; $\bar{G}_i$ is the chemical potential of the ith component; and m_2 and m_j are the concentrations of the macromolecules and the diffusible components, respectively. It can be shown that this expression is equivalent to

$$\int_{P_0}^{P_\alpha} {}_\alpha\bar{V}_i\, dP + RT \ln {}_\alpha a_i = \int_{P_0}^{P_\beta} {}_\beta\bar{V}_i\, dP + RT \ln {}_\beta a_i \qquad (1.67)$$

where ${}_\alpha a_i$ and ${}_\beta a_i$ are the activities of the ith component in the α and β phases, respectively.

Dialysis Equilibrium

If the membrane separating the two phases is so constructed as to be movable and able to completely adjust to equalize the pressure on both sides of the membrane, the activities of the dializable constituents of the system will be equal on both sides of the membrane at equilibrium. However due to the presence of a charged macromolecule or particle in the α phase, the requirement of electroneutrality in each phase will result in an unequal distribution of the ionic species. Considering the presence of only one uniunivalent dializable electrolyte in the system, Donnan (1934) was able to establish that

$$({}_{\alpha}m_f + {}_{\alpha}m_b)\, {}_{\alpha}m_b\, {}_{\alpha}y_{\pm}^2 = {}_{\beta}m_b^2\, {}_{\beta}y_{\pm}^2 \tag{1.68}$$

where ${}_{\alpha}m_f$ is the molarity of fixed charges on the particle; ${}_{\alpha}m_b$ and ${}_{\beta}m_b$ are the molarity of the byions of the electrolyte and ${}_{\alpha}y_{\pm}$ and ${}_{\beta}y_{\pm}$ are the mean activity coefficients of the electrolyte, respectively. If we neglect the activity coefficients, then

$${}_{\alpha}m_b = \frac{{}_{\alpha}m_f + \sqrt{{}_{\alpha}m_f^2 - 4\,{}_{\beta}m_b^2}}{2} \tag{1.69}$$

At low concentrations of electrolyte, the electrolyte is largely excluded from the polyelectrolyte phase and, at high concentrations of electrolyte, the differences in electrolyte concentration between both phases become negligible. If more than one dializable electrolyte is present, the distribution becomes more complicated although the same principle applies.

In practice, the large forces between the macroions and their counterions will depress the ion activity coefficient so that ${}_{\alpha}y_{\pm} < {}_{\beta}y_{\pm}$. Thus the use of dialysis equilibrium to determine the binding of ions by macromolecules requires that we either know the values of the mean activity coefficient of the salt of the ion of interest in the phase containing the macromolecule or the ionic strength of the phases must be high enough to make the difference in activity coefficients negligible. It should also be noted that, if a difference in pressure exists at equilibrium, the activities of the dializable components will no longer be equal in both phases and corrections must be made for this difference.

Osmotic Pressure

If the membrane separating the two phases is so constructed that it cannot respond completely to pressure differences between the two sides of the membrane, a pressure difference will exist at equilibrium which is defined as the osmotic pressure, ΔP.

If we assume that the partial molal volume of the solvent is constant (incompressible) and equal on both sides of the membrane, the osmotic pressure can be expressed in terms of the activity of the solvent by

$$\Delta P = \frac{-RT}{\bar{V}_1} \ln \frac{{}_\alpha a_1}{{}_\beta a_1} \tag{1.70}$$

If the macromolecules are uncharged and no dializable component other than the solvent is present, this expression, in the limit as the concentration of macromolecules approaches zero, becomes

$$\Delta P = \frac{RT}{\bar{V}_1} \frac{{}_\alpha m_2}{{}_\alpha m_1} = RT\, {}_\alpha m_2 = RT \frac{{}_\alpha c_2}{M_2} \tag{1.71}$$

where c_2 is the weight of macromolecules per liter, since by the Gibbs-Duhem relationship,

$$m_1 \text{ d} \ln a_1 + m_2 \text{ d} \ln a_2 = 0 \tag{1.72}$$

In the case of a charged macromolecule or polyion in a system containing no dializable electrolyte, the counterions of the polyion will also contribute to the osmotic pressure. Since their number, in general, will be much larger than the number of polyions, the observed osmotic pressure will not be directly related to the molecular weight.

When we add simple electrolytes to the system, the situation becomes more complex because of the Donnan distribution of the dializable ions. However if we plot the reduced osmotic pressure, $\Delta P/c_2$, versus c_2 at different concentrations of dializable electrolytes (Fig. 1.11), we obtain curves which conform to the equation

$$\Delta P/c_2 = RT\,(1/M_2 + A_2 c_2 + \ldots) \tag{1.73}$$

where A_2, the second virial coefficient, is dependent upon the concentration of dializable electrolytes. For all curves the limit as c_2 approaches zero yields a value for the intercept from which the molecular weight can be obtained. As the concentration of the dializable electrolyte is increased the slope of the curve decreases until finally the contribution of the Donnan distribution and the differences between activity coefficients between the phases become negligible and the slope approaches zero.

If the colloid is polydispersed, the values obtained for molecular weight will be the weight mean molecular weight $\bar{M}_w$.

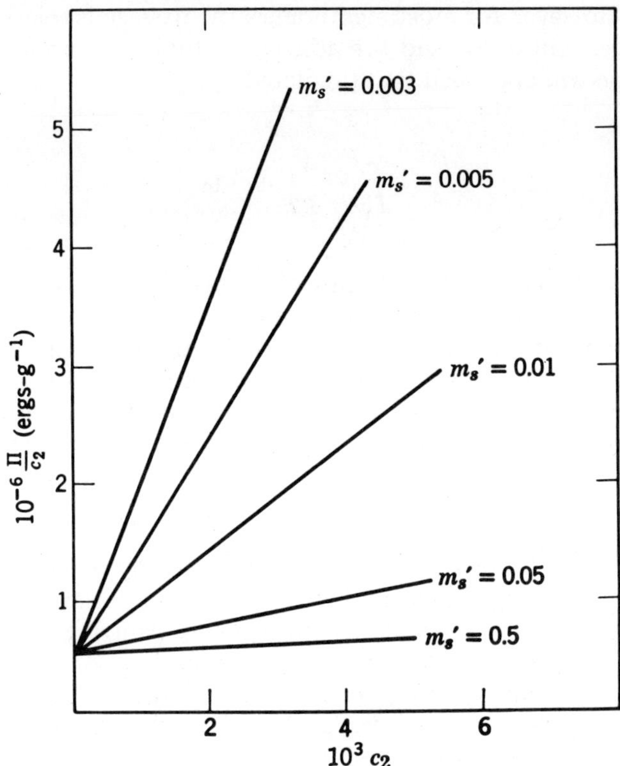

From Palls and Hermanns (1952)

FIG. 1.11. OSMOTIC BEHAVIOR OF SODIUM PECTINATE AT VARIOUS CONCENTRATIONS OF ADDED SODIUM CHLORIDE

π = Osmotic pressure; c_2 = concentration of sodium pectinate; m_s' = molal concentration of NaC1.

Ultrafiltration

When a colloidal sol is concentrated by passage through a membrane as occurs in ultrafiltration or reverse osmosis, the composition of the ultrafiltrate will differ from that of the original dispersing medium depending upon the difference in pressure across the membrane and its permeability. During the ultrafiltration process, the system will tend toward membrane equilibrium. If the passage of the dispersing medium through the membrane were rapid due to a high permeability, insufficient time would be allowed for the system to achieve equilibrium and little change would be observed in its composition. However for most membranes, the passage is relatively slow and changes will occur and the activity of the ith component in the ultrafiltrate will approach that predicted by

$$\ln {}_{\beta}a_i = \int_{P_\beta}^{P_\alpha} \frac{\bar{V}_i}{RT}\, dP + \ln {}_{\alpha}a_i \qquad (1.74)$$

Flexner (1937) investigated this phenomenon with membranes of different permeability at different pressures and compared the ultrafiltrate composition with that predicted by the Donnan equilibrium (Table 1.8). In all cases the chloride content of the ultrafiltrate was equal to or higher than that expected from the Donnan equilibrium. In all cases the composition was different from the mother liquor.

SEDIMENTATION PHENOMENA OF COLLOIDS

With the development of the ultracentrifuge by Svedberg (1926), a powerful tool was made available for the estimation of the sizes and shapes of macromolecules. With ordinary centrifuges, either the accelerations produced are too slow or convection currents disturbed the sedimentation phenomenon. However, with the ultracentrifuge, accelerations as high as 500,000 xg can be achieved. Also constancy of temperature is obtained by the use of very low pressures in the rotor chamber and currents due to centrifugation are avoided by the use of a sector-shaped cell.

TABLE 1.8

EFFECT OF MEMBRANE PERMEABILITY, PRESSURE, AND DONNAN EQUILIBRIUM ON THE COMPOSITION OF THE ULTRAFILTRATE OF A 7% CASEIN DISPERSION IN NaCl-PHOSPHATE BUFFER pH 7.4

Membrane Permeability (ml/cm)	Mother Liquor Cl^- Conc. (mg/100 ml)	Pressure (cm. Hg)	Eq. Ratio (Cl^-)	Filtrate Cl^- Conc. (mg/100 ml)	Deviation (mg/100 ml)
0.008	358	100	1.123	417	+15
0.019	362	227	1.125	416	+9
0.019	362	140	1.125	413	+6
0.019	362	40	1.125	407	0
0.010	350	50	1.130	406	+11
0.010	350	140	1.128	416	+21
0.010	350	227	1.128	422	+27

Source: Flexner (1937).

Equilibrium Ultracentrifugation

At moderate speeds in the ultracentrifuge and after reasonably long times, dependent upon the depth of the cell and the macromolecular size, a sedimentation equilibrium is achieved. In a colloidal dispersion subjected to centrifugal acceleration, the expression for chemical potential of the dispersed phase contains an additional term:

$$\bar{G}_2 = \bar{G}_2^\circ + RT \ln m_2 y_2 - M_2 \omega^2 x^2/2 \tag{1.75}$$

where ω is the angular velocity of the centrifuge and x is the distance from the center of rotation. At equilibrium,

$$\mathrm{d}\bar{G}_2/\mathrm{d}x = (\partial \bar{G}_2/\partial P)_{c_2}\, \mathrm{d}P/\mathrm{d}x + (\partial \bar{G}_2/\partial c_2)_P\, \mathrm{d}c_2/\mathrm{d}x - M_2 \omega^2 x = 0 \tag{1.76}$$

Since

$$(\partial \bar{G}_2/\partial P)_{c_2} = \bar{V}_2 = M_2 \bar{v}_2 \tag{1.77}$$

and

$$\mathrm{d}P/\mathrm{d}x = \omega^2 x \tag{1.78}$$

where ρ is the density of the dispersion, we find by substitution and rearrangement

$$\frac{\mathrm{d}c_2}{c_2} = \frac{M_2\,(1 - \bar{v}_2 \rho)\, \omega^2\, x\, \mathrm{d}x}{RT\,[1 + c_2 (\partial \ln y_2/\partial c_2)]} \tag{1.79}$$

Therefore in the case where the activity coefficient is independent of the concentration of macromolecules, a plot of the natural logarithm of c_2 against x^2 will be linear and the molecular weight can be obtained from the slope of the line provided the other values are known. In the more general case where the $(\partial \ln y_2/\partial c_2)_P$ is not equal to zero, one can relate it to the osmotic virial coefficients by means of the Gibbs-Duhem equation:

$$(\partial \ln y_2/\partial c_2)_P = 2A_2 M_2 + 3A_3 M_2 c_2 + \ldots \tag{1.80}$$

Therefore, in the nonideal case the plot of the natural logarithm of c_2 versus x^2 will not be linear. However, the molecular weight can be evaluated by substituting equation 1.80 in 1.79 and fitting the data to the resulting expression or by extrapolating the reciprocal of the apparent molecular weight (evaluated without regard for the nonideality of the system) to zero solution concentration (Williams *et al.* 1958).

If the colloidal system is polydispersed, a treatment analogous to that for the homogeneous system (provided the activity coefficients are independent of the concentration of macromolecules) yields

$$dc_x/dx = \frac{\bar{M}_{wx}\,(1 - \bar{v}\rho)\,\omega^2 x c_x}{RT} \tag{1.81}$$

or

$$\bar{M}_w = 2RT\,(c_b - c_a)/(1 - \bar{v}\rho)\,\omega^2 c_0\,(x_b^2 - x_a^2) \tag{1.82}$$

where c_a, c_x and c_b are the concentrations of macromolecules at the top of the cell, x_a, at the distance x from the center of rotation and at the bottom of the cell, x_b, respectively. $\bar{M}_{wx}$ is the weight average molecular weight at distance x and c_0 is the initial concentration of macromolecules in the system. The value of $c_b - c_a$ may be evaluated by graphical integration of $\int_{x_a}^{x_b} (dc_x/dx)\,dx$. The effect of polydispersity upon the shape of the plot of the natural logarithms of the concentration against x^2 is shown in Fig. 1.12. The existence of nonideality in the system complicates the interpretation of the data since,

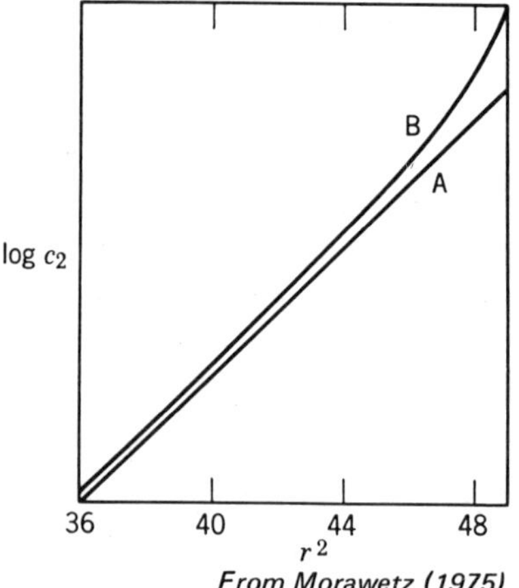

From Morawetz (1975)

FIG. 1.12. DEPENDENCE OF THE EQUILIBRIUM TOTAL MACROMOLECULAR CONCENTRATION, c_2, AS A FUNCTION OF THE DISTANCE FROM THE CENTER OF ROTATION, r, UPON THE MOLECULAR WEIGHT DISTRIBUTION OF MACROMOLECULES IN EQUILIBRIUM ULTRACENTRIFUGATION

A—Monodispersed macromolecules. B—Macromolecules with the same weight average molecular weight, containing 1 gm of a fraction with $M = 2\,\bar{M}_w$ and 5 gm of a fraction with $M = 0.8\,\bar{M}_w$.

while polydispersity results in a positive second derivative, nonideality usually results in a negative second derivative. Thus one effect may tend to mask the other. A number of different approaches to the solution of this problem have been made (Albright and Williams 1967; Fujita 1969).

Sedimentation Velocity

As the speed of the ultracentrifuge is increased, the particles of the dispersed phase will finally sediment at a speed sufficient for us to ignore the back diffusion of the macromolecules and will reach a constant velocity dependent upon the molecular frictional coefficient, the buoyancy, the centrifugal acceleration and their molecular weight,

$$v = \frac{dx}{dt} = \frac{M_2 (1 - \bar{v}_2 \rho) \omega^2 x}{Nf} \tag{1.83}$$

We may then define the sedimentation constant by rearranging this equation so that

$$s = \frac{dx/dt}{\omega^2 x} = \frac{1}{\omega^2} \, d \ln x/dt = \frac{M_2 (1 - \bar{v}_2 \rho)}{Nf} \tag{1.84}$$

Since the diffusion coefficient is related to the frictional coefficient by equation 1.47, the molecular weight can be determined by the following expression:

$$M_2 = \frac{RTs}{\mathfrak{D}(1 - \bar{v}_2 \rho)} \tag{1.85}$$

Since the frictional coefficient is a function of concentration both $1/s$ and $1/\mathfrak{D}$ must increase with concentration of the dispersed phase. Therefore these values should be evaluated by extrapolation to zero concentration.

If the molecular weight is known, sedimentation data will provide information on the shape and degree of hydration from the evaluation of the frictional coefficient as was done in translational diffusion.

While we neglected diffusion in the above treatment, we must remember that diffusion is taking place throughout the sedimentation with the resultant broadening of the boundary of the sedimenting macromolecules. This poses no problems if a single sedimenting specie is present, since the appropriate value of x is the maximum of the concentration gradient.

In applying this method to multicomponent systems, several complications arise. In a system in which one of the components of the dispersing medium other than solvent is present in high concentration (for example, urea) the centrifugal force will establish a gradient of this component in the cell. Thus the density of the system will be a function of the distance from the center of rotation and the plot of the natural logarithm of x versus time will not be linear. Suitable procedures for overcoming this have been suggested by Schachman (1959).

In considering charged macromolecules there are important charge effects since the counterions will tend to sediment more slowly than the macromolecules. Since a macroscopic separation of the polyion from its counterions is impossible, this phenomenon will add to the frictional coefficient and appreciably decrease the sedimentation rate. This effect however should decrease with increasing ionic strength. For a discussion of this effect, consult Alexandrowicz and Daniels (1968).

VISCOSITY OF COLLOIDAL SYSTEMS

The viscosity of a macromolecular dispersion is a function of the size and shape and concentration of the macromolecules. In the case of steady-state laminar flow, we can define the viscosity, or the viscosity coefficient, as the ratio of the shearing stress, the force applied per unit area, to the velocity gradient produced. If the viscosity is independent of the shearing stress, it is called Newtonian and, if it varies with the stress, it is non-Newtonian.

Defining the specific viscosity as the ratio of the difference in viscosity between the colloid and the dispersing medium to the viscosity of the dispersing medium, Einstein (1911) demonstrated that for rigid spherical particles with interparticle distances large compared to particle diameters that

$$\eta_{sp} = (\eta - \eta_0)/\eta_0 = 2.5\,\phi_2 \tag{1.86}$$

where ϕ_2 is the volume fraction occupied by the spheres. As the concentration of macromolecules increases, interaction will occur and therefore the volume fraction is best evaluated by plotting the reduced viscosity, against the concentration such that

$$\eta_{sp}/c_2 = [\eta] + k'\,[\eta]^2\,c_2 \tag{1.87}$$

where the intercept is defined as the intrinsic viscosity, $[\eta]$, and is equal to 2.5 ϕ_2/c_2 and k' is the Huggins constant. The effective hydrodynamic volume of the particles can then be calculated as,

$$V_h = \frac{\phi_2 M_2}{Nc_2} \tag{1.88}$$

The case of nonspherical rigid molecules is more complex since orientation will be possible and will be a function of the velocity gradient (Fig. 1.13). At low velocity gradients such that essentially random distribution of particle orientation occurs, Simha (1940) established the effect of the axial ratio of a rigid ellipsoid of revolution upon the ratio of specific viscosity to the volume fraction of the

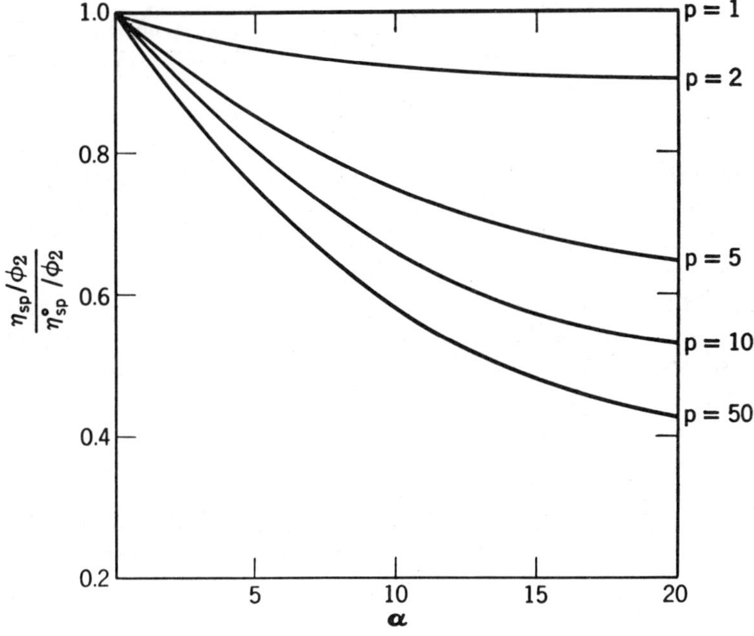

From Scheraga (1955)

FIG. 1.13. RELATIVE DECREASE IN THE SPECIFIC VISCOSITY, η_{sp}, OF DISPERSIONS OF ELLIPSOIDAL PARTICLES WITH INCREASING RATIO OF THE VELOCITY GRADIENT TO THE ROTATING DIFFUSION COEFFICIENT

$\alpha = \beta/D_r$; ϕ_2 = volume fraction of spheres.

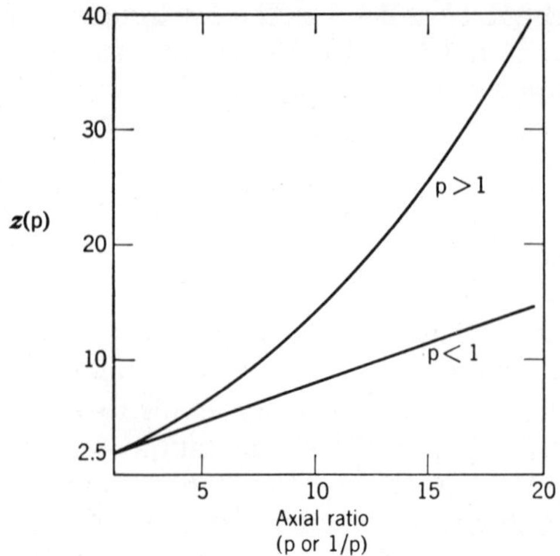

From Simha (1940)

FIG. 1.14. DEPENDENCE OF THE RATIO OF SPECIFIC VISCOSITY TO THE VOLUME FRACTION OF DISPERSED PHASE AT VERY LOW VELOCITY GRADIENTS UPON THE AXIAL RATIO OF ELLIPSOIDAL PARTICLES DISPERSED PHASE

$\alpha = \beta / D_r \rightarrow 0.$

macromolecules as shown in Fig. 1.14. Therefore in the case of the rigid ellipsoid of revolution, the value of the coefficient of the volume fraction is different from 2.5 and will depend upon the axial ratios. The situation is more complex if the particle is appreciably charged (Tanford 1961).

In the case of flexible chain molecules, the theoretical treatments of viscosities cannot compare in rigor to that of light scattering. However in practice intrinsic viscosity data may be greatly simplified by the fact that coils formed by flexible chain molecules closely approximate the behavior for the limiting case of complete impermeability. Based upon this observation Flory (1953) predicts that the intrinsic viscosity is related to the mean square radius of gyration by

$$[\eta] = \phi' \langle s^2 \rangle^{3/2} / M_2 \qquad (1.89)$$

where

$$\phi' = (10/3)\, \pi N \,(R_\eta / \langle s^2 \rangle^{1/2})^3 \qquad (1.90)$$

and R_η is the radius of the volume of a rigid sphere with the same intrinsic viscosity as the coil. ϕ' should be a universal constant independent of the nature of the macromolecule and the dispersing medium. Morawetz (1975) calculates that it should have the value 4.2×10^{24}. In dispersing media which eliminate binary interactions between segments of the macromolecule, the θ-solvents of Flory, the intrinsic viscosity will be related to the molecular weight of the macromolecule by

$$[\eta] = K' M_2^{1/2} \tag{1.91}$$

In other than θ-solvents the value of the exponent will vary.

CHROMATOGRAPHY OF MACROMOLECULAR DISPERSION

One of the currently most useful methods for the fractionation and separation of macromolecules is chromatography. It is defined as the separation of substances by their differential migration through a polyphase system as a result of a combination of the flow of the mobile phase and a selective reversible distribution of the substances between the stationary and mobile phases. For the separation of food macromolecules, column chromatography is usually employed. The macromolecular mixture is deposited at the top of a bed of the stationary phase in a column and an aqueous system of appropriate composition is passed through the column. The partitioning of the macromolecules between the two phases usually depends upon either the adsorption of the macromolecules upon the stationary phase according to their characteristic adsorption isotherm, as in the case of ion-exchange chromatography, or their distribution between the mobile and stationary phases according to their ability to permeate the pores of the stationary phase, as in gel permeation chromatography.

If we consider the ideal situation where the activity coefficients of the macromolecules are uniform throughout the system, each macromolecular specie can be considered to be distributed between the two phases according to a characteristic partition coefficient, k_i, independent of concentrations, or

$$k_i = c_{is}/c_{im} \tag{1.92}$$

where c_{is} and c_{im} are the concentrations of the ith component of the macromolecular mixture in the stationary and mobile phase, respectively. In the case of adsorption chromatography this condition would exist in the linear portion of the adsorption isotherm at low concentrations of the adsorbed specie. In the case of gel-permeation, the ith specie will, under ideal conditions, permeate those pores of the stationary phase large enough to allow it to enter, but will not enter any of the smaller pores, or

$$k_i = \frac{c_{im} \, V_i/V_s}{c_{im}} = \frac{c_{is}}{c_{im}} \tag{1.93}$$

where V_i is the volume of the pores in the stationary phase capable of holding the ith component and V_s is the total volume of the pores of the stationary phase. Defining a volume, V, such that

$$Vc_{im} = v_m \, c_{im} + v_s \, c_{is} \tag{1.94}$$

where v_m and v_s are the volume of mobile and stationary phase, respectively, in one theoretical "plate" of the column, and dividing through by c_{im}, we have

$$V = v_m + v_s \, k_i \tag{1.95}$$

If we then consider passing a volume of solvent, v, through a column which contains n theoretical "plates," it can be demonstrated (Martin and Singe 1941) that the concentration of the ith component leaving the column in the effluent mobile phase is

$$c_i = (n^{1/2}/V_R)\,[w_i/(2\pi)^{1/2}]\,e^{-n/2\,(1 - v/V_R)^2} \tag{1.96}$$

where:

$$V_R = n\,(v_m + v_s k_i) \tag{1.97}$$

and w_i is the weight of the ith component placed on the column. Plots of this function are illustrated in Fig. 1.15. From these and equation 1.97, it can be seen that as the partition coefficient increases the volume of mobile phase required to remove the component from the column increases. The maximum concentration of the ith component will leave the column when $v = V_R$ and the maximum concentration will be

$$c_{i\,max} = (n^{1/2}/V_R)\,[w_i/(2\pi)^{1/2}] \tag{1.98}$$

The peak width at the point of inflection is

$$\Delta v = 2V_R\,(1/n)^{1/2} \tag{1.99}$$

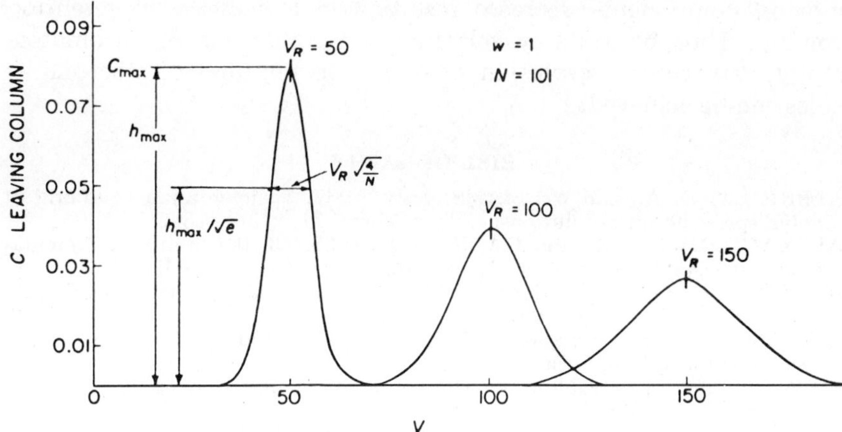

FIG. 1.15. IDEALIZED COLUMN CHROMATOGRAPH. CONCENTRATION OF THREE COMPONENTS WITH DIFFERENT RETENTION VOLUMES, V_R, AS FUNCTIONS OF THE VOLUME OF MOBILE PHASE LEAVING THE COLUMN

w = the total weight of each component; N = number of theoretical plates.

Therefore, the separation of two components by chromatography requires, since the elution curves are Gaussian, that the difference in their retention volumes, V_R, be greater than the sum of their peak widths at the point of inflection. This equation also indicates that the resolution of the components can be improved by increasing the number of theoretical plates since the retention volume increases linearly with the number of theoretical plates while the peak width will increase linearly with the square root of the number of theoretical plates.

If the conditions of chromatography deviate from the ideal situation employed in the derivation of equation 1.96, the resulting elution curves will differ from Gaussian. For example, if the macromolecules are partitioned between the phases by adsorption and the concentrations were sufficiently high to cause the partition coefficient to decrease at the higher concentrations on the column the curve will tend to tail off.

If the elution of macromolecules is performed by stepwise changes in the composition of the mobile phase or by gradient elutions in which the composition changes continuously, the partition coefficients will change and, therefore, by a suitable choice of conditions, satisfactory resolution of the components can be obtained with the passage of a reasonable volume of mobile phase since by suitable changes in the mobile phase the partition coefficients of the slower moving components decrease resulting in a decrease in retention volume. Thus by suitably selecting the conditions of chromatography, satisfactory resolution of numerous mixtures of macromolecules can be achieved.

BIBLIOGRAPHY

ALBRIGHT, D. A., and WILLIAMS, J. W. 1967. Sedimentation equilibria in polydisperse nonideal solutions. J. Phys. Chem. *71*, 2780-2786.

ALEXANDROWICZ, Z., and DANIELS, E. 1968. On the limiting sedimentation coefficients of poly-electrolytes. Biopolymers *6*, 1500-1502.

BENOIT, H. 1950. Application de l'effet Kerr a l'étude des solutions d'acide thymonucléique. J. Chem. Phys. *47*, 719-721.

BIER, M. 1959. Electrophoresis, Theory, Methods and Applications, Vol. I. Academic Press, New York.

BIER, M. 1967. Electrophoresis, Theory, Methods and Applications, Vol. II. Academic Press, New York.

BOEDER, P. 1932. Über Strömungsdoppelbrechung. Z. Physik *75*, 258-281.

CRAMER, J. L., and NEUBERGER, A. 1943. The state of tyrosine in egg albumin and in insulin as determined by spectrophotometric titration. Biochem. J. *37*, 302-310.

DEBYE, P. 1946. Light scattering in solutions. J. Appl. Phys. *15*, 338-342.

DEBYE, P. 1947. Molecular weight determination by light scattering. J. Phys. Colloid Chem. *51*, 18-32.

DEBYE, P., and BUECHE, A. M. 1948. Intrinsic viscosity, diffusion and sedimentation rates of polymers in solution. J. Chem. Phys. *16*, 573-579.

DERJAGUIN, B. 1939. A theory of interaction of particles in presence of electric double layers and the stability of lyophobe colloids and disperse systems. Acta. Physicochim. USSR *10*, 333-346.

DONNAN, F. G. 1934. Die genaue thermodynamik der membrangleichgewichte. Z. Physik. Chem. (Leipzig) Abt. A, *168*, 369-380.

DOTY, P., and STEINER, R. F. 1950. Light scattering and spectrophotometry of colloidal solutions. J. Chem. Phys. *18*, 1211-1220.

DURETTE, P. L., and HORTON, D. 1971. Conformational analysis of sugars and their derivatives. *In* Advances in Carbohydrate Chemistry and Biochemistry, Vol. 25, M. L. Wolfrom, R. S. Tipson, and D. Horton (Editors). Academic Press, New York.

EDSALL, J. T. 1943. Proteins as acids and bases. *In* Proteins, Amino Acids and Peptides, E. J. Cohn, and J. T. Edsall (Editors). Reinhold Publishing Co., New York.

EINSTEIN, A. 1911. New determination of molecular dimensions. Ann. Physik *34*, 591-592.

FLEXNER, L. B. 1937. A thermodynamic analysis of ultrafiltration. The ultrafiltration of sucrose and colloidal solutions. J. Biol. Chem. *121*, 615-630.

FLORY, P. J. 1953. Configurational and frictional properties of the polymer molecule in dilute solution. *In* Principles of Polymer Chemistry, Cornell Univ. Press, Ithaca, N.Y.

FUJITA, H. 1969. A new approximation to the sedimentation equilibrium equation for polydisperse nonideal solutions. J. Phys. Chem. *73*, 1759-1761.

GLASER, A. N., and McKENZIE, H. A. 1963. The denaturation of proteins. IV. Conalbumin and Iron (III)—Conalbumin in urea solution. Biochim. Biophys. Acta. *71*, 109-123.

GLASER, A. N., McKENZIE, H. A., and WAKE, R. G. 1963. The denaturation of proteins. II. Ultraviolet absorption spectra of bovine serum albumin and ovalbumin in urea and in acid solution. Biochim. Biophys. Acta *69*, 240-248.

GRAHAM, T. 1950. On the diffusion of liquids. Phil. Trans. Ser. A. *140*, 1-46.

HAMAKER, H. C. 1937. The London-van der Waals attraction between spherical particles. Physica *4*, 1058.

HERMANNS, J. J., and FUJITA, H. 1955. Distribution of micellar weights in soap solutions. Proc. Roy. Neth. Acad. Sci. Ser. B, *58*, 91-96.

JONES, I. D., and GORTNER, R. K. 1932. Free and bound water in elastic and non-elastic gels. J. Phys. Chem. *36*, 387-436.

KRIMM, S. 1962. Infrared spectra and chain conformation of proteins. J. Mol. Biol. *4*, 528-540.

LAL, H. 1959. Hydrogen ion equilibria and the interaction of Cu^{II} and Co^{II} with bovine serum albumin. J. Am. Chem. Soc. *81*, 844-848.

LEHNINGER, A. L. 1970. Protein: Conformation. *In* Biochemistry, Worth Publishing, New York.

LONGSWORTH, L. G. 1941. The influence of pH on the mobility and diffusion of ovalbumin. Ann. N.Y. Acad. Sci. *41*, 267-285.

MALMSTROM, B. G., VANNGARD, T., and LARSSON, M. 1958. An electron-spin resonance study of the interaction of manganous ions with enolase and its substrate. Biochim. Biophys. Acta. *30*, 1-5.

MARTIN, A. J. P., and SINGE, R. L. M. 1941. Separation of the higher mono-amino acids by counter-current liquid-liquid extraction: the amino-acid composition of wool. Biochem. J. *35*, 91-121.

McKENZIE, H. A. 1970. Effects of changes in environmental conditions on the state of association, conformation and structure. *In* Milk Proteins, Vol. I, H. A. McKenzie (Editor). Academic Press, New York.

McMEEKIN, T. L., and WARNER, R. C. 1942. The hydration of β-lactoglobulin crystals. J. Am. Chem. Soc. *64*, 2393-2398.

MOFFITT, W., and YANG, J. T. 1956. The optical rotatory dispersion of simple polypeptides I. Proc. Natl. Acad. Sci. *42*, 596-603.

MORAWETZ, H. 1975. Macromolecules in Solution, 2nd Edition. John Wiley & Sons, New York.

MOSCOWITZ, A. 1962. Theoretical aspects of optical activity. Part 1: Small molecules. Advan. Chem. Phys. *4*, 67-111.

MOY, J. H., CHAN, K. C., and DOLLAR, A. M. 1971. Bound water in fruit products by the freezing method. J. Food Sci. *36*, 498-499.

NEUGEBAUER, T. 1942. Berechnung der lichtzerstreuung von fadenkettenlösungen. Ann. Physik. Ser. 5, *12*, 509-538.

NORBÖ, R. 1939. Bestimmung der calciumionenkonzentration in ultrafiltrat von biologischen flussigkeiten. Biochem. Z. *301*, 58-60.

NOZAKI, Y., and TANFORD, C. 1967. Examination of titration behavior. *In* Methods in Enzymology, Vol. XI, C. H. W. Hirs (Editor). Academic Press, New York.

OSTWALD, W. 1907. Zur systematik der kolloide. Kolloid Z. *1*, 291-300, 331-341.

PALS, D. T. F., and HERMANNS, J. J. 1952. Sodium salts of pectin, and of carboxymethyl cellulose in aqueous sodium chloride. II. Osmotic pressure. Rec. Trav. Chim. *71*, 458-467.

PAULING, L., and COREY, R. B. 1951. The structure of hair, muscle and related proteins. Proc. Natl. Acad. Sci. *37*, 261-271.

PAULING, L., COREY, R. B., and BRANSON, H. R. 1951. The structure of proteins: Two hydrogen-bonded helical configurations of the polypeptide chain. Proc. Natl. Acad. Sci. *37*, 205-211.

PAYENS, T. A. J., and VAN MARKWIJK, B. W. 1963. Some features of the association of β-casein. Biochim. Biophys. Acta. *71*, 517-530.

PERRIN, F. 1934. Mouvement Brownien d'un ellipsoide. I. Dispersion diélectrique pour des molécules ellipsoidales. J. Phys. Radium, Ser. 7, *5*, 497-511.

RAYLEIGH, LORD, 1881. On the electromagnetic theory of light. Phil. Mag. Ser. 5, *12*, 81-101.

RAYLEIGH, LORD. 1911. The incidence of light upon a transparent sphere of dimensions comparable with the wavelengths. Proc. Roy. Soc. (London) Ser. A, *84*, 25-46.

RISEMAN, J., and KIRKWOOD, J. G. 1949. The rotatory diffusion constants of flexible molecules. J. Chem. Phys. *17*, 442-446.

SCATCHARD, G., SCHEINBERG, I. H., and ARMSTRONG, S. H. 1950. Physical chemistry of protein solutions. IV. The combination of human serum albumin with chloride ion. J. Am. Chem. Soc. *72*, 535-540.

SCHACHMAN, H. K. 1959. Ultracentrifuge in Biochemistry. Academic Press, New York.

SCHERAGA, H. A. 1955. Non-Newtonian viscosity of solutions of ellipsoidal particles. J. Chem. Phys. *23*, 1526-1532.

SCHMIDT, D. G., and PAYENS, T. A. J. 1964. Differences in aggregation behavior of the genetic variants of α_{s1}-casein. Neth. Milk Dairy J. *18*, 108-111.

SHANBAG, S., STEINBERG, M. P., and NELSON, A. I. 1970. Bound water defined and determined at constant temperature by wide-line NMR. J. Food Sci. *35*, 612-615.

SIMHA, R. 1940. The influence of Brownian movement on the viscosity of solutions. J. Phys. Chem. *44*, 25-34.

SUNDARARAJAN, N. R., and WHITNEY, R. McL. 1975. Murexide for determination of free and protein-bound calcium in model systems. J. Dairy Sci. 58, 1595-1608.

SUSI, H. 1969. Infrared spectra of biological macromolecules and related systems. *In* Structure and Stability of Biological Macromolecules, S. N. Timasheff, and G. P. Fasman (Editors). Marcel Dekker, New York.

SUSI, H., TIMASHEFF, S. N., and STEVENS, L. 1967. Infrared spectra and protein conformations in aqueous solutions. J. Biol. Chem. *242*, 5460-5466.

SVEDBERG, T. 1926. Über die bestimmung von molekulargewichten durch zentrifugierung. Z. Physik. Chem. (Leipzig) *121*, 65-77.

TANFORD, C. 1961. Physical Chemistry of Macromolecules. John Wiley & Sons, New York.

TANFORD, C. 1962. The interpretation of hydrogen ion titration curves of proteins. *In* Advances in Protein Chemistry, Vol. 17, C. B. Anfinsen, M. I. Anson, and K. Bailey (Editors). Academic Press, New York.

TEALE, F. W. J., and WEBER, G. 1959. Ultraviolet fluorescence of proteins. Biochem. J. *72,* 15P.

TIMASHEFF, S. N., and TOWNEND, R. 1964. Structure of the β-lactoglobulin tetramer. Nature *203*, 517-519.

UDENFRIEND, S. 1962. Fluorescence Assay in Biology and Medicine. Academic Press, New York.

URRY, D. W. 1968. Optical rotation. Ann. Rev. Phys. Chem. *19*, 477-530.

VERWEY, E. J. W., and OVERBEEK, J. T. G. 1948. Theory of the Stability of Lyophobic Colloids. Elsevier Publishing Co., New York.

VINK, H. 1963. Studies on membrane equilibria in multicomponent systems. Acta. Chem. Scand. *17*, 2524-2535.

VON SMOLUCHOWSKI, M. 1917. Versuch einer mathematischen theorie der koagulationskinetik kolloider Lösungen. Z. Physik. Chem. (Leipzig) *92*, 129-168.

VON WEIMAN, P. P. 1908. Zur lehre von den züstaden der materie. Kolloid Z. *2*, 199-208, 230-237, 275-284, 301-307, 326-335.

WILLIAMS, J. W., VAN HOLDE, K. E., BALDWIN, R. L., and FUJITA, H. 1958. The theory of sedimentation analyses. Chem. Rev. *58*, 715-806.

ZIMM, B. H. 1948. Apparatus and methods for measurement an interpretation of the angular variation of light scattering: preliminary results on polystyrene solutions. J. Chem. Phys. *16*, 1099–1116.

ZITTLE, C. A., DELLA MONICA, E. S., RUDD, R. K., and CUSTER, J. H. 1958. Binding of calcium to casein: Influence of pH and calcium and phosphate concentrations. Arch. Biochem. Biophys. *76*, 342–353.

CHAPTER 2

Robert McL. Whitney

Milk Proteins

INTRODUCTION

The importance of milk in human nutrition and the problems involved in its processing and preservation have resulted in extensive studies of its composition and properties. The milk proteins, particularly, have been investigated in greater detail than any other proteins except possibly those of blood. Definitive work on their physical-chemical and colloidal properties depends upon their fractionation and isolation. Braconnot (1830) first reported the separation of casein from milk by acid precipitation and later Mulder (1838) and Hammarsten (1883, 1885) proposed similar methods for the separation of this protein fraction which made it possible for these and other early workers to investigate its properties. Sebelien (1885) separated the whey proteins into globulin and albumin fractions. However, it wasn't until after the development of more sophisticated fractionation techniques, primarily in the last 20 yr, and modern methods for the investigation of milk protein colloidal properties that our knowledge of the behavior of this complex protein system advanced rapidly.

While some treatment of the nomenclature, classification, composition and primary structure of the milk proteins is necessary for the understanding of their colloidal behavior, the primary thrust of this chapter will be directed toward our knowledge of their structure and conformation and the effect of these characteristics upon their properties. Some studies have been made upon the proteins of the milk of all mammals, but bovine milk, because of its commercial importance, has been more comprehensively investigated and, therefore, the discussions in this chapter will be based largely on the milk of this genus.

MILK PROTEINS, THEIR COMPOSITION AND NOMENCLATURE

The Major Proteins

During the 19th and early 20th centuries, the fractionation of the milk proteins was limited to casein and the classical lactalbumin and lactoglobulin fractions of the whey proteins. Casein, which currently is

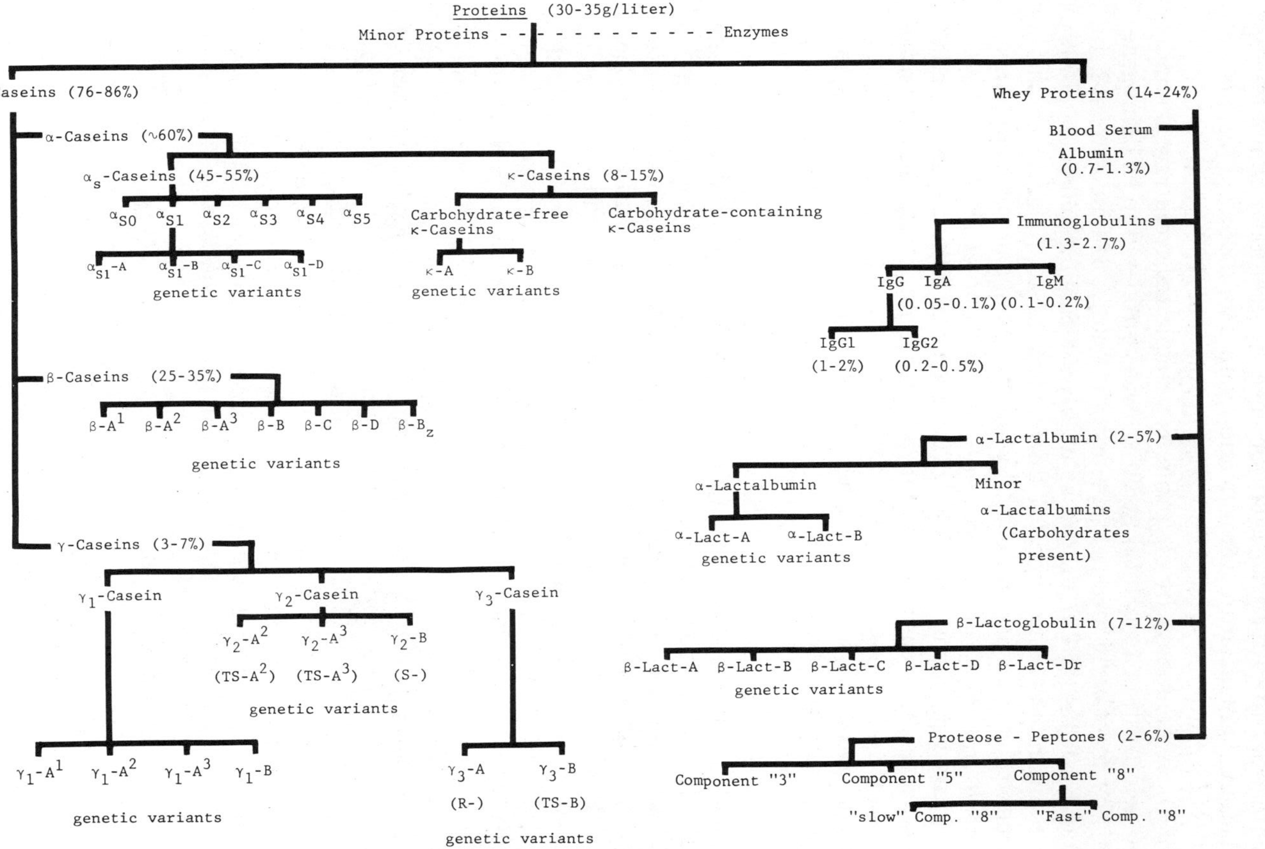

FIG. 2.1. THE DISTRIBUTION OF PROTEINS IN BOVINE MILK

defined by the American Dairy Science Association's Committee on Protein Nomenclature Classification and Methodology (Jenness *et al.* 1956; Brunner *et al.* 1960; Thompson *et al.* 1965; and Rose *et al.* 1970) as consisting of the proteins precipitated from raw skimmilk by acidification to pH 4.6 at 20° C, was considered to be homogeneous by most early investigators until the studies of Linderstrøm-Lang and Kodema (1925) on the solubility of casein. Mellander (1939) demonstrated the presence of three fractions, α-, β-, and γ-casein, in order of decreasing mobility by moving-boundary electrophoresis, and their separation was accomplished by Hipp *et al.* (1952). Subsequently, Waugh and von Hippel (1956) separated the α-casein fraction into a calcium-sensitive fraction, α_s-casein and a calcium-insensitive fraction, κ-casein.

The classical lactalbumin fraction was shown to be heterogenous by Sjögren and Svedberg (1930), upon ultracentrifugation, and 4 yr later Palmer (1934) reported the crystallization of a protein which was later called β-lactoglobulin due to its identification by Pederson (1936) with the β-peak in his ultracentrifuge patterns and its low solubility at low ionic strengths at its isoelectric point. Following this work, Sørenson and Sørenson (1939) crystallized a protein from the lactalbumin fraction which subsequent work by Gordon and Semmett (1953) demonstrated was the α-peak in Pederson's ultracentrifuge patterns, and, therefore, it was called α-lactalbumin. By repeated fractionation of the mother liquor from the crystallization of β-lactoglobulin, Polis *et al.* (1950) crystallized an albumin which they demonstrated in almost all respects to be identical to bovine blood serum albumin.

The lactoglobulin fraction was found to be a complex mixture of immunoglobulins similar to those present in blood. Smith (1946, 1948) first classified them as euglobulins and pseudoglobulins, depending upon their solubility in water. However, as knowledge of this fraction increased, they were reclassified according to their antigenic determinants in line with the system proposed by a committee of the World Health Organization (Ceppelini *et al.* 1964).

The free-boundary electrophoretic patterns of the whey or serum proteins indicated, in addition to the above, the presence of three proteins (Larson and Rolleri 1955) which were shown to be in the proteose-peptone fraction of Rowland (1938), or that fraction of the milk proteins that was not precipitated by heating at 95–100° C for 20 min followed by acidification to pH 4.7.

α_s-Caseins.—The α_s-caseins are those proteins in the α-casein fraction that are precipitated in 0.4 *M* $CaCl_2$ at pH 7 and 4°C and are capable of being stabilized by κ-casein. They consist of one major component, α_{s1}-casein and several minor α_s-caseins.

At the present time, four genetic variants of α_{s1}-casein are known to exist: A, D, B, and C in order of decreasing mobility in starch-gel electrophoresis at pH 8.6 in 7 *M* urea and a discontinuous buffer system (Thompson 1971). The phenotypes of the various genotypes have also been observed (i.e., A (A/A), AB (A/B), AC (A/C) etc.). The polymorphs are breed specific with α_{s1}-casein B predominant in Western dairy cattle and α_{s1}-casein C predominant in Zebu cattle (Aschaffenburg *et al.* 1968).

The primary structures of all of the known genetic variants have recently been established by Mercier *et al.* (1971) as shown in Fig. 2.2. As indicated, the B-variant consists of 199 amino acid residues as follows: Aspartic acid $(Asp)_7$, Asparagine $(Asn)_8$, Threonine $(Thr)_5$, Serine $(Ser)_8$, Phosphoserine $(SerP)_8$, Glutamic acid $(Glu)_{25}$, Glutamine $(Gln)_{14}$, Proline $(Pro)_{17}$, Glycine $(Gly)_9$, Alanine $(Ala)_9$, Valine $(Val)_{11}$, Methionine $(Met)_5$, Isoleucine $(Ile)_{11}$, Leucine

```
                                           10                                   20
H.Arg-Pro-Lys-His-Pro-Ile-Lys-His-Gln-Gly-Leu-Pro-Gln-[Glu-Val-Leu-Asn-Glu-Asn-Leu-
                                                         Absent in Variant A
                                           30                                   40
[Leu-Arg-Phe-Phe-Val-Ala]-Pro-Phe-Pro-Gln-Val-Phe-Gly-Lys-Glu-Lys-Val-Asn-Glu-Leu-

                                           50                                   60
Ser-Lys-Asp-Ile-Gly-Ser-Glu-Ser-Thr-Glu-Asp-Gln-[Ala]-Met-Glu-Asp-Ile-Lys-Glu-Met-
                    P       P                    ThrP (Variant D)
                                           70                                   80
Glu-Ala-Glu-Ser-Ile-Ser-Ser-Ser-Glu-Glu-Ile-Val-Pro-Asn-Ser-Val-Glu-Gln-Lys-His-
            P       P   P   P                           P
                                           90                                  100
Ile-Gln-Lys-Glu-Asp-Val-Pro-Ser-Glu-Arg-Tyr-Leu-Gly-Tyr-Leu-Glu-Gln-Leu-Leu-Arg-

                                          110                                  120
Leu-Lys-Lys-Tyr-Lys-Val-Pro-Gln-Leu-Glu-Ile-Val-Pro-Asn-Ser-Ala-Glu-Glu-Arg-Leu-
                                                        P
                                          130                                  140
His-Ser-Met-Lys-Gln-Gly-Ile-His-Ala-Gln-Gln-Lys-Glu-Pro-Met-Ile-Gly-Val-Asn-Gln-

                                          150                                  160
Glu-Leu-Ala-Tyr-Phe-Tyr-Pro-Glu-Leu-Phe-Arg-Gln-Phe-Tyr-Gln-Leu-Asp-Ala-Tyr-Pro-

                                          170                                  180
Ser-Gly-Ala-Trp-Tyr-Tyr-Val-Pro-Leu-Gly-Thr-Gln-Tyr-Thr-Asp-Ala-Pro-Ser-Phe-Ser-

                                          190                               199
Asp-Ile-Pro-Asn-Pro-Ile-Gly-Ser-Glu-Asn-Ser-[Glu]-Lys-Thr-Thr-Met-Pro-Leu-Trp.OH
                                            Gly   (Variant C)
```

From Mercier et al. 1971; Grosclaude et al. 1973

FIG. 2.2. PRIMARY STRUCTURE OF BOVINE α_{s1}-CASEIN B

The enclosed amino acid residues are the sites corresponding to the mutational differences of the genetic variants A, C, and D

(Leu)$_{17}$, Tyrosine (Tyr)$_{10}$, Phenylalanine (Phe)$_8$, Lysine (Lys)$_{14}$, Histidine (His)$_5$, Tryptophan (Trp)$_2$, and Arginine (Arg)$_6$, with a calculated molecular weight of 23,613. The other genetic variants differ from B in that C has a glycine residue substituted for the glutamic acid residue at position 192, D has a phophorylated threonine residue substituted for the alanine residue at position 53, and A lacks 13 amino acids from positions 14 to 26.

With regard to the minor α_s-caseins, Annan and Manson (1969) have isolated the component migrating in alkaline starch-gel electrophoresis immediately ahead of α_{s1}-casein which they called α_{s0}-casein. They also obtained a fraction containing at least three components, ($\alpha_{s2,\,3,\,4}$-caseins). Hoagland *et al.* (1971) isolated α_{s3}- and α_{s4}-casein by DEAE-cellulose chromatography and tentatively established their amino-acid composition (Table 2.1). The phosphorus content of α_{s4}-casein was found to be 1.03%. These investigators also isolated α_{s5}-casein which moves electrophoretically in alkaline gels just in front of the β-casein band. From its amino acid composition and the effect of mercaptoethanol upon its electrophoretic properties, they tentatively established that this protein consisted of α_{s3}- and α_{s4}-casein molecules linked together by a disulfide bond. The C-terminals of these three proteins are probably Leu–Tyr–OH. While no genetic variants have been definitely established for these caseins, Aschaffenburg *et al.* (1968) observed indications of poly-

TABLE 2.1

TENTATIVE AMINO ACID COMPOSITION OF α_{s3}- AND α_{s4}-CASEIN

	Residue Number[1]	
Amino Acid	α_{s3}-Casein	α_{s4}-Casein
Aspartic Acid	13	12
Threonine	10	10
Serine	11-12	10-11
Glutamic Acid	29	28
Proline	8	7
Glycine	2	2
Alanine	6	6
½-Cystine[2]	1	1
Valine	10	10
Methionine	3	2-3
Isoleucine	8	7
Leucine	9	9
Tryptophan	7	7
Phenylalanine	4	4
Lysine	15	16
Histidine	3	3
Arginine	4	4

Source: Thompson (1971).
[1] Calculated on the basis of four residues of arginine
[2] Not corrected for destruction.

morphism in Zebu casein and Michalek (1967) noted the absence of α_{s3}- and α_{s4}-casein in milk from some Red Danish cattle.

κ-Caseins.—The κ-caseins comprise that portion of the α-casein fraction that is soluble in 0.4 *M* $CaCl_2$ at pH 7.0 and 4°C. They are capable of stabilizing the α_s-caseins. As isolated, they apparently consist of a mixture of polymers of a monomeric unit of molecular weight of approximately 19,000 held together by intermolecular disulfide bonds (Swaisgood and Brunner 1963). However, Beeby (1964) detected free SH groups in κ-casein when all of the calcium had been removed by ethylenediaminetetraacetate (EDTA) or oxalate, suggesting the possibility that the reduced monomers rather than the disulfide-linked polymers are more nearly the native form of κ-casein. However, more work is needed before a definite conclusion can be reached.

The monomers themselves possess considerable heterogeneity since gel electrophoresis of κ-casein from pooled milk after reduction of the disulfide bonds followed by carboxymethylation yields a number of protein bands (Mackinlay *et al.* 1966). This heterogeneity arises from at least three different sources: genetic difference, variations in carbohydrate content, and a possible variation in the para-κ-casein portion of the molecule. Two genetic variants have been reported (Woychik *et al.* 1966). While the gel electrophoretic patterns of the reduced genetic variants are still complex, the component of the A variant which binds the most protein dye possesses a greater mobility than the corresponding component of κ-casein B. These components of both variants are free of carbohydrate while the other components, although they have similar amino acid and phosphorus contents, contain increasing amounts of carbohydrates with increasing electrophoretic mobility at pH 8.6. Mackinlay and Wake (1966) isolated a fraction containing the three fastest components and found it to contain approximately 10% carbohydrate compared to approximately 5% for whole κ-casein.

Investigations of the carbohydrate moiety have yielded different ratios of *N*-acetylneuraminic acid, D-galactose, *N*-acetyl galactosamine and D-manose. Wheelcock and Sinkerson (1970) fractionated the glycopeptides from the action of rennin on milk by DEAE cellulose and found them to contain varying ratios of the carbohydrates. All of the glycopeptides contain D-galactose and *N*-acetyl galactosamine. The first glycopeptide eluted was the only one containing mannose but was devoid of *N*-acetylneuraminic acid. The second contained less than 1 mole of *N*-acetylneuraminic acid per mole of glycopeptide and therefore this might be due to contamination by peptides eluted later which contained *N*-acetylneuraminic acid, D-galactose and *N*-acetyl galactosamine in the ratio of 1:2:1. In contrast, Tran and Baker (1970) proposed a trisaccharide unit, α-*N*-acetyl neuraminyl-(2→6)-β-galactosyl-(1→3 or 6)-*N*-acetyl galactosamine attached to the peptide chain through the OH groups of serines and threonines in the C-terminal portion of κ-casein.

When the disulfide-reduced form of either variant is treated with rennin, two electrophoretically different para-κ-caseins derivatives are observed moving toward the cathode in alkaline media (Mackinlay and Wake 1966). The major derivative appears to originate from the carbohydrate-free κ-casein component and the minor derivative from the κ-casein component that moves immediately in front of the carbohydrate-free component (Mackinlay *et al.* 1966). The origin of these two para-κ-caseins is still uncertain since the minor component may be an artifact resulting from cyanate in the urea used (Kim *et al.* 1969).

Recently, Jollès *et al.* (1972) and Mercier *et al.* (1973) independently reported the primary structures of the reduced forms of the carbohydrate-free components of κ-casein A and B. These two structures are in nearly total agreement as illustrated in Fig. 2.3. According to Mercier *et al.* (1973) the B-variant consists of 169 amino acid residues as follows: Asp_4, Asn_7, Thr_{14}, Ser_{12}, $SerP_1$, $Pyroglu_1$, Glu_{12}, Gln_{14}, Pro_{20}, Gly_2, Ala_{15}, 1/2-Cys_2, Val_{11}, Met_2, Ile_{13}, Leu_8, Tyr_9, Phe_4, Lys_9, His_3, Trp_1, and Arg_5, with a calculated molecular weight of 19,005. κ-casein A differs from B by the substitutions of a threonine residue for isoleucine in position 136 and an aspartic acid residue for alanine in position 148.

β-Caseins.—β-casein is the casein fraction that possesses a free-boundary electrophoretic mobility of -3.1×10^{-5} cm^2 volt^{-1} see^{-1} in veronal buffer, pH 8.6, ionic strength 0.1, at 2°C and is soluble in 3.3 *M* urea but insoluble in 1.7 *M* urea.

The known genetic variants of β-casein are more numerous and their differentiation is more complicated. Aschaffenburg (1961) reported the existence and breed specificity of β-casein A, B, and C in order of decreasing mobility in zonal electrophoresis in alkaline media. However, in 1966, Peterson and Kopfler observed that β-casein A existed in three genetic forms when subjected to acid gel electrophoresis. The confusion in nomenclature caused by this discovery was clarified by Kiddy *et al.* (1966) by arranging the genetic variants in acid gel electrophoresis in the following order of decreasing mobility: C, B, A^1, A^2 and A^3. Thus the genetic variants maintained the same letter in both alkaline and acid gel electrophoresis. In phenotyping Indian Zebu cattle, Aschaffenburg *et al.* (1968), demonstrated the existence of two new genetic variants: β-casein D, which migrates between B and C in gel electrophoresis at pH 9.1, and

```
1                                         10                                      20
PyroGlu-Glu-Gln-Asn-Gln-Glu-Gln-Pro-Ile-Arg-Cys-Glu-Lys-Asp-Glu-Arg-Phe-Phe-Ser-Asp-
        (Gln)       (Glu)   (Glu)
                                          30                                      40
    Lys-Ile-Ala-Lys-Tyr-Ile-Pro-Ile-Gln-Tyr-Val-Leu-Ser-Arg-Tyr-Pro-Ser-Tyr-Gly-Leu-

                                          50                                      60
    Asn-Tyr-Tyr-Gln-Gln-Lys-Pro-Val-Ala-Leu-Ile-Asn-Asn-Gln-Phe-Leu-Pro-Tyr-Pro-Tyr-

                                          70                                      80
    Tyr-Ala-Lys-Pro-Ala-Ala-Val-Arg-Ser-Pro-Ala-Gln-Ile-Leu-Gln-Trp-Gln-Val-Leu-Ser-

                                          90                                      100
    Asp-Thr-Val-Pro-Ala-Lys-Ser-Cys-Gln-Ala-Gln-Pro-Thr-Thr-Met-Ala-Arg-His-Pro-His-
    (Asn)                                                                         (Pro)
                    105↓106               110                                     120
    Pro-His-Leu-Ser-Phe·Met-Ala-Ile-Pro-Pro-Lys-Lys-Asn-Gln-Asp-Lys-Thr-Glu-Ile-Pro-
    (His)
                                          130                     136             140
    Thr-Ile-Asn-Thr-Ile-Ala-Ser-Gly-Glu-Pro-Thr-Ser-Thr-Pro-Thr-[Ile]-Glu-Ala-Val-Glu-
                                                                Thr(Variant A)
                    NeuNAc(2→3(6))Gal(1→3(6)GalNAc(carbohydrate containing κ-caseins)
                                148       150                                     160
    Ser-Thr-Val-Ala-Thr-Leu-Glu-[Ala]-Ser-Pro-Glu-Val-Ile-Glu-Ser-Pro-Pro-Glu-Ile-Asn-
                     (Variant A)Asp   P
                                     169
    Thr-Val-Gln-Val-Thr-Ser-Thr-Ala-Val.OH
```

FIG. 2.3. THE PRIMARY STRUCTURE OF BOVINE κ-CASEIN B

The enclosed amino acid residues are the sites corresponding to the mutational differences of the genetic variants A. The arrow indicates the point of attack of rennin (Mercier *et al.* 1973). The amino acids in parenthesis indicate the differences in the sequence observed by Jolles *et al.* (1972).

β-casein B_z, which behaves electrophoretically the same as β-casein B but possesses a different peptide map for its chymotryptic digest.

Ribadeau-Dumas *et al.* (1972) and Grosclaude *et al.* (1972) have established the primary structure of the genetic variants found in Western breeds of cattle (Fig. 2.4). As indicated, the A^2 variant consists of 209 amino acid residues with the following composition: Asp_4, Asn_5, Thr_9, Ser_{11}, $SerP_5$, Glu_{18}, Gln_{21}, Pro_{35}, Gly_5, Ala_5, Val_{19}, Met_6, Ile_{10}, Leu_{22}, Tyr_4, Phe_9, Trp_1, Lys_{11}, His_5, Arg_4, with a calculated molecular weight of 23,979. The other genetic variants differ from A^2 in that A^1 has a histidine residue substituted for the proline residue at position 67; A^3 has a glutamine substituted for histidine at position 106; B has two substitutions, histidine for proline at position 67 and arginine for serine at position 122; and C has two substitutions, lysine for glutamic acid at position 37 and histidine for proline at position 67 and, in addition, the serine at position 35 is not phosphorylated.

10 20
H.Arg-Glu-Leu-Glu-Glu-Leu-Asn-Val-Pro-Gly-Glu-Ile-Val-Glu-Ser-Leu-Ser-Ser-Ser-Glu-
P P P P

→γ_1-caseins
30 40
Glu-Ser-Ile-Thr-Arg-Ile-Asn-Lys-Lys-Ile-Glu-Lys-Phe-Gln-Ser-Glu-[Glu]-Gln-Gln-Gln-
(absent in variant C) P Lys (Variant C)

50 60
Thr-Glu-Asp-Glu-Leu-Gln-Asp-Lys-Ile-His-Pro-Phe-Ala-Gln-Thr-Gln-Ser-Leu-Val-Tyr-

70 80
Pro-Phe-Pro-Gly-Pro-Ile-[Pro]-Asn-Ser-Leu-Pro-Gln-Asn-Ile-Pro-Pro-Leu-Thr-Gln-Thr-
(variants C, A^1 and B) His

90 100
Pro-Val-Val-Val-Pro-Pro-Phe-Leu-Gln-Pro-Glu-Val-Met-Gly-Val-Ser-Lys-Val-Lys-Glu-

→γ_3-caseins (R-, TS-B)
110 120
Ala-Met-Ala-Pro-Lys-[His]-Lys-Glu-Met-Pro-Phe-Pro-Lys-Tyr-Pro-Val-Gln-Pro-Phe-Thr-
Gln (variant A^3)
→γ_2-caseins (S-, TS-A^2)

130 140
Glu-[Ser]-Gln-Ser-Leu-Thr-Leu-Thr-Asp-Val-Glu-Asn-Leu-His-Leu-Pro-Pro-Leu-Leu-Leu-
Arg (variant B)

150 160
Gln-Ser-Trp-Met-His-Gln-Pro-His-Gln-Pro-Leu-Pro-Pro-Thr-Val-Met-Phe-Pro-Pro-Gln-

170 180
Ser-Val-Leu-Ser-Leu-Ser-Gln-Ser-Lys-Val-Leu-Pro-Val-Pro-Glu-Lys-Ala-Val-Pro-Tyr-

190 200
Pro-Gln-Arg-Asp-Met-Pro-Ile-Gln-Ala-Phe-Leu-Leu-Tyr-Gln-Gln-Pro-Val-Leu-Gly-Pro-

209
Val-Arg-Gly-Pro-Phe-Pro-Ile-Ile-Val-OH

From Ribadeau-Dumas et al. 1972

FIG. 2.4. PRIMARY STRUCTURE OF BOVINE β-CASEIN A^2

The enclosed amino acid residues are the sites corresponding to the mutational differences of the genetic variants A^1, A^3, B and C. The arrows indicate the portions of the β-casein sequence believed to be identical with γ_1-, γ_2-, and γ_3-caseins.

The specific differences in the primary structure of the genetic variants of Zebu cattle, β-casein B_z and D, have not as yet been established. However, Thompson *et al.* (1969B) have provided tentative data upon the amino acid and phosphorous contents of these proteins (Table 2.2).

γ-Caseins.—The γ-caseins were first defined as those proteins in milk which occur in the slowest moving component of the free-boundary electrophoretic pattern of casein at pH 8.6, ionic strength 0.1, at 2°C. Genetic polymorphism in this fraction was first noted by Aschaffenburg (1961). More recently by employing column chromatography and gel electrophoresis in alkaline and acid media, Groves *et al.* (1962; 1968) and Groves and Kiddy (1968) defined the γ-caseins as the fraction eluted at 0.02 *M* phosphate, pH 8.3, from DEAE-cellulose columns. They noted the existence of at least four polymorphic γ-caseins, A^1, A^2, A^3 and B, and three other minor proteins designated as R-, S-, and TS-caseins. At that time, they believed that the last of these existed in two polymorphic forms, TS-casein A^2 and B based upon their sensitivity to temperature. In addition, they noted a close genetic relationship between the β- and γ-caseins in that similarly designated polymorphic forms of each consistently were observed together in milk from individual cows. In this connection, it is interesting to note that γ-casein appears to be absent in milks found to be homozygous with respect to β-casein C.

TABLE 2.2

TENTATIVE AMINO ACID COMPOSITION OF β-CASEIN VARIANTS B_z AND D (ESTIMATED WHOLE NUMBER OF RESIDUES)

Amino Acid	β-Casein B_z[1]	β-Casein D[1]
Lysine	11	12
Histidine	6	5
Arginine	5	4
Aspartic Acid	9–10	10
Threonine	10	11
Serine	15	16
Glutamic Acid	36–37	36–37
Proline	32–33	33–34
Glycine	5	5
Alanine	6	7
Methionine	5	5
Valine	17	17–18
Isoleucine	10	10
Leucine	20	20
Tyrosine	4	4
Phenylalanine	8	8
Tryptophan	1	1
Phosphorus	5	3–4

Source: Thompson *et al.* (1969B).
[1] Based on the presence of 5 glycine residues per molecule

Further work by these investigators (Groves and Gordon 1969; Gordon *et al.* 1972; Groves *et al.* 1972, 1973) secured convincing evidence that the γ-, TS-, S- and R-casein are segments of the larger β-casein molecules and that the TS-casein A and B were not polymorphic but rather that S- and TS-casein A and R- and TS-casein B are the appropriate pairs of genetic variants. Based upon this information, the A.D.S.A. Committee on the Nomenclature and Methodology of Milk Proteins (Whitney *et al.* 1976) has renamed the components of the whole γ-casein fraction according to their polypeptide chain length as derived from their sequences and indicated in Fig. 2.4. Therefore, the protein previously called γ-casein, which is identical with the β-casein segment from residue 29 to 209, is designated as γ_1-casein. Its genetic variants are then termed γ_1-A^1, γ_1-A^2, etc. The designations γ_2-casein A^2 and B are applied to those proteins previously called TS-A^2 and S-casein, respectively, since they are identical to the corresponding β-casein segments from residue 106 to 209. Finally, R- and TS-casein B are designated γ_3-casein A and B, respectively, and correspond to the β-casein segments from residue 108–209. These changes in nomenclature are summarized in Table 2.3.

TABLE 2.3

SUMMARY OF RECENT NOMENCLATURE CHANGES FOR THE γ-CASEIN FRACTION

		Relationship to β-Casein	
Recommended Nomenclature	Former Nomenclature	Genetic Variant	Residues (Inclusive)
γ_1-A^1	γ-A^1	β-A^1	29–209
γ_1-A^2	γ-A^2	β-A^2	29–209
γ_1-A^3	γ-A^3	β-A^3	29–209
γ_1-B	γ-B	β-B	29–209
γ_2-A^2	TS-A^2	*β-$A^{1\text{ or }2}$	106–209
γ_2-A^3	TS-A^3	β-A^3	106–209
γ_2-B	S-	β-B	106–209
γ_3-A	R-	**β-$A^{1\text{ or }2\text{ or }3}$	108–209
γ_3-B	TS-B	β-B	108–209

Source: Whitney *et al.* (1976).
*Genetic point mutation β-A^1, A^2 occurs prior to residue 106, hence $\gamma_2 A^1$ and $\gamma_2 A^2$ are identical.
**Genetic point mutations for β-A^1, A^2, A^3 occur prior to residue 108, hence γ_3-A^1, $\gamma_3 A^2$ and $\gamma_3 A^3$ are identical.

Blood Serum Albumin.—Since the isolation of crystalline albumin from whey by Polis *et al.* (1950) and their demonstration that it was identical in all properties investigated to blood serum albumin except in electrophoretic behavior at pH 4.0, where both samples behaved heterogeneously, very little work has been done on this protein as isolated from milk. In contrast, extensive work has been done on the protein obtained from bovine blood plasma. Considerable evidence exists that this protein is heterogeneous. For example: Ehrenpries *et al.* (1957) observed two sedimenting peaks upon ultracentrifugation suggesting the presence of a dimer and Spencer and King (1971) demonstrated several protein bands by electrophoretic focusing with the two major isoelectric components differing by one unit charge. The chemical nature of this difference is not known.

In spite of this heterogeneity, the amino acid composition of blood serum albumin has been determined by a number of investigators (Brand 1946; Henderson and Snell 1948; Velick and Ronzoni 1948; Stein and Moore 1949; McClure *et al.* 1953; Spahr and Edsall 1964; King and Spencer 1970). Since there is considerable variation in their results, those of King and Spencer (1970) on a purified 1/2-cystinyl blood plasma albumin are reported in Table 2.4 because they are the most recent and agree well with the averages of the other analyses.

Considerable progress has been made on the sequencing of the amino acid residues in bovine blood serum albumin by several investigators (Witter and Tuppy 1960; Bradshaw and Peters 1969; Brown *et al.* 1971; King and Spencer 1970, 1972). King and Spencer (1970) using CNBr followed by reduction and carboxymethylation obtained 5 peptides ranging in size from 36 to 211 amino acid residues and established the sequence of the peptides in the molecule. In a subsequent paper, King and Spencer (1972) determined the amino acid sequence in the 87 amino acid N-terminal peptide and the 36 amino acid C-terminal peptide. The molecule appears to be a single chain with one free sulfhydryl group which is located at position 34 in the N-terminal peptide and probably 17 intramolecular disulfide bonds. The N-terminal and C-terminal amino-acid residues are aspartic acid and alanine, respectively. Brown *et al.* (1971) have secured peptide sequences covering approximately 80% of the albumin molecule. It should not be long before this work is complete, provided the problems involved in the heterogeneity of this protein can be resolved.

β-Lactoglobulins.—The first observation of genetic variants in the milk proteins was made by Aschaffenburg and Drewry (1957A) by filter-paper electrophoresis of β-lactoglobulin at pH 8.6. They demonstrated the presence of β-lactoglobulins A and B in Western breeds

TABLE 2.4

TENTATIVE AMINO ACID COMPOSITION OF BOVINE BLOOD SERUM ALBUMIN (ESTIMATED WHOLE NUMBER OF RESIDUES)

Amino Acid	Number of Residues
Lysine	58
Histidine	17
Ammonia	(27)
Arginine	23
Aspartic Acid	54
Threonine	32
Serine	26
Glutamic Acid	80
Proline	28
Glycine	15
Alanine	44
½-Cystine	36
Valine	35
Methionine	4
Isoleucine	14
Leucine	58
Tyrosine	19
Phenylalanine	26
Tryptophan	2
Total (omitting ammonia)	571
Molecular weight (Calc.)	65,243

Source: King and Spencer (1970).

of cattle and established their genetic origin. In 1962, Bell noted the presence of β-lactoglobulin C in milk from some Australian Jersey cows. Two other variants were detected in 1966. Grosclaude *et al.* (1966) found β-lactoglobulin D in Montbeliarde cattle in France and Bell *et al.* (1966) isolated β-lactoglobulin Dr from the milk of some Australian Droughtmaster cattle. The various genetic variants of β-lactoglobulin differ in their electrophoretic mobility in alkaline starch of polyacrylamide gel in the following order of decreasing mobility A>B>C>D. β-lactoglobulin Dr, which has the same amino-acid composition as β-lactoglobulin A, possesses an electrophoretic mobility on starch gels at pH 8.5 slower than β-lactoglobulin C due to the presence of a carbohydrate moiety attached to it.

The primary structure of β-lactoglobulin A has been tentatively established by the work of Frank and Braunitzer (1967, 1968), Braunitzer and Chen (1972), Braunitzer *et al.* (1972) Townend (1965), Bell *et al.* (1966, 1968), and McKenzie *et al.* (1972), and is summarized in Fig. 2.5. As indicated, the A variant consists of 162 amino-acid residues with the following composition: Asp_{11}, Asn_5, Thr_8, Ser_7, Glu_{16}, Gln_9, Pro_8, Gly_3, Ala_{14}, 1/2 Cys_5, Val_{10}, Met_4,

Ile_{10}, Leu_{22}, Tyr_4, Phe_4, Trp_2, Lys_{15}, His_2, Arg_3, with a calculated molecular weight of 18,362. There is only one sulfhydryl group per molecule and, from the work of McKenzie *et al.* (1972) and the most recent primary sequence (Braunitzer *et al.* 1972), it may be postulated that it is equally distributed between positions 119 and 121 while a disulfide bridge is located either between positions 106 and 121 or 106 and 119 depending on the position of the sulfhydryl group. The other disulfide bridge in both forms links position 66 with 160. Genetic variant B differs from A in the following substitutions: alanine for valine in position 118 and glycine for aspartic acid at position 64. Variant C differs from B by the substitution of histidine for glutamine at positions 59. β-lactoglobulin D differs from B by the substitution of a glutamine residue for a glutamic acid resi-

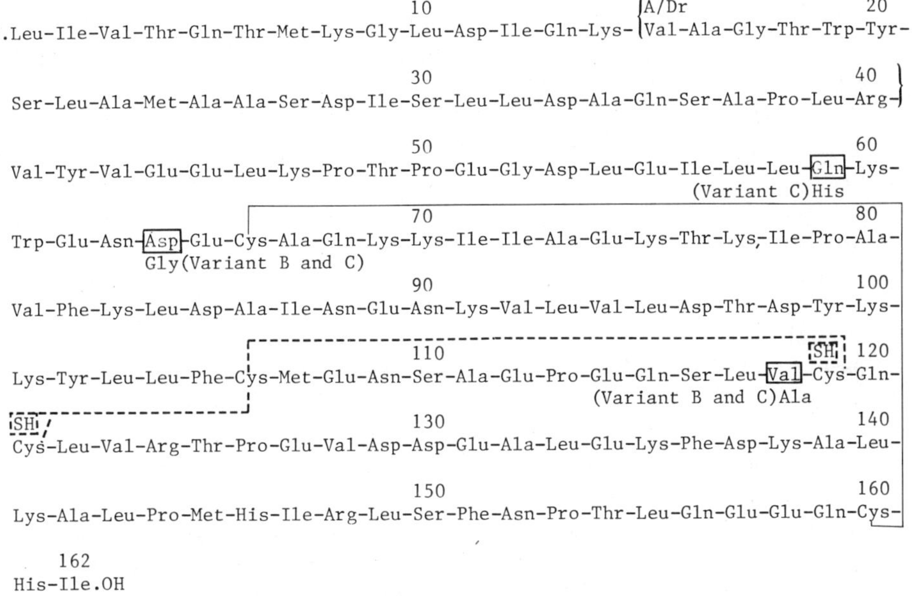

FIG. 2.5. PRIMARY STRUCTURE OF BOVINE β-LACTOGLOBULIN A

The location of the SH-group and the disulfide are deduced by applying the observations of McKenzie *et al.* (1972) to the most recent primary sequence of Braunitzer *et al.* (1972). The SH group is assumed to exist in a 50:50 distribution between positions 119 and 121 with the -SS- bridge location depending upon the position of the SH group. The enclosed amino acid residues are the sites corresponding to the differences in the genetic variants B and C. The difference peptide between variants A and Dr is indicated by brackets. In β-lactoglobulin Dr, a carbohydrate moiety is attached to this peptide.

due probably at position 51. The Droughtmaster variant has the amino acid composition as A but has a carbohydrate moiety attached to the protein which appears to consist of 1.0 *N*-acetylneuraminic acid, 4.3 hexosamine (with glucosamine to galactosamine ratio of 4:1), 1.9 mannose and 0.8 galactose residues (Bell *et al.* 1970).

α-Lactalbumins.—Western cattle yield milk with only one genetic variant of α-lactalbumin, the B-variant which is the slower moving variant in alkaline zonal electrophoresis (Aschaffenburg 1963A). Blumberg and Tombs (1958) and Aschaffenburg (1963A) noted the presence of both the A and B variants in milks from African Fulani and Zebu cattle. Bhattacharya *et al.* (1963) noted the same phenomenon in milk from Indian Zebu cattle.

Approximately 10 yr of work by several investigators (Weil and Telka 1957; Weil and Seibles 1964; Dautrivaux *et al.* 1966; Brew and Campbell 1967) culminated in the publication of the complete primary structure of α-lactalbumin by Vanaman *et al.* 1970) (Fig. 2.6). As indicated, the B variant consists of 123 amino-acid residues with the following composition: Asp_9, Asn_{12}, Thr_7, Ser_7, Glu_8, Gln_5, Pro_2, Gly_6, Ala_3, 1/2 Cys_8, Val_6, Met_1, Ile_8, Leu_{13}, Tyr_4, Phe_4, Trp_4, Lys_{12}, His_3, Arg_1 with a calculated molecular weight of 14,174. All of the 1/2-cystine residues are connected by disulfide linkages as indicated in Fig. 2.6. α-Lactalbumin A differs from B in the substitution of a glutamine residue for the arginine residue in position 10 (Gordon 1971).

In addition to the genetic variants of α-lactalbumin, several minor components have been observed by different investigators in recrystallized preparations of α-lactalbumin. Aschaffenburg and Drewry (1957B) observed a faster moving band in paper electrophoresis at pH 8.6 and isolated the protein. It was found to have the same amino acid composition as α-lactalbumin, but contained in addition one hexosamine residue per molecule (Gordon 1971). Those investigators tentatively called this protein "satellite" α-lactalbumin. In starch gel electrophoreses at pH 7.7, Hopper and McKenzie (1973) observed three minor components in their α-lactalbumin preparation. One of these moves faster than the major component and has the same amino acid composition but probably possesses one less amide group. The two others move more slowly than the major component and have the same amino acid composition but contain carbohydrates. Part of the carbohydrate moiety in the faster one of these components consists of sialic acid while the slower one contains no sialic acid. These authors call these proteins α-lactalbumin fast component (F), slow component (S1) and slow component (S2), respectively. Recently Barman (1973) has reported the isolation of a

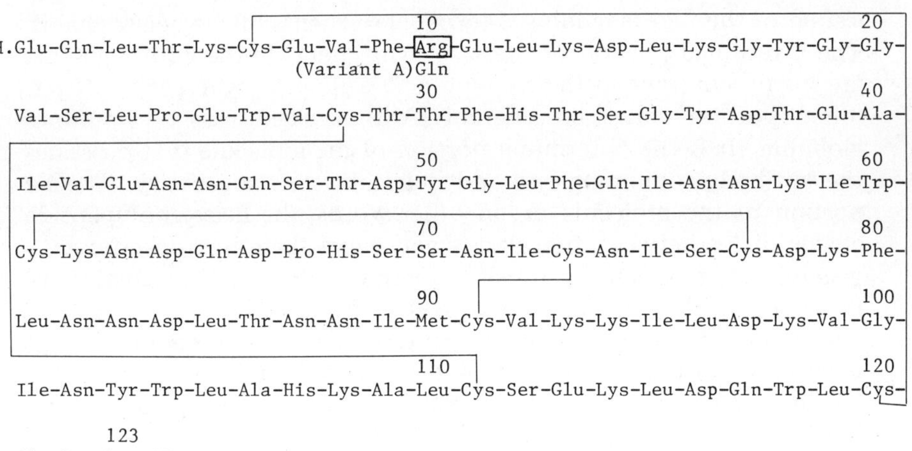

FIG. 2.6. PRIMARY STRUCTURE OF α-LACTALBUMIN B

The disulfide bridges in the molecule are as indicated between positions 6 and 120; 28 and 111; 61 and 77; and 73 and 91 (Vanaman *et al.* 1970). The enclosed amino acid residue is the site corresponding to the difference in genetic variant A.

protein from an α-lactalbumin preparation by chromatography which he called α-lactalbumin III. This protein has the kinetic properties of α-lactalbumin but a different amino acid composition and only three disulfide bonds. More work needs to be done on these proteins to confirm their composition and determine their structure.

Immunoglobulins.—Much of the work on the immunoglobulins has been done on the immunoglobulins of the serum of various species, but much of the basic information secured can be translated to the immunoglobulins of milk since considerable evidence exists for the identity of these proteins in the serum and lacteal secretion of the same species (Pierce and Feinstein 1965). However, there is some evidence that a small change may occur during transport (Kickhöfen *et al.* 1968).

The immunoglobulins of milk like those in blood serum are a complex mixture of proteins with antibody activity and closely related structures. They all appear to be either monomers or polymers of a four-chain molecule consisting of two light polypeptide chains (20,000 mol wt) and two heavy chains (50,000–70,000 mol wt) (Fig. 2.7) (Butler 1969).

Digestion of the four-chain unit with papain yields two Fab fragments, consisting of the N-terminal fragment of one of the heavy chains plus one entire light chain in each, and one Fc fragment, consisting of the two remaining C-terminal segments of the heavy chains. The C-terminal portion of the two light chains in the Fab fragments are highly constant in their amino acid sequence, whereas the N-terminal portion varies considerably within each class of immunoglobulins. It is the N-terminal portion of the molecule that possesses the antibody-combining site and determines its specificity. The Fc portion of the molecule or, in other words, the heavy polypeptide chains differ between the classes and carry their class-specific antigenetic determinants. Each class of immunoglobulin has a distinctive heavy chain which is called γ- for IgG, μ- for IgM and α- for IgA.

The current accepted nomenclature of the immunoglobulin classes is based upon their antigenetic determinants in accordance with the World Health Organizations nomenclature report (Ceppalini *et al.* 1964).

Bovine IgM as identified by immunoelectrophoresis is a pentamer of the four-chain unit linked together by disulfide bonds with a sedimentation constant of 19 S and a carbohydrate content of 12.3% (Gough *et al.* 1966; Kumar and Mickolajcik 1973).

Bovine IgG1 and IgG2, as individually characterized by immunoelectrophoresis, are normally monomers of the four-chain unit containing 2 to 4% carbohydrate (Kickhöfen *et al.* 1968). IgG1 is the principle immunoglobulin in lacteal secretions and possesses a mean sedimentation constant of approximately 6.3 S although different values have been reported. The sedimentation constant for the IgG2 subclass has a mean value of 6.6 S. Work of some investigators (Kick-

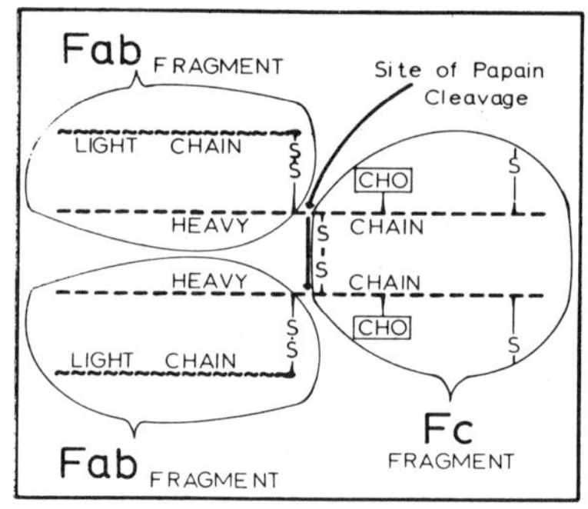

From Rose et al. (1970)

FIG. 2.7. CLASSICAL FOUR-POLYPEPTIDE-CHAIN STRUCTURAL UNIT OF AN IMMUNOGLOBULIN

The ½-cystein residues shown on the right end of the heavy chains bind the IgA and IgM subunits. CHO = carbohydrate moiety.

höfen *et al.* 1968; Pierce and Feinstein 1965) have indicated the possibility of additional subclasses.

Bovine IgA, an antigenically distinct immunoglobulin, has been isolated from milk (Gough *et al.* 1966). It has a carbohydrate content of 8–9% and a sedimentation constant of 10–12 S. It is eluted between IgM and IgG during Sephedex G-200 chromatography of whey and is considered to be a dimer of the four-chain unit (Butler and Maxwell 1972). Groves and Gordon (1967) demonstrated the presence of a glycoprotein-a in milk from individual cows and Butler *et al.* (1968) observed that it occurs in the free form and bound to IgA. Therefore glycoprotein-a and lacteal IgA are probably homologous to the "secretory (transport) piece" and "secretory IgA," respectively, that are noted in other species.

Another minor component, J-component, occurs strongly bonded to the polymeric immunoglobulins IgM and IgA (Butler 1973). It has molecular weight of 22,000–26,000 and contains carbohydrate.

While some of the immunoglobulins from milk have been subjected to amino acid and carbohydrate analysis (Groves and Gordon 1967); Kickhöfen *et al.* 1968; Milstein and Feinstein 1968; Nolan and Smith 1962; Kumar and Mikolajcik 1973) and the amino acid sequence of certain portions of the heavy chains of IgG1 and IgG2 have been established (Milstein and Feinstein 1968; Nolan and Smith 1962), the heterogeneity observed within each class of immunoglobulins (Josephson *et al.* 1972) make the use of this knowledge in the explanation of their physical behavior more difficult and, therefore, the results of these analyses have been omitted.

Proteose-peptones.—While the appropriateness of the term "proteose-peptone" for the three components, 3, 5, and 8, observed in the free-boundary electrophoretic patterns of the whey proteins, has been questioned (McKenzie 1967; Jenness 1970), this term has been retained by the A.D.S.A. Milk Protein Nomenclature Committee (Whitney *et al.* 1976) for the present.

Zonal electrophoresis of this fraction has revealed a greater degree of heterogeneity than indicated by free-boundary electrophoresis (Kolar 1967; Ng 1967; Kolar and Brunner 1969). Component 3 shows a single band but component 5 moves as a doublet and component 8 separates into two multi-zonal areas. When stained for glycoproteins additional zones are detected suggesting a total of 15 discernible bands. Four fractions: components 3, 5, 8-slow and 8-fast have been isolated and partially characterized (Jenness 1959; Kolar 1967; and Ng 1967). Their composition and some of their physical properties are reported in Table 2.5.

TABLE 2.5

COMPOSITION AND SOME PHYSICAL PROPERTIES OF PROTEOSE-PEPTONE FRACTIONS

	Component 3	Component 5	Component 8-slow	Component 8-fast
Nitrogen (%)	13.1	13.8	12.3	13.3
Phosphorus (%)	0.5	1.0	2.0	2.4
Hexoses (galactose, glucose, mannose and fucose) (%)	7.2	0.9	4.5	1.4
Hexosamines (galactosamine and glucosamine) (%)	6.0	0.2	2.5	0.3
Sialic acid (%)	3.0	0.3	3.3	0.4
Electrophoretic mobility pH 8.6 ($cm^2\ v^{-1}\ sec^{-1} \times 10^5$)	-3.48	-4.9	-9.2	-9.2
Isoelectric point	3.7			3.3
Molecular weight	163,000 40,800[1]	14,300	9,900	4,100

Source: Rose *et al.* (1970).
[1] Determined in the presence of 5 *M* guanidine-HCl.

The Minor Proteins

In addition to the major protein fractions discussed above, some minor proteins have been isolated from or identified in milk.

A small amount of the iron-binding protein transferrin which appears to be electrophoretically and immunologically identical to blood serum transferrin is present in milk (Szuchet-Derechin and Johnson 1962; Groves 1965). Disc-gel electrophoresis indicates the presence of multiple bands with transferrin isolated from either blood serum or whey and also indicates the possible existence of polymorphism (Groves 1965).

Another iron-binding protein in milk called lactoferrin was first isolated in a partially purified form by Sørenson and Sørenson (1939) as a red protein. It was originally considered to be a breakdown product of the enzyme, lactoperoxidase (Theorell and Akeson 1944; Polis and Shmukler 1953). However, with its isolation and characterization by Groves (1960) it has been established as a native protein of milk. Like transferrin, lactoferrin shows a number of bands on gel electrophoresis which can be resolved also on DEAE-cellulose columns (Groves 1965; Szuchet-Derechin and Johnson 1966). These fractions show similar absorption spectra but differ in their sedimentation coefficient versus concentration curves. Some evidence of polymorphism also has been observed (Groves 1965). Lactoferrin has an iron content of 0.12% (Gordon *et al.* 1962) which would suggest a molecular weight of 93,000 based on two moles of iron per mole of protein. This value is in good agreement with sedimentation-diffusion data. The amino acid and carbohydrate contents of both this protein and bovine serum transferrin have been compared and found to be significantly different (Gordon *et al.* 1963). They also differ immunologically (Szuchet-Derechin and Johnson 1962). Lactoferrin has also been shown to exist in milk in a colorless iron-free form (Gordon *et al.* 1962).

A crystalline protein, lactollin, has been found to be associated with the red protein in very small amounts when isolated from milk (Groves 1960). Amino acid analysis of this protein indicates the absence of methionine and only small amounts of alanine and cystine, but a large number of aromatic amino acid residues. Its molecular weight from sedimentation data is 43,000 which corresponds with its amino acid composition based on four residues of alanine and cystine per molecule.

In addition to the glycoproteins present in the κ-casein, immunoglobulin and proteose-peptone fractions, Bezkorovainy (1965, 1967) isolated what he believes to be a family of M-1 glycoproteins that are negatively charged at pH 4.5. They yield multiple bands on zonal electrophoresis and have an average molecular weight of 10,000. Upon further fractionation of the M-1 glycoproteins from colostrum, Bezkorovainy and Grohlich (1969) isolated one fraction with a molecular weight of 7,200 containing 28.4% carbohydrate and another fraction with a molecular weight of 12,000 containing 39.0% carbohydrate. Both proteins have large amounts of glutamic acid, proline and theonine but no tryptophane or cystine. Histidine, tyrosine and arginine are absent from the larger protein. The carbohydrate moiety in the M-1 glycoproteins apparently contains galactose, glucosamine, galactosamine and sialic acid. In addition to these glycoproteins, colostrum contains appreciable amounts of orosomucoid and M-2 glycoproteins.

A protein fraction rich in copper (0.19%) in undialyzable form at pH 6.5 but dialyzable at pH 3.5 was prepared from milk by Dills and Nelson (1942). They were unable to characterize it further or check its homogeneity because of lack of sufficient material. Considerably later, Hanson *et al.* (1967) demonstrated by immuno-diffusion techniques the presence in normal milk of the copperbinding blood serum protein, ceruloplasmin.

In addition to lactoferrin which was demonstrated by Blanc (1967) to act as a bacteriostatic agent, two proteins fractions called lactinins have been partially fractionated and concentrated (Auclair and Berridge 1953A, 1953B). However, electrophoretic patterns of these fractions indicated considerable heterogeneity and purer preparations are needed before they can be further characterized.

Laskowski and Laskowski (1950) found a trypsin inhibitor in bovine colostrum in relatively large amounts on the first day after parturition and in decreasing amounts thereafter. They were able to crystallize the trypsin-trypsin inhibitor complex and, finally, the inhibitor (Laskowski and Laskowski 1951). The complex has a molecular weight of 89,000 (Laskowski et al. 1952) suggesting that it is a trimer consisting of three molecules of trypsin and three molecules of inhibitor. Spectroscopic data indicates a large tyrosine-tryptophane ratio.

A kininogen which, when incubated with trypsin or snake venom, releases a material with the kinin-like ability to contract smooth muscle, has been detected in the whey fraction and concentrated by chromatography on DEAE-cellulose (Leach *et al.* 1967).

The fat globule in milk is surrounded by a lipoprotein membrane which is derived from the plasma membrane of the mammary cell (Bargmann *et al.* 1961). The protein composition of the membrane has not as yet been clearly elucidated but, in addition to a number of enzymes that have been detected, the proteins in the membrane are considered to be largely glycoproteins. Jackson *et al.* (1962) isolated a glycoprotein from the membrane which has molecular weight of 123,000 and the following composition: 11.28% nitrogen, 5.2–5.9% hexose, 3.9% hexosamine and 4.0–4.6% sialic acid. Swope *et al.* (1968) obtained two glycoproteins by gel chromatography on Biogel P-30 with molecular weights of 123,000 and 306,500. Their hexose contents were 5.4% and 7.8% respectively.

The Enzymes

To complete the picture of the protein compliment of milk, one should include the numerous enzymes which have been detected and isolated from milk. Space does not allow discussion of them in detail. Therefore, the reader is referred to the report of the American Dairy Science Associations Committee on Enzyme Nomenclature (Shahani *et al.* 1973). The enzymes recognized by the committee as being native to milk and the reaction that they catalyze are given in Table 2.6. Some additional enzymes have been detected but insufficient work has been done on them for the committee to accept them as normal components of milk.

STRUCTURE AND CONFORMATION OF MILK PROTEINS

The size, shape and configuration of the protein molecule is not only determined by its primary structure and composition but also

TABLE 2.6

ENZYMES IN BOVINE MILK

Enzyme	Reaction
Aldolase (4.1.2.13)	Fructose 1-6 phosphate $\rightleftharpoons$ dihydroxyactone + phosphoglyceric aldehyde
α-Amylase (3.2.1.1)	Starch + H_2O $\rightleftharpoons$ dextrins
β-Amylase (3.2.1.2)	Starch + H_2O $\rightleftharpoons$ Maltose
Carbonic anhydrase (?) (4.2.1.1)	$H_2CO_3 \rightleftharpoons CO_2 + H_2O$
Catalase (1.11.1.6)	$H_2O_2 \rightleftharpoons H_2O + 1/2\ O_2$
Cytochrome-c reductase (1.6.99.3)	NADH + Acceptor $\rightleftharpoons$ NAD + H-Acceptor
Diaphorase (1.6.4.3)	NADH + lipoamide $\rightleftharpoons$ NAD + reduced lipoamide
A-Esterase (3.1.1.2)	$C_6H_5OC(=O)\text{-}CH_3 + H_2O \rightleftharpoons C_6H_5OH + CH_3COOH$
B-Esterase (3.1.1.1)	Tributyrin + $H_2O \rightleftharpoons C_3H_7COOH$
C-Esterase (3.1.1.8)	$C_6H_5OC(=O)\text{-}C_2H_5 + H_2O \rightleftharpoons C_6H_5OH + C_2H_5COOH$
Lactose synthetase (2.4.1.22)	UDP galactose + αD-glucose $\rightleftharpoons$ UDP + Lactose
Lipase (3.1.1.3)	Triglyceride + H_2O $\rightleftharpoons$ Glycerol + Free fatty acids
Lysozyme (3.3.1.17)	*M. lysodeckticus* cells $\rightleftharpoons$ lysis
Phosphoprotein phosphatase (3.1.3.16)	*p*-nitrophenyl phosphate $\overset{pH\ 4.0}{\rightleftharpoons}$ *p*-nitrophenol + $PO_4^{\equiv}$
Phosphatase alkaline (3.1.3.1)	*p*-nitrophenyl phosphate $\rightleftharpoons$ *p*-nitrophenol + $PO_4^{\equiv}$
Peroxidase (1.11.1.7)	H_2O_2 + hydrogen donor $\rightleftharpoons$ H_2O + reduced donor
Protease (3.4.4.-)	Protein + H_2O $\rightleftharpoons$ peptides
Rhodanase (2.8.1.1)	$S_2O_3^{=} + CN^- \rightleftharpoons CNS^- + SO_3^{=}$
Ribonuclease (2.7.7.16)	RNA + H_2O $\rightleftharpoons$ nucleotides
Salolase (A-esterase?)	Phenyl salicylate $\rightleftharpoons$ phenol + salicylic acid
Xanthine oxidase (1.2.3.2)	Xanthine + triphenyl tetrazolium-chloride $\rightleftharpoons$ uric acid + formazan

Source: Shahani *et al.* (1973).

by the effect of secondary bonding forces such as electrostatic, hydrogen and hydrophobic bonding. These forces are influenced by the environment of the protein molecule including such factors as the temperature, pH and composition of the dispersing media.

α_{s1}-Casein

From the primary structure of the α_{s1}-caseins it can be seen that there are a large number of proline residues (8.5%) which are quite uniformly distributed throughout the molecule and would therefore considerably hinder the formation of structural coils such as the α-helix. In addition, there are a sufficiently large number of hydrophobic residues, such as proline, valine, leucine, isoleucine, phenylalanine and tryptophan, to yield a Bigelow parameter of an average hydrophobicity of 1170. Therefore the combination of these two conditions will result in a thermodynamically unstable number of nonpolar groups exposed at the surface which would encourage association or polymer formation due to hydrophobic bonding. This phenomenon is commonly observed but is very complex and dependent upon the temperature, pH, ionic strength and composition of the dispersing media. The behavior of the various genetic variants is similar but show some distinct differences (Payens and Schmidt 1966).

To obtain monomeric dispersions of α_{s1}-casein, strongly dissociating solvents are commonly used such as 6 *M* urea at pH 7.3 (McKenzie and Wake 1959), 3 *M* guanidine-HC1 at pH 7.0 (Noelken 1967), anhydrous formic acid, and 26% aqueous methanol with 0.05 *N* sodium trichloroacetate (NaTCA) (Swaisgood and Timasheff 1968). However, if sufficiently high pH values are employed, such as pH 12.0 in phosphate buffer ionic strength 0.3 (Dreizen *et al.* 1962) or pH 10.5 in glycine-NaOH buffer, ionic strength 0.5 (Schmidt *et al.* 1967), so that the charge on the protein is sufficient to overcome the hydrophobic attraction, monomers are also obtained.

While other environmental factors, such as changes in ionic strength, will modify the results, Fig. 2.8 shows the effect of pH upon the association as indicated by the change in sedimentation constant. The change occurring between pH 7 and 10.5 has been attributed to the dissociation of the hydroxyl group in the tyrosine residues and the ϵ-amino group of lysine which would greatly increase the negative charge of the α_{s1}-casein molecule.

In the alkaline range of pH values between pH 8.0 and 9.85, the behavior is complex and dependent upon ionic strength and the species of ions present. Swaisgood and Timasheff (1968) in their detailed study of this phenomenon demonstrated that, at low ionic

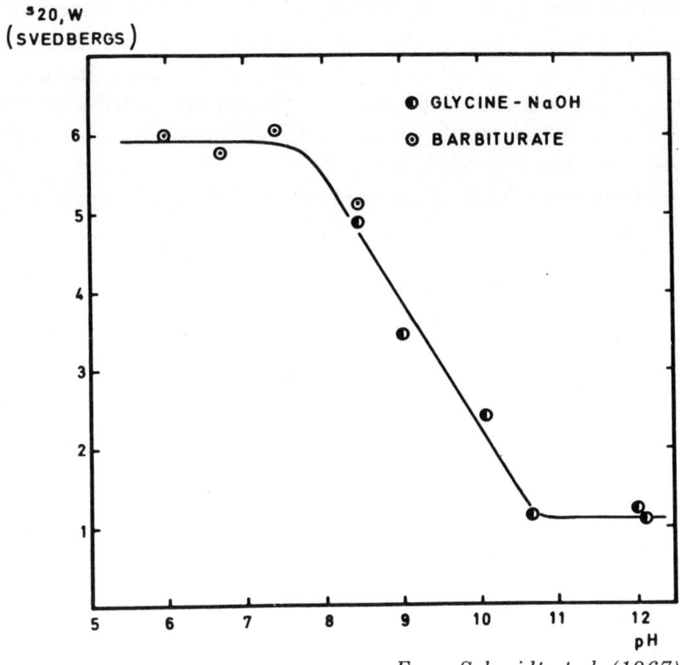

From Schmidt et al. (1967)

FIG. 2.8. THE SEDIMENTATION COEFFICIENT OF α_{s1}-CASEIN B AND C AS A FUNCTION OF THE pH. PROTEIN CONCENTRATION 7.0×10^{-3} GM/ML 2°C and IONIC STRENGTH 0.2

strengths of 0.02–0.05, a time dependent association occurs between the monomers and dimers and some higher polymers, while at higher ionic strength, (0.1–0.3) the equilibrium is rapidly attained with the concentration of polymers present increasing with increasing concentration of protein. Replacement of NaC1 with sodium tricloroacetate lowers the observed sedimentation constant of the protein between pH 8 and 10. When NaCl is replaced with Tris-Cl at 0.1 ionic strength, the sedimentation constant is also lowered, possibly due to the absence of sodium ions which have been observed to bind to α_{s1}-casein (Ho and Waugh 1965). Viscosity data suggests that each step in the aggregation is accompanied by a conformational change from a compact, possibly globular, dimer to a rigid rod or stiff coiled tetramer to a random coiled hexamer.

Schmidt and Payens (1964) (also Payens and Schmidt 1965, 1966) investigated the behavior of α_{s1}-casein in the neutral pH region. Ultracentrifugal studies of the genetic variants B and C individually at pH 6.4 in barbiturate buffer, ionic strength 0.2 at 2, 9 and 14°C indicated a positive linear dependence of the mean molecular weight upon the concentration over the range studied. The slopes increased with increasing temperature indicating a rapid endothermic association which is characteristic of hydrophobic bonding. In later work, Schmidt (1970) and Schmidt and van Markwijk (1968) investigated the effect of ionic strength on the association of α_{s1}-casein B at pH 6.6 employing the light-scattering technique which permits the use of lower concentrations of proteins (Fig. 2.9). At an ionic strength of 0.003, no association was detected but, as the ionic strength increased, considerable polymerization is observed. As the ionic strength increased the ζ-potential of the molecules was decreased. Thus the hydrophobic attraction of the molecules was able to overcome the repulsive forces due to charge on the protein. Schmidt

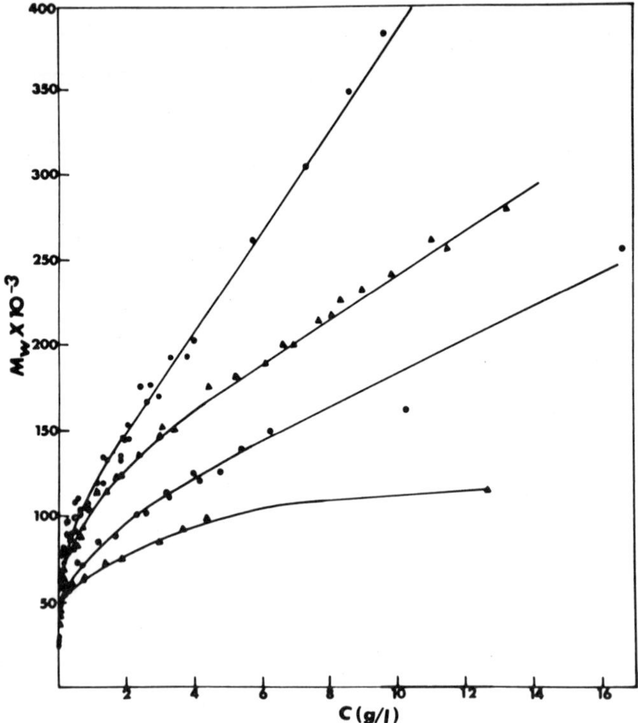

FIG. 2.9. APPARENT MOLECULAR WEIGHTS OF α_{s1}-CASEINS DETERMINED BY LIGHT SCATTERING

Upper two curves from data of Schmidt (1969) in imidazole-HC1-NaC1 buffer, pH 6.6, 0.2 ionic strength and 21°C: ● = α_{s1}-C; ▲ = α_{s1}-B. Lower two curves from data of Swaisgood and Timasheff for α_{s1}-C in Tris-NaC1 buffers, pH 8 at 25°C: ● = 0.2 ionic strength; ▲ = 0.1 ionic strength (Swaisgood 1973).

(Lyster 1972) also reports that elevated pressures reduce the tendency of α_{s1}-casein to associate.

κ-Caseins

Because of the heterogeneity of κ-casein and the difficulty in the preparation of pure and uniform samples, the interpretation of information upon its size, shape and conformation is difficult. Evidence secured in dissociating solvents indicates that κ-casein as prepared consists of a mixture of polymers of the monomeric unit linked together by intermolecular dissulfide bonds (Swaisgood *et al.* 1964; Mackinlay and Wake 1964). Dispersions of κ-casein in 5 *M* guanidine-HC1, pH 5.0; 7 *M* urea, pH 8.5; and 33% or more acetic acid in 0.15 *M* NaC1 yield heterogeneous mixtures of weight mean molecular weight of 88,000–118,000 with a minimum molecular weight of 57,200 (Swaisgood and Brunner 1963).

In aqueous salt systems, such as 1% κ-casein in 0.2 *M* phosphate buffer pH 7.0 at 20° C, κ-casein has a sedimentation constant of from 13.2–19 S depending upon the preparation (Mackinlay and Wake 1964; Waugh 1958; Swaisgood 1963) with molecular weights of the order of 650,000. The polymer size is not homogeneous although the distribution of sizes is not very broad. From a consideration of the primary structure, one can see that the monomer unit consists of hydrophobic N-terminal portion (the para-κ-casein) with a Bigelow hydrophobicity of 1,310 and a hydrophilic C-terminal portion (the macropeptide) with a hydrophobicity of 1,083. Thus the molecules could orient themselves in the association complexes with their hydrophobic ends in the interior of the complex and the hydrophilic ends on the surface (Hill and Wake 1969). The concentration dependency of their reduced viscosity supports this concept by indicating a spherical geometry, while, in dissociating solvents, similar data indicates a random chain behavior (Swaisgood 1973). Clark and Nakai (1972) secured additional evidence of this picture of the κ-casein polymer from fluorescent studies. Unlike α_{s1}- and β-casein, κ-casein does not demonstrate ionic strength or concentration dependence. Increases in temperature increase its aggregation (Yoshida 1969). However, at pH values in the vicinity of 12, the disulfide and non-covalent bonds are both broken and the monomer is released with a sedimentation constant of 1.45 S. Waugh (1958) calculated the dimensions of the monomer from frictional ratios and concluded that it existed as a coil with a diameter of 16 Å and a length of 150Å.

Reduction of the disulfide bonds in κ-casein and their subsequent alkylation does not appear to effect the polymer size (Swaisgood 1963). However, Mackinlay and Wake (1964) reduced κ-casein with sulfite in the presence of phenyl mercuric hydroxide and obtained a sedimentation constant of 9.1 S for the resulting polymeric *S*-sulpho-κ casein in 0.2 *M* phosphate buffer pH 7.0 at 20°C. In the presence of dissociating solvents, all of the various reduced and alkylated κ-caseins sediment as monomers. The following dissociating solvents have been employed: 5 *M* guanidine-HCl, pH 5.0; 7 M urea, pH 8.5; 20% or more dimethylformamide; or 45.5% dioxan in phosphate buffer pH 7.0 (Swaisgood and Brunner 1963; Mackinlay and Wake 1964; Woychik *et al.* 1966).

β-Caseins

The outstanding physical characteristic of β-casein is its temperature dependent association (Swaisgood 1973). If one examines its primary structure, it is readily apparent that most of the negative charges exist in the 21 N-terminal amino acid residues at neutral pH values and that the balance of the molecule is essentially neutral and contains a large number of hydrophobic residues. The molecule itself possesses a large Bigelow hydrophobicity, 1335 for the A^2 variant, and a large proportion of proline residues ($\sim$17%) which should result in an even greater number of nonpolar groups exposed at the surface than α_{s1}-casein. Therefore, the tendency for hydrophobic bonding should be great and an increase in association with an increase in temperature is to be expected.

β-casein exists as a monomer independent of temperature, pH, and ionic strength in dissociating solvents such as 6.6 *M* urea (Nielson and Lillevik 1957) and 3 *M* guanidine-HC1 (Noelken and Reibstein 1968). In the absence of dissociating solvents it is very dependent upon temperature but also somewhat dependent upon pH and ionic strength. Gehrke *et al.* (1964) noted that β-casein does not aggregate at pH 8.6 at 20°C, while Sullivan *et al.* (1955) observed that it is monomeric in veronal buffer pH 7.78 ionic strength 0.1 at 15°C and below but polymeric above this temperature, and Payens and van Markwijk (1963) observed that it still existed in equilibrium with polymers at 8.5°C and was completely monomeric only at 4.0°C in barbiturate-NaC1 buffer pH 7.5, ionic strength 0.2. From sedimentation and intrinsic viscosity measurements, the monomer at 4°C appears to be either highly asymmetric, axial ratio of 12.2–15.9, or a random coil. Even in dissociating solvent such as 6 *M* guanidine-HC1, the viscosity is essentially unchanged (Noelken and Reibstein 1968) suggesting little if any folded structure in the monomer.

Garnier (1966) investigated the thermal changes in conformation of β-casein from 5–45°C by spectral methods (Fig. 2.10). By measuring the optical density at 286 nm and the specific optical rotation

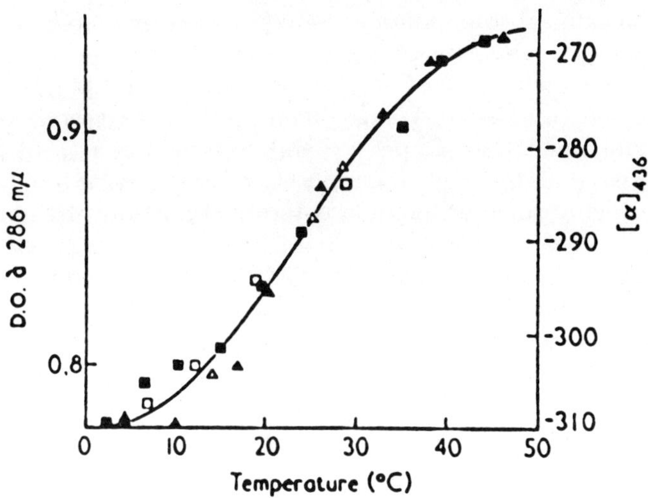

From Garnier (1966)

FIG. 2.10. THERMAL CHANGES IN CONFORMATION OF β-CASEIN, 0.9 × 10^{-4} *M* IN 0.1 *M* NaC1 AT pH 7.0

▲—Optical density at 286 nm. ■—Specific optical rotation at 436 nm. △—Optimal density and □ specific rotation from the same solution upon decreasing the temperature after reaching 45°C.

at 436 nm, he noted a rapidly reversible endothermic transition (ΔH∼30 kcal/mole) with a half transition temperature of 23–24°C. The optical rotatory dispersion data suggested a decrease in poly-*L*-proline II structure 12–5%) and a slight increase in α-helices (11–16%) during the transition from lower to higher temperatures. Studies on a carboxyacyl derivative of β-casein, which did not polymerize with increasing temperature, however, also demonstrated the optical rotatary dispersion thermal transition (McKenzie and Wake 1959). Therefore, the transition occurs independent of and, since it is rapid, probably prior to polymerization.

With increasing temperature in aqueous systems of β-casein, a boundary corresponding to a polymer appears in the sedimentation pattern in addition to the monomer boundary (Payens and van Markwijk 1963). The rate of equilibration, in contrast to the association of α_{s1}-caseins, is not rapid compared to the rate of sedimentation. Also, unlike α_{s1}-casein, β-casein does not form intermediate size polymers but possesses a rather narrow size distribution. At 8.5°C and pH 7.5 in barbiturate buffer, ionic strength 0.2, the polymer behaves, according to these authors, as a stiff rod or a random coil of ∼20 monomeric units. At 13.5°C, the degree of polymerization is greater, ∼100 monomeric units (Payens 1966). The sedimentation constant of the polymer is concentration dependent (Fig. 2.11). While the viscosity studies of Noelken and Reibstein (1968) support the conclusions of Payens and van Markwijk (1963), the sedimentation studies of Waugh *et al.* (1971) suggest a compact globular polymer at 37°C, pH 6.6 and an ionic strength of 2.0 and they suggest that the high intrinsic viscosity observed could result from the highly solvated flexible N-terminal portion of the β-casein molecules which possess the bulk of the charges and hydrophilic groups. These investigators also studied the effect of ionic strength, in the form of added NaCl, on the sedimentation constant of β-casein at 37°C and pH 6.6 and noted a high concentration dependence and a self-sharpening boundary such as observed with a highly

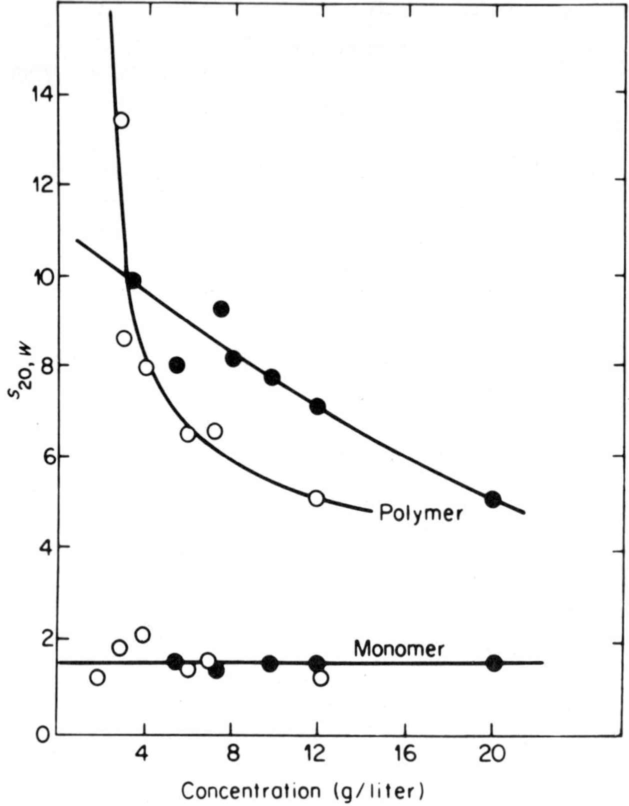

From Payens and van Markwijk (1963)

FIG. 2.11. CONCENTRATION DEPENDENCE OF THE SEDIMENTATION COEFFICIENTS OF β-CASEIN MONOMERS AND POLYMERS IN BARBITUATE BUFFER, pH 7.5 AND IONIC STRENGTH 0.2: ● AT 8.5°C; ○ AT 13.5°C

asymmetric molecule at ionic strengths less than 1.5. Above this ionic strength the s_{20w} values plateau and the self-sharpening characteristic of the boundary disappears. Archibald analysis at an ionic strength of 2.0 gives a weight average molecular weight of 820,000 corresponding to $\sim$34 monomer units. At an ionic strength of 4.5, the β-casein precipitates. The application of moderate pressures depolymerizes aggregated β-casein (Payens and Heremans 1969).

γ-Caseins

The structure and configuration of the γ-caseins have not been investigated to any great extent, but from their primary structure, one would expect them to possess a strong temperature dependent association due to their hydrophobic character. Their calculated Bigelow hydrophobicities are the highest of the caseins: γ_1-A^3, 1,400; γ_1-A^2, 1,400; γ_1-A^1, 1,386; γ_1-B, 1,390; γ_2-A^2, 1,489; γ_2-B, 1,493; γ_3-A, 1,503; and γ_3-B, 1,511. In addition their net charges at pH 6.6 are the lowest of the caseins: γ_1-A^3, -2.8; γ_1-A^2, -2.3; γ_1-A^1, 1.8; γ_1-B, -1.3; γ_2-A^2, +2; γ_2-B, +3; γ_3-A, +1; γ_3-B, +2 (Swaisgood 1973).

γ_2-caseins and γ_3-casein B are insoluble at pH 8 and 25°C, but soluble at 3°C, while γ_3-casein A is soluble at both temperatures. Sedimentation studies of these proteins (Groves and Townend 1970) indicate that they are in monomeric form at alkaline pH values (8–10.6) and low ionic strengths (0.01–0.02) except for the γ_3-casein B which shows nonideal behavior at pH 8.0. In the presence of guanidine-HC1, γ_1-casein A^3 sediments as a monomer and in the presence sodium dodecyl sulfate, the proteins of this group move as monomers in disc gel electrophoresis.

Blood Serum Albumin

The structural picture of blood serum albumin isolated from milk has not been investigated, but that obtained from bovine blood serum has been extensively studied. However, the heterogeneity observed in the protein from this source does raise the question as to whether the protein obtained from milk will have the same heterogeneous characteristics.

Bovine blood serum albumin (BSA) possesses at least three different kinds of heterogeneity: (1) that due to polymer formation, (2) that related to the sulfur linkages in the molecule and (3) a microheterogeneity in structure. Tiselius *et al.* (1956) were able to separate three fractions by stepwise elution from a hydroxapatite column with phosphate buffer pH 6.8 and 21°C. Fraction 1 was pure monomer with a sedimentation constant of 6.58 S. Fractionation of BSA on DEAE Sephedex A-50 resulted in two dimer fractions (Janatova *et al.* 1968): one relatively high in sulfhydryl content which could not be converted to the monomer by thioglycolate reduction and the other essentially free of sulfhydryl groups which could be 50% converted to monomer with thioglycolate. Small quantities of higher polymers were observed. Anderson (1966) noted that the degree of polymerization appeared to depend upon the freshness of the samples and the method of preparation and so the presence of polymers may be an artifact of the treatment of the sample. The monomer appeared to consist of at least three and usually four components: one mercaptalbumin with a free sulfhydryl group, two non-mercaptalbumins which are mixed disulfides of a modified form of mercaptalbumin and cystine or glutathione, and the fourth component is a mixed disulfide of mercaptalbumin and cystine. Anderson (1966) suggests that the modified form differs from mercaptalbumin with regard to the pairing of the disulfide bonds in the molecules. In explaining the behavior of BSA in the range of pH 3 to 5, Foster *et al.* (1965) (also Peterson and Foster 1965) demonstrated rather conclusively the microheterogeneity of BSA. They postulate that the protein consists of a collection of species which, while grossly similar, differ with respect to the pH range in which they undergo the transition from native form (N) to acid form (F). They suggest the following possible explanations of microheterogeneity: (1) small differences in amino acid composition or sequence, (2) randomization of amide groups both as to number and location, (3) subtle differences in tertiary and secondary structure that are "frozen-in" the molecule, (4) and randomization of disulfide linkages.

Due possibly to the above heterogeneity, there is some variability with regard to the conclusion reached by various workers concerning the structure and configuration of native BSA. Values for the molecular weight vary from 66,500 (Creith and Knight 1965) to 68,000 (Harrington *et al.* 1956) from sedimentation-diffusion data. The most generally accepted configuration at neutral pH values consists of a rigid linear aggregate of spheres linked together by the continuity of the peptide chain (Squire *et al.* 1968) which can be approximated by a prolate ellipsoid with a length of 138 A° and a thickness of 40 A° for the hydrated monomer. Shechter and Blout (1964) from optical rotatory dispersion data concluded that BSA at pH of 5-8 was 55% in the form of the α-helix.

The change in structure that occurs on decreasing the pH below 5 has been extensively investigated, but not all investigators are in agreement. One model that has been proposed (Riddiford and Jennings 1966) suggests that the native BSA (N state) consists of four globular subunits packed one on another and linked by the continuity of the peptide chain. The subunits held in pairs as a result of hydrophobic bonding between two members of a pair and the two pairs so obtained are bonded electrostatically into a four member globule. As the pH is lowered to pH 4.1 the electrostatic forces are first overcome and the molecule divides into two globule pairs linked by a flexible peptide chain forming the F′ form. With further lowering of the pH to 3.6 the electrostatic repulsion overcomes the hydrophobic bonding resulting in four individual but conjoint units forming the F state. Below pH 3.4, aggregation occurs. In contrast, Bloomfield's (1966) calculations, based on low angle scattering of X-rays, diffusion coefficients and intrinsic viscosities, yielded the most satisfactory agreement with the data at pH 3.6 for a rigid linear aggregate of three spheres—two with radii of 19 Å and one with a radius of 26.6 Å for an overall length of 129.2 Å. At lower pH values, substantial separation of the subunits occurs as indicated by diffusion studies. At pH values above 9, similar changes are observed.

The above changes in configuration with pH are influenced by the ionic strength of the environment. Weber (1952) found that addition of KC1 to 0.1 *M* or above prevented the change at low pH values but not at high pH, while Williams and Foster (1959) in their study of BSA by differential ultraviolet spectroscopy noted that increasing the concentration of $C1^-$ ion from 0.02–0.1 *M* or replacing it with 0.02 *M* SCN^- decreased the differential extinction coefficient at low pH. The change in this coefficient was shown to correlate with the fraction of BSA in the F-form and with the molecular radius from viscosity measurements. Apparently, the increase in ionic strength decreased the ζ-potential and the electrostatic repulsion due to the net positive charges on the BSA at low pH.

The effects of dissociating agents such as urea and guanidine HC1 upon BSA are similar to those of low pH. Jirgensons (1952) noted that the specific optical rotation and the intrinsic viscosity increased with increasing concentrations of guanidine HC1 from 1.5–5.0 *M*. The change in viscosity was reversible but the change in optical rotation was not entirely so.

β-Lactoglobulins

Since its isolation from milk, extensive studies have been made of the structure and configuration of the β-lactoglobulins. Most early workers in the field obtained values for the molecular weight of 36,000–40,000 for β-lactoglobulin in the neutral pH range (see Pederson 1936). However, Bull (1946) by surface film measurements in 20% $(NH_4)_2SO_4$ solutions obtained a molecular weight of 17,000 and concluded that, under these conditions, β-lactoglobulin dissociated into two fragments. Subsequently, Townend and Timasheff (1957; Timasheff and Townend 1961A, 1961B) noted in sedimentation studies and light scattering measurement that β-lactoglobulin A and B dissociated into monomers of molecular weight 18,000 at pH values below 3.5 at low concentrations and concluded that each variant undergoes a rapid monomer-dimer equilibrium with the dissociation increasing as the pH is lowered. They observed that the dissociation constants for the two variants are different and increase with increasing temperature. Albright and Williams (1968) made a detailed study of the equilibrium by sedimentation-equilibrium measurement (Fig. 2.12) which demonstrated not only the effect of pH on the dissociation but also that the dissociation increased with decreasing protein concentration and ionic strength. Work at Canberra (McKenzie 1967) established a similar behavior for the other variants of β-lactoglobulin. Green and Aschaffenburg (1959) determined the geometric structure of β-lactoglobulins A and B in the crystalline state and found both proteins to consist of two spheres of radius 17.9 A° joined in such a way that the centers are 33.5 A° apart with a two-fold axis of symmetry. It was originally thought that hybrid dimers of β-lactoglobulin A and B could not exist, but, when moving

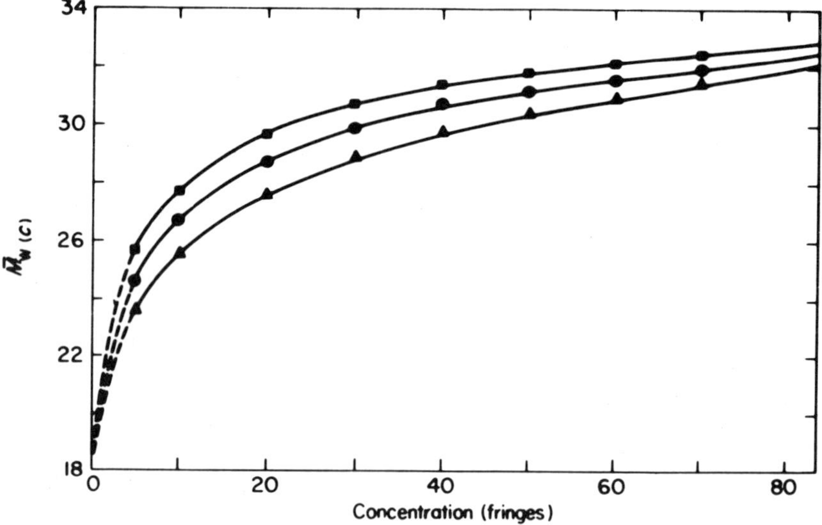

FIG. 2.12. THE DISSOCIATION OF β-LACTOGLOBULIN B AT LOW pH

Idealized curves for the weight-average molecular weight as a function of concentration from sedimentation-equilibrium measurements: ■ = pH 2.58, ionic strength, 0.15; ● = pH 2.20, ionic strength, 0.15; ▲ = pH 2.58, ionic strength, 0.10 (Albright and Williams 1968). 40.2 fringes = 10 gm/liter.

boundary theory was applied to the earlier free boundary electrophoretic results, their existence was clearly established (Gilbert 1970). While much of the evidence on the monomer dimer association has been secured at pH values below 3.5, it has been demonstrated that even at pH 5.2 there is some slight dissociation and that the dissociation increases when the pH is either increased or decreased from this value and becomes appreciable at pH 3.5 and below or pH 7.5 and above (McKenzie 1971A). While the effect of pH and ionic strength suggest that the association of monomers to form dimers is due to hydrophobic bonding, the effect of temperature does not support this postulate and the other types of noncovalent bonds can not be eliminated.

In addition to the monomer-dimer association phenomenon, β-lactoglobulin A undergoes an additional association to the octamer in the vicinity of pH 4.6. The dimer-octamer association is rapid as indicated by sedimentation velocity studies (Timasheff and Townend 1961A). The equilibrium constant for the reaction decreases with increasing temperature. On the basis of Green and Aschaffenburg's model (1959), Timasheff and Townend (1964) and Green (1964) proposed the model for the octamer shown in Fig. 2.13. While these investigators did not imply that this equilibrium involved the simultaneous collision of four dimer to form the octamer, they did assume that the dimer and the octamer were the predominate species present and that higher polymers are unlikely due to the stereochemistry of the polymerization. At 0°C only small amounts of tetramers and hexamers are observed but the quantities of these intermediates increases with increasing temperature. Adams and Lewis (1968) in their sedimentation equilibrium study of β-lactoglobulin A at pH 4.6, ionic strength 0.2 and 16°C noted that the data could be better described by an indefinite association. Recently Roark and Yphantis (1969) reexamined this system with samples of β-lactoglobulin A and found evidence of heterogeneity. On a column of Sephadex G-100 at pH 4.55 and 4°C, two fractions were observed: one with a high and one with a low association constant. The fraction with a high association constant gave results that could not be fit by the indefinite association model but rather by the Timasheff-Townend model. The cause of the heterogeneity is not known. As to the mechanism of the octamerization, the negative enthalpy and the large negative entropy suggest that hydrophobic bonding is not an important factor. The pH dependence with maximum association at pH 4.6 suggest that the carboxyl groups are involved in the reaction with the possible formation of hydrogen bonds between the protonated carboxyl groups. Examination of the primary structure of β-lactoglobu-

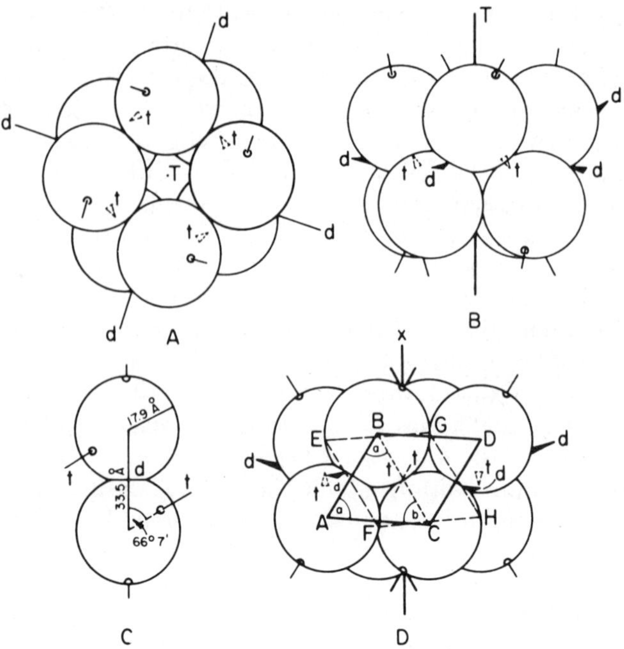

From Green (1964); Timasheff and Townend (1964)

FIG. 2.13. STAGGERED STRUCTURES FOR THE OCTAMER OF β-LACTOGLOBULIN A

A—Top view, 422 symmetry, d = dyad axis of symmetry, t = octamer bond. B—Side view, 422 symmetry, T = tetrad axis of symmetry. C—Dimer structure. D—222 symmetry, X = overall dyad axis of symmetry. The preferred structure is 422.

lin A indicates a concentration of carboxyl groups in the vicinity of the aspartic acid difference amino acid at position 64, suggesting that this portion of the molecule is involved.

With regard to the octamerization of the other genetic variants of β-lactoglobulin, the B variant can form mixed octamers with the A variant and undergoes weak octamerization by itself (Kumosinski and Timasheff 1966). The other variants do not appear to octamerize to any detectable extent. The extra aspartic acid residue (position 64) in β-lactoglobulin A apparently is important in this reaction. However, it is interesting to note that the Dr variant does not octamerize even though it possesses this aspartic acid residue. The carbohydrate moiety apparently is attached to the peptide chain close enough to the reactive site to sterically prevent octamerization (Bell *et al.* 1970).

At pH values above 8 in addition to the dissociation of the dimer to the monomer, time dependent aggregations have been reported (McKenzie and Sawyer 1966). Buchet *et al.* (1966) noted that the reaction was slowed down by the addition of disodium ethylenediaminetetraacetate suggesting that this reagent, by binding copper and other ions involved in the oxidation of SH-groups, was preventing some of the aggregation by preventing the formation of intermolecular disulfide linkages.

Conformational studies of β-lactoglobulins in solution have been numerous, but the interpretation of the results has been a subject of controversy. The experimental data of McKenzie *et al.* (1967) upon the effect of pH and temperature upon the optical rotation at 578 nm and a_0, a parameter of the Moffitt-Yang (1956) equation for optical rotatory dispersion, for the various genetic variants are shown in Fig. 2.14. The b_0 parameter of the Moffitt-Yang equation remains essentially low and constant for all genetic variants at acid and neutral pH values (-65° to -75°). These data agree with those of Timasheff *et al.* (1966) for the variants B and C, but the minimum near pH 4.6 for the variant A was not apparent in their results. However, this minimum was observed by Lontie and Préaux (1966). There appears to be little change in conformation when the dimer dissociates to the monomer at low pH. However, there is a marked change in optical rotation and the a_0 parameter in the pH range from 4–6. The minimum at pH 4.6 in the curves for variant A and its temperature dependence suggest a conformational change which may be necessary before octamerization can occur. The increases observed in variants B and C have been considered in their relationship to the ionization-linked transition (see following discussion on Electrochemical Properties of Milk Proteins) observed in their titration curves (Basch and Timasheff 1967).

Interpretation of optical rotatory dispersion and circular dichroism data by various investigators has resulted in a number of different structural pictures of the β-lactoglobulin molecule. For example, McKenzie (1967) concludes that β-lactoglobulins in pH range of 2–6 consists of approximately 33% α-helix, 33% β-conformation and 33% disordered chain while Townend *et al.* (1967) conclude that it consists of 10% α-helix, 47% β-conformation and 43% disordered chain. The presence of some β-conformation has been established by infrared spectroscopy (Timasheff and Susi 1966).

Above pH 7.5 there is a marked change in the optical rotation and optical rotatory dispersion in the variants A and B and less so in the case of C. All workers apparently associate this transition with the titration of one abnormal carboxyl group per monomer (see follow-

LIBRARY
ATHROFA GOGLEDD DDWYRAIN CYMRU
THE NORTH EAST WALES INSTITUTE
of higher education CONNAH'S QUAY

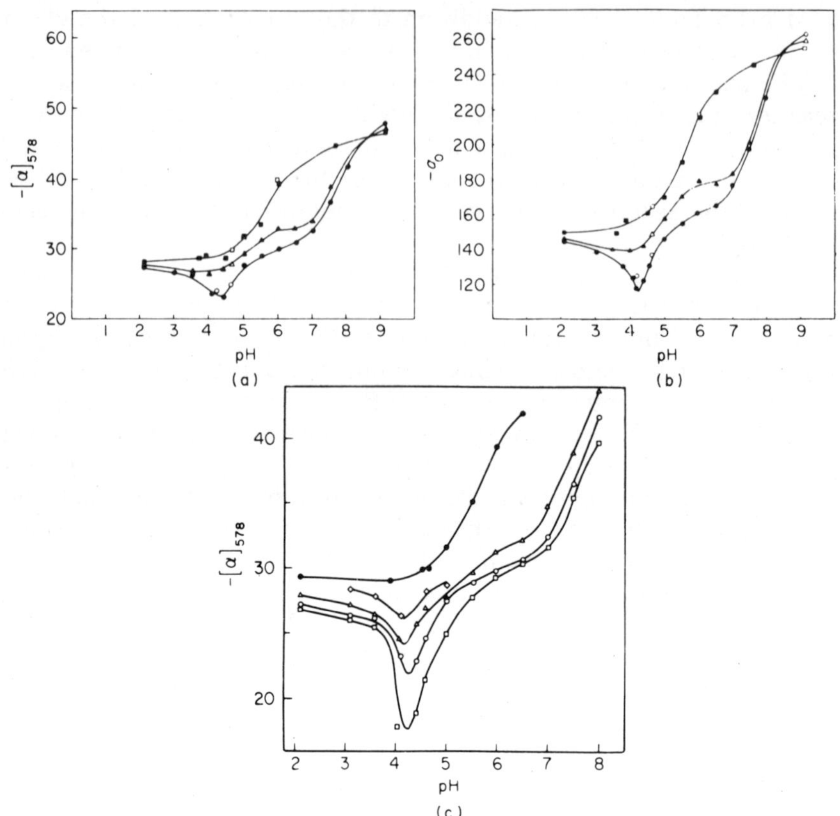

From McKenzie and Sawyer (1967); McKenzie et al. (1967)

FIG. 2.14. pH DEPENDENT CONFORMATIONAL TRANSITIONS FOR β-LACTOGLOBULINS A, B AND C

(a)—Specific rotation at 578 nm at 20°C. (b)—Parameter a_o in Moffitt Yang equation at 20°C. Symbols for (a) and (b): ● = A variant; ▲ = B-variant; ■ = C-variant. (C) The effect of temperature on specific rotation at 578 nm for the A-variant near pH 4.5: ◇ = 45°C; △ = 30°C; ○ = 20°C; □ = 10°C. The C-variant at 20°C (●) is shown for comparison.

ing discussion on Electrochemical Properties of Milk Proteins) although the behavior of the C variant is difficult to explain. This transition is also accompanied by an increase in the reactivity of the SH-group.

Above pH 8.0, the optical rotatory dispersion changes slowly with time at 20°C after the initial transition (McKenzie and Sawyer 1967). The rate of change is in the order A>B>C and proceeds more rapidly at 3°C than at 20°C. This change apparently is involved in the denaturation and the aggregation of the protein observed to occur in this pH range.

α-Lactalbumins

Sedimentation velocity studies of α-lactalbumins on the alkaline side of the isoelectric zone yielded sedimentation constants, extrapolated to zero concentration, of 1.87-1.98 S (Wetlaufer 1961; Kronman and Andreotti 1964; Szuchet-Derechin and Johnson 1965). The molecular weights obtained from these and other measurements indicated that, under these conditions, it existed in monomeric form with the measured values (14,900-16,500) somewhat higher than that obtained from the amino acid sequence, 14,176.

At pH values below the isoelectric zone, Kronman and his coworker (Kronman and Andreotti 1964; Kronman *et al.* 1964) observed that α-lactalbumin became associated to form dimers and trimer and aggregated to polymers of sedimentation constants in the range of 10-14 S. The association was rapid and reversible and was greater at 10°C than at 25°C while the aggregation was time dependent and decreased with decreasing temperature, pH and ionic strength. The aggregation was reversible and dependent upon protein concentration with little aggregation below 1% protein. Kronman and his coworkers attributed this association and aggregation to conformational changes in α-lactalbumin below pH 4 noting that, at pH 2 and at protein concentrations such that association and aggregation were absent, the sedimentation velocity revealed an increase in the frictional ratio suggesting an increase in hydrodynamic volume. This swelling of the molecule was accompanied by a change in the spectra yielding a difference spectra with maxima at 285-286 nm, 292-293 nm and 230 nm due largely to changes in environment of the tryptophan residues. The amplitude of the change was dependent upon pH and temperature, but practically independent of ionic strength and destroyed by urea (Kroman *et al.* 1965). However, by solvent perturbation studies they were able to establish that these changes were caused by conformational change in the environment of buried tryptophan residues rather than exposure of these groups at surface of the molecule. To further investigate the aggregation phenomenon under acid conditions, Robbins *et al.* (1965) amidinated the amino groups of α-lactalbumin to increase its hydrophobicity without changing its charge. Amidination did not appreciably change the configurational behavior but increased the association and aggregation phenomenon. They therefore concluded that this phenomenon was due to hydrophobic bonding resulting from the conformational change or to a decrease in the electrostatic barrier. Optical rotatory dispersion studies revealed that native α-lactalbumin exists as a tightly folded molecule with approximately 40% α-helix structure (Herskovits and Mescanti 1965), while circular dichroism spectra suggest 26% α-helix 14% β-configuration and 60% random coil (Robbins and

Holmes 1970). The acid conformational changes that occurred below pH 4 were accompanied by a decrease in the Moffitt and Yang b_0 value from 230–154° but the Cotton effects at 225 nm were identical for the protein at pH 2 and 6 indicating that the change in b_0 was not due to changes in the α-helix region but to changes in the environment of the buried tryptophan residues due to the swelling of the molecule (Kronman *et al.* 1966).

At alkaline pH, little or no association and aggregation is observed, but some change in configuration occurs as indicated by changes in the Cotton effects between 250 and 300 nm and to decreases in the Moffitt and Yang b_0 value to 182° at pH 11.5 (Kronman *et al.* 1966). Reaction of the tyrosine residues with cyanic fluoride over the pH range from 9–12 at 3° C and 25° C indicates that 3 groups react in the native α-lactalbumin and four groups react at pH 10 and above (Gorbunoff 1967).

Brown *et al.* (1969), noting the similarity of the amino acid sequences between α-lactalbumin and hen egg-white lysozyme and their functional similarity in that lysozyme was involved in the cleavage of a β-1-4-glucopyranosyl linkage and α-lactalbumin was the modifier protein of lactose synthetase proposed a structural model for α-lactalbumin based upon the known main-chain conformation of lysozyme. Changes in the internal side chains can generally be interrelated but there are some regions which could not be deduced unequivocably. The surface cleft, which is the site of substrate binding in lysozyme is shorter in α-lactalbumin. Castellino and Hill (1970) investigated the rate of carboxymethylation of α-lactalbumin and found their results compatible with model proposed by Brown *et al.* However, Krigbaum and Kügler's (1970) measurements with small angle X-ray diffraction indicated considerable differences in the overall dimension of lysozyme and α-lactalbumin in solution, 28 × 28 × 50 A°, and 22 × 44 × 57 A°, respectively. The resolution of this question awaits the determination of the native structure of α-lactalbumin by X-ray diffraction.

Immunoglobulins

The study of the structure and conformation of the immunoglobulins is complicated by their heterogeneity, but, because of the interest in immunology, considerable work has been done. Unfortunately, only a limited amount of this work has been on the bovine immunoglobulins of milk. We can, however, generalize from studies of immunoglobulins of other sources. Enzyme and chemical modifications have been employed to yield information on secondary structure of the four chain unit (Dorrington and Tanford 1970). Enzyme modification demonstrated that the N-terminal amino acids are not exposed on the surface of the native immunoglobulins, while chemical modification by selectively reducing the interchain disulphide bonds indicated that they were not involved in the specific structure of the antigen-binding sites.

From X-ray diffraction and electron microscope studies of human IgG1, the arrangement of the polypeptide chains is gradually emerging (Poljak *et al.* 1972), especially for the antigen-binding end of the molecule. The Fab fragments have 2 globular structures which are built up from both the heavy and light chains. This is consistent with the concept that both chains are involved in the antigen-binding site. Dorrington and Tanford's (1970) evidence secured from optical rotatory dispersion and circular dichroism spectra indirectly support this structure by suggesting that there is negligible α-helix in the immunoglobulins.

IgM and IgG molecules are large enough for them to be visible in the electron microscope. Green (1969) noted that IgG and the four-chain units in IgM are shaped like a Y or a T, with a flexible central hinge. In IgM the 5 Y-shaped tetramers form a ring with 5 branching arms projecting from it.

In the presence of urea or guanidine hydrochloride the antigen binding sites completely unfold (Bjork and Tanford 1971). The denaturation is completely reversed by removal of the agent in neutral solution. Acids and alkalies have a similar effect (Doi and Jirgensons 1970).

Proteose-peptones

Little is known about the effect of environment on the structure and configuration of the proteose-peptones in milk. Component 3 apparently exists in polymeric form in veronal buffer pH 8.6, $\Gamma/2 = 0.1$, at 20°C with a molecular weight of about 163,000 (Ng *et al.* 1970). Even, when 5 *M* guanidine-HC1 is added to the buffer, there is still evidence of association. However, the molecular weight in this medium is reduced to approximately 40,000. A minimum molecular weight calculated from the concentration of the limiting amino acid, tyrosine, is 22,000.

Component 5 and 8-slow and 8-fast are small proteins of low molecular weight. Contrary to component 3, they do not polymerize in veronal buffer, pH 7.0, $\Gamma/2 = 0.1$ (Kolar and Brunner 1970).

FRACTIONATION OF MILK PROTEINS

Numerous methods have been employed for the fractionation of milk proteins, some based on their differential solubility in various solvent systems and others based on their chromatographic or electrophoretic behaviors. Space does not allow for the discussion of all methods. Therefore only the more commonly used procedures will be considered. For details, the reader should consult the original references.

Whole Casein

As a starting material for the preparation of the various caseins, whole casein can be prepared by a number of procedures (McKenzie 1971B). The most commonly used method is based upon acid precipitation. Hydrochloric acid is added to skimmilk to adjust the pH to the minimum solubility of the casein. Several different temperatures have been employed: 2, 20 and 30°C. Except for precipitation at 2°C where minimum solubility is obtained at pH 4.3, the skimmilk is adjusted to pH 4.5-4.6. The precipitated casein is removed by centrifugation, washed with water, redispersed by adjusting the pH to 7.0-7.2 with NaOH and reprecipitated.

Casein can also be prepared from milk by high speed centrifugation at 105,000 xg at different temperatures both in the presence and in the absence of added calcium ion. In a modification of the procedure of von Hipple and Waugh (1955), McKenzie (1971B) added $CaCl_2$ at a final added concentrations of 0.07 *M* to skimmilk maintaining the pH at 6.6-6.8 at 3°C. After centrifugation the precipitate was resuspended in 0.08 *M* NaCl and 0.07 *M* $CaCl_2$ and the calcium removed by treatment with oxalate, citrate or ion exchange resin (Amberlite IR-120 in the sodium form). A similar procedure has been employed at 37°C (Waugh *et al.* 1962).

By adding 26.4 gm of $(NH_4)_2SO_4$ per 100 ml of skimmilk at 2°C, McKenzie (1971B) was able to precipitate the casein by centrifugation at 14,600 xg. Upon redispersion and reprecipitation with $(NH_4)_2SO_4$, the casein was dissolved in water and exhaustively dialyzed against glass-distilled water to remove residual $(NH_4)_2SO_4$. The advantages and disadvantages of these procedures are discussed by McKenzie (1971B).

Fractionation of Casein into α-, β-, and γ-Caseins

Hipp *et al.* (1952) proposed two procedures for the separation of whole acid casein into α-, β-, and γ-caseins. The first is based on their differential solubilities in 50% alcohol in the presence of salt at different pH values and temperatures. The second is based upon their differences in solubility in aqueous urea solutions. Because of its common use both in the preparation of β- and γ-casein and in the separation of α-casein as a starting material for some of the methods for the isolation of α_s- and κ-caseins, the urea procedure is outlined in some detail in Fig. 2.15.

α_s-Caseins

$\alpha_{s1,2}$-casein has been prepared by Waugh *et al.* (1962) from casein precipitated by the addition of excess calcium at 37°C as indicated above. A solution of casein is cooled to 0–1°C and $CaCl_2$ is added to 0.17 *M* at pH 7.0. After standing overnight, the precipitate is dissolved in 0.05 *M* potassium citrate at pH 6.5 and 2°C. The solution is diluted with water at 5–7°C and $CaCl_2$ added to a final concentration of 0.07 *M*. The resulting precipitate is redissolved in potassium citrate and the $\alpha_{s1,2}$-fraction can then be separated from the other α_s-caseins by stepwise elution from DEAE-cellulose columns, in 4.5 *M* urea and 0.01 *M* imidazole-HCl buffer.

Schmidt and Payens (1963) dissolved acid casein at pH 7.0 and added $CaCl_2$ to 0.4 *M*. The resulting precipitate was suspended, the calcium removed, and the casein dissolved in 6.6 *M* urea. Upon dilution to 4.6 *M* urea, a crude α_s-casein was precipitated which was further purified by redissolving in 6.0 *M* urea and diluting to 3.3 *M*. The precipitated α_s-casein was then separated into the various components by preparatory column electrophoresis on cellulose

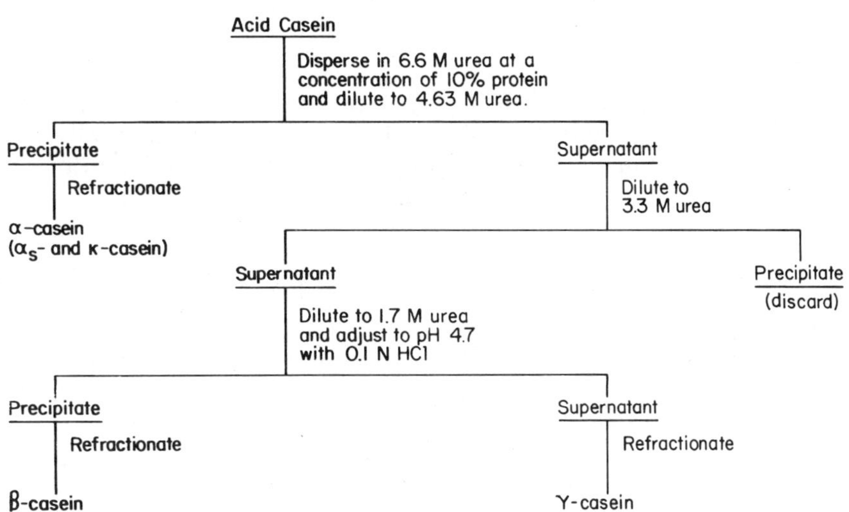

From McKenzie and Wake (1959); Hipp et al. (1959)

FIG. 2.15. UREA FRACTIONATION OF ACID CASEIN

powder at pH 6.5 ionic strength 0.05 in 5 *M* urea. The α_{s1}-casein so prepared is homogeneous by starch gel electrophoresis.

Zittle and Custer (1963) dissolved acid casein in 6.6 *M* urea and acidified the solution with H_2SO_4 to pH 1.5. Upon dilution with water, a precipitate of α_s- and β-caseins was formed which was further fractionated by the method of Zittle *et al.* (1959) in which NaCl is added in a urea fractionation procedure. The crude α_{s1}-casein obtained is further purified by precipitating the other proteins from an ethanol-water solution at pH 7.2 with ammonium acetate.

Whole α-casein was prepared from acid casein by Thompson and Kiddy (1964) by a modified urea fractionation procedure. The α-casein was then dispersed at pH 7.2 and 2–4°C and $CaCl_2$ added to 0.4 *M* to precipitate crude α_{s1}-casein. Upon removal of calcium, the crude protein was dissolved in 50% ethanol-water and the minor α_s-caseins precipitated with ammonium acetate. Final purification was accomplished by gradient chromatography from DEAE-cellulose at pH 7.0, 3.3 *M* urea.

α_{s1}-casein can also be prepared by column chromatography of whole casein by the above procedure (Thompson 1966). Rechromatography is needed in order to obtain a pure product.

The minor α_s-caseins have been isolated by Annan and Mason (1969) by chromatography on SE-Sephadex in 8 *M* urea at pH 4.0, and by Hoagland *et al.* (1971) by DEAE-cellulose chromatography in 3.3 *M* urea pH 7.0 with added mercaptoethanol.

κ-Caseins

McKenzie and Wake (1961) prepared κ-casein starting with the S fraction of Waugh and von Hippel (1956) which was essentially the supernatant obtained in the preparation of α_s-casein by addition of $CaCl_2$ at 37°C. The protein in this fraction is precipitated with Na_2SO_4, redissolved and dialyzed against 0.005 *M* NaCl. Ethanol is added to 50% and the κ-casein is precipitated with ammonium acetate. The precipitation from 50% ethanol is repeated to further purify the protein.

Starting with crude α-casein dissolved in 6.6 *M* urea, Swaisgood and Brunner (1962) cooled the system to 3°C and added trichloroacetic acid (TCA) to a final concentration of 12%. The precipitated α_s-casein is removed and the supernatant adjusted to pH 7 and the TCA and urea removed. $CaCl_2$ is then added to 0.25 *M* at 3°C and the system centrifuged to remove the precipitate. The supernatant is adjusted to pH 11.3 and the κ-casein precipitated at pH 4.4.

The supernatant from the urea sulfuric-acid method of Zittle and Custer (1963) for the preparation of α_s-casein contains κ-casein with traces of α_s- and β-casein. The κ-casein can be precipitated by the addition of $(NH_4)_2SO_4$. Further purification can be achieved by reprecipitation from aqueous ethanol.

While several chromatographic procedures have been proposed for the preparation of κ-casein, the method of Yaguchi *et al.* (1968) appears to be the only one in which the κ-casein is free of other casein components. This procedure involves gel filtration of whole casein on Sephadex G-150 columns in 6 *M* urea at pH 8.6 (0.005 *M* tris-citrate buffer) at room temperature. The κ-casein is eluted with the void volume. This procedure should prove quite useful as a final step in the purification of the κ-caseins obtained by the other procedures since they all appear to contain traces of other casein components.

Mackinlay and Wake (1971) have discussed the limitations of these and other methods and the effect of the methodology upon the κ-casein.

β-Caseins

β-casein can be prepared in relatively large amounts by the two methods of Hipp *et al.* (1952) with the urea procedure (Fig. 2.15) most commonly used. Aschaffenburg (1963B) has simplified the procedure by dispersing whole casein at pH 7.5 in 3.3 *M* urea and adjusting the pH to 4.6. The bulk of the α_s- and κ-caseins precipitate and the supernatant is adjusted to pH 4.9. After dilution to 1 *M* urea and warming to 30° C, the β-casein precipitates. This crude material is dispersed in 3.3 *M* urea at pH 7.5, adjusted to pH 4.6 at 37° C and centrifuged to remove impurities. The supernatant is adjusted to pH 4.9, diluted to 1 *M* urea at 30° C, and the precipitated β-casein removed by centrifugation.

While the above method yields β-casein which is pure by moving-boundary electrophoresis, starch gel electrophoresis reveals the presence of impurities. Therefore a number of investigators have used chromatographic procedures. One such procedure, as modified by Thompson (1966), involves the chromatography of crude β-casein upon a DEAE-cellulose column at pH 7.0, 3.3 *M* urea in the presence of mercaptoethanol, eluting with a linear NaCl gradient. The same procedure has been used on whole casein for the preparation of immunelectrophoretically pure β-casein. This procedure was also successful in the separation of the genetic variants, B and D, of β-casein.

γ-Caseins

The procedures of Hipp *et al.* (1952) have been used to prepare the γ-casein fraction (Fig. 2.15). This fraction, as expected, is electrophoretically heterogeneous (Murthy and Whitney 1958). Groves *et al.* (1962) (see also Groves and Gordon 1969) fractionated whole casein by dispersing it in 0.005 *M* phosphate buffer pH 8.3 and applying it to a fibrous DEAE-cellulose column at 3°C. The γ_2-and γ_3-caseins are eluted with the starting buffer and the γ_1-casein is removed at 0.02 *M* phosphate, while β-casein is eluted at 0.10 *M* phosphate. The γ-and β-caseins are purified by rechromatograph on microgranular DEAE-cellulose under the same conditions. The γ_2- and γ_3-casein fractions contain κ-casein and some other impurities. In the purification of these components, protease activity is noted (Groves *et al.* 1973), and therefore before further purification, they are suspended in 0.005 *M* phosphate buffer pH 8.3 and heated for 5 min at 100°C to inactivate the enzymes. After dialysis against the starting buffer, they are rechromatogrammed on microgranular DEAE-cellulose in the same buffer with or without 3 *M* urea at 3°C. Although the resolution is poorer with urea, the yields are better because of the improved solubilities. The pure proteins are obtained by again repeating the chromatography.

Blood Serum Albumin

Little attention has been paid to the isolation of blood serum albumin from milk since its first crystallization by Polis *et al.* (1950). They separated the whey into two fractions at $(NH_4)_2SO_4$ concentrations of 2.3 and 3.3 *M* at pH 6.0. Exhaustive dialysis of the 3.3 *M* salt fraction resulted in the crystallization of β-lactoglobulin at pH 5.3. The mother liquor is adjusted to pH 9.1 and treated by the addition of $(NH_4)_2SO_4$ stepwise to 2.4 *M* and 2.6 *M* removing the precipitate after each increase. Finally the protein in the supernatant is completely precipitated by adjusting the pH to 5.0 and the salt concentration to 3.4 *M*. This precipitate is redispersed and the fractionation with $(NH_4)_2SO_4$ repeated, and the albumin, so obtained, successively purified by precipitating the impurities at alcohol concentrations of 10%, 20%, and 33% at pH 5.3 and low temperature. The albumin is removed from the supernatant by adjusting the pH to 4.8 and alcohol concentration to 40%. The material is further purified by redispersing and precipitating the impurities by $(NH_4)_2SO_4$ at 2.6 *M*. Crystalline albumin is formed by adding $(NH_4)_2SO_4$ until turbid and allowing it to crystallize at 3°C.

While it has not been used appreciably for the preparation of blood serum albumin, Armstrong *et al.* (1970) developed a chromatographic procedure for the fractionation of ammonium-sulfate whey proteins upon Sephadex G-75 at pH 6.3 in imidazole-HCl buffer, ionic strength 0.043. The second fraction obtained is primarily blood serum albumin.

β-Lactoglobulin

A large variety of methods has been developed for the isolation of β-lactoglobulin, several of which have been used to simultaneously produce a fraction from which α-lactalbumin can be obtained. Their advantages and limitations have been discussed by McKenzie (1971A).

Aschaffenburg and Drewry (1957B) prepared whey by adding 200 gm per liter of Na_2SO_4 to milk at 40°C. Acidification with 11 *M* HCl to pH 2.0 at 25°C precipitated a fraction rich in α-lactalbumin while the supernatant was chiefly β-lactoglobulin. The supernatant was adjusted to pH 6.0 with 14 *M* NH_4OH and 200 gm of $(NH_4)_2SO_4$ were added per liter. The precipitated β-lactoglobulin was dispersed in pH 5.2 buffer and dialyzed against the same buffer. After removal of the insoluble material the supernatant was exhaustively dialyzed against water yielding β-lactoglobulin crystals.

Robbins and Kronman (1964), wishing to avoid the use of Na_2SO_4 and low pH values, developed a procedure using $(NH_4)_2SO_4$ and no pH values lower than 4.0. Armstrong *et al.* (1967) made a thorough study of the methods available and modified Robbins and Kronman's procedure to obtain more satisfactory results. A schematic representation of two of their procedures are shown in Fig. 2.16. The solubilities of the various genetic variants of β-lactoglobulin at pH 3.5 in method Ia vary appreciably and the yield of β-lactoglobulin A by this method is low. Therefore method IIa appears to be the better procedure for this variant.

Fox *et al.* (1967) employed the addition of trichloroacetic acid to acid-casein whey at a concentration of 34.2 gm per liter. After standing overnight the precipitate is removed and can be used as a source of α-lactalbumin (Aschaffenburg 1968). The supernatant is concentrated to 1/10 the volume of the skimmilk by negative pressure dialysis at 4°C and either dialyzed exhaustively against water and lyophilized as β-lactoglobulin or further purified by $(NH_4)_2SO_4$ additions prior to dialysis.

Armstrong *et al.* (1970) have fractionated ammonium-sulfate bovine whey into five fractions by column chromatography on Sephadex G-75 at pH 6.3 in imidazole-HCl buffer of ionic strength 0.043. Fraction III as eluted from the column contains primarily β-lactoglobulin. This material can be further purified by chromatography on DEAE Sephadex A-50. This procedure can also be used to purify the β-lactoglobulin obtained by the other methods.

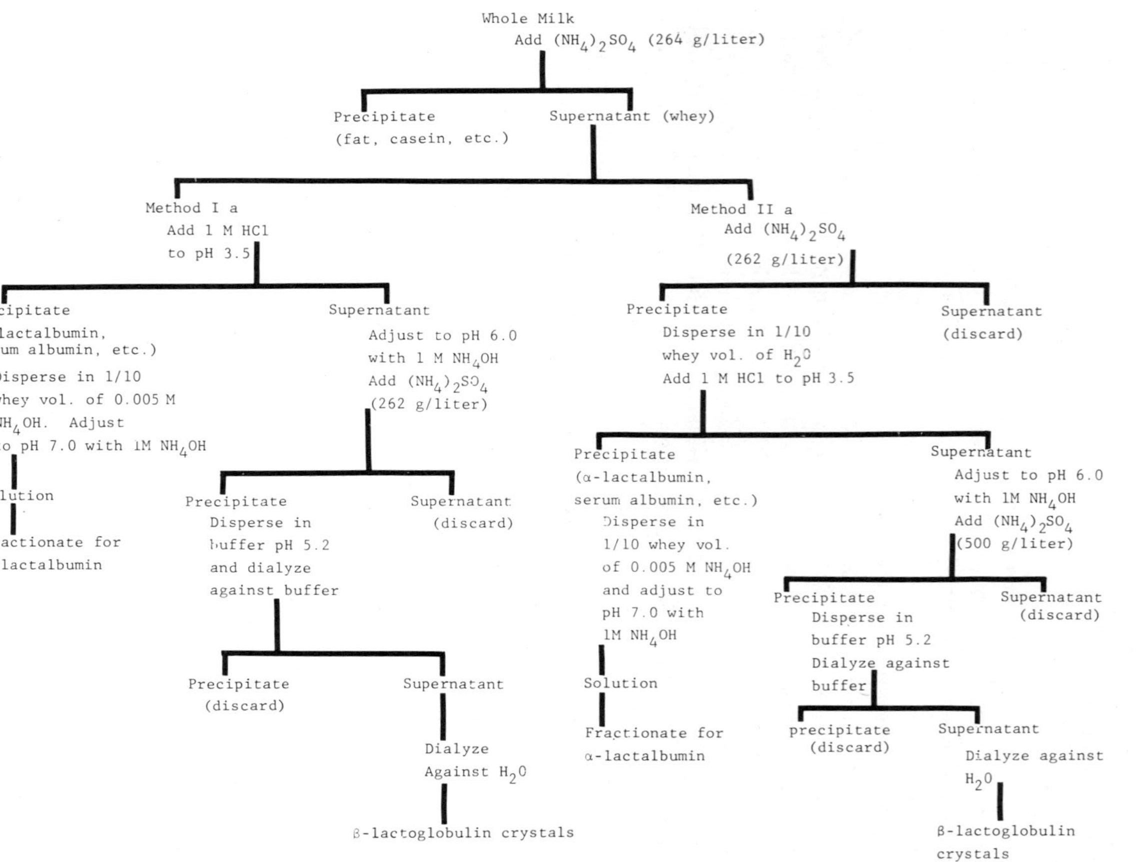

From McKenzie (1967)

Chromatography on Sephadex G-75 does not separate the variants of β-lactoglobulin. However, β-lactoglobulin A and Dr can be separated on DEAE Sephadex A-50 (Bell *et al.* 1970) and the A and C variants have been separated on DEAE cellulose at pH 5.0 in 0.05 *M* phosphate buffer with NaCl gradients (Basch *et al.* 1965).

α-Lactalbumin

The procedure of Gordon and Semmett (1953) for the isolation of α-lactalbumin as simplified and improved (Gordon and Ziegler 1955) follows the same procedure as that used by Polis *et al.* (1950) for the isolation of blood serum albumin through the crystallization of β-lactoglobulin. The mother liquor is then adjusted to pH 4.0 and $(NH_4)_2SO_4$ is added to concentration of 1.3 *M*. The precipitated α-lactalbumin is redispersed in dilute NH_4OH and again precipitated at pH 4.0 to remove additional β-lactoglobulin. The precipitated α-lactalbumin is redispersed at pH 6.6 and crystallized by slowly adding a saturated $(NH_4)_2SO_4$ solution previously adjusted to the same pH. The amount of salt required depends upon the protein concentration and varies from 0.5–0.67 saturation.

Aschaffenburg and Drewry (1957B) reworked their precipitate from the β-lactoglobulin preparation which was rich in α-lactalbumin by reprecipitating it from weakly ammoniacal solutions, first at pH 3.5 to remove residual β-lactoglobulin and then at pH 4.0 to remove serum albumin. The α-lactalbumin was then crystallized as in Gordon and Semmett's (1953) procedure.

Robbins and Kronman (1964) and Armstrong *et al.* (1967) both reworked their α-lactalbumin-rich precipitates (See Fig. 2.16) from their preparations of β-lactoglobulin by salt fractionation similar to Gordon and Semmett (1953) or by chromatography.

Aschaffenburg (1968) developed a procedure based on the observation of Fox *et al.* (1967) that α-lactalbumin was insoluble in 3% trichloroacetic acid. The precipitate from TCA-treated whey is dispersed at pH 7 with NH_4OH and reprecipitated by adjusting to pH 3.5. The precipitate is redispersed at pH 6.6 and fractionated with $(NH_4)_2SO_4$ prior to crystallization in the usual manner.

Several chromatographic procedures are available either to further purify α-lactalbumin prepared by the preceding methods or to prepare it directly from whey. Bengtsson *et al.* (1962) passed a five-times crystallized α-lactalbumin through Sephadex G-75 in order to remove traces of serum albumin. Sephadex has also been used to separate the whey proteins into fractions (Hill and Hansen 1964), one fraction of which is mainly α-lactalbumin.

Starting with the classical "lactalbumin" fraction, Szuchet-Derechin and Johnson (1965) prepared small quantities of pure α-lactalbumin by chromatography on DEAE-cellulose eluting with a shallow pH gradient decreasing from 7.28–5.0 together with increasing ionic strengths of phosphate-NaCl buffers. DEAE cellulose has also been used to purify α-lactalbumin preparations with either phosphate buffer at pH 8.2 (Groves 1965) or at pH 7.3 with a NaCl gradient (Gordon 1971).

Immunoglobulin

The immunoglobulins can be separated from acid whey or colostrum whey by salt fractionation with $(NH_4)_2SO_4$ (Smith 1946). The whey is adjusted to pH 6.5 and half saturated with solid $(NH_4)_2SO_4$. The precipitated protein is redispersed, the pH adjusted to 4.6, and solid $(NH_4)_2SO_4$ is added to 25% saturation. The precipitate is removed and the supernatant adjusted to 40% saturation at pH 6.0 to precipitate the crude immune proteins. This precipitate is reworked by dispersing in water at 1°C and adjusted to pH 4.5. The insoluble matter is removed by filtration and $(NH_4)_2SO_4$ is added to 30% saturation. The precipitate is removed and the filtrate brought to pH 6.0 and 40% with $(NH_4)_2SO_4$. The precipitate that is formed is then dialyzed exhaustively against H_2O at 2°C. The "euglobulin" fraction is insoluble in water while the "pseudoglobulin" fraction is soluble. Both are now known to be heterogeneous. The "pseudoglobulin" fraction consists mainly of IgG1 and secretary IgA while the "euglobulin" fraction contains IgG2, the slower IgG1 globulins, IgA and IgM.

The main classes of the immunoglobulins of colostrum whey can be separated by gel filtration on Sephadex G-200 (Fig. 2.17) (Mach *et al.* 1969). IgM appearing in the void volume and IgA on the shoulder of the large IgG1 peak. The free secretary piece or glycoprotein-a follows the IgG1 peak. Upon rechromatography of the IgA shoulder on Sephadex G-200, a peak is obtained which in addition to the IgA contains some aggregated IgG1.

Froese (1971) chromatographed a 1:4 dilution of whey on Bio-gel P-300 eluting with either 0.01 *M* phosphate buffer pH 8.0 containing 0.15 *M* NaCl or 0.01 *M* Tris buffer pH 8.0 containing 0.5 *M* NaCl. He observed five peaks, the first two of which were immunoglobulins. The first peak was IgM and the second was IgG with IgA probably in between.

DEAE-cellulose chromatography with stepwise fractionation has also been used to separate these proteins. Mach *et al.* (1969) observed that IgA was eluted with 0.05 *M* phosphate buffer pH 7.0

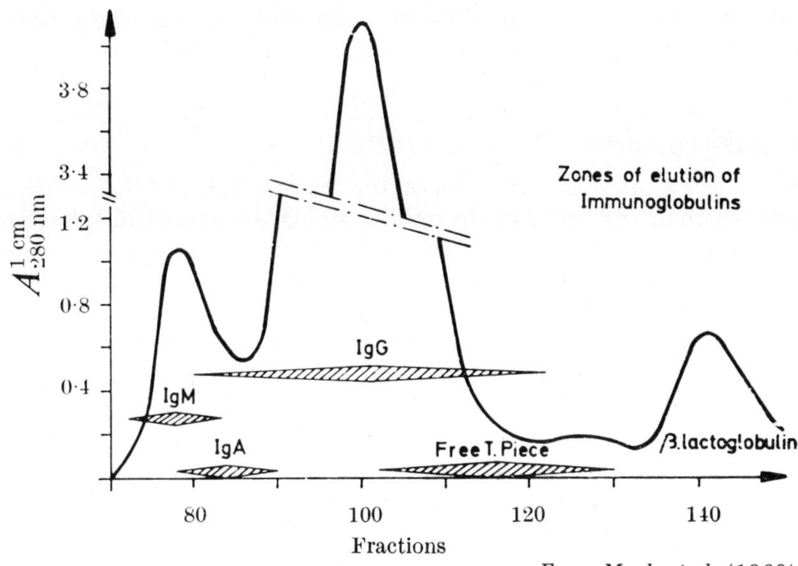

From Mach et al. (1969)

FIG. 2.17. GEL FILTRATION OF TOTAL COLOSTRUM WHEY PROTEIN ON SEPHADEX G-200 FINE (COLUMN 120 CM LENGTH, 3.7 CM DIAMETER, GEL VOLUME 1,420 CM^3, 21.6 ML/HR)

while IgM was eluted from the column with 0.2 *M* phosphate buffer pH 6.0 and purified by two Sephadex G-200 filtrations.

Groves and Gordon (1967) combined several chromatographic techniques to isolate IgG globulin and glycoprotein-a. Starting with the supernatant obtained from the exhaustively dialyzed precipitate from $(NH_4)_2SO_4$ saturated whey, they applied it to a DEAE-cellulose column and eluted it stepwise with phosphate buffer pH 5.0 at 3°C. The fraction obtained with 0.05 *M* phosphate was rechromatogrammed on phosphocellulose at pH 5 with phosphate-NaCl buffer. The fraction eluted at 0.1 *M* phosphate contains both the IgG and the glycoprotein-a. These were subsequently separated by chromatography on Sephadex G-200 at 3°C with phosphate buffer pH 7, $\Gamma/2 = 0.1$.

DEAE-Sephadex A-50 was used by Murphy *et al.* (1964) to separate the IgG1 and IgG2 globulins either with a phosphate buffer gradient with increasing ionic strength and decreasing pH from 8.0 to 4.5 or a phosphate buffer gradient at pH 8.0 from 0.02 *M* to 0.3 *M*.

It is interesting to note that specific antibodies can be separated from whey by adsorption on an immunoadsorbent. For example, Froese (1971) prepared an immunoadsorbent by coupling dinitrophenylated human serum albumin to bromacetylcellulose. After the antibodies were adsorbed, the immunoadsorbent was washed with saline and the antibodies eluted with 0.1 *M*. 2,4-dinitrophenol at pH 8.0.

Proteose-peptones

While less work has been done on the fractionation of the proteose-peptones than the other major milk proteins, Jenness (1959) has prepared component 5 approximately 90% pure. Skimmilk was saturated with NaCl to precipitate the casein, proteose-peptone, lactoperoxidase and other proteins. The precipitate was redispersed at pH 8.0 and the casein precipitated by adjusting the pH to 4.6. The filtrate was concentrated and the lactoperoxidase adsorbed on ion-exchange resin IRC-50 (H-form). Component 5 was then precipitated by carefully adjusting the pH to 4.5.

Kolar and Brunner (1969) applied several different fractionation procedures to secure proteose-peptone rich fractions. Fresh skimmilk was heated to 96–100°C for 20–30 min to denature the major whey proteins and then adjusted to pH 4.6. The precipitated casein and denatured proteins were discarded and the supernatant was dialyzed and lyophilized to yield the proteose-peptone fraction. This fraction was further separated by adjusting to pH 6.7 and adding $(NH_4)_2SO_4$ to 50% saturation. The precipitate was the sigma proteose of Aschaffenburg (1946) and the supernatant was largely component 8.

These authors also modified Jenness' procedure for the preparation of component 5 by replacing the treatment with ion-exchange resin by adjustment of pH to 8.0 and collection of the precipitate by centrifugation. The precipitate is redispersed in water, adjusted to pH 4.6 and centrifuged. The pH of the supernatant was successively adjusted from pH 6.0 to 11.5 removing the precipitate after each adjustment. These precipitates are rich in components 5 and 8.

ELECTROCHEMICAL PROPERTIES OF MILK PROTEINS

Due to the presence of acidic and basic groups on the protein molecule, the proteins are amphoteric in nature, that is, they can possess either a net positive or negative charge in solution dependent upon their environment. This charge is dependent upon the extent of dissociation of these groups and the character and number of any bound ions. The common methods for investigating the electrochemical nature of milk proteins are their titration curves and electrophoretic behavior.

Titration Curves

The shape of the titration curve is not only a function of the primary structure of the protein but reflects its configurations. For example, in the case of the $\alpha_{s1,2}$-casein isolated by the method of Waugh *et al.* (1962), it was demonstrated that, while ionic strength influences the character of the coagulation below pH 6.4, the titration curves indicated that all of the ionizable groups, in whatever form the α_s-casein occurs, are accessible and completely reversible to H^+ ion (Ho and Waugh 1965). The agreement between the ionizable groups as determined from the titration curves and from the primary structure of α_{s1}-casein is quite good (Table 2.7). The isoionic point of α_{s1}-casein is 5.16 at 20° C. The negative logarithm of the intrinsic dissociation constants, pk_i, for the side-chain carboxyl groups were found to be 5.18 and 4.88 at ionic strengths of 0.4 and 0.05, respectively. Detailed treatment of the data in this region of the curve indicated that the apparent electrostatic interaction factor, w', varies only slightly at $\Gamma/2 = 0.4$. This is in agreement with the observation that the α_{s1}-casein exists as a precipitate over the same H range. In contrast at $\Gamma/2 = 0.05$, the value of w' increases 10-fold in the region between pH 5.3 and 3.5 and then returns to approximately the original value at lower pH's. These changes coincide with the precipitation and redispersion of the α_{s1}-casein at this ionic strength. Thus

TABLE 2.7

COMPARISON OF TITRATION CURVE AND AMINO ACID ANALYSIS OF THE CHARGED GROUP IN α_{s1}-CASEINS A AND B

Group	Titration[1]	Analysis[2]
α-COOH	—	1
Side-chain COOH	35 ± 3	32
Phosphate	—	8
Imidazole	5[3]	5
Phenolic, Side-chain NH_2	24 ± 2	24
Guanidyl	6 ± 2	6
Total anionic	51 ± 3	49
Total cationic	26 ± 2[4]	26

Source: Swaisgood (1973).
[1] Recalculated from the data of Ho and Waugh (1965) for molecular weight 23,600.
[2] From sequence data for α_{s1}-casein B.
[3] Ho and Waugh's data corrected for number of groups ionizing between pH 6 and 8.5 and the number of phosphate groups.
[4] Estimated (after correction for molecular weight) from Ho and Waugh with 6 as the number of phosphate groups ionized at pH 2.

the titration curve is consistent not only with the primary structure but the changes in its degree of aggregation.

Tanford and his coworkers (Tanford 1962) demonstrated that the isoionic points of β-lactoglobulin A and B in pure water were 5.35 and 5.45, respectively. The titration curves were reversible between the acid end point and pH 9.7 (Fig. 2.18). The maximum acid-binding capacity observed indicates 20 cationic groups per monomer compared to 21 from its primary structure. The total number of carboxyl groups titrated per monomer is 26 for the A variant and 25 for the B variant. These values are one less than predicted from their primary structures. In contrast, titration indicates one more imidazole plus α-amino groups in the monomer than predicted. If the protein is denatured, the values for carboxyl, imidazole and α-amino groups obtained by titration equal those observed in the primary structure. Apparently this anomalous behavior of the native β-lactoglobulin is due to a carboxyl group with a pk_i of 7.3 which is buried in the native molecular titration range of the carboxyl groups but due to a conformational transition at pH 7.3 becomes titratable.

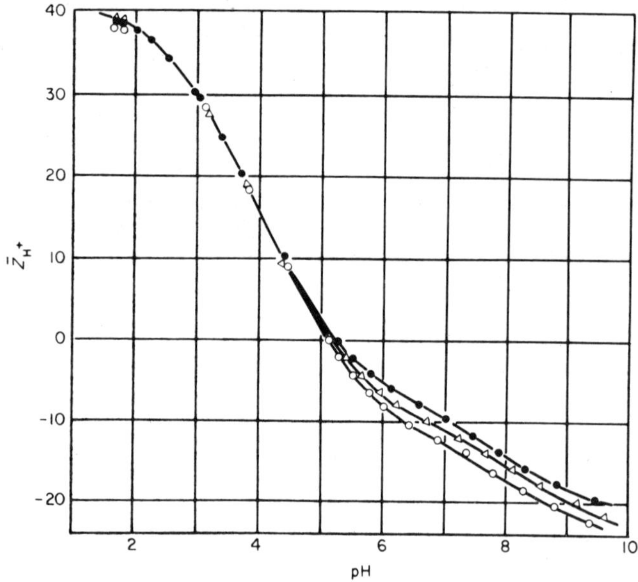

From Basch and Timasheff (1967)

FIG. 2.18. TITRATION CURVES FOR β-LACTOGLOBULINS A, B AND C IN 0.15 *M* KCl AT 25°C

○ = A-variant; △ = B-variant; ● = C-variant; $\bar{Z}_{H^+}$ expressed in terms of the number of groups per dimer (36,000 daltons).

Basch and Timasheff (1967) have considered the titration curve of β-lactoglobulin C. The maximum acid-binding capacity observed indicates two less cationic groups per monomer compared to its primary structure. Also, as in the case of the A and B variants, they observe the same anomalous carboxyl group per monomer of native protein. The discrepancy in the maximum acid-binding capacity of the C-variant has been partially explained by a conformational transition of a protonated histidine residue from the surface to the interior of the molecule either with the transfer of the proton to a carboxyl group or the formation of an imidazolium $^{+}$ ... COO^{-} ion pair. The titration curve of the denatured β-lactoglobulin C is in complete agreement with its primary structure. More work is needed on this variant.

Brignon *et al.* (1969) demonstrated that the maximum acid binding capacity of β-lactoglobulin D was the same as the other variants. The titration curves are identical to pH 4. At pH 6.5, the D-variant dissociates less protons than B as would be predicted from the substitution of a glutamine residue for a glutamic acid residue in B. As in the other variants, one anomalous carboxyl group per monomer is observed in this variant.

The isoionic points of a number of milk proteins are given in Table 2.8.

Electrophoretic Behavior

Since the development of free-boundary electrophoresis by Tiselius (1937) electrophoresis has been extensively used for the identification and characterization of the various milk proteins, as indicated previously in this chapter. With the advent of zonal electrophoresis on polyacrylimide or starch gels with their higher resolving power, this technique has largely replaced free-boundary electrophoresis for this purpose (Swaisgood 1975). The relative mobility of the

TABLE 2.8

THE ISOIONIC POINTS OF SOME MILK PROTEINS

Protein	Isoionic Points at 20°C
α_{s1}-casein A and B	5.16
β-casein	5.35
κ-casein	5.37
Bovine serum albumin	5.13
β-Lactoglobulin-A	5.35
β-Lactoglobulin-B	5.41
β-Lactoglobulin-C	5.39

Source: Ho and Waugh (1965) and McKenzie (1971 A).

proteins on these gels not only reflects the charge of the proteins but also their size and configuration in the particular medium employed. For example, Schmidt and Both (Swaisgood 1975) employ a 6 *M* urea-10% starch gel with 0.76 *M* Tris-citrate buffer, pH 8.6, for separating the genetic variants of α_{s1}-casein and β-casein, but κ-casein, because of the presence of disulfide polymers either remains in the slots or streaks. To phenotype the κ-caseins, they change their gel system to 5 *M* urea–10% starch with 0.03 *M* mercaptoethanol to reduce the disulfide bonds and form the monomers which then can be identified.

The order of the genetic variants of α_{s1}-casein upon alkaline urea gels can be explained on the basis of their relative charge and size. If we arrange them in decreasing order of their net charge at pH 8.6, they will be listed as follows: D, B, C and A. However α_{s1}-casein A contains 13 fewer amino acids than the other variants and therefore is apparently sufficiently smaller to have a comparable charge density and less frictional resistance to its movement through the gel so that the order of mobility of the α_{s1}-caseins is A>D>B>C. Comparable explanations can be proposed for the other milk proteins and their genetic variants.

The physical state of the milk proteins in their natural environment has been investigated by their free-boundary electrophorectic behavior with the natural-protein-free-milk system (NPFMS) as the buffer (Wilson *et al.* 1957; Hansen 1960). The NPFMS was prepared by dialyzing an "ocean" of milk against a limited amount of water or lactose solution in ratio of 500:1. At equilibrium, the NPFMS will contain all of the dialyzable constituents of the milk at the same activities that they possess in the milk. A typical electrophoretic pattern of skimmilk against its NPFMS is shown in Fig. 2.19 with the pattern of its ultracentrifuged fractions. These patterns indicate that most of the casein present exists in the form of micelles but at 2°C some of the caseins exist in nonmicellular form. The mobilities of the proteins vary with the sample of milk and with its previous treatment.

ASSOCIATION OF MILK PROTEINS

In addition to the association that occurs between the monomers of the various milk proteins dependent upon their environment, as indicated in the paragraphs under Structure and Conformation of Milk Protein, the milk proteins are known to form complexes with small ions and molecules, to bind water, and to form complexes with other proteins or macromolecules. The most important example of the latter is the formation of the casein micelles.

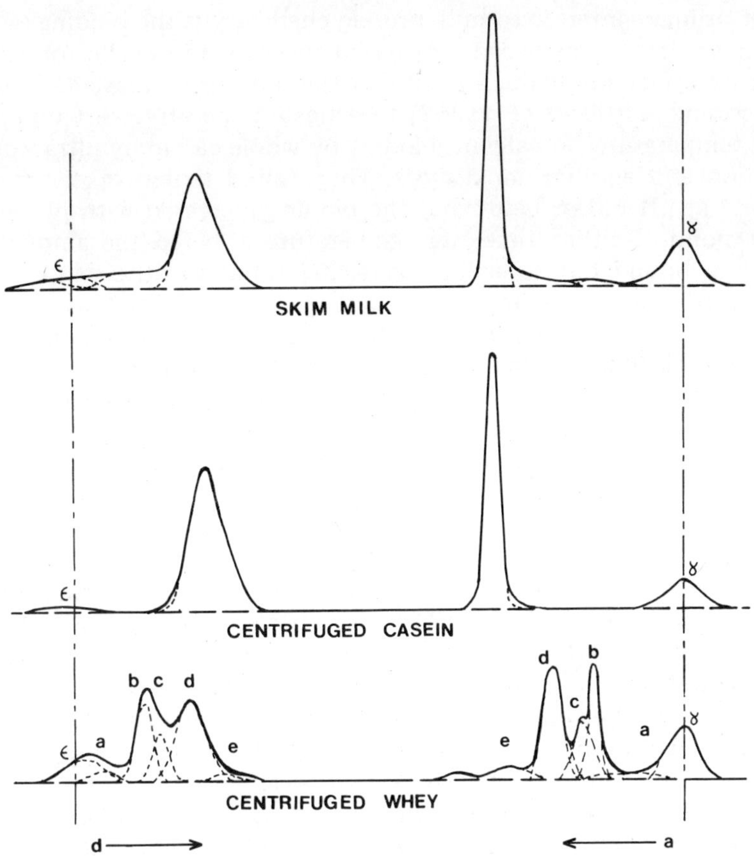

FIG. 2.19. FREE-BOUNDARY ELECTROPHORETIC PATTERNS OF A SKIM-MILK AND ITS FRACTIONS OBTAINED BY ULTRACENTRIFUGATION (55,150 X GM FOR 10 HR AT 0-1°C) WITH ITS NATURAL-PROTEIN-FREE-MILK SYSTEM AS THE BUFFER

The major peak in the skimmilk and centrifuged casein is that of the casein micelles. In the centrifuged whey the components are: a = immunoglobulins and γ-casein; b = β-casein; c = α-lactlabumin; d = β-lactoglobulin and e = serum albumin.

Association with Small Ions and Molecules

The presence in proteins of such negatively charged groups as phosphates, side-chain carboxyls, terminal carboxyls, and sulphydryl groups makes them capable of binding a large number of different cations such as calcium, barium, strontium, magnesium (Dickson and Perkins 1969), copper (Hanson *et al.* 1967), thallium (Sundararajan and Whitney 1969), potassium and sodium (Ho and Waugh 1965) and iron (Gordon *et al.* 1962).

Of primary interest to milk protein chemistry is the binding of calcium to the caseins because of its involvement in micelle formation and its effect upon the stability of the milk protein system during processing. Zittle *et al.* (1958) investigated the effects of time, pH and temperature on calcium binding by whole casein by ultracentrifugation and equilibrium dialysis. They found that no calcium was bound at pH 5.0 or below but the binding increased with pH above that point. Neither time nor temperature affected the amount of calcium bound but reversible aggregates formed as the temperature increased. More calcium was bound when phosphate was present. Rose (1968) found that the distribution of casein between micellular and nonmicellular forms was influenced by the level of calcium and phosphate present. The addition of calcium favored aggregation into micelles. When the sodium ion concentration in the system is increased, the protein-bound calcium decreases due to sodium binding (Ntailianas and Whitney 1964).

The binding of calcium by proteins follows the law of mass action. A number of investigators have attempted to investigate the number and type of binding sites on the various caseins and their association constants (Chanutin *et al.* 1942; Carr 1953; Reisfeld 1957; Gilboe 1959; Jaynes 1970; Dickson and Perkins 1971; Waugh *et al.* 1971; Sundararajan and Whitney 1975). Because of the variety of the methods and the conditions employed and possible differences in the character of the protein preparations, it is difficult to compare their results or draw firm conclusions. However, most of the results suggest that the phosphate groups have the greatest affinity for the calcium and account for most of the binding at low concentrations of calcium, but considerable evidence exists for the involvement of the carboxyl groups as well. Both α_{s1}- and β-casein have regions in their molecules that possess high concentrations of polar groups which result in strong interactions and complicate the interpretation of the results of binding studies. Waugh *et al.* (1971) noted that regardless of the ionic strength both α_s- and β-casein precipitated at the same level of sitebound calcium, 8 and 5.4 per molecule, respectively. They suggest that the calcium may form both inter- and intramolecular bridges in the caseins.

Numerous small molecules also are known to bind to milk proteins. Of primary interest for analytical purposes is the binding of various dyes such as Amido Black, Orange G, and Acid Orange 12. Ashworth (1966) and his coworkers have investigated the use of these dyes in the estimation of the protein content of dairy products. When these acid dyes are added to protein dispersions at pH 2.0, an insoluble precipitate rapidly forms. While the exact nature of the reaction is not known, it is generally assumed that the acid groups of the dye are bound by electrostatic forces to the basic groups on the protein. The amount of dye bound is dependent upon the pH, the ratio of protein to dye concentration and the character of the specific protein involved. The dye-binding capacities of several milk proteins are shown in Table 2.9.

Association with Macromolecules

Proteins are known to form association complexes with other macromolecules including polysaccharides and other proteins. Their interaction with the sulfated polysaccharides is discussed in Chapter 7 and therefore will not be discussed here. Interaction with other protein occurs in the milk system and will be treated in the following section when we consider the casein micelles and their formation. During processing of dairy products other protein complexes are formed such as the interaction of β-lactoglobulin and κ-casein.

Casein Micelles

The structure of the casein micelles in milk has long been of interest to investigators in the field of milk proteins. While much still needs to be done with regard to details and selection of the proper overall model, considerable information has been accumulated with regard to their structure. From electron microscopic (Carroll *et al.* 1968) and sedimentation and inelastic light scattering investigations (Dewan *et al.* 1974), the micelles were found to have molecular weights ranging from 1×10^7 to slightly greater than 6×10^8 with a weight average molecular weight of approximately 2.5×10^8 (Fig. 2.20). Their hydrodynamic radii range from approximately 300–

TABLE 2.9

COMPARISON OF THE DYE-BINDING CAPACITY FOR AMIDO BLACK (AB) AND ORANGE G (OG) FOR VARIOUS MILK PROTEINS AT THREE DIFFERENT LEVELS

	Total Protein (mg/mg)					
	20 mg		30 mg		40 mg	
	AB[1]	OG[1]	AB[1]	OG[1]	AB[1]	OG[1]
α_s-Casein	0.356	0.180	0.350	0.182	0.333	0.181
κ-Casein	0.365	0.190	0.359	0.193	0.350	0.194
β-Casein	0.311	0.170	0.318	0.170	0.305	0.169
β-Lactoglobulin	0.470	0.252	0.456	0.252	0.422	0.252
Proteose-peptone	0.360	0.126	0.350	0.128	0.340	0.127

Source: Ashworth and Chaudry (1962).
[1] Dye used.

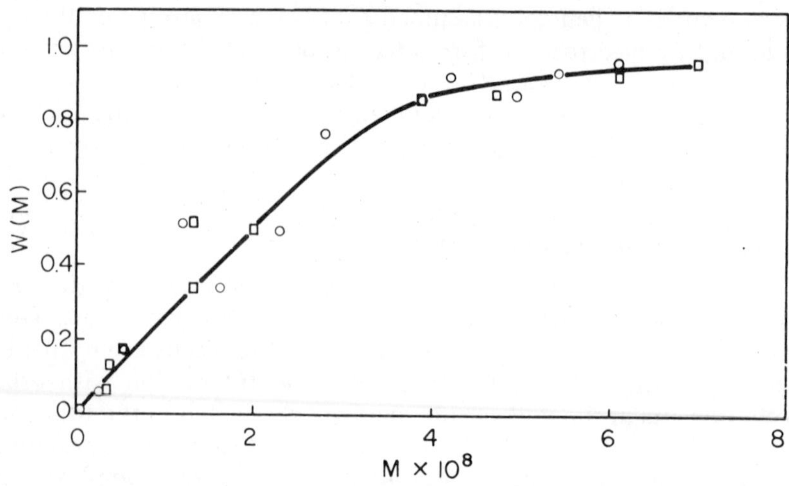

From Dewan et al. (1974)

FIG. 2.20. WEIGHT FRACTION OF CASEIN MICELLES WITH MOLECULAR WEIGHTS, DETERMINED BY SEDIMENTATION AND DIFFUSION, LESS THAN OR EQUAL TO M. TWO DIFFERENT POOLED MILK SAMPLES WERE INVESTIGATED AS INDICATED ○ AND □

1300 Å with a median of 840 Å and their sedimentation constants vary from approximately 100–1500 Svedberg units with a median of 500. While their composition is a function of temperature and the natural variability of milk composition, they have been estimated to contain 50% α_{s1}-casein, 33% β-casein and 15% κ-casein (Rose *et al.* 1969). In addition to proteins they are known to contain colloidal calcium phosphate with a calcium phosphate ratio of 1.5–1.8 (Farrell and Thompson 1974). Considerable evidence exists that indicates that their structure is porous and highly solvated, approximately 1.9 gm of water per gram of protein (Thompson *et al.* 1969A). Ribadeau-Dumas and Garnier (1970) have demonstrated that carboxypeptidase A (mol. wt. 40,000) is able to penetrate into the center of the casein micelle since it quantitatively removes the carboxyl-terminal residues from all of the caseins in the micelle.

To understand why the casein micelles appear to be stable in milk, one must examine the possible bonding forces that may be involved. We have seen earlier that the caseins are highly hydrophobic and quite capable of association through hydrophobic bonds not only between like molecules but between the various caseins. Several investigators have noted that β-casein and to a lesser extent κ- and α_{s1}-casein diffuse out of the micelle at low temperature (1°C) (Rose 1968; Downey and Murphy 1970). This behavior is consistent with the hydrophobic character of these proteins since the outstanding feature of β-casein is its temperature dependent association. Nakai and coworkers (Clark and Nakai 1971) have demonstrated by polarization fluorescence that a 1:1 association which is largely hydrophobic can occur between α_{s1}- and κ-casein. Dissociating agents, such as sodium dodecyl sulfate, guanidine-HCl, and urea, which overcome the hydrophobic association of the individual caseins, tend to disrupt the casein micelles reducing the micelles to subunits approaching 100 Å (Carroll *et al.* 1971). Thus the stability of the casein micelles must be due in part to hydrophobic bonding.

The action of inter- and intra-molecular ionic bonds is less certain. If we consider the polar groups in the sequences of the caseins, the possibility for ion-pair bonds in the apolar environments in the interior of the micelle is apparent. Pepper *et al.* (1970) demonstrated that carbamylation of 5 of the 9 lysine residues of κ-casein destroyed its ability to stabilize α_{s1}-casein. Thus, in addition to the hydrophobic forces, ion-pair bonds may be a factor in micelle stability. The association of calcium with the caseins discussed previously must also be involved in micelle stability though it is difficult to separate its effect from that of the colloidal calcium phosphate.

As indicated previously, the caseins in their monomeric form possess little secondary structures but upon association conformational changes occur. Undoubtedly conformational changes take place upon micelle formation and therefore must be involved in the stabilization of the micelle. Since numerous proteins are partially stabilized in their configurations by hydrogen bonds, as in the case of a α-helices and β-configurations, the possibility of hydrogen bonds between the components of the micelles can not be eliminated. However, as yet, no direct evidence of their existence has been obtained.

The presence of disulfide bonds in the polymer forms of κ-casein undoubtedly are present in the casein micelles and therefore must contribute to their general structure. However, since Woychik (1965) has demonstrated that the reduced and alkylated form of κ-casein also has the ability of stabilizing α_{s1}-casein against precipitation by calcium, their contribution to the formation of the casein micelle may not be great.

One of the most interesting and important characteristics of the casein micelles is the presence of colloidal calcium phosphate. Pyne and McGann (1960) demonstrated that, as the pH of milk is decreased to 5.0 to 5°C, the colloidal calcium phosphate content is reduced to essentially zero. If a small volume of this milk is dialyzed against several large volumes of the original milk, the pH is restored to the original value but the colloidal calcium phosphate is not reformed. Further studies of this colloidal calcium phosphate-free milk (CPF milk) (McGann and Pyne 1960) indicated that it has a greater viscosity than ordinary milk and is more translucent. It is much more sensitive to the addition of calcium ions than ordinary milk, precipitating at 25 *mM* calcium ion, but is slightly more heat stable. Gel chromatography of normal and CPF milk on Sepharose-2B in synthetic milk serum demonstrated a decrease in the size or a change in shape of the micelles upon the removal of colloidal calcium phosphate. (Downey and Murphy 1970). The elution of the normal casein micelles occurred with the void volume indicating a molecular weight greater than 10^8 while the elution of the CPF micelles indicated a molecular weight of approximately 2×10^6. While washing the normal casein micelles by centrifugation removes some of the calcium and phosphate from the micelles, the bulk of the calcium phosphate is not readily removed (Farrell and Thompson 1974) suggesting that the bulk of the colloidal calcium phosphate is rather firmly held as a calcium-phosphate-caseinate complex. However when milk is treated with excess cation exchange resin buffered at pH 6.7 (Jaynes 1970) or with ethylenediamintetraacetate (EDTA), Jenness *et al.* (1966) the calcium content of the micelles is readily reduced and a marked increase in nonmicellular casein is observed. On the basis of these studies of colloidal calcium phosphate, it is evident that the presence of a calcium-phosphate-caseinate complex in the casein micelles is one of the major factors in their formation and stabilization. However, the exact nature of the complex is not yet known.

On the basis of available evidence a number of models have been proposed for the casein micelles. Waugh (1971) notes that α_{s1}- and κ-casein associate to form low weight ratio complexes. Upon the addition of calcium ion to a mixture of α_{s1}- and β-casein, the proteins aggregate as limiting polymers in the form of a rosette with charged phosphate loops on the surface. These limiting polymers are spherical, having a maximum degree of association near 30 and a radius of approximately 100 Å. If κ-casein is present, the precipitation of these core polymers is prevented by the formation of the α_{s1}- κ-casein complexes which coat the core aggregates with the κ-casein on the surface. The size of the micelle is then dictated by the amount of κ-casein available. This model as proposed by Waugh is capable of explaining the lyophilic nature of the micelles as well as the availability of the κ-casein to the action of rennin.

However Parry and Carroll (1969) attempted to identify this surface layer of κ-casein by electronmicroscopy with the aid of a ferritin-labeled anti-κ-casein immunoglobulin and found little or no concentration of κ-casein on the surface. Therefore Parry proposed that the κ-casein served as a point of nucleation about which the other caseins aggregated to be subsequently stabilized by colloidal calcium phosphate. He explained the action of rennin by showing that the nonmicellular κ-casein can be involved in the rennin-coagulation of the micelles possibly by the formation of bridges. Both of these models attribute to κ-casein the responsibility for the size of the casein micelles. However, Parry's model does not predict a subunit structure and Waugh's does not include colloidal calcium phosphate. An additional objection to these models was raised by Ashoor *et al.* (1971) when they demonstrated that papain which had been immobilized by cross-linking with gluteraldehyde was capable of hydrolyzing all three of the major components of isolated casein micelles. Therefore the κ-casein can not be preferentially localized in the micelle.

Garnier and Ribadeau-Dumas (1970) consider that trimers of κ-casein are linked to three chains of α_{s1}- and β-caseins which radiate in the form of a Y from the κ-casein. These chains are linked with other κ-casein trimers to form a loose network. Such a structure allows for the observed porosity of the micelles and the availability of all the casein components to carboxypeptidase A. Such a model would require a fixed and rigid structure for κ-casein and the presence of only disulfide trimers. The involvement of colloidal calcium phosphate in the micellular structure is not considered.

Rose (1969) proposed a mechanism for the formation of the micelles rather than a definite structure. Based upon the hydrophobic character of β-casein, he proposed that the β-casein monomers self-associated into chains to which α_{s1}-monomers were hydrophobically bound. The κ-casein in turn interacts with the α_{s1}-casein. As the micelle is formed it is stabilized by the formation of colloidal calcium phosphate bridges. Such a mechanism would account for a more or less random distribution of the casein components and the porous structure. However, Waugh (1971) has demonstrated that α_{s1}- and β-casein form randomly mixed polymers and α_{s1}- and κ-casein can form micelles without β-casein.

Shimmin and Hill (1964) and Morr (1967B) proposed that the casein micelles consisted of subunits containing all the casein components held together by hydrophobic bonds and calcium caseinate bridges. These subunits, in turn, are held in the micellular structure by colloidal calcium phosphate. Shimmin and Hill observed subunits of 100 Å diameter in electronmicrographs of ultra-thin sections of embedded casein micelles while Morr estimated, on the basis of sedimentation studies, that their diameter was approximately 300 Å. In support of this picture of the casein micelles, Carroll *et al.* (1971)

disrupted micelles with EDTA, urea, sodium lauryl sulfate and sodium fluoride and obtained particles 100±20Å in diameter and Schmidt and Bucheim (1970) disrupted the casein micelles by dialyzing the milk free of calcium in the cold and by the use of high pressure and obtained 100 Å subunits. However, the reported correlation between micellular size and κ-casein (Rose *et al.* 1969; and Waugh 1971) argues against the concept of uniform subunits.

A final conclusion as to the detailed structure of the casein micelles awaits further research.

The Fat Globule Membrane

Since van Leewenhoeck (1674) first observed the fat globule in milk, investigators have attempted to discover the nature of the substance which envelops the fat globule and its origin. However, it wasn't until 1961 that Bargmann *et al.* (1961) were able to demonstrate by electronmicroscopy that, during milk secretion, the fat globules are gradually enveloped by the plasma membrane of the lactating cell and are then extruded into the aveolar lumen. Later Wooding (1971) noted that the Golgi vesicles and endoplasmic reticulum of the lactating cell were incorporated into the plasma lemma prior to secretion. He also demonstrated that the nature of the milk fat globule membrane changes with time after the globule is excreted from the cell, changing from the initial dark-light-dark pattern of the plasma membrane with a layer of cytoplasmic material 100–200Å thick between the globule and the membrane to a final structureless membrane with adsorbed lipoprotein micelles (Fig. 2.21). He suggests that the liproprotein micelles are fragments of the initial membrane. Bauer (1972), however, by electron-microscopic examination of freeze-etched specimens observed as much as 50% of the original unit membrane surrounding the globule and suggested that at least part of the loss of unit membrane was due to the preparation of the specimen for electronmicroscopic examination. Henson *et al.* (1971) observed that, while the general surface of the globule was covered by a single layered membrane, a trilaminar membrane occurred when material was occluded between the membrane and the globule. Enzymatic analysis of the membrane material indicated the presence of marker enzymes for the Golgi apparatus, plasma membrane and endoplasmic reticulum confirming the involvement of these sources in the membrane (Martel-Pradal and Got 1972).

Chemical analysis of various fractions of the membrane material separated by different procedures indicated that the fractions had markedly different composition. For example, Alexander and Lusena (1961) investigating a desoxycholate dissociated membrane by differential centrifugation obtained two high-density fractions contain-

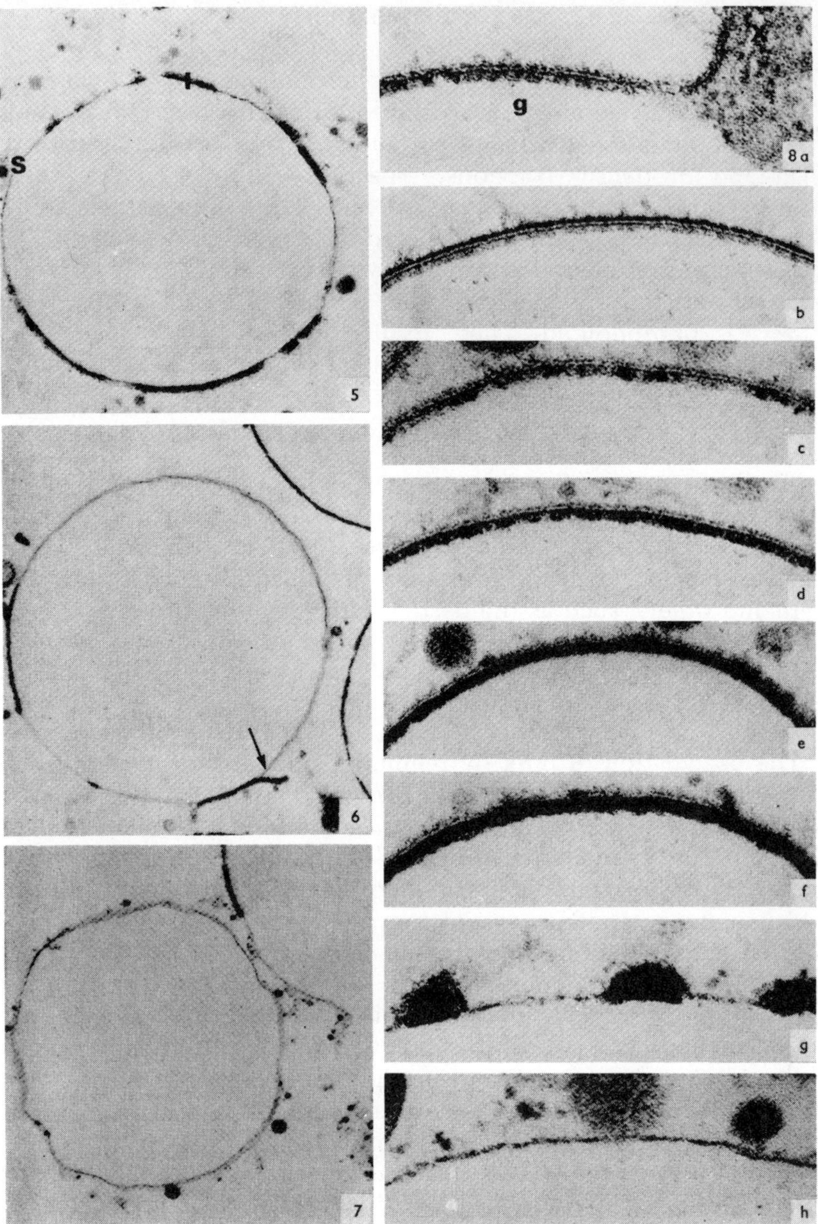

From Wooding (1971)

FIG. 2.21. MILK FAT-GLOBULE MEMBRANE (MFGM) AT DIFFERENT STAGES AFTER SECRETION

(5)—Globule showing secondary MFGM linking parches of initial MFGM × 37,000. (6)—Globule with very little initial MFGM remaining, part of which is coming away from the underlying secondary MFGM × 29,000. (7)—Globule with secondary MFGM × 18,000. (8)—Portions of transverse section of globule membrane from the goat secretory alveolus showing the suggested transition from initial (a) to secondary (h) MFGM × 150,000.

ing 32% lipid, 1/2 of which was phospholipid and a low-density fraction which contained 65% lipid, 1/4 of which was phospholipid. Xanthine oxidase and alkaline phosphotase were associated with the high- and low-density fractions, respectively. Brunner and Thompson (1961) centrifugally separated the membrane lipoproteins into a mucoidal-appearing supernatant and a reddish brown pellet. The supernatant contained lipoproteins with 60% lipids, 23% of which were phospholipid, while the pellet contained 11% lipid, of which 38% was phospholipid. These results indicate that the membrane possesses a very complex structure which remains to be elucidated.

EFFECT OF PROCESSING ON MILK PROTEINS

The milk proteins undergo numerous changes in their structure and configuration during processing which are of great importance to the acceptability of the product. Some are beneficial like the changes that occur in the formation of the curd in cheese manufacture and others, such as the destabilization of milk proteins by heat, are considered to be defects to be avoided in the product.

Effect of Rennin

The coagulation of milk by rennin in the preparation of curd is a two-stage process. The first stage involves the splitting of the κ-casein at the peptide bond between phenylalanine and methionine (residues 105 and 106) as indicated in Fig. 2.3. Thus forming para-κ-casein and either the glycomacropeptide (GMP) or the macropeptide (MP) depending upon whether or not the κ-casein molecule attacked possessed a carbohydrate moiety (MacKinley and Wake 1971). The porous nature of the micelles allows the rennin molecules to diffuse throughout the micelles and react with the κ-casein.

The second step in the process involves an aggregation of the micelles which have been altered by the rennin action to form a three dimensional network of intact micelles which retain their subunit structure (Ruegg *et al.* 1974). Since para-κ-casein has a net positive charge at neutral pH values, the removal of the GMP or MP from the micelles reduces the ζ-potential of the micelles to about 1/2 of that of the native micelles and increases their hydrophobicity. Both of these changes facilitate their closer approach and aggregation. The micelles maintain their individual structures during coagulation and syneresis for 40 min at 40° C.

Heating milk prior to rennin action inhibits the overall reaction possibly due to the formation of the β-lactoglobulin-κ-casein complex which may hinder the rennin action (Morr 1975). Also heating is known to influence the distribution of the milk salts between the colloidal and solution phases and thus might increase the charge of the casein micelles and in this manner inhibiting coagulation.

Upon the development of acid and proteolysis by the cheese culture, changes occur in the character of the rennin curd. The colloidal calcium phosphate becomes soluble as the pH approaches 5 with the resulting dissipation of the individual micelles (McGann and Pyne 1960). However, the casein components become more strongly complexed as the pH approaches the isoelectric point pH 4.5 and the curd shrinks and syneresis occurs with the whey being expelled.

Effect of Heating

Denaturation of Whey Proteins.—When the whey proteins are subjected to temperatures greater than 60°C, they change from their globular structure to a more random open conformation. This denaturation can be followed by the Harland and Ashworth (1945) procedure based upon their loss of solubility at pH 4.6 (Fig. 2.22). However, the rate of denaturation of the individual protein varies. Larson and Rolleri (1955) followed the denaturation of the individual whey proteins by free-boundary electrophoresis (Fig. 2.23). From these results, it may be seen that the whey proteins can be arranged in the following order according to the ease of denaturation: immunoglobulins > serum albumin > β-lactoglobulin > α-lactalbumin. Lyster (1970) observed that the denaturation of α-lactalbumin is a first order reaction between 90 and 155°C but β-lactoglobulin is denatured according to second order reaction kinetics. The rate constant for β-lactoglobulin follows two equations one between 68 and 90°C and the other between 90 and 135°C. There are apparently three forms of denatured β-lactoglobulin: one involving a disulfide interchange and the others involving the hydrophobic, hydrogen and ionic bonds. The susceptibility of whey proteins in skimmilk to denaturation is dependent upon the total solids content of the milk (McKenna and O'Sullivan 1971). The percentage of denaturation at 80°C for 20 min decreased from 80% to 59% to 39% as the total solids increased from 9 to 28 to 44%. At heating times greater than 5 min at 75–80°C, the denaturation follows a first order reaction, but at 5 min or less it appears to be zero order. Comparable rates of whey protein denaturation are observed upon heating either skimmilk, ultracentrifugal whey or cottage cheese whey which has been adjusted to the original pH of the skimmilk (Kenkare *et al.* 1964). In contrast to these results, a recent report suggests that casein micelles protect whey proteins against denaturation (Sabarwal and Ganguli 1972).

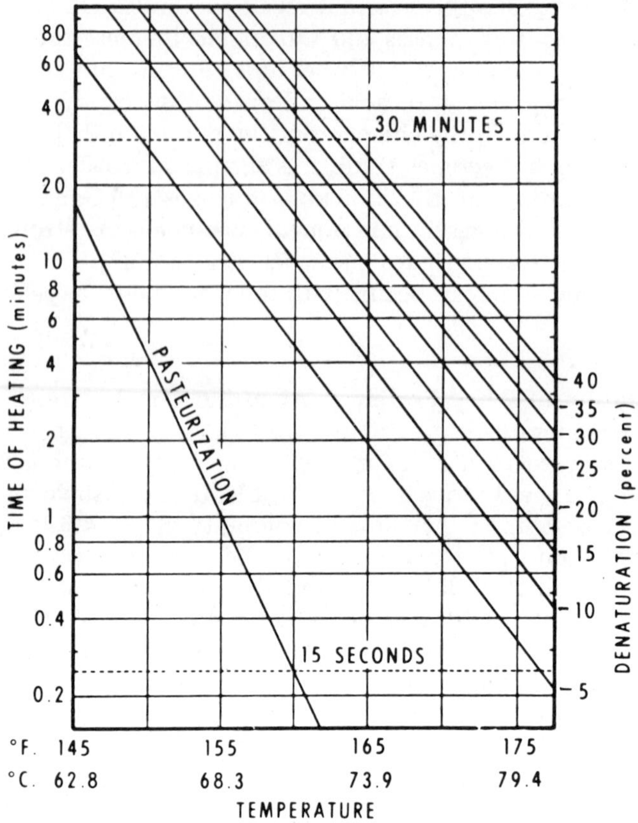

From Harland et al. (1952)

FIG. 2.22. THE HEAT DENATURATION OF THE SERUM PROTEINS IN SKIMMILK AS DETERMINED BY THE SALT PRECIPITATION METHOD

Following denaturation, the whey proteins in heated whey systems aggregate in intermediate-sized complexes of molecular weight equal to or greater than 200,000 which do not sediment at 1,000 xg (Morr and Josephson 1968). Then in the presence of calcium ions they aggregate to form large particles that precipitate at 1,000 xg. When caseins or casein micelles are added, they prevent the formation of these large particles in heated whey. The stabilization by casein is believed to be due to calcium bridges between casein and the whey proteins. When the whey proteins are denatured by heating skimmilk, they also form the intermediate-sized aggregates of molecular weight equal to or greater than 100,000. However ultra-high temperature heating at 146°C for up to 16 sec causes less aggregation than heating at 90°C for 10–30 min.

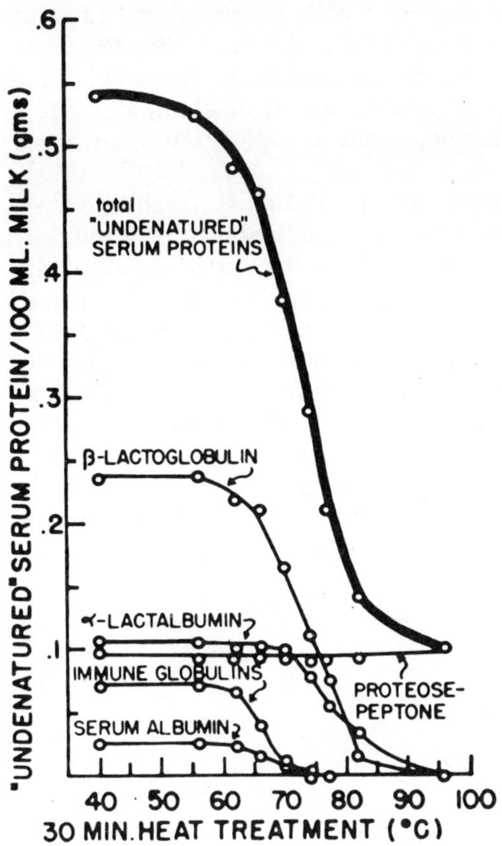

From Larson and Rolleri (1955)

FIG. 2.23. THE EFFECT OF HEATING TEMPERATURE UPON THE EXTENT OF DENATURATION OF THE TOTAL AND INDIVIDUAL SERUM PROTEINS IN MILK AS DETERMINED ELECTROPHORETICALLY

Heating skimmilk also results in the formation of a β-lactoglobulin-κ-casein complex linked by a disulfide interchange (Sawyer 1969). Studies on the isolated proteins indicate that complex formation is inhibited when sulfhydryl blocking agents or reducing agents are present. McKenzie *et al.* (1971) observed that when heat denatured β-lactoglobulins and κ-caseins were mixed at room temperature, no interaction took place. When the mixture was heated at 75° for 7.5 hr, 50% of the κ-casein interacted. The complex once formed is relatively stable and does not dissociate even in 6 *M* guanidine-HCl. Addition of mercaptoethanol, however, completely dissociates the complex. This reaction appears to be a major factor in the heat stability of milk. Tessier and Rose (1964) differentiated between two types of milk: one (A) with a maximum and minimum heat stability in the pH range of 6.4–7.1; the other (B) with a gradually increasing heat stability throughout the range (Fig. 2.24). Addition of κ-casein to type A milk reduced or eliminated the minimum while addition of β-lactoglobulin to type B milk converted it to a type A milk.

Heat Coagulation of Milk Proteins.—In addition to the above reaction which influences the heat stability of a milk, other factors are known to be involved. The salt balance is one of these factors. For example, the higher level of ionic calcium in colostral milk is blamed for its low heat stability. Milks of normal heat stability have a calcium ion activity in their ultrafiltrates of 2–4 *mM* while those with abnormally low heat stability range from 4–7 (Boogaerdt 1954). Adjustments to a more alkaline pH, addition of calcium sequestering agents such as polyphosphates, and heat treatment above 76°C reduce the free ionic calcium and thus increase the protein stability.

The influence of pH upon heat coagulation has long been known. Miller and Sommer (1940) notes a sudden drop in stability between

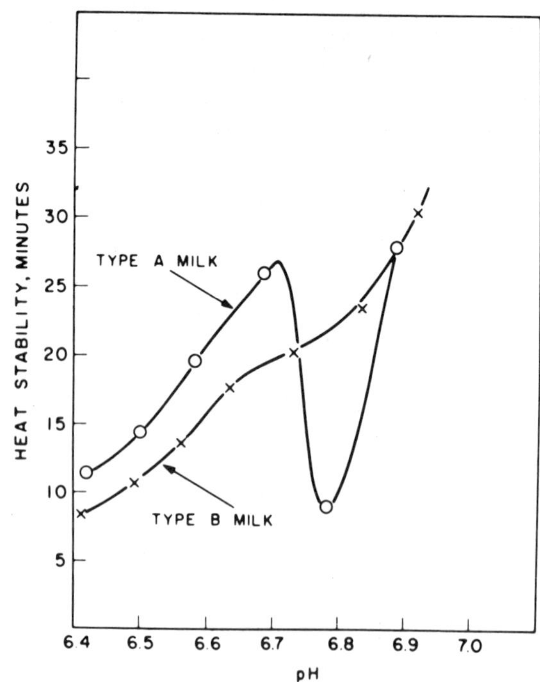

From Tessier and Rose (1964)

FIG. 2.24. THE EFFECT OF pH ON THE HEAT STABILITY OF MILK FROM INDIVIDUAL COWS, TYPE A AND TYPE B

pH 6.4 and 6.2 and a more gradual reduction at lower pH values down to pH 5.4 below which the drop is again precipitous. Each milk exhibits a pH-stability maximum. The pH of milk decreases with increasing temperature due to changes in the salt equilibria, especially the calcium and phosphate equilibrium. With prolonged heat treatment such as 120° C for 90 min, acidity is developed due to the decomposition of lactose and casein dephosphorylation.

The heat coagulation of milk is also a function of the milk solids concentration. The product coagulates at lower temperatures and shorter times as the concentration of nonfat milk solids increases. For example, while fresh milk at 130° C coagulates in about 1 hr, an evaporated milk (26% total solids) will coagulate in 10 min at that temperature. The relationship between the temperature and the logarithm of the coagulation time is linear at each level of milk solids.

Previous forewarming of milk prior to concentration and heat treatment influences the time required for coagulation. If the nonfat solids content is below 13.5%, forewarming decreases the heat stability of the product to subsequent heating. Above this value, forewarming increases the heat stability. Therefore, in the manufacture of sterile milk concentrates it is essential to forewarm the milk to 90° C or above (Morr 1975). This treatment produces the following effects at low concentrations of solids and therefore reduces their involvement in heat coagulation: (1) denaturation of the whey proteins, (2) formation of the β-lactoglobulin-κ-casein complex, and (3) precipitation of calcium phosphate.

Gelation of Sterile Concentrated Milks.—When concentrated milks are sterilized at 140° C or higher for 2 sec or less (High temperature-short time method, HTST), large casein particles are formed. The higher the solids concentration the larger the particle size (Wilson *et al.* 1961). While the predominant size on a volume basis is 2–3 μ in diameter, particles as large as 20 μ are observed (Wilson 1973). These large particles tend to dissociate when stored at 4° C and at a faster rate when stored at 21° C and 37° C. During storage the particles tend to form short chains and finally form a network or gel. According to Morr (1967A) gelation is favored by a high total solids content, and the addition of compounds that reduce the availability of calcium ion such as alkali, citrate and orthophosphate. Addition of calcium has the opposite effect of retarding gelation but reducing heat stability. In contrast, Wilson *et al.* (1963) observed that addition of sodium polyphosphate, which should complex calcium, to the concentrated milk prior to sterilization delayed the onset of gelation. The action of polyphosphates is not completely understood. They do not prevent the formation of the large protein complexes during sterilization but they do inhibit the formation of the particle chains. κ-casein is also involved in the gelation phenomenon since pretreatment of milk with rennin or neuraminidase to split the κ-casein into the MP and para-κ-casein reduces the heat stability but retards the gelation (Morr 1975).

Effect of Freezing.—The freezing of milk has little effect on protein stability. However, if the frozen milk is stored for several weeks, the proteins are gradually destabilized and form a precipitate upon thawing, especially if the storage temperature is allowed to fluctuate and low temperatures (-20°C) are not employed (Morr 1975). Upon freezing, part of the water is converted to ice, and the lactose, salts and proteins become highly concentrated in the remaining aqueous phase. The concentration of solutes in the aqueous phase is independent of the total solids in the original product and is a function of the temperature of freezing (Table 2.10). With a concentration of the lactose in the remaining aqueous phase, it becomes supersaturated and crystallizes. This crystallization permits more water to freeze, increasing the concentration of the casein micelles and salts until equilibrium is restored. At this final concentration of casein micelles, some water of solvation is probably removed by a salting-out effect destabilizing the micelles. The calcium ion is also important since removal of colloidal phosphate and the addition of hexametaphosphate improves the stability of the milk to freezing. Concentration of the milk prior to freezing effects the stability due to its effect on the salt balance. Addition of sugars and hydrocolloids and hydrolysis of the lactose also have been used to stabilize the casein micelles in frozen milk.

TABLE 2.10

PERCENTAGE WATER FROZEN IN FLUID AND CONCENTRATED SKIMMILK AT VARIOUS TEMPERATURES

	Skimmilk (9.3 % TS)		Concentrated Skimmilk (26 % TS)	
Temperature (°C)	Water Frozen (%)	Solids in Unfrozen Solution (Calc. %)	Water Frozen (%)	Solids in Unfrozen Solution (Calc. %)
-24	96.0	72.0	88.0	74.5
-20	95.5	69.5	86.0	71.5
-16	95.0	67.1	84.5	69.4
-12	94.5	65.2	81.0	64.8
-8	92.5	57.8	74.0	57.5
-4	87.5	45.1	53.0	42.8
-2	75.0	29.0	20.0	30.5

Source: Morr (1974).

The Effect of Drying.—During the spray-drying of milk, the temperature of the milk droplets increases from the bulk temperature of the sprayed product to the temperature of the drying gas. Under properly controlled conditions the milk proteins are only slightly altered. The extent of denaturation of the whey proteins is essentially the same as in the concentrated milk from which it is made. (Harland *et al.* 1952). The loss of solubility of dry milk during processing and storage is largely due to changes in the stability of the caseinate-phosphate complex rather than the whey proteins.

In roller drying of milk, the solubility of the casein is altered considerably. The film of milk remains on the drum for several seconds during which the solids content increases progressively. At a solids level above 60% the casein is rapidly insolubilized and at 80% coagulation is almost instantaneous. At a fixed moisture level, the time required to produce a constant degree of insolubility is a logarithmic function of the temperature (Fig. 2.25) (Wright 1932). At moisture levels below 13% the rate of insolublization is decreased so that the rate exhibits a maximum at this moisture level.

The solubility of dry-heated powder varies with the temperature of reconstitution while in high moisture powders the protein is irreversibly insolubilized. The solubility of a milk powder in 20° and 50°C water can therefore be used to differentiate between dry-heat and moist-heat insolubility (Howatt and Wright 1933). Dehydration

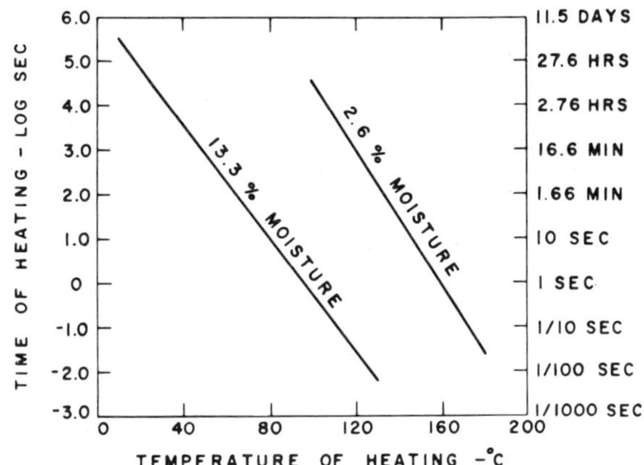

From Wright (1932)

FIG. 2.25. THE TIME-TEMPERATURE RELATIONSHIP FOR THE DEVELOPMENT OF 50% PROTEIN INSOLUBILITY IN HEATED NONFAT MILK POWDER AT TWO MOISTURE CONTENTS. SOLUBILITY MEASURED AT 20°C

at low moisture levels is suggested as a possible cause of dry-heat insolubility. These changes in dry-heated milk powder are wholly related to the caseinate complex with no further detectable changes in the whey proteins. Most of the knowledge of the behavior of the proteins in dry milk was obtained approximately 20 yr ago. It would be interesting to see what information could be secured in this area with the newer protein techniques.

BIBLIOGRAPHY

ADAMS, E. T. JR., and LEWIS, M. S. 1968. Sedimentation equilibrium in reacting systems. VI. Some applications to indefinite self-association. Studies with β-lactoglobulin A. Biochemistry 7, 1044-1053.

ALBRIGHT, D. A., and WILLIAMS, J. W. 1968. A study of the combined sedimentation and chemical equilibrium of β-lactoglobulin B in acid solution. Biochemistry 7, 67-68.

ALEXANDER, K. M., and LUSENA, C. V. 1961. Fractionation of the lipoproteins of the fat globule membrane from cream. J. Dairy Sci. *44*, 1414-1419.

ANDERSSON, L. O. 1966. The heterogeneity of bovine serum albumin. Biochim. Biophys. Acta. *117*, 115-133.

ANNAN, W. D., and MANSON, W. 1969. Fractionation of the α_s-casein complex of bovine milk. J. Dairy Res. *36*, 259-268.

ARMSTRONG, J. M., HOPPER, K. E., McKENZIE, H. A., and MURPHY, W. H. 1970. On the column chromatography of bovine whey proteins. Biochim. Biophys. Acta. *214*, 419-426.

ARMSTRONG, J. M., McKENZIE, H. A., and SAWYER, W. H. 1967. On the fractionation of β-lactoglobulin and α-lactalbumin. Biochim. Biophys. Acta. *147*, 60-72.

ASCHAFFENBURG, R. 1946. Surface activity and proteins of milk. J. Dairy Res. *14*, 316-329.

ASCHAFFENBURG, R. 1961. Inherited casein variants in cow's milk. Nature *192*, 431-432.

ASCHAFFENBURG, R. 1963A. Milk protein polymorphism. *In* Man and Cattle, A. E. Mourant, and F. E. Zeuner (Editors). Royal Anthropological Institute (London), Occasional Paper *18*, 50-54.

ASCHAFFENBURG, R. 1963B. Preparation of β-casein by a modified urea fractionation method. J. Dairy Res. *30*, 259-260.

ASCHAFFENBURG, R. 1968. Preparation of α-lactalbumin from cow's or goat's milk: a method for improving the yield. J. Dairy Sci. *51*, 1295-1296.

ASCHAFFENBURG, R., and DREWRY, J. 1957A. Genetics of the β-lactoglobulins of cow's milk. Nature *180*, 376-378.

ASCHAFFENBURG, R., and DREWRY, J. 1957B. Improved method for the preparation of crystalline β-lactoglobulin and α-lactalbumin from cow's milk. Biochem. J. *65*, 273-277.

ASCHAFFENBURG, R., and SEN, A., and THOMPSON, M. P. 1968. Genetic variants of casein in Indian and African Zebu cattle. Comp. Biochem. Physiol. *25*, 177-184.

ASHOOR, S. H., SAIR, R. A., OLSON, N. F., and RICHARDSON, T. 1971. Use of a papain superpolymer to elucidate the structure of bovine casein micelles. Biochim. Biophys. Acta. *229*, 423-430.

ASHWORTH, U. S. 1966. Determination of protein in dairy products by dye-binding. J. Dairy Sci. *49*, 133-137.

ASHWORTH, U. S., and CHAUDRY, M. A. 1962. Dye-binding capacity of milk proteins for Amido Black 10B and Orange G. J. Dairy Sci. *45*, 952-957.

AUCLAIR, J. E., and BERRIDGE, N. J. 1953A. The inhibition of microorganisms by raw milk. J. Dairy Res. *20*, 370-374.

AUCLAIR, J. E., and BERRIDGE, N. J. 1973B. Séparation par précipitation acétonique des deux constituants de la lacténine. Biochim. Biophys. Acta. *10*, 196-197.

BARGMANN, W., FLEISCHHAUER, K., and KNOOP, A. 1961. Über die morphologie der milchsekretion. II. Zugleich eine kritik am schema der sekretionsmorphologie. Z. Zellforsch. Mikrosk. Anat. *53*, 545-568.

BARMAN, T. E. 1973. The isolation of an α-lactalbumin with three disulfide bonds. European J. Biochem. *37*, 86-89.

BASCH, J. J., KALAN, E. B., and THOMPSON, M. P. 1965. Preparation of β-lactoglobulin C. J. Dairy Sci. *48*, 604-606.

BASCH, J. J., and TIMASHEFF, S. N. 1967. Hydrogen ion equilibria of the genetic variants of bovine β-lactoglobulins. Arch. Biochem. Biophys. *118*, 37-47.

BAUER, H. 1972. Ultrastructural observations on the milk fat globule envelope of cow's milk. J. Dairy Sci. *55*, 1375-1387.

BEEBY, R. 1964. The presence of sulphydryl groups in κ-casein. Biochim. Biophys. Acta. *82*, 418-419.

BELL, K. 1962. One-dimensional starch-gel electrophoresis of bovine skim-milk. Nature *195*, 705-706.

BELL, K., McKENZIE, H. A., and MURPHY, W. H. 1966. Isolation and properties of β-lactoglobulin Droughtmaster, Australian J. Sci. *29*, 87.

BELL, K., McKENZIE, H. A., MURPHY, W. H., and SHAW, D. C. 1970. β-lactoglobulin (Droughtmaster): A unique protein variant. Biochim. Biophys. Acta. *214*, 427-436.

BELL, K., McKENZIE, H. A., and SHAW, D. C. 1968. Amino acid composition and peptide maps of β-lactoglobulin variants. Biochim. Biophys. Acta. *154*, 284-294.

BENGTSSON, C., HANSON, L. Å., and JOHANSSON, B. C. 1962. Immunological studies of modified and enzymatically degraded β-lactoglobulin and α-lactalbumin. Acta. Chem. Scand. *16*, 127-134.

BEZKOROVAINY, A. 1965. Comparative study of the acid glycoproteins isolated from bovine serum, colostrum and milk whey. Arch. Biochem. Biophys. *110*, 558-567.

BEZKOROVAINY, A. 1967. Physical and chemical properties of bovine milk and colostrum whey M-1 glycoproteins. J. Dairy Sci. *50*, 1368.

BEZKOROVAINY, A., and GROHLICH, D. 1969. Separation of the bovine colostrum M-1 glycoprotein into two components. Biochem. J. *115*, 817-822.

BHATTACHARYA, S. D., ROYCHAUDHURY, A. K., SINHA, N. K., and SEN, A. 1963. Inherited α-lactalbumin and β-lactoglobulin polymorphism in Indian Zebu cattle. Comparison of Zebu and buffalo α-lactalbumin. Nature *197*, 797-799.

BJORK, I., and TANFORD, C. 1971. Gross conformation of free polypeptide chains from rabbit immunoglobulin G. I. Heavy chain. Biochemistry *10*, 1271-1280.

BLANC, B. 1967. La lactotransferrine, ses relations avec les aspects physiologiques du metabolisme du fer. *In* Protides of the Biological Fluids, Proceedings of the 14th Colloquium, 1966, H. Peters (Editor). Elsevier, Amsterdam, The Netherlands.

BLOOMFIELD, V. 1966. The structure of bovine serum albumin at low pH. Biochemistry *5*, 684-689.

BLUMBERG, B. S., and TOMBS, M. P. 1958. Possible polymorphism of bovine α-lactalbumin. Nature *181*, 683-684.

BOOGAERDT, J. 1954. Instability of milk due to a high content of calcium ions. Nature *174*, 884.

BRACONNOT, H. 1830. Mémoire sur le caséium et sur le lait; nouvelles ressources qu'ils peuvent affrir à la société. Ann. Chim. Phys. *43*, 337-351.

BRADSHAW, R. A., and PETERS, T., JR. 1969. The amino acid sequence of peptides (1-24) of rat and human serum albumins. J. Biol. Chem. *244*, 5582-5589.

BRAND, E. 1946. Amino acid composition of simple proteins. Ann. N. Y. Acad. Sci. *47*, 187-228.

BRAUNITZER, G., and CHEN, R. 1972. Die Spaltung des β-lactoglobulins AB mit Bromcyan. Hoppe-Seyler's Z. Physiol. Chem. *353*, 674-676.

BRAUNITZER, G., CHEN, R., SCHRANK, B., and STANGL, A. 1972. Automatische sequenzanalyse eines proteins (β-lactoglobulin AB). Hoppe-Seyler's Z. Physiol. Chem. *353*, 832-834.

BREW, K., and CAMPBELL, P. N. 1967. The characterization of the whey proteins of guinea-pig milk. Biochem. J. *102*, 258-264.

BRIGNON, G. *et al.* 1969. Chemical and physico-chemical characterization of genetic variant D of bovine β-lactoglobulin. Arch. Biochem. Biophys. *129*, 720-727.

BROWN, J. R. *et al.* 1971. Amino acid sequence of bovine and porcine serum albumin. Federation Proc. *30*, Pt. II, 1241.

BROWNE, W. J., NORTH, A. C. T., and PHILLIPS, D. C. 1969. A possible three-dimensional structure of bovine α-lactalbumin based on that of hen's egg-white lysozyme. J. Mol. Biol. *42*, 65-86.

BRUNNER, J. R. *et al.* 1960. Nomenclature of the proteins of bovine milk, first revision. J. Dairy Sci. *43*, 901-911.

BRUNNER, J. R., and THOMPSON, M. P. 1961. Characteristics of several minor-protein fractions isolated from bovine milk. J. Dairy Sci. *44*, 1224-1237.

BUCHET, J.-P., PREAUX, G., and LONTIE, R. 1966. Influence de l'abaissement de la temperature et de l'ethylenediamine tetraacetate de sodium sur la transformation des β-lactoglobulins à pH 8.6. Arch. Intern. Physiol. Biochim. *74*, 721-722.

BULL, H. B. 1946. Monolayers of β-lactoglobulin. II. Film molecular weight. J. Am. Chem. Soc. *68*, 745-747.

BUTLER, J. E. 1969. Bovine immunoglobulins: A review. J. Dairy Sci. *52*, 1895-1909.

BUTLER, J. E. 1973. Immunoglobulins of the mammary secretions. *In* Lactation: A Comprehensive Treatise, Vol. III, B. Larson, and V. Smith (Editors). Academic Press, New York.

BUTLER, J. E., COULSON, E. J., and GROVES, M. L. 1968. Identification of glycoprotein-a as a probable fragment of bovine IgA. Federation Proc. *27*, 617.

BUTLER, J. E., and MAXWELL, C. F. 1972. Preparation of bovine immunoglobulins and free secretary component and their specific antisera. J. Dairy Sci. *55*, 151-164.

CARR, C. W. 1953. Studies on the binding of small ions in protein solutions with the use of membrane electrodes. IV. The binding of calcium ions in solutions of various proteins. Arch. Biochem. Biophys. *46*, 424-431.

CARROLL, R. J., FARRELL, H. M., JR., and THOMPSON, M. P. 1971. Electron microscopy of the casein micelle. Forces contributing to its integrity. J. Dairy Sci. *54*, 752.

CARROLL, R. J., THOMPSON, M. P., and NUTTING, G. C. 1968. Glutaraldehyde fixation of casein micelles for electron microscopy. J. Dairy Sci. *51*, 1903-1908.

CASTELLINO, F. J., and HILL, R. L. 1970. The carboxymethylation of bovine α-lactalbumin. J. Biol. Chem. *245*, 417-424.

CEPPELLINI, R. *et al.* 1964. Nomenclature for human immunoglobulins. Bull. World Health Organ. *30*, 447-450.

CHANUTIN, A., LUDWIG, S., and MASKET, A. V. 1942. Studies of the calcium-protein relationship with the aid of the ultracentrifuge. J. Biol. Chem. *143*, 737-751.

CLARK, R. F. L., and NAKAI, S. 1971. Investigation of κ-α_{s1}-casein interaction by fluorescence polarization. Biochemistry *10*, 3353-3357.

CLARK, R. F. L., and NAKAI, S. 1972. Fluorescent studies of κ-casein with 8-anilinonaphthalene-I-sulfonate. Biochim. Biophys. Acta. *257*, 61-69.

CREITH, J. M., and KNIGHT, C. G. 1965. On the estimation of the shape of macromolecules from sedimentation and viscosity measurements. Biochim. Biophys. Acta. *102*, 549-558.

DAUTRIVAUX, M., CROUWY, F., MOSCHETTO, Y., and BISERTE, G. 1966. Étude de l'α-lactalbumine du lait de vache. II. Hydrolysats trypisiques. Bull. Soc. Chim. Biol. *48*, 1111-1117.

DEWAN, R. K. *et al.* 1974. Molecular weight and size distribution of bovine milk casein micelles. Biochim. Biophys. Acta. *342*, 313-321.

DICKSON, J. R., and PERKINS, D. J. 1969. Interaction between the alkaline-earth metal ions and purified bovine caseins. Biochem. J. *113*, 7P.

DICKSON, J. R., and PERKINS, D. J. 1971. Studies on the interaction between purified bovine caseins and alkaline earth-metal ions. Biochem. J. *124*, 235-240.

DILLS, W. L., and NELSON, J. M. 1942. Isolation of a copper bearing protein from cow's milk. J. Am. Chem. Soc. *64*, 1616-1618.

DOI, E., and JIRGENSONS, B. 1970. Circular dichroism studies on the acid denaturation of γ-immunoglobulin G and its fragments. Biochemistry *9*, 1066-1073.

DORRINGTON, K. L., and TANFORD, C. 1970. Molecular size and conformation of immunoglobulins. Advan. Immunol. *12*, 333-381.

DOWNEY, W. K., and MURPHY, R. F. 1970. The temperature-dependent dissociation of the β-casein from bovine casein micelles and complexes. J. Dairy Res. *37*, 361-372.

DREIZEN, P., NOBLE, R. W., and WAUGH, D. F., 1962. Light scattering studies of $\alpha_{s1,2}$-caseins. J. Am. Chem. Soc. *84*, 4938-4943.

EHRENPREIS, S. P., MAURER, P. H., and RAM, J. S. 1957. Modified bovine serum albumin. I. Preparation and physiocochemical studies of some derivatives. Arch. Biochem. Biophys. *67*, 178-195.

FARRELL, H. M., JR. and THOMPSON, M. P. 1974. Physical equilibria: proteins. *In* Fundamentals of Dairy Chemistry, 2nd Edition, B. H. Webb, A. H. Johnson, and J. A. Alford (Editors). Avi Publishing Co., Westport, Conn.

FOSTER, J. F., SOGAMI, M., PETERSON, H. A., and LEONARD, W. J., JR. 1965. The microheterogeneity of plasma albumins. J. Biol. Chem. *240*, 2495-2502.

FOX, K. K., HALSINGER, V. H., POSATI, L. P., and PALLANSCH, M. J. 1967. Separation of β-lactoglobulin from other milk serum proteins by trichloracetic acid. J. Dairy Sci. *50*, 1363-1367.

FRANK, G., and BRAUNITZER, G. 1967. Zur primärstruktur der β-lactoglobuline. Hoppe-Seyler's Z. Physiol. Chem. *348*, 1691-1692.

FRANK, G., and BRAUNITZER, G. 1968. Gewinnung alles tryptischen peptide von β-lactoglobulin A und B. Hoppe-Seyler's Z. Physiol. Chem. *349*, 1456-1462.

FROESE, A. 1971. Isolation of dinitrophenyl-specific antibodies from bovine colostrum. Can. J. Biochem. *49*, 522-528.

GARNIER, J. 1966. Conformation de la caséine β en solution. Analyse d'une transition thermique entre 5 et 40°C. J. Mol. Biol. *19*, 586-590.

GARNIER, J., and RIBADEAU-DUMAS, B. 1970. Structure of the casein micelle. J. Dairy Res. *37*, 493-504.

GEHRKE, C. W., FREEARK, C. W., OH, Y. H., and CHUN, P. W. 1964. Isolation of electrophoretically pure β-caseins. Anal. Biochem. *9*, 423-430.

GILBERT, G. A. 1970. Mixed hemoglobulin molecules. Biochem. J. *119*, 32P.

GILBOE, D. D. 1959. Studies in ion binding and ion exchange of proteins. Ph.D. Thesis, University of Wisconsin, Madison. Diss. Abstr. *19*, 1548.

GORBUNOFF, M. J. 1967. Exposure of tyrosine residues in protein reaction of cyanuric fluoride with ribonuclease, α-lactalbumin and β-lactoglobulin. Biochemistry *6*, 1606-1614.

GORDON, W. G. 1971. α-Lactalbumin. *In* Milk Proteins, Vol. II. H. A. McKenzie (Editor). Academic Press, New York. pp. 331-365.

GORDON, W. G., GROVES, M. L., and BASCH, J. J. 1963. Bovine milk "Red Protein": Amino acid composition and comparison with blood transferrin. Biochemistry *2*, 817-820.

GORDON, W. G. *et al.* 1972. Probable identification of γ-, TS-, R- and S-caseins as fragments of β-casein. J. Dairy Sci. *55*, 261-263.

GORDON, W. G., and SEMMETT, W. F. 1953. Isolation of crystalline α-lactalbumin from milk. J. Am. Chem. Soc. *75*, 328-332.

GORDON, W. G., and ZIEGLER, J. 1955. α-Lactalbumin. Biochem. Prep. *4*, 16-22.

GORDON, W. G., ZIEGLER, J., and BASCH, J. J. 1962. Isolation of an iron-binding protein from cow's milk. Biochim. Biophys. Acta. *60*, 410-411.

GOUGH, P. M., JENNESS, R., and ANDERSON, R. K. 1966. Characterization of bovine immunoglobulins. J. Dairy Sci. *49*, 718.

GREEN, D. W. 1964. *Cited by* S. N. Timasheff, and R. Townend. Structure of the β-lactoglobulin tetramer. Nature *203*, 517-519.

GREEN, D. W., and ASCHAFFENBURG, R. 1959. Two-fold symmetry of the β-lactoglobulin molecule in crystals. J. Mol. Biol. *1*, 54-64.

GREEN, N. M. 1969. Electron microscopy of the immunoglobulins. Advan. Immunol. *11*, 1-30.

GROSCLAUDE, F., MAHÉ, M.-F., MERCIER, J.-C., and RIBADEAU-DUMAS, B. 1972. Caractérisation des variants génétique des caséines α_{s1} et β bovines. European J. Biochem. *26*, 328-337.

GROSCLAUDE, F., MAHÉ, M.-F., and RIBADEAU-DUMAS, B. 1973. Structure primaire de la caséine α_{s1} de la caséine β bovines. European J. Biochem. *40*, 323-324.

GROSCLAUDE, F., PUJOLLE, J., GARNIER, J., and RIBADEAU-DUMAS, B. 1966. Mise en évidence de deux variants supplémentaires des protéines du lait de vache: α_{s1}-Cn D et Lg D. Ann. Biol. Animale Biochim. Biophys. *6*, 215-222.

GROVES, M. L. 1960. The isolation of a red protein from milk. J. Am. Chem. Soc. *82*, 3345-3350.

GROVES, M. L. 1965. Preparation of some iron-binding proteins and α-lactalbumin from bovine milk. Biochim. Biophys. Acta. *100*, 154-162.

GROVES, M. L., and GORDON, W. G. 1967. Isolation of a new glycoprotein-a and a γG-globulin from individual cow milks. Biochemistry *6*, 2388-2394.

GROVES, M. L., and GORDON, W. G. 1969. Evidence from amino acid analysis for a relationship in the biosynthesis of γ- and β-caseins. Biochim. Biophys. Acta. *194*, 421-432.

GROVES, M. L., GORDON, W. G., KALAN, E. B., and JONES, S. B. 1972. Composition of bovine γ-caseins A^1 and A^{3x} and further evidence for a relationship in biosynthesis of γ- and β-caseins. J. Dairy Sci. *55*, 1041-1049.

GROVES, M. L., GORDON, W. G., KALAN, E. B., and JONES, S. B. 1973. TS-A^2, TS-B, R- and S-caseins: Their isolation, composition and relationship to the β- and γ-casein polymorphs A^2 and B. J. Dairy Sci. *56*, 558-568.

GROVES, M. L., GORDON, W. G., and KIDDY, C. A. 1968. Polymorphism of electrophoretically slow-moving casein and their relationship to the γ-casein and β-casein variants. J. Dairy Sci. *51*, 946.

GROVES, M. L., and KIDDY, C. A. 1968. Polymorphism of γ-casein in cow's milk. Arch. Biochem. Biophys. *126*, 188-193.

GROVES, M. L., McMEEKIN, T. L., HIPP, N. J., and GORDON, W. G. 1962.

Preparation of β- and γ-casein by column chromatography. Biochim. Biophys. Acta. *57*, 197-203.

GROVES, M. L., and TOWNEND, R. 1970. Molecular weight of some human and cow caseins. Arch. Biochem. Biophys. *139*, 406-409.

HAMMARSTEN, O. 1883. Zur frage, ob das casein ein einheitlicher stoff sei. Z. Physiol. Chem. 7, 227-273.

HAMMARSTEN, O. 1885. Über den gehalf des caseins an schwefel und über die bestimmung des schwefels in proteinsubstanzen. Z. Physiol. Chem. *9*, *273*-309.

HANSEN, P. M. T. 1960. Investigation of the physical state of the milk proteins in undiluted skim milk. Ph.D. Thesis, Univ. of Illinois, Urbana.

HANSON, L. Å., SAMUELSSON, E. G., and HOLMGREN, J. 1967. Detection of Caeruloplasmin in bovine milk and blood serum. J. Dairy Res. *37*, 493-504.

HARLAND, H. A., and ASHWORTH, U. S. 1945. The preparation and effect of heat treatment on the whey proteins of milk. J. Dairy Sci. *28*, 879-886.

HARLAND, H. A., COULTER, S. T., and JENNESS, R. 1952. The effect of the various steps in the manufacture on the extent of serum protein denaturation in nonfat dry milk solid. J. Dairy Sci. *35*, 363-368.

HARRINGTON, W. F., JOHNSON, P., and OTTEWILL, R. H. 1956. Bovine serum albumin and its behavior in acid solution. Biochem. J. *62*, 569-582.

HENDERSON, L. M., and SNELL, E. E. 1948. A uniform medium for determination of amino acids with various microorganisms. J. Biol. Chem. *172*, 15-29.

HENSON, A. F., HOLDSWORTH, G., and CHANDRAN, R. C. 1971. Physiochemical analyses of the bovine milk fat globule membrane. II. Electron microscopy. J. Dairy Sci. *54*, 1752-1763.

HERSKOVITS, T. T., and MESCONTI, L. 1965. Conformation of proteins and polypeptides. J. Biol. Chem. *240*, 639-644.

HILL, R. D., and HANSEN, R. R. 1964. The separation of milk protein on dextran gel. J. Dairy Res. *31*, 291-295.

HILL, R. J., and WAKE, R. G. 1969. Amphiphile nature of κ-casein as the basis for its micelle stabilizing properties. Nature *221*, 635-639.

HIPP, N. J., GROVES, M. L., CUSTER, J. H., and McMEEKIN, T. L. 1952. Separation of α-, β- and γ-casein. J. Dairy Sci. *35*, 272-281.

HO, C., and WAUGH, D. F. 1965. Interaction of bovine α_s-casein with small ions. J. Am. Chem. Soc. *87*, 110-117.

HOAGLAND, P. D., THOMPSON, M. P., and KALAN, E. B. 1971. Amino acid composition of α_{s3}-, α_{s4}- and α_{s5}-caseins. J. Dairy Sci. *54*, 1103-1110.

HOPPER, K. E., and McKENZIE, H. A. 1973. Purification and properties of bovine milk glyco-α-lactalbumin. Biochim. Biophys. Acta. *214*, 242-244.

HOWATT, G. R., and WRIGHT, N. C. 1933. Factors affecting the solubility of milk powders. J. Dairy Res. *4*, 265-272.

JACKSON, R. H., COULSON, E. J., and CLARK, W. R. 1962. The mucoprotein of the fat/plasma interface of cow's milk. I. Chemical and physical characterization. Arch. Biochem. Biophys. *97*, 373-377.

JANATOVA, J., FULLER, J. K., and HUNTER, M. J. 1968. The heterogeneity of bovine albumin with respect to sulfhydryl and dimer content. J. Biol. Chem. *243*, 3612-3622.

JAYNES, H. O. 1970. A study of the resin-contact-time method for the determination of bound calcium. Ph.D. Thesis, Univ. of Illinois, Urbana.

JENNESS, R. 1959. Characterization of milk serum protein component 5. J. Dairy Sci. *42*, 895.

JENNESS, R. 1970. Protein composition of milk. *In* Milk Proteins, Vol. I, H. A. McKenzie (Editor). Academic Press, New York.

JENNESS, R. *et al.* 1956. Nomenclature of the proteins of bovine milk. J. Dairy Sci. *39*, 536-541.

JENNESS, R., MORR, C. V., and JOSEPHSON, R. V. 1966. Some properties of colloidal phosphate free skim milk. J. Dairy Sci. *49*, 712.

JIRGENSONS, B. 1952. Optical rotation and viscosity of native and denatured proteins. I. Influences of pH, and the stability of various proteins toward several denaturing reagents. Arch. Biochem. Biophys. *39*, 261-270.

JOLLÈS, J. F., SCHOENTGEN, F., ALAIS, C., and JOLLÈS, P. 1972. Studies on the primary structure of cow κ-casein: The primary sequence of cow para-κ-casein. Chimia *26*, 645-646.

JOSEPHSON, R. V., MIKOLAJCIK, E. M., and SINGH, V. K. 1972. Isoelectric focusing of bovine colostrum immunoglobulins. J. Dairy Sci. *55*, 1050-1057.

KENKARE, D. B., MORR, C. V., and GOULD, I. A. 1964. Factors affecting the heat aggregation of proteins in selected skim milk sera. J. Dairy Sci. *47*, 947-953.

KICKHÖFEN, B., HAMMER, D. K., and SCHEEL, D. 1968. Isolation and characterization of γG-type immunoglobulin from bovine serum and colostrum. Hoppe-Seyler's Z. Physiol. Chem. *349*, 1755-1773.

KIDDY, C. A., PETERSON, R. F., and KOPFLER, F. C. 1966. Genetic control of the variants of β-casein A. J. Dairy Sci. *49*, 742.

KIM, Y. K., YAGUCHI, M., and ROSE, D. 1969. Isolation and amino acid composition of para-kappa-casein. J. Dairy Sci. *52*, 316-320.

KING, T. P., and SPENCER, M. E. 1970. Structural studies and organic ligand-binding properties of bovine plasma albumin. J. Biol. Chem. *245*, 6134-6148.

KING, T. P., and SPENCER, M. E. 1972. Amino acid sequences of the amino and the carboxyl terminal cyanogen bromide peptides of bovine plasma albumin. Arch. Biochem. Biophys. *153*, 627-640.

KOLAR, C. K. 1967. Isolation and characterization of proteose-peptone components 8-fast, 8-slow, and 5 from cow's milk. Ph.D. Thesis, Michigan State Univ., East Lansing.

KOLAR, C. K., and BRUNNER, J. R. 1969. Proteose-peptone fraction of bovine milk: Distribution in the protein system. J. Dairy Sci. *52*, 1541-1546.

KOLAR, C. K., and BRUNNER, J. R. 1970. Proteose-peptone fraction of bovine milk: Lacteal serum components 5 and 8-casein-associated glycoproteins. J. Dairy Sci. *53*, 997-1008.

KRIGBAUM, W. R., and KÜGLER, F. R. 1970. Molecular conformation of egg-white lysozyme and bovine α-lactalbumin in solution. Biochemistry *9*, 1216-1223.

KRONMAN, M. J., and ANDREOTTI, R. E. 1964. Inter- and intramolecular interactions of α-lactalbumin. I. The apparent heterogeneity at acid pH. Biochemistry *3*, 1145-1151.

KRONMAN, M. J., ANDREOTTI, R. E. and VITALS, R. 1964. Inter- and intramolecular interactions of α-lactalbumin. II. Aggregation reactions at acid pH. Biochemistry *3*, 1152-1160.

KRONMAN, M. J., BLUM, R., and HOLMES, L. G. 1966. Inter- and intramolecular interactions of α-lactalbumin. VI. Optical rotation dispersion properties. Biochemistry *5*, 1970-1978.

KRONMAN, M. J., CERANKOWSKI, L., and HOLMES, L. G. 1965. Inter- and intramolecular interactions of α-lactalbumin. III. Spectral changes at acid pH. Bio-chemistry *4*, 518-525.

KRONMAN, M. J., and HOLMES, L. G. 1965. Inter- and intramolecular interactions of α-lactalbumin. IV. Location of tryptophan groups. Biochemistry *4*, 526-532.

KUMAR, S., and MIKOLAJCIK, E. M. 1973. Selected physico-chemical characteristics of bovine colostrum immunoglobulin. J. Dairy Sci. *56*, 255-258.

KUMOSINSKI, T. F., and TIMASHEFF, S. N. 1966. Molecular interactions in β-lactoglobulin. X. The stoichiometry of the β-lactoglobulin mixed tetramerization. J. Am. Chem. Soc. *88*, 5635-5642.

LARSON, B. L., and ROLLERI, G. D. 1955. Heat denaturation of specific serum proteins in milk. J. Dairy Sci. *38*, 351-360.

LASKOWSKI, M. JR., and LASKOWSKI, M. 1950. Trypsin inhibitor in colostrum. Federation Proc. *9*, 194.

LASKOWSKI, M., JR., and LASKOWSKI, M. 1951. Crystalline trypsin inhibitor from colostrum. J. Biol. Chem. *190*, 563-573.

LASKOWSKI, M., JR., MARS, P. H., and LASKOWSKI, M. 1952. Comparison of trypsin inhibitor from colostrum with other crystalline trypsin inhibitors. J. Biol. Chem. *198*, 745-752.

LEACH, B. E., BLALOCK, C. R., and PALLANSCH, M. J. 1967. Kinin-like activity in bovine milk. J. Dairy Sci. *50*, 763-764.

LINDERSTRØM-LANG, K., and KODEMA, S. 1925. On the solubility of casein in hydrochloric acid. Compt. Rend. Trav. Lab. Carlsberg, Ser. Chim. *16*, 1-62.

LONTIE, R., and PRÉAUX, G. 1966. Polarimetric investigation of β-lactoglobulin A and B and of the reactivity of their thiol groups. *In* Protides of Biological Fluids: Proceedings of the 14th Colloquium, 1966, H. Peters (Editor). Elsevier, Amsterdam, The Netherlands.

LYSTER, R. L. J. 1970. The denaturation of α-lactalbumin and β-lactoglobulin in heated milk. J. Dairy Res. *37*, 233-243.

LYSTER, R. L. J. 1972. Reviews of the progress of dairy science. Section C. Chemistry of milk proteins. J. Dairy Res. *39*, 279-318.

MACH, J. P., PAHUD, J. J., and ISLIKER, H. 1969. IgA with "Secretary Piece" in bovine colostrum and saliva. Nature *223*, 952-954.

MACKINLAY, A. G., HILL, R. J., and WAKE, R. G. 1966. The action of rennin on κ-casein. The heterogeneity and origin of the insoluble products. Biochim. Biophys. Acta. *115*, 103-112.

MACKINLAY, A. G., and WAKE, R. G. 1964. The heterogeneity of κ-casein. Biochim. Biophys. Acta *93*, 378-386.

MACKINLAY, A. G., and WAKE, R. G. 1966. Fractionation of *S*-carboxymethyl-κ-casein and characterization of the components. Biochim. Biophys. Acta. *104*, 167-180.

MACKINLAY, A. G., and WAKE, R. G. 1971. κ-casein and its attack by rennin (chymosin). *In* Milk Proteins, Vol. II, H. A. McKenzie (Editor). Academic Press, New York.

MARTEL-PRADAL, M. B., and GOT, R. 1972. Presence d'enzymes marquers des membranes plasmiques, de l'appareil de golgi et du reticulum endoplasmique dans les membranes des globules lipidiques de lait maternel. Fed. Eur. Biochem. Soc. Letters *21*, 220-222.

McCLURE, L. E., SCHIELER, L., and DUNN, M. S. 1953. The free amino groups of crystalline bovine plasma albumin. J. Am. Chem. Soc. *75*, 1980-1982.

McGANN, T. C. A., and PYNE, G. T. 1960. The collodial phosphate of milk. III. Nature of its association with casein. J. Dairy Res. *27*, 403-417.

McKENNA, B. M., and O'SULLIVAN, A. C. 1971. Whey protein denaturation in concentrated skimmilks. J. Dairy Sci. *54*, 1075-1077.

McKENZIE, G. H., NORTON, R. S., and SAWYER, W. H. 1971. Heat induced interaction of β-lactoglobulin and κ-casein. J. Dairy Res. *38*, 343-351.

McKENZIE, H. A. 1967. Milk proteins. *In* Advances in Protein Chemistry, Vol. 22, C. B. Anfinsen, M. L. Anson, and J. T. Edsall (Editors). Academic Press, New York.

McKENZIE, H. A. 1971A. β-lactoglobulins. *In* Milk Proteins, Vol. II, H. A. McKenzie (Editor). Academic Press, New York.

McKENZIE, H. A. 1971B. Whole casein: Isolation, properties, and zone electrophoresis. *In* Milk Proteins, Vol. II, H. A. McKenzie (Editor). Academic Press, New York.

McKENZIE, H. A., RALSTON, G. B., and SHAW, D. C. 1972. Location of sulfhydryl and disulfide groups in bovine β-lactoglobulins and effects of urea. Biochemistry *11*, 4539-4547.

McKENZIE, H. A., and SAWYER, W. H. 1966. Zone electrophoresis of β-lactoglobulins. Nature *212*, 161-163.

McKENZIE, H. A., and SAWYER, W. H. 1967. Effect of pH on β-lactoglobulins. Nature *214*, 1101-1104.

McKENZIE, H. A., SAWYER, W. H., and SMITH, M. B. 1967. Optical rotary dispersion and sedimentation in the study of association-dissociation: Bovine β-lactoglobulin near pH 5. Biochim. Biophys. Acta. *147*, 73-92.

McKENZIE, H. A., and WAKE, R. G. 1959. Studies of casein. III. The molecular size of α-, β-, and κ-casein. Australian J. Chem. *12*, 734-742.

McKENZIE, H. A., and WAKE, R. G. 1961. An improved method for the isolation of κ-casein. Biochim. Biophys. Acta. *47*, 240-242.

MELLANDER, O. 1939. Electrophoretische untersuchung von casein. Biochem. Z. *300*, 240-245.

MERCIER, J.-C., BRIGNON, G., and RIBADEAU-DUMAS, B. 1973. Structure primaire de la caséine κB bovine, Séquence complète. European J. Biochem. *35*, 222-235.

MERCIER, J.-C., GROSCLAUDE, F., and RIBADEAU-DUMAS, B. 1971. Structure primaire de la caséine α_{s1} bovine. Séquence complète. European J. Biochem. *23*, 41-51.

MICHALEK, W. 1967. Anomalous electrophoretic pattern of milk proteins. J. Dairy Sci. *50*, 1319-1320.

MILLER, P. G., and SOMMER, H. H. 1940. The coagulation temperature of milk as affected by pH, salts, evaporation and previous heat treatment. J. Dairy Sci. *23*, 405-421.

MILSTEIN, C. P., and FEINSTEIN, A. 1968. Comparative studies of two types of bovine immunoglobulin G heavy chains. Biochem. J. *107*, 559-564.

MOFFITT, W., and YANG, J. T. 1956. The optical rotatory dispersion of simple polypeptides. I. Proc. Natl. Acad. Sci. *42*, 596-603.

MORR, C. V. 1967A. Some effects of pyrophosphate and citrate ions upon the colloidal caseinate-phosphate micelle and ultrafiltrate of raw and heated skimmilk. J. Dairy Sci. *50*, 1038-1044.

MORR, C. V. 1967B. Effect of oxalate and urea upon ultracentrifugation properties of raw and heated skimmilk casein micelle. J. Dairy Sci. *50*, 1744-1751.

MORR, C. V. 1975. Chemistry of milk proteins in food processing. J. Dairy Sci. *58*, 977-984.

MORR, C. V., and JOSEPHSON, R. V. 1968. Effect of calcium, *N*-ethylmaleimide and casein upon heat-induced whey protein aggregation. J. Dairy Sci. 1349-1355.

MULDER, J. G. 1838. Zusammensetzung von fibrin, albumin, leimzucker, leucin, u. s. w. Ann. Pharm. *28*, 73-82.

MURPHY, F. A., AALUND, O., OSEBALD, J. W., and CARROLL, E. J. 1964. Gamma globulins of bovine lacteal secretions. Arch. Biochem. Biophys. *108*, 230-239.

MURTHY, G. K., and WHITNEY, R. McL. 1958. A comparison of some of the chemical and physical properties of γ-casein and immune globulins of milk. J. Dairy Sci. *41*, 1-12.

NG, W. C. 1967. The isolation and physical-chemical characterization of a glycoprotein from the protose-peptone fraction as cow's milk. Ph.D. Thesis. Michigan State Univ., East Lansing.

NG, W. C., BRUNNER, J. R., and RHEE, K. C. 1970. Proteose-peptone fraction of bovine milk: Lacteum serum component 3- a whey glycoprotein. J. Dairy Sci. *53*, 987-996.

NIELSON, H. C., and LILLEVIK, H. A. 1957. Molecular weight determination by osmotic pressure measurements of casein in 6.6 *M* urea. J. Dairy Sci. *40*, 598.

NOELKEN, M. E. 1967. The molecular weight of α_{s1}-casein B. Biochim. Biophys. Acta. *140*, 537-539.

NOELKEN, M., and REIBSTEIN, M. 1968. Conformation of β-casein B. Arch. Biochem. Biophys. *123*, 397-402.

NOLAN, C., and SMITH, E. L. 1962. Glycopeptides. J. Biol. Chem. *237*, 453-458.

NTAILIANAS, H. A., and WHITNEY, R. McL. 1964. Dialysis equilibrium for determining binding of calcium by milk proteins. J. Dairy Sci. *48*, 773.

PALMER, A. H. 1934. The preparation of a crystalline globulin from the albumin fraction of cow milk. J. Biol. Chem. *104*, 359-372.

PARRY, R. M., JR., and CARROLL, R. J. 1969. Location of κ-casein in milk micelles. Biochim. Biophys. Acta. *194*, 138-150.

PAYENS, T. A. J. 1966. Association of caseins and their possible relation to structure of the casein micelle. J. Dairy Sci. *49*, 1317-1324.

PAYENS, T. A. J., and HEREMANS, K. 1969. Effect of pressure on the temperature-dependent association of β-casein. Biopolymers *8*, 335-345.

PAYENS, T. A. J., and SCHMIDT, D. G. 1965. The thermodynamic parameters of the association of α_{s1}-casein C. Biochim. Biophys. Acta. *109*, 214-222.

PAYENS, T. A. J., and SCHMIDT, D. G. 1966. Boundary spreading of rapidly polymerizing α_{s1}-casein B and C during sedimentation. Arch. Biochem. Biophys. *115*, 136-145.

PAYENS, T. A. J., and VAN MARKWIJK, B. W. 1963. Some features of the association of β-casein. Biochim. Biophys. Acta. *71*, 517-530.

PEDERSON, K. O. 1936. Ultracentrifugal and electrophoretic studies on the milk proteins. I. Introduction and preliminary results with fraction from skim milk. Biochem. J. *30*, 948-960.

PEPPER, L., HIPP, N. J., and GORDON, W. G. 1970. Effects of modification of ϵ-amino groups on the interaction of κ- and α_{s1}-caseins. Biochim. Biophys. Acta. *207*, 340-346.

PETERSON, A., and FOSTER, J. F. 1965. The microheterogeneity of plasma albumins. J. Biol. Chem. *240*, 2503-2507.

PETERSON, R. F., and KOPFLER, F. C. 1966. Detection of new types of β-casein by polyacrylamide gel electrophoresis at acid pH: A proposed nomenclature. Biochem. Biophys. Res. Commun. *22*, 388-392.

PIERCE, A. E., and FEINSTEIN, A. 1965. Biophysical and immunological studies on bovine immune globulins with evidence for selective transport within the mammary gland from maternal plasma to colostrum. Immunology *8*, 106-123.

POLIS, B. D., and SHMUKLER, H. W. 1953. Crystalline Lactoperoxidase. J. Biol. Chem. *201*, 475-500.

POLIS, B. D., SHMUKLER, H. W., and CUSTER, J. H. 1950. Isolation of a crystalline albumin from milk. J. Biol. Chem. *187*, 349-354.

POLJAK, R. J. *et al.* 1972. Structure of Fab[1] New at 6 Å resolution. Nature New Biol. *235*, 137-140.

PYNE, G. T., and McGANN, T. C. A. 1960. The colloidal phosphate of milk. II. Influence of citrate. J. Dairy Res. *27*, 9-17.

REISFELD, R. A. 1957. A study of the calcium-binding properties of whole casein, alpha-casein and beta-casein. Ph.D. Thesis, Ohio State Univ., Columbus. Diss. Abstr. *17*, 1204.

RIBADEAU-DUMAS, B., BRIGNON, G., GROSCLAUDE, F., and MERCIER, J.-C. 1972. Structure primaire de la caséine β bovine. Séquence complète. European J. Biochem. *25*, 505-514.

RIBADEAU-DUMAS, B., and GARNIER, J. 1970. Structure of the casein micelle. J. Dairy Res. *37*, 269-278.

RIDDIFORD, C. L., and JENNINGS, B. R. 1966. The Kerr effect study of the low pH configuration changes of bovine plasma albumin in aqueous solution. J. Am. Chem. Soc. *88*, 4359-4364.

ROARK, D. E., and YPHANTIS, D. A. 1969. Studies of self-associating systems by equilibrium ultracentrifugation. Ann. N. Y. Acad. Sci. *164*, 245-278.

ROBBINS, F. M., and HOLMES, L. G. 1970. Circular dichroism spectra of α-lactalbumin. Biochim. Biophys. Acta. *221*, 234-240.

ROBBINS, F. M., and KRONMAN, M. J. 1964. A simplified method for preparing α-lactalbumin and β-lactoglobulin from cow's milk. Biochim. Biophys. Acta. *82*, 186-188.

ROBBINS, F. M., KRONMAN, M. J., and ANDREOTTI, R. E. 1965. Inter- and intramolecular interactions of α-lactalbumin. V. The effect of amidination on association and aggregation. Biochim. Biophys. Acta. *109*, 223-233.

ROSE, D. 1968. Relation between micellar and serum casein in bovine milk. J. Dairy Sci. *51*, 1897-1902.

ROSE, D. 1969. A proposed model of micelle structure in bovine milk. Dairy Sci. Abstr. *31*, 171-175.

ROSE, D. *et al.* 1970. Nomenclature of the proteins of cow's milk, third revision. J. Dairy Sci. *53*, 1-17.

ROSE, D., DAVIES, D. T., and YAGUCHI, M. J. 1969. Quantitative determination of the major components of casein mixtures by column chromatography on DEAE-cellulose. J. Dairy Sci. *52*, 8-11.

ROWLAND, S. J. 1938. Determination of nitrogen distribution in milk. J. Dairy Res. *9*, 42-46.

RUEGG, M., LUSCHER, M., and BLANC, B. 1974. Hydration of native and rennin coagulated caseins as determined by differential scanning calorimetry and gravimetric sorption measurements. J. Dairy Sci. *74*, 387-393.

SABARWAL, P. F., and GANGULI, N. C. 1972. Effect of micellar casein on heat induced denaturation of whey proteins in milk. J. Dairy Sci. *55*, 765-767.

SAWYER, W. H. 1969. Complex between β-lactoglobulin and κ-casein. J. Dairy Sci. *52*, 1347-1355.

SCHECHTER, E., and BLOUT, E. R. 1964. An analysis of the optical rotatory dispersion of polypeptides and proteins. Proc. Natl. Acad. Sci. *51*, 695-702.

SCHMIDT, D. G. 1969. Ph.D. Thesis, Univ. of Utrecht, Utrecht, The Netherlands.

SCHMIDT, D. G. 1970. The association of α_{s1}-casein B at pH 6.6. Biochim. Biophys. Acta. *207*, 130-138.

SCHMIDT, D. G., and BUCHEIM, W. 1970. Electronenmikroskopische untersuchung der feinstruktur von caseinmiscellen in kuhmilch. Milchwissenschaft *25*, 596-600.

SCHMIDT, D. G., and PAYENS, T. A. J. 1963. The purification and some properties of a calcium-sensitive α-casein. Biochim. Biophys. Acta. *78*, 492-499.

SCHMIDT, D. G., and PAYENS, T. A. J. 1964. Differences in aggregation behavior of the genetic variants of α_s-casein. Neth. Milk Dairy J. *18*, 108-111.

SCHMIDT, D. G., PAYENS, T. A. J., VAN MARKWIJK, B. W., and BRINKHUIS, J. A. 1967. On the subunit of α_{s1}-casein. Biochem. Biophys. Res. Commun. *27*, 448-455.

SCHMIDT, D. G., and VAN MARKWIJK, B. W. 1968. Further studies on the associating subunit of α_{s1}-casein. Biochim. Biophys. Acta. *154*, 613-614.

SEBELIEN, J. 1885. Beitrag zur kenntniss der eiweisskörper der kuhmilch. Z. Physiol. Chem. *9*, 445-468.

SHAHANI, K. M. *et al.* 1973. Enzymes in bovine milk: A review. J. Dairy Sci. *56*, 531-543.

SHIMMIN, P. D., and HILL, R. D. 1964. An electron microscope study of the internal structure of casein micelles. J. Dairy Res. *31*, 121-123.

SJÖGREN, B., and SVEDBERG, T. 1930. The molecular weight of lactalbumin. J. Am. Chem. Soc. *52*, 3650-3654.

SMITH, E. L. 1946. Isolation and properties of immune lactoglobulins from bovine whey. J. Biol. Chem. *165*, 665-676.

SMITH, E. L. 1948. The isolation and properties of the immune proteins of bovine milk and colostrum and their role in immunity: A review. J. Dairy Sci. *31*, 127-138.

SØRENSON, M., and SØRENSEN, S. P. L. 1939. The proteins in whey. Compt. Rend. Trav. Lab. Carlsberg, Ser. Chim. *23*, 55-59.

SPAHR, P. E., and EDSALL, J. T. 1964. Amino acid composition of human and bovine serum mercaptalbumins. J. Biol. Chem. *239*, 850-854.

SPENCER, E. M., and KING, T. P. 1971. Isoelectric heterogeneity of bovine plasma albumin. J. Biol. Chem. *246*, 201-208.

SQUIRE, P. G., MOSER, P., and O'KONSKI, C. T. 1968. The hydrodynamic properties of bovine serum albumin monomer and dimer. Biochemistry *7*, 4261-4272.

STEIN, W. H., and MOORE, S. 1949. Amino acid composition of β-lactoglobulin and bovine serum albumin. J. Biol. Chem. *178*, 79-91.

SULLIVAN, R. A. *et al.* 1955. The influence of temperature and electrolytes upon the apparent size and shape of α- and β-casein. Arch. Biochem. Biophys. *55*, 455-468.

SUNDARARAJAN, N. R., and WHITNEY, R. McL. 1969. Binding of thallous ion by casein. J. Dairy Sci. *52*, 1445-1448.

SUNDARARAJAN, N. R., and WHITNEY, R. McL. 1975. Murexide for determination of free and protein-bound calcium in model systems. J. Dairy Sci. *58*, 1575-1608.

SWAISGOOD, H. E. 1963. The isolation and physical-chemical characterization of kappa-casein from cow's milk. Ph.D. Thesis, Michigan State Univ., East Lansing.

SWAISGOOD, H. E. 1973. The caseins. *In* CRC Critical Reviews in Food Technology. The Chemical Rubber Company, Cleveland, Ohio.

SWAISGOOD, H. E. 1975. Methods of Gel Electrophoresis. American Dairy Science Association, Champaign, Ill.

SWAISGOOD, H. E., and BRUNNER, J. R. 1962. Characterization of κ-casein obtained by fractionation with trichloroacetic acid in a concentrated urea solution. J. Dairy Sci. *45*, 1-11.

SWAISGOOD, H. E., and BRUNNER, J. R. 1963. Characteristics of kappa-casein in the presence of various dissociating agents. Biochem. Biophys. Res. Commun. *12*, 148-151.

SWAISGOOD, H. E., BRUNNER, J. R., and LILLEVIK, H. A. 1964. Physical parameters of κ-casein from cow's milk. Biochemistry *3*, 1616-1623.

SWAISGOOD, H. E., and TIMASHEFF, S. N. 1968. Association of α_{s1}-casein C in the alkaline pH range. Arch. Biochem. Biophys. *125*, 344-361.

SWOPE, F. C., RHEE, K. C., and BRUNNER, J. R. 1968. The isolation and partial characterization of a fat globule membrane glycoprotein. Milchwissenschaft *23*, 744-746.

SZUCHET-DERECHIN, S., and JOHNSON, P. 1962. Red proteins from bovine milk. Nature *194*, 473-474.

SZUCHET-DERECHIN, S., and JOHNSON, P. 1965. The "Albumin" fraction of bovine milk. I. Overall chromatographic fractionation on DEAE-cellulose. European Polymer J. *1*, 271-281.

SZUCHET-DERECHIN, S., and JOHNSON, P. 1966. The "Albumin" fraction of bovine milk III. The micro-heterogeneity of the red protein. European Polymer J. *2*, 29-35.

TANFORD, C. 1962. The interpretation of hydrogen ion titration curves of proteins. *In* Advances in Protein Chemistry, Vol. 17, C. B. Anfinsen, K. Bailey, and J. T. Edsall (Editors). Academic Press, New York.

TESSIER, H., and ROSE, D. 1964. Influence of κ-casein and β-lactoglobulin on the heat stability of skimmilk. J. Dairy Sci. *47*, 1047-1051.

THEORELL, H., and ÅKESON, Å. 1944. Highly purified milk peroxidase. Arkiv. Kimi. Mineral. Geol. *17B*, No. 7, 1-6.

THOMPSON, M. P. 1966. DEAE-cellulose-urea chromatography of casein in the presence of 2-mercaptoethanol. J. Dairy Sci. *49*, 792-795.

THOMPSON, M. P. 1971. α_s- and β-caseins. *In* Milk Proteins, Vol. II, H. A. McKenzie (Editor). Academic Press, New York.

THOMPSON, M. P., GORDON, W. C., BOSWELL, R. T., and FARRELL, H. M., JR. 1969A. Solubility, solvation and stabilization of α_{s1}- and β-caseins. J. Dairy Sci. *52*, 1166-1173.

THOMPSON, M. P., GORDON, W. C., PEPPER, L., and GREENBERG, R. 1969B. Amino acid composition of β-caseins from the milks of *Bos Indicus* and *Bos Taurus* cows: a comparative study. Comp. Biochem. Physiol. *30*, 91–98.

THOMPSON, M. P., KALAN, E. B., and GREENBERG, R. 1967. Properties of casein modified by treatment with carboxypeptidase. J. Dairy Sci. *50*, 767–769.

THOMPSON, M. P., and KIDDY, C. A. 1964. Genetic polymorphism in caseins of cow's milk. II. Isolation and properties of α_{s1}-caseins A, B and C. J. Dairy Sci. *47*, 626–632.

THOMPSON, M. P. *et al.* 1965. Nomenclature of the proteins of cow's milk, second revision. J. Dairy Sci. *48*, 159–169.

TIMASHEFF, S. N., MESUNTI, L., BASCH, J. J., and TOWNEND, R. 1966. Conformational transitions of bovine β-lactoglobulins A, B, and C. J. Biol. Chem. *241*, 2496–2501.

TIMASHEFF, S. N., and SUSI, H. 1966. Infrared investigation of the secondary structure of β-lactoglobulins. J. Biol. Chem. *241*, 249–251.

TIMASHEFF, S. N., and TOWNEND, R. 1961A. Molecular interactions in β-lactoglobulin. V. The association of the genetic species of β-lactoglobulin below the isoelectric point. J. Am. Chem. Soc. *83*, 464–469.

TIMASHEFF, S. N., and TOWNEND, R. 1961B. Molecular interactions in β-lactoglobulin. VI. The dissociation of the genetic species of β-lactoglobulin at acid pH's. J. Am. Chem. Soc. *83*, 470–473.

TIMASHEFF, S. N., and TOWNEND, R. 1964. Structure of the β-lactoglobulin tetramer. Nature *203*, 517–519.

TISELIUS, A. 1937. A new apparatus for electrophoretic analysis of colloidal mixtures. Trans. Faraday Soc. *33*, 524–531.

TISELIUS, A., HJERTEN, S., and LEVIN, O. 1956. Protein chromatography on calcium phosphate column. Arch. Biochem. Biophys. *65*, 132–155.

TOWNEND, R. 1965. β-lactoglobulin A and B: The environment of the Asp/Gly difference residue. Arch. Biochem. Biophys. *109*, 1–6.

TOWNEND, R., KUMOSINSKI, T. F., and TIMASHEFF, S. N. 1967. The circular dichroism of variants of β-lactoglobulin. J. Biol. Chem. *242*, 4538–4545.

TOWNEND, R., and TIMASHEFF, S. N. 1957. The molecular weight of β-lactoglobulin. J. Am. Chem. Soc. *79*, 3613–3614.

TRAN, V. D., and BAKER, B. E. 1970. Carbohydrate moiety of κ-casein. J. Dairy Sci. *53*, 1009–1012.

VANAMAN, T. C., BREW, K., and HILL, R. L. 1970. The disulfide bonds of bovine α-lactalbumin. J. Biol. Chem. *245*, 4583–4590.

VAN LEEWENHOECK, A. 1674. Some observations made by a microscope continued by M. Leewenhoeck in Holland lately communicated by Dr. Regnius de Graaf. Phil. Trans. Roy. Soc. (London) *2*, 66–67.

VELICK, S. F., and RONZONI, E. 1948. The amino acid composition of aldolase and D-glyceraldehyde phosphate dehydrogenase. J. Biol. Chem. *173*, 627–639.

VON HIPPEL, P. H., and WAUGH, D. F. 1955. Casein: monomers and polymers. J. Am. Chem. Soc. *77*, 4311–4319.

WAUGH, D. F. 1958. The interactions of α_s-, β- and κ-caseins in micelle formation. Discussions Faraday Soc. *25*, 186–192.

WAUGH, D. F. 1971. Formation and structure of casein micelles. *In* Milk Proteins, Vol. II, H. A. McKenzie (Editor). Academic Press, New York.

WAUGH, D. F., CRAEMER, L. K., SLATTERY, C. W., and DRESDNER, G. W. 1971. Core polymers of casein micelles. Biochemistry *9*, 786–795.

WAUGH, D. F. *et al.* 1962. The α_s-caseins of bovine milk. J. Am. Chem. Soc. *84*, 4929–4938.

WAUGH, D. F., and VON HIPPEL, P. H. 1956. κ-casein and the stabilization of casein micelles. J. Am. Chem. Soc. *78*, 4576–4582.

WEBER, G. 1952. Polarization of the fluorescence of macromolecules. Biochem. J. *51*, 155-167.
WEIL, L., and SEIBLES, T. S. 1964. On the structure of α-lactalbumin. II. Amino acid sequences of four peptides formed in the peptic hydrolysis of the native protein. Arch. Biochem. Biophys.*105*, 457-464.
WEIL, L., and TELKA, M. 1957. Tryptic digestion of native and chemically modified α-lactalbumin. Arch. Biochem. Biophys. *71*, 473-478.
WETLAUFER, D. B. 1961. Osmometry and general characterization of α-lactalbumin. Comp. Rend. Trav. Lab. Carlsberg, Ser. Chim. *32*, 125-138.
WHEELCOCK, J. V., and SINKERSON, G. 1970. Carbohydrates of bovine κ-casein glycopeptides. Biochem. J. *119*, 13P.
WHITNEY, R. McL. *et al.* 1976. Nomenclature of the proteins of cow's milk, fourth revision. J. Dairy Sci. *59*, 795-815.
WILLIAMS, E. J., and FOSTER, J. F. 1959. An investigation of bovine plasma albumin by differential ultraviolet spectroscopy. J. Am. Chem. Soc. *81*, 865-870.
WILSON, H. K. 1973. Measurements and counts of large protein particles in sterile concentrated skimmilk. J. Dairy Sci. *56*, 69-75.
WILSON, H. K., MURTHY, G. K., and WHITNEY, R. McL. 1957. The electrophoretic pattern of undiluted skimmilk with its native protein-free milk system as a buffer. J. Dairy Sci. *40*, 598-599.
WILSON, H. K., VETTER, J. L., SASAGO, K., and HERREID, E. O. 1963. Effects of phosphates added to concentrated milks before sterilization at ultra-high temperatures. J. Dairy Sci. *46*, 1038-1043.
WILSON, H. K., YOSHINO, U., and HERREID, E. O. 1961. Size of protein particles in ultra-high-temperature sterilized milk as related to concentration. J. Dairy Sci. *44*, 1836-1842.
WITTER, A., and TUPPY, H. 1960. *N*-(4-dimethylamino-3,5-dinitrophenyl) maleimide: A coloured sulfhydryl reagent. Biochim. Biophys. Acta. *45*, 429-442.
WOODING, F. B. C. 1971. The structure of milk fat globule membrane. J. Ultrastructure Res. *37*, 388-400.
WOYCHIK, J. H. 1965. Preparation and properties of reduced κ-casein. Arch. Biochem. Biophys. *109*, 542-547.
WOYCHIK, J. H., KALAN, E. B., and NOELKEN, M. E. 1966. Chromatographic isolation and partial characterization of reduced κ-casein components. Biochemistry *5*, 2276-2282.
WRIGHT, N. C. 1932. Factors affecting the solubility of milk powders. J. Dairy Res. *4*, 122-141.
YAGUCHI, M., DAVIES, D. T., and KIM, Y. K. 1968. Preparation of κ-casein by gel filtration. J. Dairy Sci. *51*, 473-477.
YOSHIDA, S. 1969. Reversible transconformation of casein by heating and cooling. J. Agr. Chem. Soc. Japan *43*, 514-520.
ZITTLE, C. A., and BINGHAM, E. W. 1959. Action of purified milk phosphatase on phosphoserine and on casein. J. Dairy Sci. *42*, 1772-1780.
ZITTLE, C. A., and CUSTER, J. H. 1963. Purification and some of the properties of α_s-casein and κ-casein. J. Dairy Sci. *46*, 1183-1188.
ZITTLE, C. A., DELLA MONICA, E. S., RUDD, R. K., and CUSTER, J. H. 1958. Binding of calcium to casein: Influence of pH and calcium and phosphate concentration. Arch. Biochem. Biophys. *76*, 342-353.

CHAPTER 3

Frank Sosulski

Concentrated Seed Proteins

INTRODUCTION

Until recently, seed proteins were utilized in their natural state and food technologists were concerned primarily with conditions in the intact cellular tissues during maturation, storage, processing and cooking. With the exception of enzymes, the cell proteins were studied as one component in a complex biological system and not as a concentrated or isolated substance.

Modern food technologists have adopted processes which are familiar in the research laboratory to extract and concentrate the principal food components from plant storage organs such as seeds. The extraction of oil, protein, starch, pentosans, lecithin, pectins and micro-constituents has become a major industry and the utilization of these food components in the resynthesis of new improved food products has become the principal challenge facing the processor.

Knowledge of the physical and chemical properties of isolated proteins is no longer of only theoretical value and has become indispensible in the development of new technological processes and preparation of finished products. Since most concentrated seed proteins are utilized in aqueous food systems, their colloidal properties and molecular interactions with other food components will determine the success or failure of many food products. Research on functional properties and utilization of concentrated seed proteins is being conducted in many advanced and developing countries. However, there is need for greater exchange of information on seed proteins other than those from soybean. Also, the standardization of procedures for the determination of functional properties and the preparation of experimental food products such as comminuted meats has become a critical necessity. The American Association of Cereal Chemists, which has established international standards for the evaluation of wheat flour and cereal products (Am. Assoc. Cereal Chemists 1972), has undertaken to develop more acceptable criteria for the determination of protein solubility, water absorption and emulsification characteristics of oilseed proteins. The work of these committees is seriously hampered by the lack of agreement on what constitutes a model pH and ionic system in which the prediction tests should be conducted. More knowledge of the aqueous environment within the food product during processing and food preparation would greatly assist in defining the optimum conditions for the experimental evaluation of protein functionality.

Soybean and wheat are the principal sources of extracted and concentrated seed proteins which have been utilized widely as protein supplements in traditional and modern food products. During the past 15 yr there has been a substantial increase in the usage of soybean protein in human and pet food. This has been a result of the introduction of highly purified protein products on the market, greater awareness of their functional properties and nutritive value, and competitive pricing in relation to milk, egg and meat proteins. The estimated use of soy flour and protein products in the United States is shown in Table 3.1. Current estimates of soybeans used for human consumption in the United States and foreign markets are about 1.0 million metric tons per year (Butz 1974). The soy protein food market is projected to be 25-30 million tons by 1980.

Large quantities of extracted and concentrated soy proteins have been consumed by the Chinese and Japanese for many generations. About 1.0 million tons of soybean and soybean meal are used annually in Japan for traditional foods such as miso, soy sauce, tofu and natto (Table 3.2). Several companies are now producing new protein foods from soybean and wheat. Soybean and wheat proteins are being used as flours, concentrates and isolates in bakery, milk and meat products as well as in texturized protein foods.

While soybean appears to have certain natural advantages over other sources of plant proteins, research is underway to develop alternative sources of concentrated seed protein with a wider range of

TABLE 3.1

USES OF SOYBEAN PROTEIN IN THE UNITED STATES, 1970

Use Category	Million Pounds
Flours and grits	
Baked goods	60
Meat products	38
Beverages	12
Dry cereals and infant foods	8
Brewers' flakes	4
Pasta and macaroni products	1
Miscellaneous	7-12
Export	12
Corn-soy-milk blends (CSM)	90
Total	232-237
Concentrates	20-35
Isolates	25-40

Source: Wolf and Cowan (1971).

TABLE 3.2

PRODUCTION OF PLANT PROTEIN FOODS IN JAPAN, 1972

Food	1,000 Metric Tons		
	Whole Soybean	Defatted Soybean	Wheat Gluten
Traditional foods			
Tofu and fried tofu	380	80	
Kori-tofu	50		
Miso	185	10	
Natto	50		
Shoyu (soy sauce)	15	165	
Other	78	50	
Modern foods			
Bakery, milk, meat		14[1]	14

Source: Wanatabe (1974).
[1] Including texturized vegetable protein in fresh and frozen products.

functional properties. Much work on potential food uses for oilseed proteins from peanut, cottonseed, sunflower and rapeseed has already been done. Recently, the success of fine grinding and air classification in non-oilseed grain legumes has expanded the opportunity for concentrating the seed proteins in field peas, field beans, fababean and other starchy legumes. The extraction and utilization of cereal proteins other than wheat gluten is also under active investigation and oat protein concentrate appears to be particularly promising to food scientists.

SOURCES OF SEED PROTEIN

Plants supply about 75% of the world food supply, and starchy cereals and root crops constitute about 88% of the total contribution from plants. The cereals are not only the major source of calories for human nutrition but also supply 50% of the food proteins. The major cereal crops—wheat, rice and corn—are particularly low in protein content (Table 3.3) and show serious deficiencies in several essential amino acids, especially lysine. Legumes contain almost double the protein content of cereals and most species are rich in lysine. However, their productivity is low and the expansion in cereal production during the "green revolution" has been, in part, at the expense of legume acreage. Because of the expanding markets for vegetable oils, the world production of oilseed meals has increased markedly and made available a particularly rich source of seed protein. Some oilseed meals are good sources of lysine (soybean and rapeseed) while others are intermediate in lysine, but most oilseed proteins are more than adequate in sulfur-containing amino acids

TABLE 3.3

PROTEIN CONTENT OF VEGETABLE AND ANIMAL PROTEINS

Vegetable	Protein (%)	Animal	Protein (%)
Cereals	7-15	Whole milk	3.5
Legumes	20-25	Eggs	13
Oilseeds (defatted)	45-55	Meat (red)	16-22
Concentrates (soy, cottonseed)	60-80	Fish	18-25
		Meat (poultry)	20-25
Isolates (soy, wheat)	90-95	Nonfat dry milk	36

Source: Horan (1974A).

(cottonseed, sunflower, peanut). To meet the modern food processor requirements for supplemental and functional proteins, technologies have been developed for the removal of soluble carbohydrates to produce protein concentrates or to extract and isolate the proteins. These products are particularly rich in protein content and are relatively free of organoleptic and antinutritive factors which may have been present in the seed.

The majority of animal products are only intermediate in protein content (13-25%) but have excellent balances of essential amino acids (Table 3.3). Among the disadvantages of animal products as protein sources on a world basis are: high feed conversion ratios for most animal species, high cost per pound or per unit of protein, and problems associated with safe, long-term storage of these perishable products. The characteristically high prices for meat products would preclude any significant increase in the per capita consumption of animal protein foods in developing countries or among those with limited fixed incomes in advanced nations. The challenge has been for protein chemists and technologists to develop new protein foods from plant sources which can serve as replacements or supplements to the conventional protein foods derived from animals.

Cereals

In general, cereal proteins are bland in flavor and are not associated with undesirable pigments or toxic constituents in the seed. A few species contain allergens but most are low in allergenicity. However, cereals are poor sources of concentrated seed proteins for the food industry because of low protein content, deficiencies in essential amino acids and poor extractability of seed components. Concentration of seed proteins is difficult because the seed coat is fused to the endosperm, which contains the bulk of the storage proteins. The proteins structures in the cell may be amorphous and are closely associated with starch and other cell constituents. Mechanical separation of seed components by grinding and air classification has been only partially successful. Extraction of the proteins by wet-milling techniques is hampered by the insolubility of the storage proteins in aqueous solvents.

The hard spring wheats and oats are higher in protein content than other cereals and have functional properties which are of interest to the food industry. Triticale and rye are alternate sources of gluten-type proteins which have the cohesive-elastic properties desired by the baking industry and, more recently, in meat extenders. Rice and corn proteins are available in concentrated form as by-products of the milling industry. Barley, sorghum and millet have limited potential as protein concentrates for food or industrial use.

Wheat

Early in this century, Osborne (1907) fractionated wheat proteins sequentially with water, neutral salt solutions, 70% ethanol and dilute acid or alkali. The proteins extracted by these solvents were classed as albumins, globulins, prolamines (wheat gliadin) and glutelins (wheat glutenin). The latter two relatively insoluble proteins were found to include almost 90% of the wheat protein in Osborne's study. Pence *et al.* (1954) found the range of these protein classes in five wheat varieties were 6–12% albumin, 5–11% globulin and 78–85% gliadin plus glutenin. Comparative data obtained by Chen and Bushuk (1970) illustrate that one bread wheat variety was relatively high in glutenin (acid-soluble plus residue protein) while the durum variety contained about equal proportions of gliadin and glutenin (Table 3.4). The residual glutenin may be extracted with alkali or reducing and dissociating agents.

The inherent characteristics of viscoelasticity and loaf volume in bread doughs are caused primarily by the gluten developed when water, gliadin, glutenin and small amounts of albumin, globulin, lipids, etc. are mixed. When hydrated, the gliadin becomes a viscous mass while hydrated glutenin is tough and cohesive but less elastic than the whole gluten. Thus the rubberlike properties of dough are a result of the blend of these two major wheat proteins in a structure called gluten. The unique properties in hydrated gluten of cohesiveness, extensibility and elasticity serve to maintain the structure of the dough against the pressures caused by gases from yeast fermentation or chemical reactions. By proper manipulation of the dough during gas formation, a cellular, airy structure is retained in the final baked goods which is universally prized for its desirable mouth feel and texture. By the simple process of dough formation and gas ex-

TABLE 3.4

PERCENTAGE OF THE TOTAL PROTEIN EXTRACTED WITH THE OSBORNE SEQUENCE OF SOLVENTS

Endosperm Proteins	Water Soluble	Salt Soluble	Ethanol Soluble	Acid Soluble	Residue Protein
HRS wheat	11.9	5.2	28.5	16.6	34.0
Durum wheat	12.2	4.7	40.7	18.3	23.2
Triticale	26.4	6.5	24.4	17.3	19.0
Rye	34.3	10.7	19.0	9.4	20.6
Rice[1]	2.7–4.6	3.7–6.4	1.8–2.9	—	78.0
Corn[2]	—	12.4	33.9	36.8	16.9
Oat flour[3]	13	37	—	18	18

Source: Chen and Bushuk (1970).
[1] Cagampang *et al.* (1966).
[2] Mertz and Bressani (1957).
[3] Wu *et al.* (1972).

pansion, an unpalatable powder is converted into the world's most popular food products such as bread, cakes, donuts, pancakes, chapattis, etc. Pasta products like spaghetti, noodles and macaroni are made from doughs without fermentation or gas expansion.

Early studies indicated that gliadin contained at least eight components (Woychik *et al.* 1961) with a molecular weight range of 42,000–47,000 (Jones *et al.* 1961). The glutenin was considered by Neilsen *et al.* (1962) to have a basic molecular weight of 20,000 but the subunits were bonded together by disulfide bonds into macromolecules with molecular weights in the millions. Recent research has demonstrated 40–50 different gliadin proteins (Wrigley 1970) with molecular weights of 36,000 and 44,000. Glutenin appears to have at least 15 subunits ranging in molecular weight from 11,000-133,000 (Bietz and Wall 1972).

The largest group of amino acids in wheat gluten proteins have nonpolar (hydrophobic) side chains that result in extensive apolar bonding and water insolubility (Dimler 1963; Krull and Wall 1969). Wheat proteins also contain large amounts of the amino acid, glutamine, as indicated by the high levels of glutamic acid and ammonia in the acid hydrolysate of wheat protein (Table 3.5). Most of the carboxyl groups of glutamic and aspartic acids are present in the amide form and are not free to ionize. Since the contents of basic amino acids such as lysine, histidine and arginine are also low, the gluten proteins have a low total ionic capacity. Although glutamine is neutral, the polar terminal group engages in hydrogen bonding between glutamine residues and other amino acids to cause extensive aggregation of protein chains. Covalent bonds due to cystine crosslinks and hydrogen bonds contribute to the substantial water insolubility of gluten proteins.

TABLE 3.5

AMINO ACID COMPOSITION OF CEREAL GRAINS

Amino Acid	gm amino acid per 16 gm sample N					
	HRS Wheat	Triticale	Rye Flour	Rice[1]	Corn[2]	Oat Groat
Tryptophan	1.5	1.6	1.2	(1.1)[3]	0.7	1.6
Lysine	2.3	3.0	2.9	3.7	3.0	3.7
Histidine	2.2	2.5	2.1	2.3	2.9	2.2
Ammonia	3.6	3.0	2.9	2.3	2.3	2.6
Arginine	4.0	4.9	4.2	7.7	5.4	6.6
Aspartic Acid	4.7	5.9	6.5	10.4	8.1	7.8
Threonine	2.8	3.1	3.3	4.1	4.3	3.3
Serine	5.0	4.6	4.3	5.2	5.1	4.8
Glutamic acid	33.1	30.9	27.5	20.4	22.4	21.0
Proline	11.1	10.7	10.4	4.8	14.9	4.7
Glycine	3.8	4.0	3.6	5.0	4.6	4.8
Alanine	3.3	3.6	3.7	6.0	7.9	4.6
½ cystine	2.6	2.8	2.3	1.1	1.2	3.3
Valine	4.5	5.0	4.9	5.7	5.6	5.6
Methionine	1.7	1.9	1.2	2.4	2.6	1.8
Isoleucine	3.8	4.1	3.6	3.9	3.6	4.2
Leucine	6.7	6.7	6.0	8.0	13.3	7.2
Tyrosine	2.7	2.3	1.9	3.3	3.9	3.0
Phenylalanine	4.8	4.8	4.5	5.2	4.1	4.9
Total	104.2	105.4	97.0	102.6	115.9	97.7
Protein content (N X 5.7)	16.3	17.9	14.5	11.1	9.0	17.8

Source: Tkachuk and Irvine (1969).
[1] Juliano *et al.* (1964).
[2] Wall *et al.* (1971).
[3] Value estimated from another literature source.

The net effect of the covalent, hydrogen and hydrophobic bonding is the insoluble, viscoelastic protein system which is capable of retaining gases during bread making. The strong cohesive-elastic properties of glutenin are related to the high molecular weight of its molecules, which result from covalent linkages of small polypeptide chains through disulfide bonds (Huebner and Wall 1974). In contrast, the more viscous gliadin molecules consist of single chain polypeptides containing only intramolecular disulfide bonds.

Rye

Rye grain is used primarily in bread-making and as a component of distiller's mash. The kernels of rye contain alkyl resorcinols and other antimetabolites that contribute to unpalatability and poor feed efficiency in livestock. Due to the potential toxic constituents and ergot contamination of the grain, rye has not been evaluated as a source of concentrated seed protein.

The gluten of rye is less elastic than that of wheat and produces a darker and denser loaf than wheat bread. The gluten characteristic may be associated with a low gliadin content, as well as a high level of water-soluble protein (Table 3.4). Relative to wheat, the rye proteins are lower in glutamic acid and proline but higher in lysine (Table 3.5).

Triticale

Triticale, a new synthetic species (wheat × rye hybrid), is not grown extensively but much research is being devoted to the developments of the crop (Burrows *et al.* 1972). The new species does not appear to contain the antinutritive factors which occur in rye but resembles rye more than wheat in its amino acid composition (Table 3.5). As a result of shrivelled seeds and low starch contents, the published values for protein content may be higher than would be the case for normally filled kernels. Chen and Bushuk (1970) demonstrated that the protein solubility properties of triticale are intermediate to the parent species. At the present time it is difficult to predict what the potential of this crop will be in terms of world production or protein utilization.

Rice

Rice is comparatively low in protein content, 7.5-8.8% (McCall *et al.* 1953), but has a fairly good balance of essential amino acids (Table 3.5). The glutelin, oryzenin, is the principal protein fraction of milled rice while the albumins and globulins are concentrated in the aleurone layers and germ (Table 3.4). The natural glutelin had a molecular weight of 600,000 and, after reduction with mercaptoethanol, the principal glutelin fraction was 1/10 of that size (Tecson *et al.* 1971). As indicated by the amino acid composition in Table 3.5, rice proteins are relatively water insoluble because of low acidic and basic amino acid side-chains but lack the gluten characteristic because of low glutamine and cystine levels.

Fine grinding and air classification have been very satisfactory for separation of high-protein fractions from soft wheat but were less successful on rice flour (Stringfellow *et al.* 1961). This can be attributed to the vitreous nature of rice, the very small starch granules and the intimate dispersion of protein bodies among the starch granules.

Houston *et al.* (1964) produced high-protein rice flour by surface or peripheral abrasion and found the protein content was more than twice that of the original milled rice. The product is useful as a protein supplement in breakfast and infant cereals. The potential for further concentration of rice protein above the 20% level does not appear promising at the present time.

Corn

Protein represents about 10% of the corn kernel but the large germ has 20% protein and the endosperm contains only 8% protein (Hamilton *et al.* 1951). The corn grain has significant levels of albumins and globulins but they arise primarily from the germ (Table 3.4). The distribution of protein fractions in the endosperm are 8% albumins and globulins, 38% zein and 44% glutelin (Paulis *et al.* 1969). Zein, the principal alcohol-soluble protein in corn, is rich in glutamine and asparagine plus leucine, alanine and proline residues but is deficient in basic and acidic amino acids, especially lysine (Table 3.5). The high proportion of nonpolar amino acids causes zein to be soluble in alcoholic solvents while the hydrogen bonding of the amide groups contribute the property of water insolubility to this protein (Reiners *et al.* 1973). The average molecular weight of zein is 45,000, but the glutelins have not been fully characterized.

Zein is currently extracted from corn gluten meal, a by-product of the corn wet-milling industry. The alcohol-extracted protein is utilized as an industrial protein because of its film-forming properties (Reiners *et al.* 1973). The principal use is as a binder in formulations for coating tablets, but small amounts are also used as a protective coating on nuts and confections. In general, the unusual solubility properties of zein have limited its use in the food field, even though adequate quantities of raw material for zein extraction are available.

Glutelin is the major protein in the mutant corn variety, opaque-2, (53%) and contains about 30 times the lysine concentration of zein (Paulis *et al.* 1969). Glutelin is cross-linked through disulfide bonds which render the protein insoluble in both saline and ethanolic solutions. The molecular weights or subunit structure of corn glutelin has not been determined.

Oats

Hischke *et al.* (1968) obtained protein contents of 18.1–22.2% (N × 6.25) in groats from seven cultivars while Wu *et al.* (1972) reported a range of 16.4–20.5% for three varieties. The globulin content of groat proteins is very high for a cereal and the prolamine level is usually very low (Table 3.4). These salt and ethanol solubility characteristics are associated with high levels of basic amino acids but low contents of glutamine and proline in the protein (Table 3.5).

Cluskey *et al.* (1973) developed an alkaline wet-milling process to produce protein concentrates, starch and a residue fraction. The bland concentrates varied between 59 and 75% in protein (N × 6.25) and showed a nitrogen dispersibility curve with good solubility at pH 2.5 and 8 (Fig. 3.1). The concentrates were reported to have reasonable hydration capacity and emulsion stability (Wu *et al.* 1973). Air classification of oat groats and flour has given satisfactory yields of protein concentrate which may find food uses because of the good amino acid balance (Table 3.5) and bland flavor of the products (Wu and Stringfellow 1973).

Grain Legumes

Oilseed and starchy legumes have high levels of proteins that are rich in several essential amino acids. Only oilseed meals are presently utilized as sources of concentrated seed protein. However, procedures have been developed for the air classification of the starch and protein fractions in the non-oilseed legumes (Youngs 1970). This dry milling technique offers a low cost procedure for upgrading the protein content of legumes which have less than 25% protein. However,

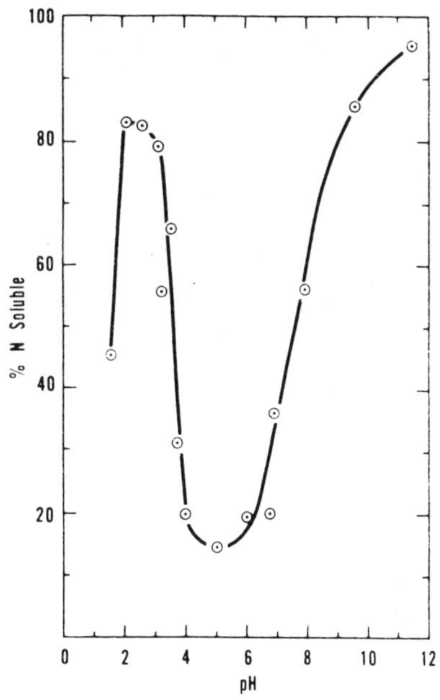

From Wu et al. (1973)

FIG. 3.1. NITROGEN SOLUBILITY OF DEFATTED GARLAND GROAT PROTEIN CONCENTRATE AT VARIOUS pH VALUES

the economics of fine grinding and air classification of pea and bean crops is dependent on the availability of markets for the starch-rich fraction.

The protein contents of dehulled legume flours are typically about 40% in soybean and lupine, 25–30% in fababean and pea bean, 20–25% in mung bean and field pea and 20% in lima bean, lentil and chickpea (Table 3.6). Significant levels of oil are present in soybean (23%), lupine (8%) and chickpea (6%) flours but variations in the low levels of fiber and ash are relatively small. Peanuts contain 25–30% protein and 50% oil and are discussed as an oilseed crop.

Osborne and Campbell (1898A, 1898B) demonstrated that, unlike the cereal proteins, the seeds of the legume family contain high levels of albumin and globulin. The high water solubility of soybean proteins has been utilized traditionally in the Orient in the preparation of foods and beverages. The two soybean varieties in Table 3.8 of the following section gave water dispersibility values of 69.0 and 75.7%. Baker *et al.* (1961) and Zarkadas *et al.* (1965) reported that 56% and 85% of pea protein were water- and salt-soluble, respectively.

Investigations of Danielsson (1949) into the globulin proteins of 34 legumes indicated that nearly all species contained the two major fractions, legumin and vicilin. Based on ultracentrifuge studies, the molecular weights were 186,000 for vicilin and 331,000 for legumin, corresponding to the 7–8 S and 11–12 S sedimentation coefficients. Estimates of the legumin to vicilin ratios vary from 2:1–4:1 in the various species.

In his review of plant proteins, Danielsson (1956) noted that seed globulins of molecular weight 300,000–400,000 readily dissociated into half molecules. Earlier, Johnson and Shooter (1950) demonstrated a reversible dissociating system for arachin, a globulin of peanut, and its half-molecule which was affected by pH and salt concentration. Conarachin, the second globulin of peanut, has a more complex dissociating system based on 6 or 7 components. Joubert (1957) has used ultracentrifuge techniques to show that globulins from lupine and several bean species are also capable of reversible association-dissociation under conditions of varying pH and ionic strength of the solution.

Due to the presence of the same major globulins in most legume species, large variations in amino acid composition would not be expected. Holt and Sosulski (1974B) found that, except for lupine, the distributions of amino acids in the proteins of ten legume species were very similar (Table 3.7). Substantial variations were noted in only glutamic acid and arginine which, along with aspartic acid, con-

TABLE 3.6

FLOUR COMPOSITION, NITROGEN EXTRACTABILITY AND YIELD OF PROTEIN ISOLATE FROM GRAIN LEGUMES, DRY BASIS

Legume Flour	Protein (%) N × 5.7	Fat (%)	Fiber (%)	Ash (%)	Protein Isolate				Whey N (% of Total N)
					Yield (gm/100 gm flour)	Nitrogen (%)	Yield (% of Total Protein)	Color	
Soybean	39.7	23.1	2.2	4.8	36.6	15.0	78.9	cream	10.1
Lupine	40.8	7.9	1.5	3.1	30.8	15.2	65.6	white	21.5
Fababean	30.0	1.5	1.4	2.9	28.2	14.9	80.2	tan	18.4
Pea bean	28.6	1.6	1.7	4.0	28.6	13.1	74.6	white	23.0
Mung bean	24.7	0.6	0.9	3.7	26.9	14.1	87.6	yellow	11.4
Field pea	22.7	1.0	1.5	2.9	22.7	14.0	79.8	cream	20.5
Lima bean	20.0	0.9	2.1	3.8	17.9	12.5	64.3	white	32.7
Lentil	19.8	1.1	1.1	3.4	19.0	13.3	72.8	cream	18.2
Chickpea	19.2	5.6	1.3	2.6	18.5	13.6	74.6	cream	17.9

Source: Fan and Sosulski (1974).

TABLE 3.7

AMINO ACID COMPOSITION OF TEN LEGUME SPECIES

	Grams of Amino Acid per 16 gm Flour N									
	Soy-bean	Lupine	Pea bean	Field pea	Faba-bean	Lath-yrus	Mung bean	Lima bean	Lentil	Chick pea
% N	6.18	6.17	4.61	4.57	4.47	4.47	4.25	3.65	3.64	3.36
Tryptophan	1.3	0.5	1.1	1.0	1.0	0.5	1.1	1.3	0.9	1.2
Lysine	6.4	4.5	6.7	7.2	6.3	5.5	7.4	6.4	7.5	7.0
Histidine	2.2	2.2	2.8	2.4	2.5	2.5	3.0	2.8	2.4	2.5
Amide-N	1.5	1.4	1.3	1.2	1.3	1.3	1.3	1.2	1.1	1.2
Arginine	7.4	12.3	6.8	9.7	8.9	8.3	8.4	4.7	6.8	8.4
Aspartic acid	11.5	9.5	11.9	12.2	11.2	10.0	11.9	12.1	11.2	11.9
Threonine	3.7	3.1	4.0	3.7	3.5	3.3	3.5	4.1	3.6	3.8
Serine	4.5	4.7	5.7	4.5	4.7	4.2	5.3	6.2	4.7	5.2
Glutamic acid	18.3	21.8	16.1	18.1	17.0	16.5	18.5	13.2	15.7	17.0
Proline	5.1	3.8	3.8	4.0	4.1	3.6	4.6	4.1	3.5	4.2
Glycine	4.2	3.8	3.7	4.5	4.3	3.4	3.9	3.8	4.1	4.0
Alanine	4.3	3.2	3.8	4.4	4.1	3.5	4.3	4.0	4.4	4.1
½ Cystine	1.1	0.4	0.7	1.1	1.0	0.4	1.0	1.0	1.0	1.2
Valine	4.9	3.0	4.8	4.8	4.6	3.6	5.0	4.6	4.9	4.2
Methionine	1.4	1.4	0.9	0.9	0.9	0.7	0.7	1.2	1.2	1.6
Isoleucine	4.7	3.8	4.4	4.0	4.0	3.5	4.2	4.6	4.3	4.3
Leucine	7.4	6.8	8.3	7.2	7.4	6.7	8.5	8.4	7.2	7.8
Tyrosine	3.4	3.4	3.4	3.0	3.0	2.7	2.9	3.6	2.9	2.6
Phenylalanine	4.9	3.1	5.6	4.6	4.1	3.8	5.7	5.3	4.8	5.5
Total	97.1	91.6	94.7	97.5	92.6	83.0	99.8	91.4	91.2	96.5

Source: Holt and Sosulski (1974B).

stituted 30–35% of the total amino acids. The high levels of acidic and basic groups, and low levels of amide nitrogen, would contribute high water and salt dispersibility to these legume proteins.

In the absence of salts, globulins show widely varying solubilities in water, depending on the pH of the slurry. This is due to the insolubility of globulin proteins at their isoelectric points. For most seed globulins, the isoelectric points occur in the pH range of 3–6 where the molecules have no net charge and no electrostatic repulsions exist between neighboring molecules. Associative attraction between protein molecules prevent the dispersion of proteins, and those in solution will rapidly aggregate and precipitate. If the pH is above or below the isoelectric point, the globulins dissolve in the aqueous medium, even if salts are not present.

These phenomena are illustrated by the nitrogen extraction and precipitation curves which were determined over the pH range of 2–11 on the flour proteins of 9 legume species (Fig. 3.2). The proteins in mung bean, field pea and soybean were highly extractable except at the narrow isoelectric pH of 4–5 and, after extraction with dilute alkali (pH 11–12), showed a similar precipitation pattern when the pH was adjusted to pH 2. Chickpea, lupine, lentil and pea bean showed progressively wider ranges of insolubility at the isoelectric pH, indicating greater heterogeneity in isolectric points among constituent globulins. These legume flours showed wide variations in solubility at pH 6–7, which is characteristic of most food products, and significant differences in functional properties could be predicted from this data.

There were some relationships between amino acid composition of the legumes and their extraction profiles. The highly soluble mung bean and field pea proteins contained almost 50% of acidic and basic amino acids while the more insoluble lentil and pea bean proteins had less than 45% of these ionic amino acids. Lima bean was very low in total ionic amino acids (39%) but exhibited rather high solubility in the isoelectric range. However, this protein showed poor solubility at pH 6–8, as occurred for lentil and pea bean.

About 30% of the lima bean nitrogen was soluble at the isoelectric point and 50% remained in solution when the alkali-extracted proteins were precipitated at pH 4–5 (Fig. 3.2). Concentrated proteins available to the food industry, especially casein and soybean, are very insoluble at pH 4–5. A highly dispersible protein is needed for certain acidic foods and beverages, and lima bean protein should be investigated for this purpose.

The protein precipitation curves (Fig. 3.2) gave good prediction of the yields of protein isolate from each legume flour (Table 3.6).

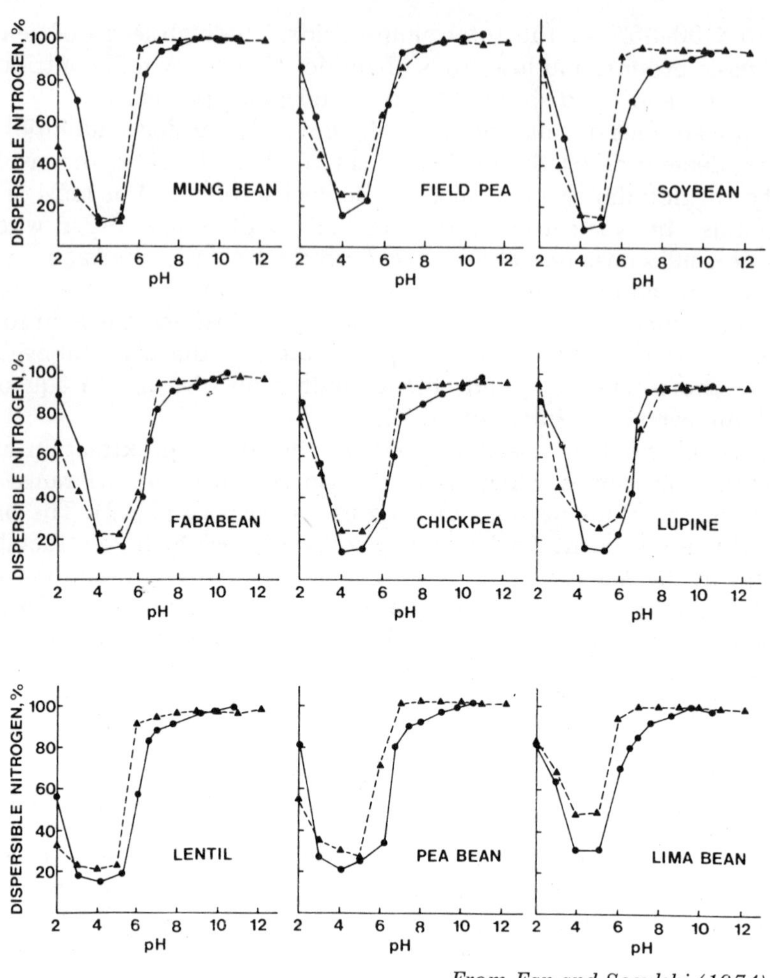

From Fan and Sosulski (1974)

FIG. 3.2. NITROGEN EXTRACTION (—●—●—●—●—) AND PRECIPITATION (—▲—▲—▲—) PROFILES FOR 9 LEGUME FLOURS

Soybean flour was the best source of protein isolate because of its high protein content, high extractability and low losses of whey protein. Lupine and fababean also provided high yields of relatively pure protein isolate. The protein isolates of pea bean, lima bean and lentil were low in nitrogen content and whey losses were high.

Oilseeds

As the name implies, oilseeds are grown primarily for their extractable lipids. However, the by-product meals are the richest sources of protein available to man. Most oilseeds proteins have essential amino acid compositions which complement those of cereals but some are deficient in lysine. To manufacture a food-grade flour, the seed is dehulled and low temperature extraction of the oil must be employed. Defatted soybean flours and grits are the major products of commerce and are available in a variety of specialized forms (full-fat, high-fat, low-fat, defatted, lecithinated) and degrees of protein denaturation. Protein concentrates and isolates represent higher levels of protein concentration and greater freedom from adverse flavor and color components (Fig. 3.3). These refined products are also modified to provide a wide range of functional properties to meet the requirements of food formulators.

Large quantities of peanut, cottonseed, rapeseed, coconut and sesame meals are produced in protein-deficient countries and concerted efforts are being made to utilize these sources of supplementary protein for human feeding. The chemical composition and processing methods for the new protein products will not be identical to that of

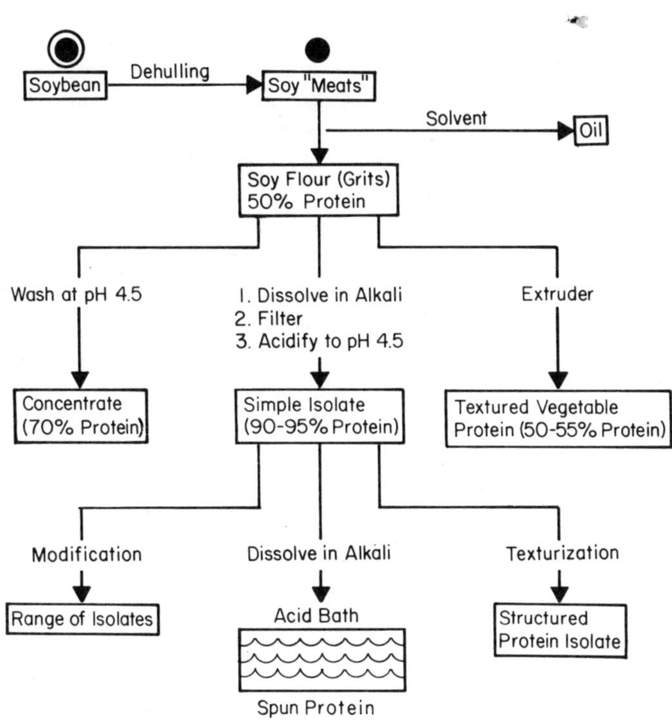

From McCleary (1973)

FIG. 3.3. TYPES OF PROTEIN PRODUCTS AVAILABLE FROM SOYBEAN

soybean. These new products will be categorized as flours, concentrates or isolates but their specifications and characteristics may be quite different from the present market classes of soybean products in the United States.

Soybean

Soybean globulin constitutes about 85% of soybean proteins with the remainder being albumins, proteose and other conjugate proteins (Wolf *et al.* 1962). Ultracentrifugation of the water-extractable soybean proteins gave four fractions with the approximate sedimentation values of 2, 7, 11 and 15 S (Table 3.8). These fractions constituted about 22, 37, 31 and 11% of the water-soluble proteins. Some of the components of these fractions have been identified and their molecular weights determined by several investigators (Wolf and Cowan 1971). The 7 S globulin is a glycoprotein and makes up over 1/2 of the 7 S fraction while 11 S globulin constitutes over 1/3 of the total soybean protein. These two major proteins have been purified and characterized with respect to their subunit structure.

The complex interactions which determine protein solubility under varying salt concentrations have been explained in part by Koshiyama (1968) using the soybean globulins. The 7S globulin, and to a lesser extent the 11 S globulin, undergo reversible association-dissociation over a critical range of ionic strength. The 7 S globulin has a molecular weight of about 200,000 (monomer form) at 0.5 ionic strength (pH 7.6) but at 0.1 ionic strength, the protein dimerizes to a molecular weight of 370,000 with a sedimentation coefficient of 9 S.

TABLE 3.8

APPROXIMATE AMOUNTS AND COMPONENTS OF ULTRACENTRIFUGE FRACTIONS OF WATER-EXTRACTABLE SOYBEAN PROTEINS

Fraction	% of Total	Components	Molecular Weight
2 S	22	Trypsin inhibitors	8,000–21,500
		Cytochrome c	12,000
7 S	37	Hemagglutinins	110,000
		Lipoxygenases	102,000
		Beta-amylase	61,700
		7 S globulin	180,000–210,000
11 S	31	11 S globulin	350,000
15 S	11		600,000

Source: Wolf and Cowan (1971).

Under acid and alkaline conditions, there may be disruption of the 11 S molecule into slower sedimenting species (Wolf and Cowan 1971). These molecular changes are particularly significant in foods which have sodium chloride or acids as natural or added ingredients. The salt concentration in most food products would, however, be sufficiently low and the pH near neutrality to favor the associated form of the two proteins.

Circle (1950) reported that low concentrations of calcium and magnesium salts suppress the solubility of soy protein to 15–30% over a pH range from 4–9. The ancient process for making tofu in Japan involves the calcium coagulation of protein from soy milk with calcium sulfate (Smith and Circle 1972). Recently, Saio *et al.* (1973) proposed the commercial fractionation of 7 S and 11 S proteins on the basis of the solubility of 7 S proteins in dilute calcium chloride. The 11 S protein was then extracted with water and found to give calcium- and heat-induced gels with higher water absorption and tensile strength than the 7 S preparation or soybean protein isolate.

Because of the low amide content, the soybean proteins are higher in total ionic (acidic and basic) groups on the amino acids than cereals (Table 3.7). The low level of hydrophobic groups is consistent with the high water dispersibility of these proteins. The cystine content is also low, indicating less protein aggregation through cross-linkages than in cereal proteins.

Peanuts

Although the world production of peanut meal protein exceeds three million tons, only a negligible portion is utilized for human consumption (Rhee *et al.* 1973). The Central Food Technological Research Institute, Mysore, India, has developed processes for the production of aflatoxin-free peanut flour and protein isolate and these processes are now being exploited commercially (Krishnaswamy 1970). Rhee *et al.* (1972, 1973) have described aqueous processes for the simultaneous extraction of oil and a 67% protein concentrate or an 89% protein isolate. Because of their moisture content at harvest, peanuts are susceptible to mold growth and development of aflatoxin. Aqueous oil-milling of peanuts makes it possible to inactivate aflatoxins by means of alkaline or oxidizing agents.

About 90% of the proteins in peanut cotyledons are soluble in neutral buffer or 5% sodium chloride solution (Dechary and Altschul 1966). Fractionation of the salt-soluble protein by precipitation with ammonium sulfate yields about 73% arachin and 25–30% alpha-conarachin, each of which is composed of many components. At high ionic strength the arachin is in the associated form (14.6 S and molecular weight 330,000) and at low ionic strength the dissociated form has 9.0 S and molecular weight 180,000. At high ionic strength, the alpha-conarachin dissociates (7.8 S and molecular weight 142,000) and associates at low ionic strength (12.6 S and molecular weight 295,000).

The protein extraction curve for peanut protein has been determined by Rhee *et al.* (1972) and shows a broad range of insolubility between pH 3–6 (Fig. 3.4). The point of minimum solubility was estimated to be pH 4.0. Above pH 8.0, more than 97% to the total proteins in peanuts were extracted. Sodium and calcium chloride suppressed protein extractability under neutral and alkaline conditions or below pH 3.0.

Peanut proteins are high in the basic amino acid, arginine, but contain low levels of lysine and the sulfur amino acids (Table 3.9). Arachin is unique in being almost devoid of methionine, as well as low in cystine and lysine. Conarachin contains three times the sulfur amino acids and two times the lysine of arachin. Conarachin is the more soluble of the two peanut proteins but the yield of the more nutritive protein is too low (20%) for commercial application.

Cottonseed

Cottonseed is one of the principal oilseeds of the world and, at 20% protein, represents about 6% of the world's total supply of edible protein (Ridlehuber and Gardner 1974). Because of deficiencies

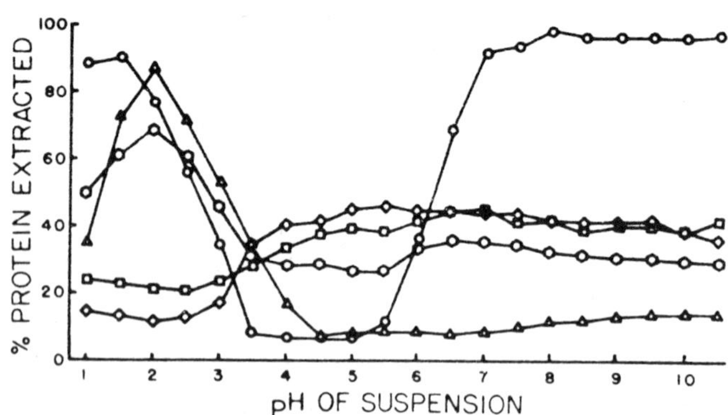

From Rhee et al. (1972)

FIG. 3.4. EFFECT OF $CaCl_2$ (OR $MgCl_2$) AT DIFFERENT LEVELS OF CONCENTRATIONS ON THE EXTRACTION OF PROTEIN AT VARIOUS pH LEVELS

○—Water. △—0.01 *M* $CaCl_2$. ○—0.01 *M* $CaCl_2$. □—0.25 *M* $CaCl_2$. ◇—$CaCl_2$.

in yield and cotton fiber quality, the glandless varieties of cottonseed that are free of gossypol have not been readily accepted by the industry. A liquid cyclone process to produce cottonseed flour from glanded seeds has proven feasible in pilot plant studies at the Southern Regional Research Center and a commercial operation has been initiated. A mixed solvent process using hexane, acetone and water to remove most of the gossypol from the meal along with the oil (Gastrock *et al.* 1965) has undergone extensive developmental research. Recently, oil and protein have been recovered simultaneously from oilseed materials by aqueous extraction of the comminuted seed, followed by a centrifugal separation of the slurry into oil, solid and aqueous phases (Cater *et al.* 1974). In this new oilseed milling process, the proteins are recovered as concentrates or isolates in the aqueous or solid phase. Protein concentrates have also been produced experimentally by air classification (Martinez *et al.* 1967) and aqueous extraction with ethanol or calcium chloride (Martinez *et al.* 1970A, 1970B).

For cytological reasons, cottonseed proteins are only 10% water soluble but 80-90% is dispersible in neutral salt or alkaline solutions.

TABLE 3.9

AMINO ACID COMPOSITION OF OILSEED MEALS AND FLOURS

	Grams of Amino Acid per 16 gm Meal N				
Amino Acid	Soybean Meal	Peanut[1] Meal	Cottonseed[2] Flour	Sunflower Four	Rapeseed Meal
Tryptophan	1.3	(1.2)[3]	(1.6)	1.2	1.6
Lysine	6.5	3.3	4.4	3.2	6.3
Histidine	2.0	2.9	2.9	2.2	3.0
Ammonia		2.0			
Arginine	7.5	10.4	12.4	8.5	6.1
Aspartic acid	11.3	11.4	9.1	8.9	7.0
Threonine	3.7	2.6	3.0	2.9	3.7
Serine	4.2	4.8	4.1	2.9	3.0
Glutamic acid	17.2	18.6	20.4	20.8	18.6
Proline	4.1	3.5	3.6	4.6	7.9
Glycine	4.3	6.1	4.1	6.2	5.2
Alanine	7.5	4.0	3.7	4.4	4.3
½ Cystine	1.9	1.2	(1.1)	2.8	3.2
Valine	5.0	3.7	4.6	5.4	5.0
Methionine	1.4	0.8	1.3	2.2	1.9
Isoleucine	4.9	3.3	3.4	4.8	4.1
Leucine	7.4	6.4	5.8	6.2	7.2
Tyrosine	3.5	4.1	3.1	2.3	2.5
Phenylalanine	4.8	5.1	5.5	4.3	3.5
Total	98.5	98.3	94.1	93.8	94.1

Source: Sosulski and Sarwar (1973).
[1] Rosen (1958).
[2] Martinez *et al.* (1970A).
[3] Values in parentheses were estimated from other literature sources.

Unlike those of soybean and peanut, the protein bodies of cottonseed do not rupture on suspension in water (Martinez *et al.* 1970A, 1970B). Alkali or salt is needed to rupture the protein-body membrane and to solubilize the storage proteins of the cottonseed. For this reason dilute sodium hydroxide is used to determine the nitrogen solubility of cottonseed protein products. Based on these extraction characteristics, the proteins of cottonseed can be fractionated into water dispersible and water nondispersible (Smith 1971). The water dispersible proteins are predominantly the metabolic proteins of the cytoplasm and the nondispersible proteins are essentially the storage proteins in the protein bodies. The metabolic proteins are low in molecular weight with a sedimentation coefficient of 2 and a minimum solubility at pH 4. The storage proteins are high in molecular weight with sedimentation coefficients of 7 and 12 and show least solubility at pH 7. Therefore, the protein extractability curves for cottonseed meal show a broad range of insolubility between pH 2 and 7 but over 80% is extractable at pH 8 (Mattil 1971).

A selective two-step extraction procedure has been used to separate and isolate the two classes of cottonseed protein (Berardi *et al.* 1969). The metabolic protein isolate represented 16% of the flour nitrogen and the storage protein 53% of the nitrogen. The two isolates differed from each other in average molecular weight, and in protein, amino acid and nonprotein nitrogen composition. Each protein isolate was highly soluble except at the isoelectric points of pH 4 and 7, respectively (Martinez *et al.* 1970A). The metabolic protein fraction contained a number of components but the major storage protein globulin had a sedimentation coefficient of 9.2 and molecular weight of 180,000 (Rossi-Fanelli *et al.* 1964).

Consistent with the high solubility of the protein fractions, the acidic and basic amino acid contents of the proteins were high, especially in arginine (Table 3.9). The metabolic protein contained about twice the level of lysine but much less phenylalanine than the storage protein but the concentration of other amino acids were similar in the two fractions.

Sunflower

Sunflower ranks as third in importance among vegetable oil crops and the meal has great potential as a protein supplement in foods. Although sunflower has a thick hull (22–43% of seed weight) the defatted flours from dehulled kernels contain 55–65% protein (Sosulski and Bakal 1969). Although deficient in lysine, the protein is rich in other essential amino acids and the protein is highly digestible (Sosulski and Sarwar 1973; Sarwar *et al.* 1973). Sunflower contains no toxic or antinutritive factors and the flours are white in color with no adverse flavors. Sunflower flours contain 3-4% chlorogenic acid which, under alkaline conditions, oxidizes and complexes with soluble proteins to give green and brown colors to the product (Cater *et al.* 1972; Sabir *et al.* 1974A)

Gheyasuddin *et al.* (1970A) prepared a colorless isolate by treating the soluble protein with sodium sulfite and washing the precipitated protein with 50% isopropanol. Recently a process for the aqueous diffusion of chlorogenic, caffeic and quinic acids, and simple sugars, from whole kernels of dehulled sunflower has been described (Sosulski *et al.* 1973). Plant breeding to develop low chlorogenic acid strains is also underway.

Sosulski and Bakal (1969) reported that the solubilities of sunflower proteins in the Osborne series of solvents were about 20, 50-60 and 3% in water, salt and ethanol solutions, respectively (Table 3.10). Globulins constituted 70-79% of the sunflower proteins of which 50% had a molecular weight of about 330,000 and 25% were only 20,000 in molecular weight (Sabir *et al.* 1973). The sedimentation coefficient for the major globulin was 12.1 S.

The poor solubility of sunflower proteins in water is confirmed in

TABLE 3.10

PEPTIZATION OF PROTEINS IN SOYBEAN, RAPESEED, FLAX AND SUNFLOWER MEALS BY THE OSBORNE SERIES OF FOUR SOLVENTS

Crop and Variety	% of total meal N soluble in H_2O	5% NaCl	70% EtOH	0.2% NaOH	N in Residue (%)
Soybean, dehulled					
Portage	69.0	8.2	4.5	5.4	12.9
Altona	75.7	6.0	3.9	4.1	10.3
Rapeseed					
Argentine	51.3	20.5	3.9	8.1	16.2
Target	50.6	20.5	4.0	9.1	15.8
Oro	48.4	22.4	3.3	8.5	17.4
Turnip rape					
Polish	44.6	24.0	4.1	6.3	21.0
Echo	44.5	25.0	4.4	6.6	19.5
Zero erucic	44.7	24.7	4.3	5.5	20.8
Flax					
Redwing	41.6	46.5	1.2	3.2	7.5
Redwood	52.1	34.4	2.0	3.5	8.0
Noralta	43.6	43.6	1.2	2.9	8.7
Sunflower, dehulled					
Commander	19.2	59.8	3.1	11.5	6.4
Advent	16.9	60.2	3.5	11.6	7.8
Peredovik	22.9	50.9	4.1	11.9	10.2

Source: Sosulski and Bakal (1969).

the extractability curve (Fig. 3.5) which shows sunflower as being only 20–30% soluble below pH 8 (Gheyasuddin *et al.* 1970B). The unique feature about sunflower protein is its high solubility at low or high concentration of sodium and calcium chloride which is not true for soybean and peanut protein (Mattil 1971). These salts are common ingredients in many foods and in imitation milk.

Although dark in color, exceptionally high yields of protein isolate have been obtained (Table 3.11). Whey losses of about 20% could be expected from the extractability curve (Fig. 3.5) and the high proportion of low molecular weight protein in sunflower (Sabir *et al.* 1973).

Sunflower proteins are low in acidic and basic amino acids (Table 3.9) and the glutamic acid level in the low molecular weight globulin

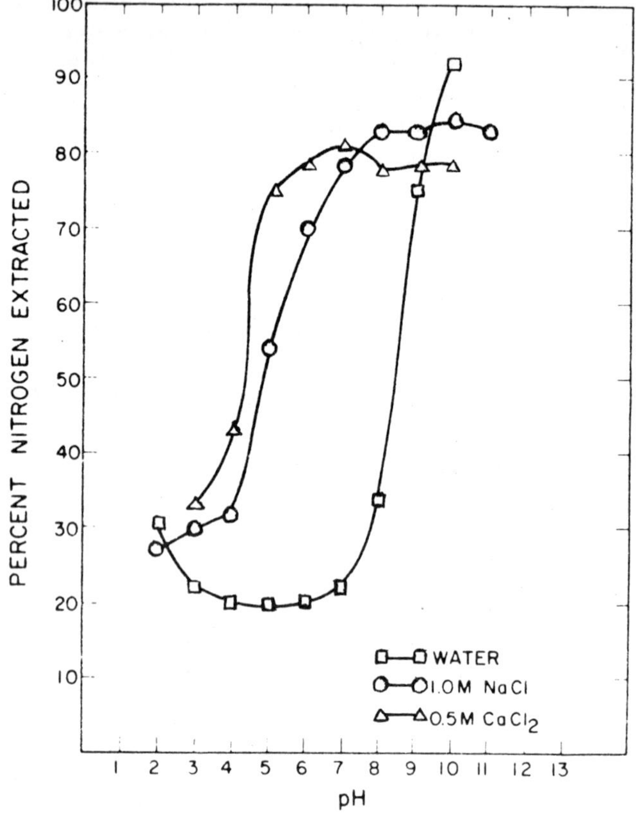

From Mattil (1971)

FIG. 3.5. EFFECT OF 1.0 *M* NaCl AND 0.5 *M* $CaCl_2$ ON THE PROTEIN SOLUBILITY PROFILE OF SUNFLOWER SEED MEAL

was 50% higher than in the 12.1 S protein (Sabir *et al.* 1973). In addition, the proportion of amino acids with hydrophobic sidechains was comparatively high, which may account, in part, for the poor water solubility of this protein.

Rapeseed

This crop ranks fifth in total world production among oilseeds. Rapeseed meal, which contains 40-45% protein (Table 3.11), has good potential as a food supplement because of its nutritionally balanced amino acids. Food uses have not been developed because of the high fiber levels in the meal and the lack of practical methods for dehulling the seed. Glucosinolates are also a nutritional problem, having antithyroid activity and adverse effects on feed consumption and efficiency. Aqueous extraction of the glucosinolates from the ground seed (Tape *et al.* 1970) and whole seed (Sosulski *et al.* 1972) have been described. Low glucosinolate varieties are now being grown in

TABLE 3.11

SPECIES AND VARIETY DIFFERENCES IN NITROGEN SOLUBILITY OF THE MEAL, YIELD AND PROPERTIES OF ISOLATED PROTEIN, AND WHEY NITROGEN CONTENT

			Isolated Protein from Meal			
Species and Variety	N Content of Meal[1] (%)	N Solubility in Alkali (%)	Yield of Isolate (gm/100 gm Meal)	N Content (%)	Color of Product	N Content of Whey (% of Total)
Soybean, dehulled						
Portage	7.7	81.2	36.4	14.6	cream	10.3
Altona	8.2	86.2	40.2	15.0	cream	12.3
Rape						
Argentine	7.1	82.3	25.2	13.7	brown	32.3
Target	6.7	83.4	25.3	13.7	brown	30.3
Oro	6.4	85.7	26.6	12.8	brown	31.3
Turnip rape						
Polish	6.2	81.3	24.5	13.4	brown	27.4
Echo	6.1	83.2	26.3	12.9	brown	25.8
Zero erucic	6.9	79.5	24.5	13.6	brown	29.2
Flax						
Redwing	7.1	94.8	40.0	12.6	tan	23.2
Redwood	3.8	96.2	36.3	12.7	tan	27.2
Noralta	6.7	91.1	36.8	11.2	tan	29.4
Sunflower, dehulled						
Commander	10.3	92.4	52.6	14.5	green	18.7
Advent	10.0	95.9	54.0	14.1	green	19.4
Peredovik	9.3	95.3	52.3	13.4	green	19.8

Source: Sosulski and Bakal (1969).
[1] Dry basis.

Canada but the residual glucosinolate level in flours is probably too high for safe consumption by humans.

Up to 50% of the proteins from several species and varieties of rapeseed are water soluble and 20–25% of the remainder was extracted with 5% sodium chloride (Table 3.10). A large portion (15–20%) was insoluble in dilute alkali. The isolation procedure indicated that over 80% of the nitrogen was alkali-soluble but low yields of isolate (50% of meal nitrogen) were obtained because of the high proportion of whey nitrogen (Table 3.11). Dark colors in the rapeseed protein isolates suggested the presence of phenolic acids and other pigment-forming compounds.

The total reserve proteins in rapeseed constitute about 70% of the total nitrogen and 20% of the remainder were low molecular weight polypeptides (Finlayson *et al.* 1969). About 35% of the seed nitrogen was a high molecular weight globulin, 12 S, which irreversibly dissociates to 7 S and 3 S in acid media. Another 7–10% of the seed nitrogen was a strongly basic globulin, 1.7 S, with a molecular weight of 13,500.

Rapeseed proteins are relatively low in acidic and basic amino acids while amino acids with hydrophobic groups tended to predominate (Table 3.9). Proline and cystine levels were particularly high.

FUNCTIONALITY

Concentrated seed proteins have essentially no gastronomic appeal or palatability when in the form of dry flours or grits. The transformation of these dry powders into attractive foods requires some type of aqueous processing in which the protein is brought into colloidal dispersion. Functionality can be defined as "the physical, chemical and organoleptic properties of the colloidally suspended protein which affect the structure, texture, flavor and color of the formulated food product."

The number of desirable functional properties associated with the behavior of proteins in aqueous colloidal systems, in the presence of carbohydrates, fats, minerals and other food ingredients, can be extremely large. However, many of these properties are difficult to measure or have only limited significance in the food system. Johnson (1970) has summarized the characteristics desired in a functional protein ingredient as: easy wettability, good water dispersibility with no settling, clear dispersions over a wide pH range, desirable viscosity characteristics, possible reactions or synergistic effects with other colloids to obtain desired viscosity, gel formation, water holding, controlled water absorption rates, emulsifying, stabilizing, thickening, dough forming, elasticity, film forming, adhesion, cohesion, aeration properties and low or compatible flavor. Other useful characteristics include fat binding, texturizing, fiber formation, pigmentation and color control (Wolf and Cowan 1971).

Specific functional and nutritional characteristics are required for each food system (Horan 1974A). In bakery products, water absorption is an important functional property. For meat extenders, the binding of fat and water through efficient emulsification is essential. Textural characteristics are highly important in meat analogs. Color, flavor, mouth-feel and dispersibility must be considered in milk replacers. For an egg white replacer, foaming ability and heat gelation are critical characteristics.

In the following sections, these properties will be discussed under the headings: flavor, color, water binding, water absorption and retention, swelling and viscosity, gelation, dough formation, emulsification, fat absorption, foaming, protein concentration and denaturation.

Flavor

For broad applications of enriched protein foods, the protein ingredients should be neutral in taste and odor, and should not develop unusual flavors when processed with other food components. Most raw proteinaceous materials from seeds have peculiar flavors which must be masked or extracted for many food applications. The beany flavor and astringency of soybean products and the characteristic taste of cereals have varying degrees of importance, depending upon the food system in which the products are to be used.

Generally, the flavors and odors arise from low molecular weight components which are usually present in low concentration. While most are natural compounds in the seed, some are generated during aqueous processing by the action of natural enzymes such as lipoxidases and lipases. Rapid drying temperatures will entrap the natural and added flavor compounds while slow drying caused a certain percentage of reversible and irreversible bonding between these volatile constitutents and proteins (Gremli 1974). When used in foods that require mastication, e.g. meat substitutes, the bound flavors may be released gradually during consumption.

In soybean, the characteristic beany and bitter flavors of the raw flours and grits can be partially removed by the application of moist heat. However, toasting also causes an undesirable flavor profile and protein dispersibility will be reduced. Soybean concentrates and isolates were developed to reduce the flavor levels but a recent study of commercial concentrates and isolates has shown the presence of beany and bitter flavor in all products (Kalbrener *et al.* 1971). This occurs despite the vigorous competition among manufacturers to produce the most flavorless product in each class of soybean product (Mattil 1974).

About 35 flavor components have been identified in soybean products and the beany flavors are attributed to 3-methyl-1-butanal, *n*-hexanol, *n*-heptanol and *n*-hexanal which have a low flavor threshold (Arai *et al.* 1967). Of nine phenolic acids found in soybean flour, syringic acid was the major compound responsible for bitterness and astringency (Arai *et al.* 1966).

The acceptability of soy milk outside Asian countries is seriously limited by another type of beany bitter flavor which develops during milk preparation. Apparently the action of lipoxygenase on lipids during soaking and wet milling is largely responsible for these undesirable flavor compounds. Flavor improvement can be achieved by enzyme inactivation during seed grinding or steam distillation of the volatile constituents from the soy milk. Ethyl vinyl ketone and 1-octen-3-ol are major flavor constituents in soy milk (Badenhop and Wilkins 1969; Mattick and Hand 1969).

Other legumes also display varying degrees of beany flavors. The grassy taste of field peas appears to concentrate with the protein in air classified concentrates. These flavor components are readily volatilized by moist heat as during baking. Fababean has a less intense flavor than soybean or field peas.

Sunflower flours are essentially bland in flavor and, on toasting, develop a pleasant nutty taste. Of the ten phenolic compounds isolated from the flour, only chlorogenic acid (2.2% of flour) and caffeic acid (0.2%) were unbound (Sabir *et al.* 1974B). Chlorogenic acid has not presented a flavor problem in sunflower products, due probably to a high flavor threshold.

The characteristic mustard flavor is present in rapeseed and most volatile constituents are removed with the glucosinolates during aqueous processing. Flours from the new low-glucosinolate varieties retain some undesirable flavor tones which are still present in food products after cooking.

Color

The colors of the flour from most dehulled seeds range from white to cream or grey and are frequently assumed to have widespread utility in processed foods. However, the presence of dark pigments may not be detected until the product is suspended in an aqueous medium for 1 or 2 hr. Phenolic compounds develop dark colors under alkaline pH conditions and are usually detected during the preparation of protein isolates when extractions are done at pH

8-11. Pigmentation is only partly reversed when the pH is adjusted to the isoelectric point for protein precipitation. For example, Sosulski and Bakal (1969) reported that rapeseed and sunflower isolates developed dark brown and green colors while flax protein isolate was tan-colored and soybean isolate remained a light cream. The color constituents in rapeseed are relatively complex but chlorogenic acid was the principal color-forming compound in sunflower flour (Sabir *et al.* 1974A). About 30% of the chlorogenic acid was covalently bonded to flour constituents and the soluble fraction was closely associated with low molecular weight polypeptides during Sephadex column chromatography. Removal of the chlorogenic acid resulted in a sunflower protein concentrate which was near-white in appearance at pH 10 (Sosulski *et al.* 1973).

Raw soybean flour with water dispersible protein of over 90% has high enzyme activity and is used as a bleaching agent or to promote color formation in baked goods (Johnson 1970). The enzyme-active soy flour contains the enzyme lipoxidase which oxidizes polyunsaturated fats, and the oxidized fats, probably hydroperoxides, bleach the yellow carotenoids in the wheat flour dough into a colorless form (Wolf and Cowan 1971). During baking the enzyme activity is destroyed and undesirable flavor components of the soy flour are lost so that the finished baked products have an acceptable flavor as well as a white crumb.

The browning of bread crusts and other baked goods or pancakes during frying is attributed to reactions between the proteins, sugars and oligosaccharides in the wheat flour and dough ingredients. The addition of about 0.5% soy flour to the dough or batter will also enhance the brown color of bread crusts, breading mixes, pancakes and waffles, and this effect is due to complexes formed between the soy protein and wheat flour carbohydrates (Wolf and Cowan 1971).

Water Binding

Water binding is defined as the water vapor absorbed by a dried protein flour after equilibration against water vapor of known relative humidity. Mellon *et al.* (1947) investigated the significance of the various sections of the sigmoid curve obtained when the amount of water vapor absorbed was plotted against relative humidity. The initial water uptake of 24-33% of casein weight was absorbed by amino groups and the remainder was by other polar groups in the molecule. The initial step in the binding of water by amino groups was a sharing of one water molecule between two amino groups at relative humidities below 6%. The second step was a linear increase in absorbed water with increase in relative humidity up to 60%. At relative humidities above 60–70% the amount of absorption increased more rapidly and appeared to be a condensation of water on water molecules already attached to the amino groups. The epsilon amino group in lysine accounted for over 80% of the total free amino groups in casein.

Kilara *et al.* (1972) confirmed the linear nature of the moisture absorption curve at relative humidities of 11–60% for soybean and sunflower protein products (Fig. 3.6). The moisture absorption isotherms showed sharp increases above 70% RH. At 70–95% RH, soy protein isolate was superior to sunflower isolates and concentrate, which paralleled the relative differences in lysine concentration in each product (Table 3.9). The highest water absorptions were obtained with sunflower flour, which probably reflected the influence of carbohydrates rather than protein effects.

Hagenmaier (1972) found that the pH (of an aqueous suspension of the protein) and protein solubility were not directly associated with water-binding capacity, which was determined at 84% RH. However, animal proteins were generally lower in amide nitrogen and bound more water than oilseed proteins. The total moles of hydrophilic groups, taken as the sum of hydroxyl, carboxyl and basic groups less the moles of amide groups from glutamine and asparagine, were closely associated with the water binding of animal and oilseed proteins. This relationship was first noted by Bull and Breese (1968) and was especially significant for oilseed proteins because of the significant level of amide nitrogen in the globulin proteins. About 1/2 of the glutamic and aspartic acids in globulins are in the amine form. Hagenmaier suggested that deamidation of oilseed proteins might increase their water binding potential and enhance their value, relative to animal proteins, for food applications that require a highly hygroscopic protein.

Water Absorption and Retention

The rate of hydration is an important characteristic when water is being incorporated with dry ingredients such as occurs during baking or the preparation of extended meat products. Organoleptic characteristics associated with the degrees of hydration include dryness, juiciness and mouth feel. The same functional properties of proteins which determine the total water absorption may control water retention after baking or shrinkage during cooking.

Proteins are capable of binding large amounts of water due to the hydrogen bonding of water molecules to polar groups on the protein chains. Numerous sidechain groups are involved: carboxyl, amino,

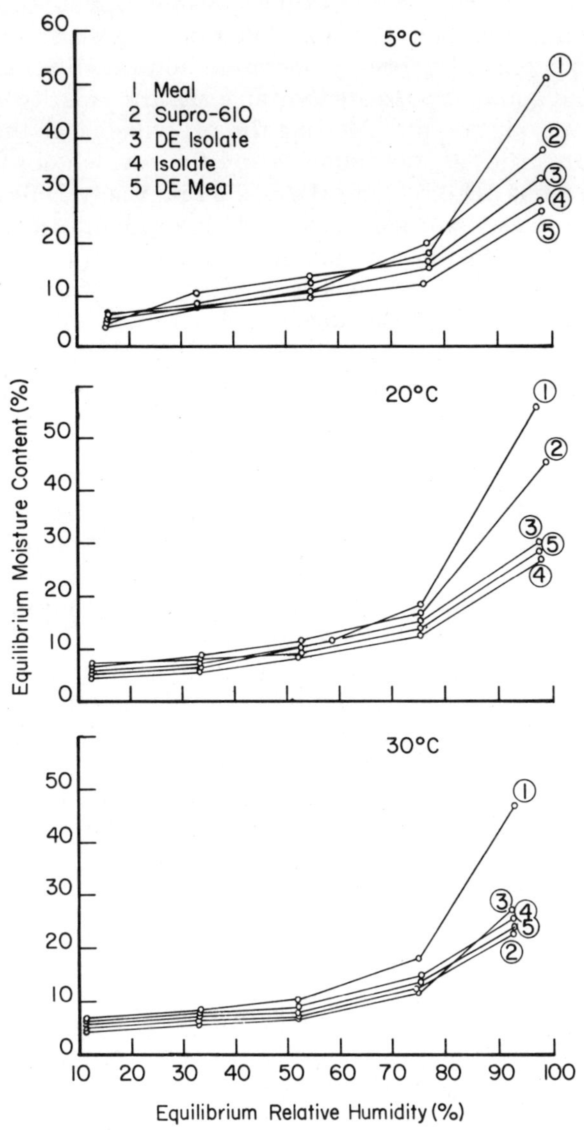

From Kilara et al. (1972)

FIG. 3.6. MOISTURE ADSORPTION ISOTHERMS FOR THE FLOURS AND ISOLATES AT VARIOUS TEMPERATURES AND RELATIVE HUMIDITIES

imidazol, carbonyl, sulfhydryl, hydroxyl and guanidino (Briskey 1970). Water binding is diminished as polar groups are blocked as by amidation of the carboxyl groups. Water binding is also greatly influenced by protein solubility, temperature and pH. For example, the carboxyl and amino groups are ionizable and the polarity can be controlled by varying the pH. Altering the pH will change the water absorption properties as was found in soy flour which absorbed twice as much water at pH 8.5 as at pH 4.5-6.3 (McAnelly 1964).

Globulin proteins are generally more hydrophilic than prolamine and glutelin proteins such as gluten because they contain more polar side chains. Therefore proteins such as soybean which contain 70–85% globulins will absorb relatively high levels of water and retain it in the finished product (Wolf and Cowan 1971). If soy flours are substituted for nonfat dry milk solids, more water must be added to the dough. Dough yield and handling characteristics are also improved and the shelf-life of bread and cakes is prolonged.

The usual hydration levels are 2:1 for soybean flours, 2.5:1 for protein concentrates and 3-4:1 for isolates (Rakosky 1974). Soybean protein concentrates prepared by alcohol leaching of defatted flakes have water absorptions of 3.4-3.8 times their dry weights (Mustakas *et al.* 1962). Particle size has a marked influence on the rate of water absorption by these products.

Water absorption is carefully controlled when incorporating the dry texturized proteins into ground or simulated meat products. Depending on the raw material and particle size the water absorption of extruded flour products vary from 2.4-3.4 times their dry weight (Wolf and Cowan 1971). Spun fibers absorb 1.4 times their weight of water.

Swelling and Viscosity

One of the most useful properties of hydrated and colloidally suspended seed proteins is their ability to provide texture and structure to viscous, gelled or doughy foods. The swelling properties of ingredients in the aqueous phase of food systems involves interactions among water, proteins, carbohydrates and lipids. The role of proteins in the swelling and viscosity characteristics of foods is less well known than those of starch.

Swelling can be defined as the volume expansion of a macromolecule in response to water uptake which progresses until a colloidal suspension is achieved or until further expansion and uptake is prevented by intermolecular forces in the swollen particle. Many foods products have insufficient moisture to achieve full protein solvency and most concentrated seed proteins are partially but not completely, soluble at the pH level of any specific food system. As a result of limited moisture availability or protein solubility, the water uptake and swelling of a protein would normally increase to an equilibrium level at which moisture content, molecular volume and viscosity would remain relatively constant.

Hermansson (1972) evaluated this hypothesis by observing the swelling of soybean protein isolate, sodium caseinate and whey protein particles during hydration. Soybean isolate reached maximum swelling and an equilibrium after 3 hr while the swelling of caseinate reached a maximum after 1 hr and then rapidly lost its ability to hold water. Whey protein concentrate showed very poor swelling ability. The particles of soybean isolate appeared to resist swelling and maintained their structure for several hours, but the structure of the milk proteins was quickly destroyed by water uptake. Whey proteins were rapidly solubilized in the experiment while caseinate showed a rapid increase in soluble protein after the particles lost the ability to hold water. Heating caused an increase in swelling, followed by a decrease. Swelling increased with pH and ionic strength. Viscosity measurements were higher for soybean isolate than the milk proteins and the viscosity increased with concentration of protein in the slurry.

The relationship between the degree of protein hydration and viscosity of soybean and sunflower slurries was investigated by Fleming *et al.* (1974B). Water absorptions of flours, concentrates and isolates were generally proportional to their protein contents, but the water uptakes of sunflower proteins were much lower than those of soybean proteins (Table 3.12). Presumably intermolecular forces due to disulfide linkages in sunflower (Table 3.9) exceeded those in soybean particles to restrict water absorption. Soybean proteins contain more hydrophilic groups than sunflower protein and this may be another influential factor.

For most products, the viscosities of the slurries increased exponentially with slurry concentration (Table 3.12). In soybean, the differences in viscosity profile were not proportional to protein content, but were predicted by water absorption.

In sunflower, the partially denatured protein concentrates were much higher in viscosity and the isolate was much lower in viscosity than would be predicted from protein content or water absorption. There were obviously other major factors controlling the water uptake, swelling and viscosity.

Gelation

Circle *et al.* (1964) found that gels were formed by heating 8% or higher dispersions of sodium soy proteinate to temperatures of 80–125°C in sealed containers for 30 min. According to Catsimpoolas

TABLE 3.12

WATER ABSORPTION AND APPARENT VISCOSITY OF SLURRIES OF SOYBEAN AND SUNFLOWER FLOURS, CONCENTRATES AND ISOLATES AT ROOM TEMPERATURE

	Protein Content (% N × 5.7)	Centrifuge Water Absorption (gm water/ gm product)	Brookfield Apparent Viscosity (Cps) Concentration of Product (W/W)			
			5	10	15	20
Soybean						
Flour	50.2	2.60	—	25	230	2000
Isopro	63.7	2.75	10	200	330	28,320
Supro 610	85.8	6.25	160	10,500	83,300+	83,300+
Promide D	86.1	4.15	1300	3200	7500	25,200
Sunflower						
Flour	53.0	1.55	90	135	200	300
Concentrate-60[1]	69.9	2.92	300	2200	13,000	83,300+
Concentrate-80	68.6	2.65	50	400	10,800	83,300+
Isolate-60	87.7	3.05	10	80	500	3780

Source: Fleming *et al.* (1974B).
[1] Temperature treatment during processing (°C).

and Meyer (1970), the slurry of soybean globulins (the sol) was heated to a progel state and, when cooled, formed a reversible gel. Excess heat produced a metasol which did not gel on cooling. By this procedure, Hermansson (1972) produced gels with 10% protein dispersions of Promine D and whey protein concentrate, but not with sodium caseinate. Recently, Fleming *et al.* (1974A) found that fababean and field pea protein concentrates developed gels with medium viscosity characteristics which were somewhat less elastic than that produced by soybean isolate. While the albumin fraction from these legumes showed no thickening ability, the globulins from soybean, fababean and field pea produced gels of variable consistency to support the theory that gelation occurred in the globulin fraction.

According to Briskey (1970), the ability of gel structures to act as a matrix to hold moisture, lipids, polysaccharides and other ingredients is of significance in many food applications. In particular, sausage and weiner emulsions acquire gel-like properties when heated. Soybean protein gels contribute texture, moisture and fat-holding capacity, and chewiness to frankfurters and luncheon meats (Wolf and Cowan 1971). In simulated sausage-type products, the gel structure is supplied entirely by the soybean protein.

Temperature, time of treatment, pH and ionic strength were factors which determined the strand thickness and pore size of the gels (Circle *et al.* 1964). In addition to the effects of electrostatic bonds, disulfide bonds are considered to be important in stabilizing the gel network (Wolf and Cowan 1971). Heating the gels to 125°C will disrupt the quaternary structure of the gels. The effect has been attributed to the dissociation of globulins into subunits, especially the 11S glycinins (Catsimpoolas and Meyer 1970).

Dough Formation

It is well known that gluten proteins have the unique characteristic of retaining gases during dough expansion and maintaining loaf volume during baking. Intermolecular covalent linkages, such as the disulfide bond of the cystine residue are primarily responsible for the viscoelastic properties of gluten. Sulfhydryl-disulfide interchange reactions take place during dough mixing, gas expansion and dough manipulation stages to permit the redevelopment of gluten structures after each step in the bread-making process. The glutenin fraction of gluten is mainly responsible for the mixing behavior of the dough but the gliadin fraction has a controlling influence on the final loaf volume.

Bread, durum and soft wheats have strong gluten-forming proteins while those in triticale and rye proteins are only intermediate to weak in these properties. None of the other cereal, legume and oilseed species contain seed proteins which exhibit the true characteristics of gluten. Most proteins, when mixed with limited amounts of water, will form dough-like masses but these lack the elastic and cohesive properties of gluten.

The increasing demands for breads throughout the world has led to the development of blended or nonwheat composition flours with much improved nutritional quality. These nonbread flours usually require more water to develop a pliable dough and modified dough handling procedures to produce a palatable loaf. In blends with wheat flour, proteins which have been heat treated to decrease protein solubility have less adverse effects on loaf volume and bread texture. Also, the level of oxidizing agents such as bromate can be increased to overcome the reducing action of the foreign protein on the sulfhydryl-disulfide ratios in the gluten.

It is of interest to note that the thermoplastic extrusion of soybean flours involves the development of a dough-like mass by the action of moist heat and pressure on the hydrated flour. The functional properties of improved adhesion, cohesion and elasticity are also characteristic of soybean-meat blends in ground meat products.

Emulsification

A wide range of ground seed, flour and concentrated seed proteins are used as emulsifiers in ground meat products, baked goods, soups, etc. In meat emulsions such as frankfurters and sausage, the dispersed or discontinuous phase is fat, the continuous phase is water and the emulsifying agent is the soluble protein from meat and seed sources. When the two immiscible phases are agitated, cohesive forces result in droplet formation by both phases. The soluble proteins from the meat and seed sources are surface active and migrate to the oil-water interfaces to lower surface tension and enhance droplet formation. Since the concentration of water exceeds that of the oil in ground meat products (Table 3.13), the oil-in-water emulsion is more stable and, with further agitation, gradually replaces the water-in-oil emulsion. The development of the oil-in-water emulsion is enhanced by the presence of an emulsifying agent such as protein which is soluble in the aqueous phase. The proportion of soluble protein in typical ground meat products (Table 3.13) may be marginal as collogen may constitute a significant proportion of the constituents containing nitrogen. Therefore, cereal and legume flours, concentrates and isolates are frequently required to aid in the formation of the emulsion.

In addition to the formation of emulsions, soluble proteins function to maintain and stabilize the emulsified fat droplets. The soluble

TABLE 3.13

PROXIMATE COMPOSITION (%) OF GROUND MEAT PRODUCTS FROM SASKATOON SUPERMARKETS

	Ground Beef		Wieners		Sausage	
	Mean	S.D.	Mean	S.D.	Mean	S.D.
Water	61.4	5.4	52.0	2.2	51.4	3.8
Fat	19.4	7.3	27.2	2.3	33.0	3.5
Protein	17.1	1.4	12.0	0.3	9.4	1.0
Ash	0.7	0.2	3.4	0.6	2.0	0.5
No. of Samples	10		6		8	

Source: Holt and Sosulski (1974A).

protein molecules collect on the surface of the droplet with the apolar hydrocarbon chains oriented towards the center of the oil droplet and the polar ionized groups positioned in the direction of the water (Aurand and Woods 1973). As a result of the carboxyl group orientation, the droplet takes on a negative surface charge which is balanced by sodium ions in the surrounding water. Each droplet possesses the same negative charge and repulsion forces serve to stabilize the emulsion and prevent the coalescence of oil droplets. De-emulsification can be achieved by neutralization of the electric charge carried by the dispersed phase.

The emulsifying capacity of soluble proteins is based on the hydrophilic-lipophilic balance in the molecule which determines the affinities for oil and water. The amino acid composition and protein configuration in solution plus the modifying effects of pH and ionic strength of the aqueous phase will influence the emulsifying characteristics of the protein. These properties are difficult to measure in a model system and the interactions which occur in a complex meat emulsion have not been defined.

Lin *et al.* (1974) demonstrated that wheat, soybean and sunflower flours had relatively good oil emulsification properties when compared to concentrates and isolates. The oil emulsification capacities were unrelated to water or fat absorption characteristics but high protein solubility index was associated with the percentages of oil emulsified in a model system.

Fat Absorption

Fat absorption has been equated with fat emulsification properties but there is no supporting evidence to confirm that these characteristics are related. Lin *et al.* (1974) found that soybean and sunflower products with good oil emulsifying properties tended to be low in fat absorption. They concluded that soybean proteins, which had oil absorption values of 84.4–154.5%, were less lipophilic than sunflower products which absorbed 207.8–256.7% oil. Although the mechanism of fat absorption was unclear, the authors concluded that sunflower proteins contained more nonpolar sidechains than soybean proteins which retained oil by associative bonding.

Because of this characteristic, soybean flours are added to donut mixes to control fat absorption during frying (Johnson 1970). A wheat flour mix absorbed 27.6 gm of fat per 100 gm of dry mix and the substitution of 4% soybean flour (NSI = 80) reduced the fat absorption to 11.1 gm of fat per 100 gm of mix. Soybean flours are added to breading mixes, pancakes and waffles in order to retard fat absorption during frying (Wolf and Cowan 1971).

Foaming Ability

A foam may be defined as a colloidal system consisting of gases suspended in a very viscous liquid (Aurand and Woods 1973). The foaming agent serves to lower surface tension at the interface between the gas and the liquid, and colloidal substances such as proteins may also give good stability to the foam by preventing the coalescence of gas bubbles dispersed in the liquid. Foam expansion, foam stability and foam stiffness are the important functional characteristics. When used as egg-white substitutes in bakery and confectionery goods, the protein foams must be firm enough to support the flour and/or sugar. The protein foaming agent must also have heat coagulation properties to maintain the aerated matrix during cooking.

In general, the whipping properties of oilseed proteins are related to their solubility and are influenced by pH, ionic strength and protein denaturation. Lin *et al.* (1974) demonstrated that initial foaming volumes are not related to foam stability. The latter property was closely associated with protein solubility and oil emulsification characteristics of soybean and sunflower products. The final foam volumes were very low for denatured or insoluble proteins such as the sunflower concentrates, and certain commercial soybean concentrates and isolates. Soybean proteins are used in whipped toppings and frozen desserts (Wolf and Cowan 1971). Pepsin hydrolyzates of soybean proteins serve as whipping agents in confections, chiffon mixes and angel food cake mixes. These hydrolyzates are soluble in the isoelectric range for the protein (pH 4–5). Alcohol extracted soybean isolates have very stable foam characteristics but lack the heat coagulation properties of egg albumin (Eldridge *et al.* 1963).

Protein Concentration and Particle Size

The final protein composition of seed products is determined by the type and extent of processing (Smith 1971). If the seed is dehulled to remove a portion of the crude fiber and ground, a flour or, in the case of oilseeds, a full-fat flour is produced with an increase in protein concentration of 5–10%. By defatting an oilseed, the protein content is increased from 25–30% to 40–50% in the defatted flour. Further upgrading to protein concentrates of 65–70% is done by two methods. Alcohol or water leaching of low-fat flour or flakes removes the soluble and low molecular weight constituents from the insolubilized protein. This type of wet-milling process has the disadvantage of costly effluent disposal, and crude fiber concentrates with the protein. Fine grinding and separation of protein-rich constituents from the fiber-rich or starchy fraction by liquid cyclone or air classification may be useful alternatives to the wet-milling processes. Protein isolates are essentially pure protein extracted and precipitated by a more complex procedure. The protein is solubilized in an aqueous medium and is separated from the fibrous seed residue by filtering and centrifugation. The pH of the clarified extract is then adjusted with food-grade acid to precipitate the major proteins.

Particle size and particle size distribution have an important influence on the rate of water absorption, total absorption, the rate of development of viscosity and final viscosity, fat absorption and emulsification characteristics and mouth feel of the final product (Johnson 1970). In some processes, it may be desirable to have a final product of high viscosity, but a low viscosity would be useful during processing to make the product easier to handle. Soybean grits may be useful for these applications because of their slow water absorption characteristics and low initial viscosity. During heating and processing, the slurry of soybean grits will develop higher final viscosities with firmer texture characteristics and less pastiness than the very fine material. Soybean grits have limited application in the baking field but are used extensively in pet foods and in ground meat products. Usually flours are ground to pass through a U.S. 100 mesh sieve although most commercial soybean flour products are finer than U.S. 200 mesh. Soybean grits are separated into a number of particle size ranges by a screening operation. For commercial purposes these are designated as, for example, a 10/40 grit in which all of the product will pass a U.S. 10 mesh sieve and most of the material is held on a U.S. 40 mesh sieve.

Protein concentrates are also available in the granular and fine forms while isolated soybean protein is commonly utilized as a flour (Rakosky 1974).

Denaturation

Denaturation is defined as the modifications in the secondary, tertiary and quaternary structure of the protein molecule which are usually associated with changes in the physical-chemical and functional properties. The common environmental factors which cause denaturation of proteins during processing include heat, pH, salt concentration and organic solvents.

Because of variations in manufacturing conditions, soybean flours are available with water-dispersible protein from less than 10% to over 90%, soybean protein concentrates have NSI values ranging from 0-70% (Johnson 1970) and isolates from 0-90%. Dippold (1961) demonstrated that water absorption of soybean flours increased from 290-385% as NSI's increased from 15-70% but at 80% NSI the water absorption decreased to 270%. Emulsifying capacity (Johnson 1970) and whipping properties (Yasumatsu *et al.* 1972) of soybean protein concentrates improve with increases in water-dispersible protein. Soybean protein isolates with water-dispersible protein levels of 50-60% have better gelling properties than those with solubilities of 80% or more.

Denaturation of wheat gluten is measured by the loss of solubility in dilute acetic acid as a function of time. Little decrease of solubility was obtained at 70°C for 60 min but soluble nitrogen decreased rapidly at higher temperatures, and at 90°C most of the gluten became insoluble in 30 min (Pence *et al.* 1953). The denaturation of the wet gluten in these experiments adversely affected loaf volume of reconstituted doughs in direct proportions to the heat treatment within the 70-90°C range of the experiment.

Fukushima (1969) found the denaturing ability of organic solvents depended on their hydrophobicities and their degree of dilution by water. Highly hydrophobic solvents such as hexane, which are immiscible with water, possessed slight abilities to denature soybean proteins, even at high temperatures. The denaturing power of organic solvents increased with addition of water and that of water increased with addition of organic solvents, especially with the lower alcohols such as ethanol. The denaturing ability of water and aqueous-organic solvents occurred rapidly in a few minutes and the degree of denaturation was very temperature dependent.

Fukushima (1969) reported that the major soybean protein molecules are compactly folded around the inner hydrophobic region. The major structures of the polypeptide chains are beta or disorganized forms, and only a small amount of alpha-helix is present. The chains are only slightly susceptible to proteinase hydrolysis unless there is disruption of the internal structures.

Most of the hydrophobic side-chain residues are located towards the center of the molecule and the hydrophilic residues are positioned towards the surface of the molecule. Therefore, water-immiscible hydrophobic solvents will not penetrate the hydrated outer shell but water, owing to its strong hydrogen bond-forming ability, destroyed the hydrophilic shells easily (Fig. 3.7). While water destroys only the hydrophilic shell, organic solvents which possess both hydrophobic and hydrophilic radicals will penetrate into the hydrophobic region as shown in Fig. 3.7.

SEED PROTEINS AS FOOD INGREDIENTS

Protein-enriched Blends

For a majority of the world's population, starchy cereals, tubers and fruits furnish 2/3 or more of their total protein intake. When the staple is corn, cassava or fibrous cereals, malnutrition is common among adults, and young children suffer severely. Even cereals of

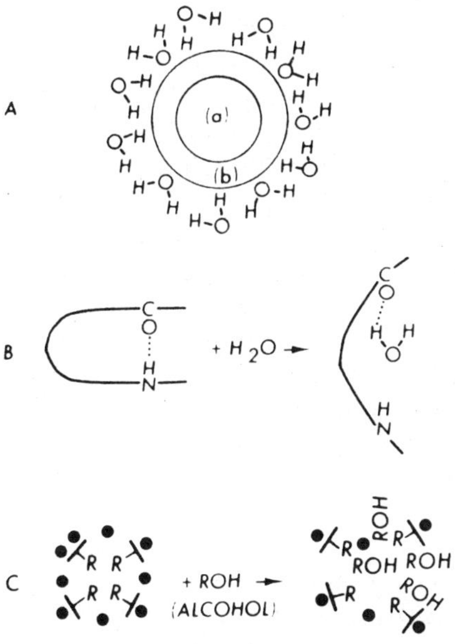

From Fukushima (1969)

FIG. 3.7. SCHEMATIC REPRESENTATION OF THE MECHANISM OF PROTEIN DENATURATION WITH WATER AND ALCOHOLS

A—One globular protein molecule surrounded by water (HOH) in which a hydrophobic region (a) and a hydrophilic region (b) are present. B and C—Disruption of hydrogen bond by water and of the hydrophobic bond by alcohol. Thick lines in B and C are polypeptide backbone of proteins; in C there is one molecule of water, and —R represents amino acid side chains.

better protein content (wheat) or protein quality (rice) do not provide adequate levels of essential amino acids for weaning diets. Most children go directly from breast feeding to starchy gruels in which the deficiency of protein relative to calories is so great that serious physical and mental retardation is a common occurrence.

About 25 yr ago the FAO/WHO Expert Committee on Nutrition recommended that protein-calorie malnutrition be investigated in all areas where it occurred and began to develop principles for protein-rich food programs to alleviate the problem. Subsequently, a Protein Advisory Group, with leading nutritionists and pediatricians as members, determined that the protein-deficient diets were low in lysine, threonine, sulfur amino acids and tryptophan. Legume proteins were generally good sources of lysine but were limited in sulfur amino acids. Oilseed proteins were rich in one or more of these essential amino acids. In developing protein-rich weaning foods, a combination of two or more concentrated seed proteins along with the cereal flour was necessary to provide the desired balance of essential amino acids. Frequently, the nutritive value of the blend was enhanced by adding skimmilk powder to upgrade the protein quality. The basic formula was usually 25–30% legume or oilseed flour, 5–10% dry skimmilk and 60–70% cereal flour plus minerals and vitamins. The first applications of these seed protein mixtures were for distribution to infants after weaning and to pre-school children in countries affected by protein malnutrition. The processes developed were simple to ensure that these products could be manufactured locally and prepared for consumption by adding water and cooking for a few minutes.

Numerous formulae have been developed for food distribution programs in many countries, especially throughout the tropics. While great care was taken to formulate the desired nutritive balance, less attention was often given to appearance, taste and texture of the cooked product. Lack of acceptability on the local level may explain why most of these distribution programs have not achieved great success. Exceptions would be Pronutro and Incaparina, which are commercially successful (De Muelenaere 1969), and CSM (corn-soy-milk) has been widely distributed by U.S. AID in over 90 developing countries (Senti 1969).

With present knowledge of functional properties and available technology to modify seed proteins, it should be possible to gain consumer acceptance of cereal-legume or oilseed blends. The first approach is to incorporate the new protein flours and extruded products into existing foods without altering their taste or texture. This is being achieved in protein-fortified breadfast cereals, bread, doughnut, macaroni, cake, cereal-fruit, snack and meat products. A later stage would involve more sophisticated technology to develop entirely new products such as texturized protein products or high protein beverages with superior nutrition and palatability.

Protein-enriched Bread and Pasta

Bread in its various forms is almost universally popular as a food. In the United States soybean flours are commonly incorporated into bread at a 3% level for their functional effects on loaf color, water absorption and dough-handling properties. In recent years, soybean flours have become a regular ingredient in bread formulations as a partial replacement for nonfat dry milk which has been rising in cost.

Recently there has been greater interest in the protein fortification of breads with high lysine protein sources. For an adequate nutritional balance, the protein supplementation must be at the 10–20% levels of wheat flour replacement in the bread formula. At these levels the nongluten protein will have adverse effects on dough-handling characteristics and the quality of the final product. All legume or oilseed proteins have deleterious effects because of excess water absorption, weak gluten strength, poor loaf volume and crumb texture, and unacceptable organoleptic properties.

Defatted soybean flours appear to have the least adverse effects on gluten among concentrated seed proteins and levels of 5–7% have been used with only moderate deterioration in loaf characteristics. Breads containing up to 12% soybean flour have been rendered acceptable by the incorporation of natural and synthetic emulsifiers at low concentrations. Natural products such as lecithin and plant gums may be used, but most emulsifiers are synthetic esters of polyalcohols with fatty acids or fats. Some function primarily as bread softeners and antistaling agents while others act as strong dough conditioners which permit greater tolerance in dough formulation and handling. Emulsifiers which have been effective as dough conditioners include glycerol esters of mono- and diglycerides, sucroglyceride esters, glycolipids and stearyl-2-lactylate salts (Tsen 1974). Tsen has proposed that a glutenin-surfactant-gliadin complex was formed which counteracted the adverse effects of the glutenin-soy globulin-gliadin complex.

Interest in the development of breads from nonwheat flours, usually starchy tuber flours and defatted oilseed flours, has arisen in developing countries. Jongh (1961) developed the concept of using a binding agent such as glyceryl monostearate to make the starch granules sufficiently coherent to retain the gases and maintain volume during baking. Kim and de Ruiter (1968) made a bread-type product using an 80:20 ratio of cassava flour to soybean flour with added glyceryl monostearate (2.0%). Rice, corn, sorghum and barley flours have been used successfully in these nongluten bread systems.

The fortification of macaroni products presents fewer problems than that of bread. Pasta manufacture is a simpler process mechanically than bread fermentation and concentrated seed proteins can be added at up to 20% of the formulation without altering the workability and shaping of the dough. Matsuo *et al.* (1972) concluded that the supplementation of 10% egg albumin, glutenin or gliadin improved the cooking quality of spaghetti, soybean flour had little effect, while rapeseed flour and fish protein concentrate had slight deleterious effects. Other workers have found that vegetable gums such as carrageenan help to maintain the tensile strength and monoglycerides reduce the stickiness of the dough in pasta products.

The incorporation of seed proteins in soft wheat products, such as cookies, cakes and pancake mixes, is also common. The presence of other components in the formula, i.e. sugar, shortening and flavoring, will serve to minimize any adverse physical or flavor effects of the protein.

Ground Meats

The most popular comminuted meat products are the emulsion-type frankfurters and sausages, followed by loaf-type products which are often canned and, lastly, the coarse-ground hamburgers, patties and chili-type products (Rakosky 1974). The muscle proteins in the lean meat of the formulation are added for their water-holding capacity and fat emulsification properties. Factors such as rigor mortis, freezing, ageing, cations, pH and salts will affect the ability of the microfibrillar protein structure to hold water and also the solubility of sarcoplasmic proteins which are the efficient emulsifiers. Due to the numerous factors which can cause failure in the meat emulsion, it is a common practice in the meat industry to add functional additives to control variation in meat protein properties and to maintain uniform quality.

Soybean proteins are added to the full range of ground meat products for reasons of economy, to prevent shrinkage during cooking and to retain juiciness and meat flavor through their water holding capacity. Soybean flours are used in sausage-type products but the granular grade of protein concentrate is preferred in coarse ground meat products because of its blander flavor and greater hydration capacity. Up to 20% supplementation of meat patties is possible without flavor adjustment. Protein isolates have specific application in canned meats where their heat stability appears to protect the meat structures against the effects of heat. All of the soybean products are used in sausage manufacture as emulsifying agents and for their gelling properties which act as a matrix to bind fat and water. The usage of the various soybean products in meat systems is summarized in Table 3.14.

In general, the trend in the meat industry has been to utilize the more concentrated protein products. This is conditioned in part by the government restrictions on the amount of meat extender that can be added. The purified protein sources have better flavor and functional properties and their addition enhances protein content where cereal additives would dilute the protein content. The addition of protein powders to ground meats is self-limiting from the aspect of texture, as well as flavor, and 10–20% of meat replacement is the maximum which can be tolerated by the system.

Compared to a low NSI (nitrogen solubility index) isolate, Smith *et al.* (1973) rated a high NSI soybean isolate as superior in process

TABLE 3.14

BASIC SOYBEAN PROTEIN PRODUCTS USED IN VARIOUS MEAT SYSTEMS[1]

Meat System	50% Protein		70% Protein Dry Basis Soybean Protein Concentrate		90% Protein Dry Basis Isolated Soybean Protein
	Soybean Flour	Soybean Grits	Coarse	Fine	
Ground (coarse)					
Patties	B	A	A	B	B
Meat balls	B	A	A	B	B
Meat loaves	B	A	A	B	B
Chili	A	A	A	B	B
Sloppy joe	B	A	A	B	B
Tacos	A	A	A	B	B
Salisbury steak	A	A	A	B	B
Sausage	AB	A	A	A	A
Emulsion					
Sausage	A			A	A
Bologna	A			A	A
Loaves	A			A	A
Canned	A			A	A
Other					
Baby foods	A	A	A	A	A
Soups					
Canned	A	A	A	A	A
Dry					A
Sauces and gravies	A	A	A	A	A
Pet foods	A	A	A	A	A
Poultry rolls	B			B	B

Source: Rakosky (1974).
[1] A = major additive, B = minor additive.

shrinkage and cooking yield of frankfurters when added at the 3.5% level of final processed weight. Highly variable cooking yields and organoleptic ratings were obtained among the soybean flours, concentrates and isolates in this finished meat product. According to Mattil (1974), control of organoleptic and functional properties continues to be the major challenge for the seed protein industry. The results obtained by Smith *et al.* (1973) with cottonseed flour was sufficiently promising, especially in organoleptic characteristics, to recommend the evaluation of cottonseed concentrates and isolates in emulsion-type meats. Alternate sources of concentrated seed proteins, alone or in blends with soybean proteins, may provide the desired blend of functional properties which the meat industry requires for a competitive, discriminating market. In addition, these second and third generation blends of concentrated seed proteins might permit much higher levels of protein supplementation and nutritional quality than is presently possible because of organoleptic considerations.

Textured Proteins

The recent development of structured protein products by extrusion cooking and by spinning has extended the opportunities for utilizing the various grades of seed proteins in meat formulations. While less functional than powdered forms of proteins, they possess textural properties of chewiness which allows their incorporation into processed meats at much greater levels or to replace the meat component entirely in simulated meat products. The formulation of textured products from seed protein powders is based on water uptake, water retention and dough formation during moist heat denaturation, acid precipitation or coagulation of protein. A comprehensive review of historical and recent technology and applications of meat analogs has been published by Horan (1974B).

In the extrusion-cooking process, defatted soybean flours are moistened to 30–40% and subjected to high temperatures (over 100°C) and high pressures for a short period (Wuhrmann 1972). The gelatinized protein mass is forced through orifices at reduced (atmospheric) pressure. The sudden release of pressure and steam causes the extruded protein mass to expand and become oriented into plate- or strand-like structures. The spongy structure of denatured hydrated protein contains air cells which are elongated in the direction of extrusion (Fig. 3.8). The protein dough is then cut into chunks or pellets and dried. The density of the product is controlled by adjusting the moisture, pressure, temperature, and duration of extrusion. Coloring and flavor are added to the flour before extrusion or to the expanded product before drying.

From Horan (1974B)

FIG. 3.8. A CHUNK OF TEXTURED SOYBEAN PROTEIN MADE BY THERMOPLASTIC EXTRUSION PROCESS

The dried textured pieces contain 50% protein and 32% carbohydrate. Before use, the extruded product is rehydrated by adding about 2.5–3 times its weight of water. This provides about 16% protein on a hydrated, as is, basis for blending in beef-, pork-, chicken- and fish-type products, or added directly to chili, casseroles and soups.

The outstanding characteristic of this type of product is that the raw material is a soybean flour or flake which is substantially cheaper than the meat product or soybean isolate required in the spun-fiber process. Extruded soybean products still retain residual flavors and flatulence factors which limit the use of the products in some applications.

The fiber spinning process employs more expensive raw material, complicated equipment and skilled operations than thermoplastic extrusion (Fig. 3.9). Slurries of 10-30% soybean isolate are adjusted to pH 10-12 where the viscosity of the spinning dope increases markedly. The protein dispersion or dope is pumped through the holes of a spinneret into a coagulating bath containing acid and salts. The spinneret has 15,000 or more holes, 50-250 μ in diameter, and the coagulated and wrapped filament which passes out of the spinneret is called the tow. The tow passes to a second bath for stretching on rolls and is neutralized. The tow is then impregnated with a heat coagulable protein binder, fats, flavoring, spices and coloring. The impregnated fibers are heated to coagulate the protein, usually egg albumin, and to bind the fibers and ingredients. After drying, the protein content of the finished product is 50-60%, dry basis.

Products manufactured from spun soybean include bacon bits, bacon strips and the analogs of beef, ham, chicken and seafood. Many of these analogs are marketed in the ready-to-eat state through the food service industry—hotels, restaurants, armed services and institutions (Wolf and Cowan 1971). Nearly 40% of all meals consumed in the United States are eaten outside of the home and the meat analogs provide the convenience and economy required in institutional food preparation.

Protein Beverages

Soy milk is a traditional food in Eastern Asia which is prepared by soaking beans overnight and then wet milling to provide a slurry. After heating the slurry to improve flavor and nutritional value, the milky suspension is separated from the insoluble residue by filtration. Beany flavors, astringency and bitterness have limited the acceptability of soy milk outside the orient. The lipoxidase activity can be controlled by wet milling at temperatures above 80°C or under acid conditions. The taste of the milk can also be improved by heat treatments which steam distill the undesirable flavor com-

---

Courtesy of General Mills, Inc.

FIG. 3.9. MANUFACTURE OF SOY PROTEIN SPUN FIBERS

Liquid protein has been fed into a spinning machine and emerges as bands, or ribbons, of tiny white fibers. To produce these ribbons, soy isolate is extracted from soy meal, precipitated and washed. It is then spun out (upper photo). Each of these ribbons of soy protein tow contains 16,000 monofilaments (lower photo). Coloring and flavor are added to the tow during the process. It is then dried, cut into the desired shapes and sized, packaged and sold.

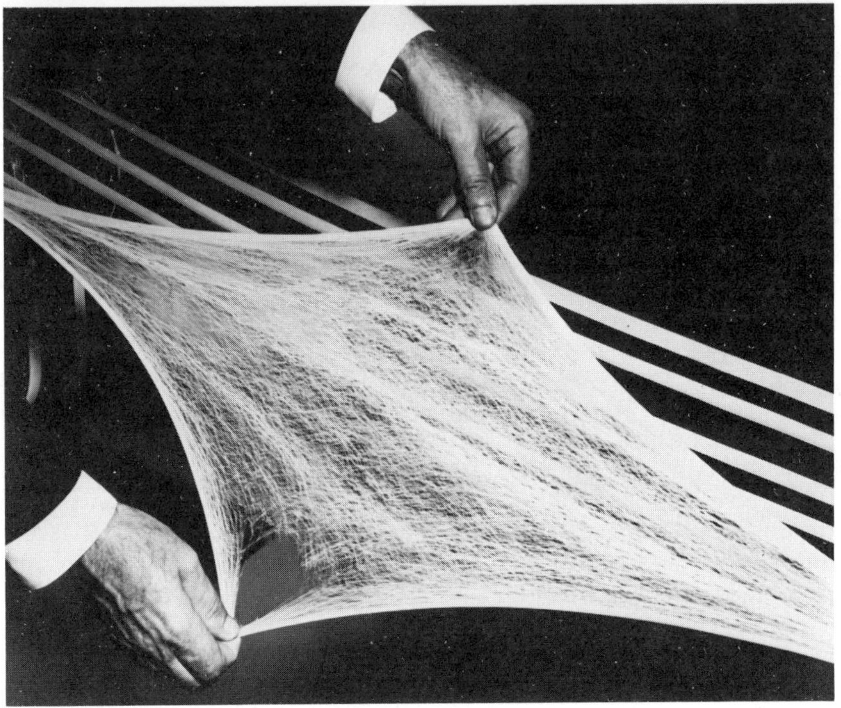

pounds. Therefore, modern soy beverages have improved flavor, texture and product stability.

The solids content of the soy milk is about 10% with 4% protein and 2% fat. After the addition of sugars, minerals, vitamins, and flavoring, the stable suspension can be either sterilized in suitable containers or concentrated and spray dried. In the manufacture of dried soy milk, problems are encountered with high viscosity in 20–30% slurries and the final products may lack good miscibility and dispersibility.

Other protein sources have been evaluated for improved functionality in milk-based products. Cottonseed protein bodies do not rupture in an aqueous suspension to provide the water dispersible protein required for a stable oil-protein-water milk emulsion. The majority of oilseed proteins are too insoluble at pH 6–7 to give a high yield of soluble protein for milk manufacture. Mattil (1971) has discussed the potential of sunflower and coconut proteins in imitation milk. Their high solubility in dilute solutions of calcium chloride over a broad range of pH (Fig. 3.5) may qualify these proteins for diverse uses in imitation and fortified milk, carbonated soft drinks and proteinated fruit juices. The lima bean protein may have particular application in a broad spectrum of beverages (Fig. 3.2).

CONCLUSIONS

The technology for conversion of concentrated seed protein into palatable and acceptable food items for infants and adults has been achieved in several food systems. The present utilization of flours, concentrates and isolates is, however, only a fraction of the potential production which could alleviate the serious shortage of quality protein in the world. The major challenge facing the food industry is to develop nutritional products which are compatible with the food habits of the majority of consumers.

There is a wide diversity of plant protein sources available to the food processor. Most protein crops are being essentially ignored due to a lack of knowledge of their chemistry and functional properties. Plant proteins are complex macromolecules which are sensitive to their physical-chemical environment during extraction and food formulation. A better understanding of their colloidal properties and interactions is required before the optimum conditions for extraction, aqueous processing and utilization can be defined. At the present time, the evaluations of functionality are based on empirical tests which are not representative of conditions in the food system. It is the task of the food industry to encourage and support mission-oriented research on new and original methods for the utilization of protein raw materials in foods. If the present trends are an indication of the potential advances that can be made, the concerted efforts of industry, government and consumer groups will be amply rewarded in terms of new and exciting food products that are economical, nutritional and convenient to prepare, and yet pleasing to the palate.

BIBLIOGRAPHY

AM. ASSOC. CEREAL CHEMISTS. 1972. Cereal Laboratory Methods, 7th Edition. American Association of Cereal Chemists, St. Paul, Minn.

ARAI, S., KOYANAGI, O., and FUJIMAKI, M. 1967. Studies on flavor components in soybean. IV. Volatile neutral compounds. Agr. Biol. Chem. (Tokyo) *31*, 868-873.

ARAI, S., SUZUKI, H., FUJIMAKI, M., and SAKURAI, Y. 1966. Studies on flavor components of soybeans. II. Phenolic acids in defatted soybean flour. Agr. Biol. Chem. (Toyko) *30*, 364-369.

AURAND, L. W., and WOODS, A. E. 1973. Food Chemistry. The Avi Publishing Co., Westport, Conn.

BADENHOP, A. F., and WILKENS, W. F. 1969. The formation of 1-octen-3-ol in soybeans during soaking. J. Am. Oil Chemists' Soc. *46*, 179-182.

BAKER, B. E., PAPACONSTANTINOU, J. A., CROSS, C. K., and KHAN, N. A. 1961. Protein and lipid constitution of some Pakistani pulses. J. Sci. Food Agr. *12*, 205-207.

BERARDI, L. C., MARTINEZ, W. H., and FERNANDEZ, C. J. 1969. Cottonseed protein isolates: two step extraction procedure. Food Technol. *23*, No. 10, 75-81.

BIETZ, J. A., and WALL, J. S. 1972. Wheat gluten subunits: Molecular weights determined by sodium dodecyl sulfate-polyacrylamide gel electrophoresis. Cereal Chem. *49*, 416-430.

BRISKEY, E. J. 1970. Functional evaluation of protein in food systems. In Evaluation of Novel Protein Products, A. E. Bender, R. Kihlberg, B. Lofqvist, and L. Munck (Editors). Pergamon Press, New York.

BULL, H. B., and BREESE, V. 1968. Protein hydration: 1. Binding sites. Arch. Biochem. Biophys. *128*, 488-496.

BURROWS, V. D. *et al.* 1972. Food Protein From Grains and Oilseeds—A Development Study Projected to 1980. Office of the Minister Responsible for the Canadian Wheat Board, House of Commons, Ottawa.

BUTZ, E. L. 1974. World protein markets—a supplier's view. J. Am. Oil Chemists' Soc. *51*, 57A—58A.

CAGAMPANG, G. B. *et al.* 1966. Studies on the extraction and composition of rice protein. Cereal Chem. *43*, 145-155.

CATER, C. M., GHEYASUDDIN, S., and MATTIL, K. F. 1972. The effect of chlorogenic, quinic and caffeic acids on the solubility and color of protein isolates, especially from sunflower seed. Cereal Chem. *49*, 508-514.

CATER, C. M., RHEE, K. C., HAGENMAIER, R. D., and MATTIL, K. F. 1974. Aqueous extraction—An alternative oilseed milling process. J. Am. Oil Chemists' Soc. *51*, 137-141.

CATSIMPOOLAS, N., and MEYER, E. W. 1970. Gelation phenomena of soybean globulins. 1. Protein-protein interactions. Cereal Chem. *47*, 559-570.

CHEN, C. H., and BUSHUK, W. 1970. Nature of proteins in Triticale and its parental species. 1. Solubility characteristics and amino acid composition of endosperm proteins. Can. J. Plant Sci. *50*, 9-14.

CIRCLE, S. J. 1950. Proteins and other nitrogenous constituents. *In* Soybean and Soybean Products, K. S. Markley (Editor). John Wiley & Sons, New York.

CIRCLE, S. J., MEYER, E. W., and WHITNEY, R. W. 1964. Rheology of soy protein dispersions. Effect of heat and other factors on gelation. Cereal Chem. *41*, 157-172.

LIBRARY
ATHROFA GOGLEDD DDWYRAIN CYMRU
THE NORTH EAST WALES INSTITUTE
of higher education CONNAH'S QUAY

CLUSKEY, J. E., WU, Y. V., WALL, J. S. and INGLETT, G. E. 1973. Oat protein concentrates from a wet-milling process: preparation. Cereal Chem. *50*, 475-481.

DANIELSSON, C. E. 1949. Seed globulins of the Gramineae and Leguminosae. Biochem. J. *44*, 387-440.

DANIELSSON, C. E. 1956. Plant proteins. Ann. Rev. Plant Physiol. *7*, 215-236.

DECHARY, J. M., and ALTSCHUL, A. M. 1966. Major seed proteins and the concept of aleurins. *In* World Protein Resources, R. F. Gould (Editor). American Chemical Society, Washington, D.C.

DE MUELENAERE, H. J. H. 1969. Development, production and marketing of high-protein foods. *In* Protein-enriched Cereal Foods For World Needs, M. Milner (Editor). American Association of Cereal Chemists, St. Paul, Minn.

DIMLER, R. J. 1963. Gluten—the key to wheat's utility. Bakers Digest *37*, 52-57.

DIPPOLD, M. W. 1961. Soyflour and meal composition, characteristics and uses. Proc. Symp. What's New in Products from the Soybean. Midwest Section, American Association of Cereal Chemists. April 1961.

ELDRIDGE, A. C., HALL, P. K., and WOLF, W. J. 1963. Stable foams from unhydrolyzed soybean protein. Food Technol. *17*, 1592-1595.

FAN, T. Y., and SOSULSKI, F. W. 1974. Dispersibility and isolation of proteins from legume flours. Can. Inst. Food Sci. Technol. J. 7, 256-259.

FINLAYSON, A. J., BHATTY, R. S., and CHRIST, C. M. 1969. Species and varietal differences in the proteins of rapeseed. Can. J. Botany *47*, 679-685.

FLEMING, S. E., SOSULSKI, F. W., and HAMON, N. W. 1974A. Gelation and thickening phenomena of vegetable protein products. J. Food Sci. *40*, 805-807.

FLEMING, S. E., SOSULSKI, F. W., KILARA, A., and HUMBERT, E. S. 1974B. Viscosity and water absorption characteristics of slurries of sunflower and soybean flours, concentrates and isolates. J. Food Sci. *39*, 188-191.

FUKUSHIMA, D. 1969. Denaturation of soybean proteins by organic solvents. Cereal Chem. *46*, 156-163.

GASTROCK, E. A. *et al.* 1965. A mixed solvent-extraction process for cottonseed. Cereal Sci. Today *10*, 572-576.

GHEYASUDDIN, S., CATER, C. M., and MATTIL, K. F. 1970A. Preparation of a colorless sunflower protein isolate. Food Technol. *24*, 242-243.

GHEYASUDDIN, S., CATER, C. M., and MATTIL, K. F. 1970B. Effect of several variables on the extractability of sunflower seed proteins. J. Food Sci. *35*, 453-456.

GREMLI, H. A. 1974. Interaction of flavor compounds with soy protein. J. Am. Oil Chem. Soc. *51*, 95A-97A.

HAGENMAIER, R. 1972. Water binding of some purified oilseed proteins. J. Food Sci. *37*, 965-966.

HAMILTON, T. S., HAMILTON, B. C., JOHNSON, B. C., and MITCHELL, H. H. 1951. The dependence of the physical and chemical composition of the corn kernel on soil fertility and cropping system. Cereal Chem. *28*, 163-176.

HERMANSSON, A. M. 1972. Functional properties of proteins for foods—swelling. Lebensm.-Wiss. u. Technol. *5*, No. 1, 24-29.

HISCHKE, H. H. JR., POTTER, G. C., and GRAHAM, W. R. JR. 1968. Nutritive value of oat protein. 1. Varietal differences as measured by amino acid analysis and rat growth responses. Cereal Chem. *45*, 374-378.

HOLT, N. W., and SOSULSKI, F. W. 1974A. Proximate and amino acid composition of ground meat products. Can. Inst. Food Sci. Technol. J. 7, 144-147.

HOLT, N. W., and SOSULSKI, F. W. 1974B. Private communication.

HORAN, F. E. 1974A. Nutritional cereal blends—from conception to consumption. Cereal Sci. Today *19*, 112-117.

HORAN, F. E. 1974B. Meat analogs. *In* New Protein Foods, Vol. 1A, A. M. Altschul (Editor). Academic Press, New York.

HOUSTON, D. F., MOHAMMAD, A., WASSERMAN, T., and KESTER, E. B. 1964. High-protein rice flours. Cereal Chem. *41*, 514-523.

HUEBNER, F. R., and WALL, J. S. 1974. Wheat glutenin subunits. 1. Preparative separation by gel-filtration and ion-exchange chromatography. Cereal Chem. *51*, 228-240.

JOHNSON, D. W. 1970. Functional properties of oilseed proteins. J. Am. Oil Chemists' Soc. *47*, 402-407.

JOHNSON, P., and SHOOTER, E. M. 1950. The globulins of ground-nut (*Arachis hypogaea*). Investigation of arachin as a dissociation system. Biochem. Biophys. Acta. *5*, 361-375.

JONES, R. W., BABCOCK, G. E., TAYLOR, N. W., and SENTI, F. R. 1961. Molecular weights of wheat gluten fractions. Arch. Biochem. Biophys. *94*, 483-488.

JONGH, G. 1961. The formation of dough and bread structures. The ability of starch to form structures, and the improving effect of glyceryl monostearate. Cereal Chem. *38*, 140-152.

JOUBERT, F. J. 1957. Ultracentrifuge studies of seed proteins of the family Leguminosae. Part III. Proteins from different bean species. J. South African Chem. Inst. *10*, 16-19.

JULIANO, B. O., BAUTISTA, G. M., LUGAY, J. C., and REYES, A. C. 1964. Studies on physicochemical properties of rice. J. Agr. Food Chem. *12*, 131-138.

KALBRENER, J. E., ELDRIDGE, A. C., MOSER, H. A., and WOLF, W. J. 1971. Sensory evaluation of commercial soy flours, concentrates and isolates. Cereal Chem. *48*, 595-600.

KILARA, A., HUMBERT, E. S., and SOSULSKI, F. W. 1972. Nitrogen extractability and moisture adsorption characteristics of sunflower seed products. J. Food Sci. *37*, 771-773.

KIM, J. C., and de RUITER, D. 1968. Bread from non-wheat flours. Food Technol. *22*, No. 7, 55-66.

KOSHIYAMA, I. 1968. Factors influencing conformation changes in a 7S protein of soybean globulins of ultracentrifugal investigations. Agr. Biol. Chem. *32*, 879-887.

KRISHNASWAMY, P. R. 1970. Limitations and scope of oilseed proteins. Proc. SOS/70 3rd Intern. Cong. Food Sci. Technol., Institute of Food Technologists, Washington, D.C.

KRULL, L. H., and WALL, J. S. 1969. Relationship of amino acid compositions and wheat protein properties. Bakers Digest *43*, 30-39.

LIN, M. J. Y., HUMBERT, E. S., and SOSULSKI, F. W. 1974. Certain functional properties of sunflower meal products. J. Food Sci. *39*, 368-370.

MARTINEZ, W. H., BERARDI, L. C., and GOLDBLATT, L. A. 1970A. Potential of cottonseed; products, composition and use. Proc. SOS/70 3rd Intern. Cong. Food Sci. Technol., Institute of Food Technologists, Washington, D.C.

MARTINEZ, W. H., BERARDI, L. C., and GOLDBLATT, L. A. 1970B. Cottonseed protein products—composition and functionality. J. Agr. Food Chem. *18*, 961-968.

MARTINEZ, W. H., BERARDI, L. C., PFEIFER, V. F., and CROVETTO, A. J. 1967. The production of protein concentrates by air classification of defatted cottonseed flour. J. Am. Oil Chemists, Soc. *44*, 139A.

MATSUO, R. R., BRADLEY, J. W., and IRVINE, G. N. 1972. Effect of protein content on the cooking quality of spaghetti. Cereal Chem. *49*, 707-711.

MATTICK, L. R., and HAND, D. B. 1969. Identification of a volatile component in soybeans that contributes to the raw bean flavor. J. Agr. Food Chem. *17*, 15-17.

MATTIL, K. F. 1971. The functional requirements of proteins for foods. J. Am. Oil Chemists' Soc. *48*, 477-480.

MATTIL, K. F. 1974. Composition, nutritional and functional properties and quality criteria of soy protein concentrates and soy protein isolates. J. Am. Oil Chemists' Soc. *51*, 81A-84A.

McANELLY, J. K. (Swift and Co.) 1964. U.S. Pat. 3,142,571. July 28.

McCALL, E. R. *et al.* 1953. Composition of rice: Influence of variety and environment on physical and chemical composition. J. Agr. Food Chem. *1*, 988-993.

McCLEARY, C. W. 1973. Vegetable proteins. Part 1. Food in Canada *33*, No. 11, 23-25.

MELLON, E. F., KORN, A. H., and HOOVER, S. R. 1947. Water absorption of proteins. 1. The effect of free amino groups in casein. J. Am. Chem. Soc. *69*, 827-831.

MERTZ, E. T., and BRESSANI, R. 1957. Studies on corn proteins. 1. A new method of extraction. Cereal Chem. *34*, 63-69.

MUSTAKAS, G. C., KIRK, L. D., and GRIFFIN, E. L., JR. 1962. Flash desolventizing defatted soybean meals washed with aqueous alcohols to yield a high protein product. J. Am. Oil Chemists' Soc. *39*, 222-229.

NIELSEN, H. C., BABCOCK, G. E. and SENTI, F. R. 1962. Molecular weight studies on glutenin before and after disulfide bond splitting. Arch. Biochem. Biophys. *96*, 252-258.

OSBORNE, T. B. 1907. The Proteins of the Wheat Kernel. Publ. *84*, Carnegie Institute, Washington, D.C.

OSBORNE, T. B., and CAMPBELL, G. F. 1898A. Proteids of the soybean (*Glycine hispida*). J. Am. Chem. Soc. *20*, 419-428.

OSBORNE, T. B., and CAMPBELL, G. F. 1898B. Proteids of the pea. J. Am. Chem. Soc. *20*, 348-362.

PAULIS, J. W., JAMES, C., and WALL, J. S. 1969. Comparison of glutelin proteins in normal and high-lysine corn endosperms. J. Agr. Food Chem. *17*, 1301-1305.

PENCE, J. W., MOHAMMAD, A., and MECHAM, D. K. 1953. Heat denaturation of gluten. Cereal Chem. *30*, 115-126.

PENCE, J. W., WEINSTEIN, N. E., and MECHAM, D. K. 1954. The albumin and globulin contents of wheat flour and their relationship to protein quality. Cereal Chem. *31*, 303-311.

RACKIS, J. J., SMITH, A. K. BABCOCK, G. E., and SASAME, H. A. 1957. An ultracentrifugal study of the association—dissociation of glycinin in acid solution. J. Am. Chem. Soc. *79*, 4655-4658.

RAKOSKY, J., JR. 1974. Soy grits, flour, concentrates and isolates in meat products. J. Am. Oil Chemists' Soc. *51*, 123A-127A.

REINERS, R. A., WALL, J. S. and INGLETT, G. E. 1973. Corn proteins: Potential for their industrial use. *In* Industrial Uses of Cereals, Y. Pomeranz (Editor). American Association of Cereal Chemists, St. Paul, Minn.

RHEE, K. C., CATER, C. M., and MATTIL, K. F. 1972. Simultaneous recovery of protein and oil from raw peanuts in an aqueous system. J. Food Sci. *37*, 90-93.

RHEE, K. C., MATTIL, K. F., and CATER, C. M. 1973. Recovers protein from peanuts. Food Eng. *45*, 82-90.

RIDLEHUBER, J. M., and GARDNER, H. K., JR. 1974. Production of food-grade cottonseed protein by the liquid cyclone process. J. Am. Oil Chemists' Soc. *51*, 153-157.

ROSEN, G. D. 1958. Groundnuts (peanuts) and groundnut meal. *In* Processed Plant Protein Foodstuffs, A. M. Altschul (Editor). Academic Press, New York.

ROSSI-FANELLI, A. *et al.* 1964. Isolation of monodisperse protein fraction from cottonseeds. Biochem. Biophys. Res. Commun. *15*, 110-115.

SABIR, M. A., SOSULSKI, F. W., and FINLAYSON, A. J. 1974A. Chlorogenic acid-protein interactions in sunflower. J. Agr. Food Chem. *22*, 575-578.

SABIR, M. A., SOSULSKI, F. W., and KERNAN, J. A. 1974B. Phenolic constituents in sunflower flour. J. Agr. Food Chem. *22*, 572-574.

SABIR, M. A., SOSULSKI, F. W., and MacKENZIE, S. L. 1973. Gel Chromatography of sunflower proteins. J. Agr. Food Chem. *21*, 988-993.

SAIO, K., WATANABE, T., and KAJI, M. 1973. Food use of soybean 7S and 11S proteins. Extraction and functional properties of their fractions. J. Food Sci. *38*, 1139-1144.

SARWAR, G., SOSULSKI, F. W., and BELL, J. M. 1973. Nutritional evaluation of oilseed meals and protein isolates by mice. Can. Inst. Food Sci. Technol. *6*, 17-21.

SENTI, F. R. 1969. Formulated cereal foods in the U.S. Food for peace program. *In* Protein-Enriched Cereal Foods for World Needs, M. Milner (Editor). American Association of Cereal Chemists, St. Paul, Minn.

SMITH, K. J. 1971. Nutritional framework of oilseed proteins. J. Am. Oil Chemists' Soc. *48*, 625-628.

SMITH, A. K., and CIRCLE, S. J. 1972. Protein products as food ingredients. *In* Soybeans: Chemistry and Technology, Vol. 1. A. K. Smith and S. J. Circle. The Avi Publishing Co., Westport, Conn.

SMITH, G. C. *et al.* 1973. Efficacy of protein additives as emulsion stabilizers in frankfurters. J. Food Sci. *38*, 849-855.

SOSULSKI, F. W., and BAKAL, A. 1969. Isolated proteins from rapeseed, flax and sunflower meals. Can. Inst. Food Technol. J. *2*, 28-32.

SOSULSKI, F. W., and SABIR, M. A., and FLEMING, S. E. 1973. Continuous diffusion of chlorogenic acid from sunflower kernels. J. Food Sci. *38*, 468-470.

SOSULSKI, F. W., and SARWAR, G. 1973. Amino acid composition of oilseed meals and protein isolates. Can. Inst. Food Sci. Technol. J. *6*, 1-5.

SOSULSKI, F. W., SOLIMAN, F. S., and BHATTY, R. S. 1972. Diffusion extraction of glucosinolates from rapeseed. Can. Inst. Food Sci. Technol. J. *5*, 101-104.

STRINGFELLOW, A. C., PFEIFER, V. F., and GRIFFIN, E. L., JR. 1961. Air classification of rice flours. Rice J. *64*, No. 7, 30-32.

TAPE, N. W., ZABRY, Z. I., and EAPEN, K. E. 1970. Production of rapeseed flour for human consumption. Can. Inst. Food Technol. J. *3*, 78-81.

TECSON, E. M. S., ESMAMA, B. V., LONTOK, L. P., and JULIANO, B. O. 1971. Studies on the extraction and composition of rice endosperm glutelin and prolamin. Cereal Chem. *48*, 168-181.

TKACHUK, R., and IRVINE, G. N. 1969. Amino acid compositions of cereals and oilseed meals. Cereal Chem. *46*, 206-218.

TSEN, C. C. 1974. Fatty acid derivatives and glycolipids in high-protein bakery products. J. Am. Oil Chemists' Soc. *51*, 81-83.

WALL, J. S., JAMES, C., and CAVINS, J. F. 1971. Nutritive value of protein in hominy feed fractions. Cereal Chem. *48*, 456-465.

WATANABE, T. 1974. Government role and participation in development and marketing of soy protein food. J. Am. Oil Chemists' Soc. *51*, 111A-115A.

WOLF, W. J., BABCOCK, G. E., and SMITH, A. K. 1962. Purification and stability studies of the 11 S component of soybean proteins. Arch. Biochem. Biophys. *99*, 265-274.

WOLF, W. J., and COWAN, J. C. 1971. Soybeans as a food source. The Chemical Rubber Co. Press, Cleveland, Ohio.

WOYCHIK, J. H., BOUNDY, J. A., and DIMLER, R. J. 1961. Starch gel electrophoresis of wheat gluten proteins with concentrated urea. Arch. Biochem. Biophys. *94*, 477-482.

WRIGLEY, C. W. 1970. Protein mapping by combined gel electrofocusing and electrophoresis: Application to the study of genotypic variation in wheat gliadins. Biochem. Genetics *4*, 509-516.

WU, Y. V., CLUSKEY, J. E., WALL, J. S., and INGLETT, G. E. 1973. Oat protein concentration from a wet-milling process. Composition and properties. Cereal Chem. *50*, 481-488.

WU, Y. V., SEXSON, K. R., CAVINS, J. F., and INGLETT, G. E. 1972. Oats and their dry-milled fractions. Protein isolation and properties of four varieties. J. Agr. Food Chem. *20*, 757-761.

WU, Y. V., and STRINGFELLOW, A. C. 1973. Protein concentrates from oat flours by air classification of normal and high-protein varieties. Cereal Chem. *50*, 489-496.

WUHRMANN, J. J. 1972. Utilization of novel proteins for human food. Nestlé Research News 1972.

YASUMATSU, K. *et al.* 1972. Whipping and emulsifying properties of soybean products. Agr. Biol. Chem. (Tokyo) *36*, 719-727.

YOUNGS, C. G. 1970. Peas for food and feed. *In* The Feed Protein Market. S. C. Hudson, Office of the Minister Responsible for the Canadian Wheat Board, House of Commons, Ottawa.

ZARKADAS, C. G., HENNEBERRY, G. D., and BAKER, B. E. 1965. The constitution of leguminous seeds. Field peas (*Pisum sativum*). J. Sci. Food Agr. *16*, 734-738.

CHAPTER 4

William J. Stadelman

Egg Proteins

The hen's egg is a complex mixture of nutrients. It consists of three well-defined portions: shell, albumen and yolk. The composition of each of the portions differs widely from the others. The shell is largely calcium carbonate with a small amount of protein matrix. The albumen is primarily a mixture of water and proteins with only a trace of carbohydrates. Finally the yolk is a mixture of water, lipids and protein. The membrane separating the yolk and albumen is known as the vitelline membrane. The solids content of the yolk is about 53% while the albumen has only about 13% solids. Amazingly, even after prolonged storage the water content of the two mixtures separated only by the vitelline membrane maintains approximately the same difference. An extensive review of all aspects of eggs was prepared by Romanoff and Romanoff (1949).

This chapter will briefly review the reports covering composition, functional and colloidal properties of the several proteins of the egg of the domestic chicken. The chemical composition and characteristics of egg proteins have been extensively studied. Discussions and reviews of these studies were reported by Hektoen and Cole (1928), Schmidt (1938), Romanoff and Romanoff (1949), Fevold (1951), Warner (1954), Parkinson (1966), Baker (1968B), Cook (1968), and Powrie (1973).

EGG ALBUMEN COMPOSITION

Egg white was separated into several fractions as early as 1900. The globulin fraction is precipitated by addition of ammonium sulfate to 1/2 saturation. The albumin fraction is crystallized by saturating the solution with ammonium sulfate and adjusting the pH to the isoelectric point, 4.8, by addition of dilute sulfuric acid or acetic acid solution. This early attempt at fractionation was reported by Hektoen and Cole (1928) who classified five proteins. They used antigen reactions to relate serum proteins of the chicken's blood to egg white proteins. Their conclusion was that a serum albumin was likely the same protein as egg white conalbumin. A much more extensive separation was reported by Longsworth *et al.* (1940) by combining the chemical procedures with moving boundary electrophoresis. They suggested that ovalbamin contained two fractions, conalbumin two forms, globulins three entities and ovomucoid several components. They reported egg white to contain 60% ovalbumin, 13.8% conalbumin, 14% ovomucoid, 2.8% globulin G_1, 4.6% globulin G_2 and 4.3% globulin G_3 by electrophoretic analysis. Warner (1954) modified their procedure to produce a simple comprehensive scheme for fractionation of egg white into the following proteins: ovalbumin, conalbumin, ovomucoid, lysozyme, globulin, ovomucin and ovidin. He developed a flow chart for the separation based primarily on the use of ammonium sulfate.

The list of proteins of egg white as reported by Warner (1954) was expanded by Baker (1968B) by breaking globulins into three fractions and adding ovoinhibitor, flavoprotein and ovoglycoprotein. The separation of flavoprotein is possible using the procedures listed by Warner (1954) according to Baker (1968B). The proteins isolated and identified thus far from egg white are listed in Table 4.1.

Using starch-gel electrophoresis Baker and Manwell (1962) resolved egg white into 22 distinct bands. They had bands for esterase, catalase, possibly two conalbumins, apoprotein, and several bands for ovalbumin. Qualitatively there were no differences in proteins found in each of the four layers of egg albumen but quantitative differences did exist. The thick white and chalaziferous layers had larger quantities of ovomucin than either of the thin whites. As the quantity of

TABLE 4.1

PROTEINS IN EGG ALBUMEN

Protein	Relative Amount in Albumen (%)	Iso-electric Point	Molecular Weight	Characteristics
Ovalbumin	54.0	4.6	45,000	Phosphoglycoprotein
Conalbumin	13.0	6.6	80,000	Binds metals especially iron
Ovomucoid	11.0	3.9-4.3	28,000	Inhibits trypsin
Lysozyme (G_1 globulin)	3.5	10.7	14,600	Lyses some bacteria
G_2 globulin	4.0?	5.5	30,000-45,000	—
G_3 globulin	4.0?	5.8	?	—
Ovomucin	1.5	?	?	Sialoprotein
Flavoprotein	0.8	4.1	35,000	Binds riboflavin
Ovoglycoprotein	0.5?	3.9	24,000	Sialoprotein
Ovomacroglobulin	0.5	4.5-4.7	760,000-900,000	?
Ovoinhibitor	0.1	5.2	44,000-	Inhibits several proteases
Avidin	0.05	9.5	53,000	Binds biotin

Source: Powrie (1973).

chalaza varies widely among eggs from different hens (Baker and Stadelman 1958) a similar variation in mucin might be expected.

Chung and Stadelman (1965) reported a change in protein content of eggs during a laying year. This confirmed the report of Cunningham *et al.* (1960). The variation is apparently due to age of the hens rather than season of the year. In general, percentage of protein in egg white declines as hens age but the change is small.

The amino acid composition of whole egg protein (Table 4.2) is such that it has often been used as a reference standard in nutrition studies. An attempt was made by May (1959) to increase threonine content by feeding higher levels of this amino acid to laying hens, without success. While there are hen-to-hen variations in protein content, the amino acid content of the proteins remains relatively constant (May and Stadelman 1960). Feeney *et al.* (1963A) reported that globulins can be modified genetically and expressed some concern for possible nonuniformity of total egg protein quality if genetic selection were used to modify globulin content. Smith *et al.* (1954) reported that the composition of individual hen's eggs was uniform but noted variations between hens.

The reported variation in protein and possibly amino acid composition among eggs has led to some concern as to using egg proteins as the standard for protein studies. Baker (1960) reviewed reports of

TABLE 4.2

TOTAL AMINO ACID CONTENT OF CHICKEN EGGS, BASED ON 100-GM, EDIBLE PORTION

Amino Acids	Fresh or Stored			Dried		
	Whole gm	Whites gm	Yolks gm	Whole gm	Whites gm	Yolks gm
Alanine	—	0.816	—	—	6.487	—
Arginine	0.840	0.634	1.132	3.070	5.044	2.167
Aspartic acid	0.897	0.850	0.897	3.280	6.762	1.717
Cystine	0.299	0.263	0.274	1.093	2.089	0.524
Glutamic acid	1.583	1.521	1.951	5.788	12.095	3.734
Glycine	0.453	0.404	0.571	1.655	3.216	1.093
Histidine	0.307	0.233	0.368	1.123	1.855	0.704
Isoleucine	0.850	0.698	0.996	3.108	5.553	1.907
Leucine	1.126	0.950	1.372	4.118	7.559	2.626
Lysine	0.819	0.648	1.074	2.995	5.154	2.057
Methionine	0.401	0.420	0.417	1.468	3.340	0.799
Phenylalanine	0.739	0.689	0.717	2.703	5.484	1.373
Proline	0.543	0.408	0.722	1.984	3.244	1.383
Serine	1.075	0.848	1.314	3.931	6.748	2.516
Threonine	0.637	0.477	0.827	2.329	3.793	1.582
Tryptophan	0.211	0.164	0.235	0.771	1.306	0.449
Tyrosine	0.551	0.449	0.756	2.014	3.573	1.448
Valine	0.950	0.842	1.121	3.474	6.693	2.147

Source: Orr and Watt (1957).

degree of heritability of egg quality factors and found that albumen and yolk weight were both genetically controlled. The albumen height, and with it ovomucin content of the egg, was also highly heritable. Genetic selection for any of these characteristics could therefore alter the composition of the egg with respect to amino acid and protein content. Similar reports were made by Feeney *et al.* (1960A), Lush (1961), Cochrane and Annau (1962) and Baker (1968A). The actual differences in commercially available eggs are likely too small to be determined in feeding trials.

The methodology for separation of egg white proteins has been extensively studied. As indicated, the earliest studies utilized precipitation techniques (Hektoen and Cole 1928) with electrophoresis being applied somewhat later (Longsworth *et al.* 1940). Forsythe and Foster (1950) proposed an ethanol fractionation scheme for separation of six egg white proteins. The application of slightly differing procedures were suggested by Rhodes *et al.* (1958) and Mandeles (1960). Disc gel electrophoresis has been applied most recently by Chang *et al.* (1970) and Butts and Cunningham (1972). Each of the methods has claimed advantages for particular applications.

Egg white is known to consist of four rather distinct layers in fresh eggs (Romanoff and Romanoff 1949). A number of workers have investigated the variation in protein composition of the layers. Forsythe and Foster (1949) suggested that the three outer layers were likely the same. They did not consider the chalaziferous layer adjacent to the yolk. Feeney *et al.* (1952) reported that fresh thick and thin white and deteriorated thin white were very uniform except for extra ovomucin and antihemagglutinin in fresh thick egg white.

Feeney *et al.* (1951) studied the deterioration of thick and thin white in intact eggs and as separated entities. They found that the separated components thin the same as intact shell eggs.

Ovalbumin

As shown in Table 4.1 ovalbumin makes up about 54% of the egg white. In earlier reports Longsworth *et al.* (1940) reported ovalbumin made up 70% of egg white proteins by chemical fractionation and 60% by electrophoretic analysis. Forsythe and Foster (1950) used a fractionation procedure employing ethanol. The ovalbumin could then be readily crystallized using ammonium sulfate.

Ovalbumin was found to consist of two fractions, A_1 and A_2, by Longsworth *et al.* (1940). A single fraction was found when the pH was high or low but at intermediate pH levels two bands were separated electrophoretically. Cann (1949) reported that electrophoretic separation of egg white components was possible only in dilute solutions. Linderstrom-Lang and Otteson (1947) suggested a different form of ovalbumin that crystallized as plates rather than the familiar needles.

Ovalbumin is classified as a glycoprotein. The carbohydrate component consists of 3 moles of D-glucosamine and 6 moles of D-mannose per 45,000 gm (1 mole) of ovalbumin according to Lee and Montgomery (1961). Montgomery *et al.* (1965) identified the glucosamine as 2-amino-2-deoxy-D-glucose. The carbohydrate units in ovalbumin are always linked to aspartic acid (Lee and Montgomery 1962). They suggested an alignment of amino acids near the carbohydrate of -glucose-lysine-tyrosine-aspartic acid-carbohydrate-leucine-threonine-serine-valine-leucine-. Montgomery (1970) reported that ovalbumin contains only one carbohydrate group per mole.

According to MacDonnell *et al.* (1951) ovalbumin contains most of the sulhydryl groups in egg white. Winzor and Creeth (1962) reported that ovalbumin has 4 sulfhydryl and 2 disulfide groups. According to Fernandez-Diez *et al.* (1964) native chicken ovalbumin had 3 weakly reactive (masked) sulfhydryl groups but after denaturation there were 4 reactive groups. Ovalbumin is somewhat unique among the egg white proteins in that it does not contain sialic acid according to Lee and Montgomery (1961) and Feeney *et al.* (1960B).

Neuberger and Marshall (1966) reported that ovalbumin contains four tryptophan residues per molecule and that the C terminal amino acid was proline with the N terminal being blocked by an acetyl group. The amino acid composition of ovalbumin is given in Table 4.3.

It was suggested by Cunningham and Lineweaver (1967) that the sulfhydryl groups of ovalbumin react with lysozyme during pasteurization to inactivate the lytic response. Lineweaver *et al.* (1967) reported that 3–5% of the ovalbumin is denatured by heating egg white at pH 9–62°C for 3.5 min. With the pH reduced to 7.0 less than 0.1% of the ovalbumin is denatured with the same pasteurizing treatment. Smith (1964) reported a more heat resistant form of ovalbumin, called S-ovalbumin. The level of the S-ovalbumin in eggs increases as storage of shell eggs is extended. It is only 5% of the total ovalbumin in fresh eggs but 81% in eggs that were in storage for 6 months. Smith and Back (1965) listed procedures for converting ovalbumin to S-ovalbumin. No further reports were found relative to this form of ovalbumin.

Marshall and Deutsch (1951) found traces of ovalbumin in blood serum of the chicken. McKenzie *et al.* (1963) indicated that ovalbumin and bovine serum albumin were very similar in behavior in urea at low pH but near neutrality and in the alkaline pH range, ovalbumin reacts much more slowly than the serum albumen. They suggest a random-coil polymer arrangement for ovalbumin. The molecular weight of ovalbumin is about 45,000 according to Warner (1954) as derived from osmotic pressure and light scattering measurements.

Conalbumin

Blood albumin is the same protein as egg white conalbumin according to Hektoen and Cole (1928). Conalbumin was reported as a single fraction by Longsworth *et al.* (1940) even though there was some suggestion of two fractions depending on pH. The preparation of crystalline conalbumin was described by Warner and Weber (1951) as being possible by saturation of the filtrate from ovalbumin preparation with ammonium sulfate. The purification procedures for conalbumin precipitate were given by Alderton *et al.* (1946).

Conalbumin is unique among the proteins of egg white in that it binds multivalent metal ions. Ferric, cupric and zinc ions combine with conalbumin at two specific sites in each protein molecule and bind one bicarbonate ion per metal ion in forming the complex according to Warner and Weber (1953). The complex dissociates very slowly. Fraenkel-Conrat and Feeney (1950) reported that iron and copper are bound by the same sites. This was confirmed by Windle *et al.* (1963). According to Fraenkel-Conrat (1950) the metal binding is not mediated by a hydroxylamine group. Warner and Weber (1953) state that phenolic groups of tyrosine are an essential part of the binding sites. The specific binding sites of conalbumin are not identical according to Aisen *et al.* (1973).

The relationship between serum proteins and conalbumin was investigated by Williams (1962B). He reported that transferrin, a serum protein, and conalbumin of hen's egg white differed only in their respective carbohydrate prosthetic groups. Marshall and Deutsch (1951) reported conalbumin to be present in chicken blood serum. Azari and Baugh (1967) reported the amino acid composition of serum transferrin and egg white conalbumin to be the same.

The stability of conalbumin has been studied under a wide range of conditions. Wishnia and Warner (1961) reported an unstable environment to exist at pH 3.2 where denaturation occurred in a few seconds. At a pH of 4.2 an equal degree of denaturation took days. At a pH of 7.0 conalbumin is 100% denatured when heated at 62°C for 3.5 min. When the pH is raised to 9.0 the same heat treatment denatures only 50% of the conalbumin according to Lineweaver *et al.* (1967).

Gaffield *et al.* (1966) indicated that ordered structure of conalbumin is highly susceptible to an acidic environment and is sensitive to detergents at neutral pH. They reported that 13 of 18 tyrosyls are buried in the molecule and that 6 of the 13 are hydrogen bonded. Several side chain chromophores (tyrosyls, tryptophyls, and possibly cystines and phenylalanines) which exist in an interior portion of conalbumin are specially oriented in such a manner as to enhance optical activity of their long wave length absorption bands. The N terminal group of conalbumin is alanine according to Fraenkel-Conrat and Porter (1952). They report a single peptide chain with a molecular weight of 87,000 for conalbumin. Conalbumin contains no sialic acid (Feeney *et al.* 1960B). Wenn and Williams (1968) reported that conalbumin exists in three states: metal free, one metal atom per molecule and two metal atoms per molecule. The same was reported by Stratil (1967). He reported ten electrophoretic bands for conalbumin in its three forms. The extra bands represented multiple molecular forms of conalbumin. The variation in conalbumin forms was also reported by Baker (1967); Clark *et al.* (1963); and Ogden *et al.* (1962). The amino acid composition of conalbumin as reported by Lewis *et al.* (1950) is given in Table 4.3.

The metal complexes of conalbumin are much more resistant to heat than the metal-free protein (Azari and Feeney 1958). In a following report Azari and Feeney (1961) state that iron complexing of conalbumin appears to stabilize the molecule so that modification of other parts of the molecule could occur without denaturation. This information led to a patent for pasteurizing egg white involving pH adjustment to 7.0 and complexing of conalbumin with aluminum salts (Lineweaver and Cunningham 1966). Aluminum salts complexed with conalbumin yield a white preciptiate. When conalbumin and iron complex a red color is produced and with copper a yellow color results.

Lysozyme

Lysozyme has been variously classified as a globulin, G_1, by Longsworth *et al.* (1940) and as an enzyme by early workers (Warner 1954) based on its ability to lyse bacterial cells. Alderton *et al.* (1945) identified the lytic enzyme as the same compound as Longsworth *et al.* (1940) had called globulin G_1. The lytic action of lysozyme was attributed to a solubilizing effect by Abraham (1939).

The lysozyme content of eggs from an individual hen was found to be relatively constant with rather wide differences among hens in the total lysozyme in their eggs. (Wilcox and Cole 1954; Cotterill and Winter 1954). Lysozyme is rather uniformly distributed throughout the egg white except for the chalaziferous layer and the chalazae

TABLE 4.3

COMPOSITION OF SOME PROTEINS IN EGG WHITE WITH SUGARS AND AMINO ACIDS AS GRAMS PER 100 GM PROTEIN

Substance	Oval-bumin[1]	Conal-bumin[2]	Lyso-zyme[3]	Avidin[4]	Ovo-mucin[5]	Ovo-mucoid[6]
Alanine	6.72	4.4	12	2.73	1.8	2.6
Arginine	5.72	7.6	11	8.40	2.6	2.1
Aspartic acid	9.30	13.3	21	11.74	5.6	9.8
Cystine	0.51[7]	3.8	8	0.86[7,8]	4.1[7]	5.5
Glutamic acid	16.50	11.9	5	9.76	7.0	4.4
Glycine	3.05	5.7	12	5.13	1.8	5.1
Histidine	2.35	2.57	1	0.97	1.7	1.38
Hydroxyproline	—	Nil	—	—	—	—
Isoleucine	7.00	5.0	6	6.35	2.6	1.1
Leucine	9.20	8.8	8	5.75	4.0	3.9
Lysine	6.30	10.0	6	8.03	4.3	4.1
Methionine	5.20	2.03	2	1.87	1.4	0.64
Phenylalanine	7.66	5.7	3	7.00	3.1	1.76
Proline	3.60	4.9	2	1.58	3.3	2.36
Serine	8.15	6.3	10	5.74[8]	4.3	4.0
Threonine	4.03	5.9	7	14.53[8]	4.6	4.6
Tryptophan	1.20	3.0	6	5.50	0.9	0.15
Tyrosine	3.68	4.6	3	1.15	2.5	1.76
Valine	7.05	8.2	6	5.20	3.3	5.1
Cysteine	1.35	—	—	—	—	—
Hexosamine	1.2	—	—	3.83	8.5	13.6
Mannose	2.0	—	—	5.80	4.6	6.8
Galactose	—	—	—	Trace	1.8	2.3

[1] Neuberger and Marshall (1966).
[2] Lewis *et al.* (1950).
[3] Canfield (1963A) listed as amino residues per molecule of Lysozyme
[4] Melamed and Green (1963).
[5] Robinson and Monsey (1971).
[6] Stevens and Feeney (1963).
[7] Reported as ½ cystine residues.
[8] Corrected for destruction.

where it is concentrated at 2–3 times the levels found in other egg white (Baker *et al.* 1959).

Cotterill and Winter (1955) suggested that a lysozyme-ovomucin complex resulted in the thick egg white. The complex is pH dependent as it becomes almost nonexistant at pH 9.0–9.5. Garabaldi *et al.* (1968) and Dam (1971) studied the lysozyme-ovomucin complex. The breakdown of the complex might explain the thinning of egg white in shell eggs (Hawthorne 1950). Dam (1971) reported that sialic acid is not involved in the lysozyme-ovomucin complex. The fact that available lysozyme increased as chalazae cords were incubated was taken as further proof of the lysozyme-ovomucin complex by Baker *et al.* (1959).

The relationship of lysozyme content of eggs to height of the thick albumen was investigated by Sauter *et al.* (1966). No relationship was found between lysozyme content of the egg and albumen height but eggs with a high lysozyme content lost albumen height more slowly than eggs with lesser amounts of lysozyme. Sauter and Peterson (1969) found that eggs with high lysozyme content at the time of lay actually had lower albumen heights than hens with lower lysozyme levels. The quality as measured in Haugh units (Romanoff and Romanoff 1949) lost by eggs during 2 weeks was 5.5 Haugh units for high lysozyme eggs as compared to 13 Haugh units for low lysozyme eggs (Sauter and Peterson 1972). Sauter *et al.* (1970) reported lysozyme contents of eggs for hens selected as high lysozyme egg producers to be up to 13.5% compared to a low lysozyme line with only 4.4% lysozyme. With this wide variation in the lysozyme level in egg white a difference in amino acid content of the total egg white would be a distinct possibility. The amino acid content of lysozyme is listed in Table 4.3. According to this analysis the molecular weight of lysozyme is 14,307 (Canfield 1963B). To aid in this detailed analysis Canfield and Anfinsen (1963A) used phosphorylated cellulose in the separation of peptides derived from chymotrypsin and pepsin digests of carboxymethylated egg white lysozyme. Another approach was the use of radioactive components. Canfield and Anfinsen (1963B) reported evidence that supports the theory that the lysozyme polypeptide chain grows unidirectionally, initiated at the N terminal and terminating at the C terminal.

According to Beychok and Warner (1959) the extreme stability of lysozyme at acid pH levels does not depend on hydrogen bonding but on disulfide cross-linking. Canfield and Lin (1965) positively located eight 1/2-cystine residues. Jollès *et al.* (1963) detailed the chemical structure. The polypeptide chain contains the 129 amino acid residues listed in Table 4.3 and is folded by 4 disulfide bridges of cystine. Blake *et al.* (1967) confirmed the structure and gave special arrangements with angles for each peptide linkage. Jollès (1967) stated that lysozyme must have disulfide bridges to remain biologically active. Native lysozyme has 4 disulfide bridges and is still active with 3 disulfide bridges. With 2 or less disulfide bridges there is no lytic acitvity. Cystine, tryptophan and histidine are all involved in biological activity of lysozyme.

Lysozyme is selective in materials it will react with according to Klotz and Walker (1948). Pepsin precipitates lysozyme but trypsin and chymotrypsin do not. Gandhi *et al.* (1968) found that 3,3-dimethyl-glutaric anhydride reacted with egg white proteins, particularly lysozyme, to change the charges on the proteins. Galyean *et al.* (1972) reported that yolk components reacted with lysozyme to inactivate the biological lytic activity.

Purified lysozyme salts upon standing in dry form change into two separable peaks both of which showed lytic activity according to Tallan and Stein (1951). Later, the same workers, Tallan and Stein (1953), reported three chromatographically distinct lysozymes.

The reactions of lysozyme during pasteurization have been studied. Seideman *et al.* (1963) reported that lysozyme is denatured at high pH values, 9.0 and above. At a pH of 7.0 only 6% of lysozyme was denatured at 62°C for 3.5 min according to Lineweaver *et al.* (1967). Cunningham and Lineweaver (1967) reported that sulfhydryl groups of ovalbumin react with lysozyme during pasteurization to inactivate the lytic response.

Ovomucoid

Ovomucoid is a heterogeneous glycoprotein consisting of three major and two minor components according to Bier *et al.* (1953). They claim the carbohydrate fraction to be mannose, galactose and glucosamine with 9% mannose and galactose at a 3:1 ratio and 13% of the glucosamine. Feeney *et al.* (1960B) reported 0.7% sialic acid in ovomucoid all bound to the protein. Jutisz *et al.* (1957) indicated they obtained a homogeneous ovomucoid by subjecting an ovomucoid preparation to column electrophoresis at a pH near the isoelectric point of ovomucoid followed by immunoelectrophoresis. Beeley and Jevons (1961) were still reporting ovomucoid preparations of that time to be heterogeneous. Rhodes *et al.* (1960) reported chicken eggs to contain 11% ovomucoid but cassowary eggs had 30% of this protein.

Fredericq and Deutsch (1949) reported that ovomucoid contained 13.2% nitrogen, 9.7% mannose and 17% glucosamine. The native protein is resistant to heat denaturation but loses its antitryptic activity. Their protein contained eight antitryptic units per milligram of protein. Melamed (1966) reported that ovomucoid was first reported in 1873 and was isolated in 1894.

Two major fractions of ovomucoid were isolated by Melamed (1967) using gel electrophoresis. One fraction was virtually free of sialic acid, the other contained 0.4 M of sialic acid per mole of protein. The ovomucoid was found to be equally distributed in thick and thin white by Lineweaver and Murray (1947). A unique property of ovomucoid is that it is a trypsin inhibitor (Feeney *et al.* 1963).

The carbohydrates of ovomucoid are present as several small units, each individually linked to the protein according to Neuberger and Papkoff (1963). The linkage to each of three carbohydrate units is through aspartic acid according to Montgomery and Wu (1963). They found threonine adjacent to the aspartic acid residue and that sialic acid was present. The molecular weight of the carbohydrates in ovomucoid was determined to be 2700–2800 and calculated to be 2567. Fraenkel-Conrat and Porter (1952) found alanine to be the N terminal group of ovomucoid with a single peptide chain containing 13 lysine groups. The amino acid composition as listed in Table 4.3 for ovomucoid was reported by Stevens and Feeney (1963).

Avidin

Avidin is differentiated from other proteins of egg white by its unique ability to complex with biotin. Eakin *et al.* (1940) found that avidin binds biotin against dialysis over a pH range of 2–10.5. Steam sterilization of the complex completely releases biotin. The molecular weight of avidin was estimated to be 87,000 by Eakin *et al.* (1941). Pennington *et al.* (1942) reported that avidin contained a large carbohydrate moiety. The carbohydrates present were reported to be mannose and hexosamine by Fraenkel-Conrat *et al.* (1952B). They estimated the molecular weight to be 58,000.

Avidin was found to occur in egg white in three forms (Fraenkel-Conrat *et al.* 1952A). One difference among forms is the addition of galactose to the carbohydrate component. The composition of avidin was reported by Melamed and Green (1963) and is listed in Table 4.3. They confirmed that each avidin can bind three molecules of biotin, as reported earlier by Fraenkel-Conrat *et al.* (1952A). Each bound biotin protects four tryptophan residues from oxidation according to Green (1963). One tryptophan could be lost to oxidation with only slight inactivation of the binding site. Loss of two tryptophans reduced biotin binding activity to less than 10%.

Avidin consists of three peptide chains with alanine as the N terminal group (Fraenkel-Conrat and Porter 1952). This protein has been specifically purified by chromatography over a biotin-cellulose using a linear gradient from high to low concentration of ammonium carbonate by McCormick (1965). The biotin-cellulose is an avidin specific adsorbent.

Ovomucin

The ovomucin content of egg white was found to vary greatly between thick and thin white portions. McNally (1933) found nine times as much in the thick white portion as in thin white. The relatively low nitrogen content, 12.9%, suggested a conjugated protein. The sialic acid content was found to be 2.6% by Feeney *et al.* (1960B) the highest of any of the egg white proteins.

The high ovomucin content of thick white and the long mucin

fibers (several centimeters according to Forsythe and Bergquist 1951) was suggested as the basis for the high viscosity of the thick white. Brooks and Hale (1959, 1961) reported that ovomucin content does not explain the rigidity of thick white. The thinning of the thick egg white could not be explained by a depolymerization of an ovomucin network.

Crude ovomucin was prepared by dilution of thick egg white with four volumes of distilled water by Robinson and Monsey (1964). They suggested that the insolubility of ovomucin fibers was due to disulfide bonds. Robinson and Monsey (1971) reported on the composition of ovomucin; a summary of their analysis is listed in Table 4.3. It wasn't until 1972 that a simple and rapid method for producing egg white ovomucin in pure form was reported by Young and Gardner (1972).

The reactions of ovomucin and lysozyme to form complexes were discussed in the section of this chapter on lysozyme.

Ovoinhibitor

For many years the small fraction of ovoinhibitor was classified with ovomucoid. Feeney *et al.* (1963B) discovered that ovomucoid inhibits only trypsin, whereas ovoinhibitor from the chicken egg inhibits trypsin, chymotrypsin and a fungal proteinase. The previously reported chymotrypsin inhibition attributed to ovomucoid had been caused by ovoinhibitor contamination.

Matsushima (1958) differentiated ovoinhibitor and gave details for its separation from egg white ovomucoid. Tomimatsu *et al.* (1966) reported ovoinhibitor to have a molecular weight between 44,000 and 49,000. They also gave a comparative analysis with ovomucoid which is reproduced in Table 4.4. The values of this amino acid analysis are in close agreement with values listed for ovomucoid in Table 4.3.

Flavoprotein

The source of flavoprotein and its importance to the egg in reproduction is discussed by Baker (1968B). Rhodes *et al.* (1959) reported that egg white normally contains about 0.9% flavoprotein. The molecular weight is from 32,000–36,000 according to these researchers.

The composition and structure of flavoprotein has not been elucidated beyond the fact that riboflavin is held by this protein as a normal component. Feeney *et al.* (1960B) reported that flavoprotein contains 0.5% sialic acid.

TABLE 4.4

COMPARATIVE COMPOSITION OF OVOINHIBITOR AND OVOMUCOID

Component	Ovoinhibitor	Ovomucoid
Alanine	3.37	3.56
Arginine	3.32	1.92
Aspartic acid	7.68	9.52
Cystine	4.13	5.84
Glutamic acid	6.25	4.71
Glycine	5.45	4.94
Histidine	2.30	1.56
Isoleucine	2.87	1.07
Leucine	3.71	3.66
Lysine	3.90	4.10
Methionine	0.54	0.70
Phenylalanine	1.60	1.51
Proline	2.82	2.03
Serine	4.29	3.55
Threonine	5.20	4.53
Tryptophan	0.0	0.15
Tyrosine	2.15	1.90
Valine	4.30	5.00
Nitrogen %	15.4	13.3
Hexosamine	2.7	13.6
Hexoses	3.5	5.1

Source: Tomimatsu *et al.* (1966).

Globulins

Two globulins different from other egg white proteins were reported by Kaminski (1954). Using starch-gel electrophoresis Feeney *et al.* (1963A) separated two genetically controlled dimorphic forms of globulins termed A_1 and A_2. Globulin A_1 contained 13.4% nitrogen, 0.80% tryptophan and 3.35% tyrosine. The molecular weight was over 30,000 but less than 45,000. Globulin A_2 was separated but not characterized.

Ovoglycoprotein

Ketterer (1965) reported a protein in egg white containing 13.8% hexosamine, 3% sialic acid and 13.6% hexose (mannose and galactose in a 2:1 ratio). He indicated a molecular weight of 24,400 and that the N terminal amino acid was threonine. He offered a name, ovoglycoprotein.

Ovomacroglobulin

This protein was isolated by Miller and Feeney (1966) and its chemical properties as given in Table 4.1 were determined. Little else is known about it except for its antigenic activity. There is a possibility that it might be the same protein reported by Lanni *et al.* (1949). This protein was reported as similar to ovomucin but with a strong inhibitor of influenza virus hemagglutination.

Enzymes

An extensive report on enzymes found in fresh hens' eggs was made by Lineweaver *et al.* (1948). In the egg white they reported finding peptidases and catalase. Beta-*N*-acetyl-glucosidase and alpha-mannosidase were reported to be present in egg white by Lush and Conchie (1966). Ball and Cotterill (1970) reported that catalase was present in the fraction of egg white usually classified as unidentified globulins.

Free Amino Acids

At least 16 free amino acids were found in albumen and yolk by Ducay *et al.* (1960). They included all of the amino acids listed in Tables 4.3 and 4.4 except cystine or cysteine. The total concentration of free amino acids was about 40 *mM* per milliliter. They found a migration of free amino acids from the yolk to the white during storage of eggs.

FUNCTIONAL PROPERTIES OF EGG WHITE

The functional properties of eggs in foods were classified by Baldwin (1973) as coagulating, foaming, emulsifying and providing nutrients. The evaluation of functional properties could be evaluated by the use of standardized procedures for determining the degree of protein denaturation according to Melnick and Oser (1949). The functional properties of egg whites are limited to coagulation, foam formation and stability, and nutritional contribution.

Coagulating

MacDonnell *et al.* (1955) reported that the ovalbumin is the egg white protein primarily responsible for coagulation and denaturation to establish a structure for food products. To test the coagulating properties of eggs, baked custards are often used. Baldwin *et al.* (1967) found tenderness of gels was influenced by pH of the egg white. Adjusting pH to an acidic range resulted in more tender gels. They also reported that coagulation temperature was 57.2°C with microwave heating and 61.8°C with conventional convection heating.

The presence of salts at low concentrations accelerates coagulation of proteins during heating according to Lepeschkin (1922). Sauter *et al.* (1954) reported that oil treating of shell eggs tended to result in improved functional performance of eggs in custards, cakes, poached and soft-cooked eggs. Such oil treating would result in lower pH egg whites.

Problems in the preparation of soft meringues are beading and leakage. The beading resulted from over-coagulation and leakage from undercoagulation in studies by Hester and Personius (1949).

Foaming

Foam quality is judged by volume and stability. According to MacDonnell *et al.* (1955) the ovoglobulins are especially important to foam volume and ovomucin is of prime importance for foam stability. The presence of ovomucin will not ensure satisfactory formation of egg white foams according to Forsythe and Bergquist (1951). They found best whipping when the long ovomucin fibers were cut to about 300 μ lengths. St. John and Flor (1931) reported that thin whites gave greater foam volume than thick whites. The breakdown of mucin fibers results in a thin white.

The influence of chemical and physical factors on egg white foam were studied by Barmore (1934). He found that beating time of the egg white was positively correlated with drainage or lack of stability of the foam and negatively correlated with specific gravity and volume of the foam. He reported that adding of potassium acid tartrate (cream of tartar) to the egg white before whipping improved foam stability.

Attempts were made by Garibaldi *et al.* (1968) to modify whipping properties of egg white by adjusting protein fractions. Modifying either ovomucin or lysozyme resulted in a decline in whipability of the resultant mixture. Sauter and Montoure (1972) working with eggs from genetic lines of hens selected for lysozyme content found that foam volume was significantly less for high lysozyme content eggs than for eggs low in lysozyme. They studied the composition of foam drainage and observed that ovalbumins drain from the foam whereas lysozyme is held in the foam.

It was recognized early that heating of egg white for pasteurization resulted in a loss of foaming properties (Slosberg *et al.* 1948). They noted that lowering of the pH to 6.5 reduced the damage and also that adding of sugar reduced damage but that the two effects were not additive. Cunningham and Lineweaver (1965) reported that adjusting of pH of egg white to 7.0 stabilized ovalbumin, lysozyme and ovomucoid but conalbumin is almost completely denatured during pasteurization unless it was previously complexed with metal, iron or aluminum, ions. They reported that volume and texture of cakes made from stabilized albumen are normal but whipping time necessary for best volume is increased.

Spray drying of albumen is a common method of processing.

Galyean and Cotterill (1973) found a change in electrophoretic and chromatographic patterns of the spray-dried albumen as compared to native albumen. Prior to spray drying, egg white must be desugared to prevent browning reactions between hexoses and free amino acids present. Kline *et al.* (1954) reported that using glucose oxidase for desugaring left 5-6% of the original carbohydrate, largely as mannose and galactose. Use of yeast fermentation resulted in less carbohydrate residue in the desugared egg white. Both desugaring methods yielded products about equal as far as foaming characteristics, provided the enzyme treated product had the proper pH adjustment prior to dehydration. Grunden *et al.* (1974) reported enzymatic treatments for desugaring resulted in egg white with improved foam volume but inferior foam stability to yeast-fermented egg white. Chang (1973) reported that addition of polyphosphates to liquid white before drying resulted in dried albumen giving angel food cakes of greater volume. Lysozyme reacts with the polyphosphates to inactivate its antifoaming characteristics.

Other Properties

During pH adjustments in egg white Cunningham and Cotterill (1962) observed that at pH 11.9 a translucent gel forms which liquifies during prolonged holding. Tung *et al.* (1970) reported egg white to have flow behavior of a pseudoplastic at 2°C between shear rates of 8.1 and 147 sec^{-1}. A characteristic of thick egg white that is somewhat unique is the "Weissenberg effect" according to Muller (1961). This property is displayed when a vertical rod is rotated in a liquid, and the liquid climbs up the rod.

Nutritional Contribution

The amino acid pattern of whole egg is close to the requirement of man. The egg white is a major contributor of amino acids particularly the amino acids containing sulfur (Tables 4.2, 4.3 and 4.4). Mauron (1972) reported that raw egg white is about 50% digested, but heat denatured egg white is 100% digested. This would be expected with trypsin and chymotrypsin inhibitors present in the native egg white.

EGG YOLK COMPOSITION

The egg yolk components are confined by the vitelline membrane. The composition of the membrane was studied electrophoretically by McCready and Roland (1972). They found it to consist of 70% lysozyme, 24.8% of an immobile protein and 5% of an unidentified peak. The composition of yolk solids is given in Table 4.5.

TABLE 4.5

COMPARITIVE CHEMICAL COMPOSITION OF YOLK SOLIDS

Chemical Class or Compound	% Total Yolk Solids: Proportions of Compounds	Proportions of Classes
Proteins		33
Livetins	4–10	
Phosvitin	5–6	
Vitellin (lipid-free)	4–15	
Vitellenin (lipid-free)	8–9	
Lipids		63
Triglycerides	41.0	
Phospholipids		
Lecithin	14.8	
Cephalin	3.2	
Sphingomyelin	0.5	
Cholesterol	3.5	
Free glucose		0.4
Inorganic elements		2.1
Other compounds (amino acids)		1.5

Source: Shenstone (1968).

Fresh egg yolk contains about 53% solids. An excellent review of macromolecules in egg yolk was prepared by Cook (1968).

Nearly all of the yolk lipids are found in the proteins, according to Weinman (1956). Stevens (1961) reported that starch-gel electrophoresis is well suited to the study of egg yolk proteins. The nomenclature used for egg yolk proteins was criticized by Schjeide and Urist (1960). The nomenclature was defended by Cook (1961) who classified them as low density factor lipoprotein, lipovitellenin, which contains about 40% lipid; high density factor lipoproteins, α- and β-lipovitellin, each of which contain about 20% lipid; and other high density factor proteins, phosvitin, α-, β- and γ-livetins, which are essentially free of a lipid component. This classification and nomenclature was confirmed by McCully *et al.* (1962). They also gave detailed procedures for fractionation of egg yolk into its several protein components. Chargaff (1942) was one of the early researchers working with yolk lipoproteins. He reported the phosphotides of lipovitellin to be mainly lecithin and cephalin. Schmidt *et al.* (1956) reported that approximately 40% of the yolk proteins are in the particulate matter fraction which normally occupies 11–13% of the yolk volume. As the phosphorus content of particulate matter is only 2.05%, it can be assumed there is very little phosvitin in this fraction.

The structure of the yolk was studied by electron microscopy by

Bellairs (1961). Proteins were found to have round profiles about 250 Å in diameter.

Lipoproteins

The lipoproteins were classified by Cook (1961) into low density factor (LDF) component and high density factor (HDF) components. The LDF protein is vitellenin. Cook and Martin (1962) classified lipoproteins as low protein lipids all containing less than 33% protein and the high protein lipids those containing over 33% protein. The low protein lipids had neutral lipid phospholipid ranges from 1:1-1:10. The high protein lipids have a constant 1:1 ratio of neutral lipid to phospholipid. Parkinson (1966) reported that lipovitellenin contains 36-41% lipid, 10% nitrogen, 1.6% phosphorus and 0.6% sulfur. The protein, vitellenin, contains 15.5% nitrogen, 0.3% phosphorus and 0.9% sulfur and has a molecular weight of about 93,000. The molecular weight of the lipoprotein is 170,000 according to Augustyniak *et al.* (1964). They reported LDF contains about 90% lipid with rather exact proportions of phospholipid and neutral lipid to maintain the LDF proteins in an orientation that confers water solubility on the LDF fraction. They were able to fractionate LDF into two components differing in particle size. Sugano and Watanabe (1961) also reported two fractions of LDF. They reported the percentages of the various proteins by ultracentrifugal estimation as LDF_1, 24%; LDF_2, 48%; HDF_1, 8%; HDF_2, 8%; livitens, 7.5%; and phosvitins, 4.5%.

Fevold and Lausten (1946) compared high and low density lipoproteins and found them to differ in ratio of combined protein and lipid, in solubility behavior, in general stability and in composition of the proteins. Both are phosphoproteins, but they differ significantly in phosphorus content. The LDF lipovitellenin makes up about 40% of the total lipoprotein of the egg yolk and 12-13% of total egg solids.

The two fractions of low density fraction were labeled LDF_1 and LDF_2 by Augustyniak *et al.* (1964). Martin *et al.* (1964) reported differences between the two fractions to be in lipid content and protein particle size. LDF_1 has 86.8% lipid and a protein molecular weight of 1,400,000. The LDF_2 component has 83.2% lipid and a protein molecular weight of about 600,000. Saari *et al.* (1964) found that LDF_1 had 89% lipid and LDF_2 had 86% lipid. The LDF_2 was more heat sensitive than LDF_1.

As listed by Cook (1961), the lipovitellenin contains only about 40% lipid. Turner and Cook (1958) studied a lipovitellenin with about 55% lipid. Much of the difficulty in studying the LDF is that they lose their identity when centrifuged. Nowak *et al.* (1966) found dye-binding capacity of LDF varied with pH, declining as pH increased.

The amino acid composition of vitellenin was reported by Cook *et al.* (1962). These values are given in Table 4.6.

The high density fraction lipoproteins are α-lipovitellin and β-lipovitellin (Martin *et al.* 1963). According to Parkinson (1966) these lipoproteins contain 17% lipid, 13% nitrogen and 1.5% phosphorus. The protein, vitellin, contains 15.5% protein, 1–2% phosphorus and 1% sulfur. Alderton and Fevold (1945) found the lipid fraction of the lipovitellins to be largely lecithin. Martin *et al.* (1963) reported that the phospholipids in all lipoproteins are similar, being made up of 74% lecithin, 18% cephalins and 8% minor phospholipids.

The high density fraction HDF represents 31% of the nondialysable egg yolk solids according to Bernardi and Cook (1960A). It contains 6 proteins consisting of 15% phosvitin, 7% α-livetin, 17% β-livetin, 10% γ-livetin, 30% α-lipovitellin and 21% β-lipovitellin. Earlier Joubert and Cook (1958A) had listed only three proteins: lipovitellin phosvitin and livetin.

The two lipovitellins were characterized by Bernardi and Cook

TABLE 4.6

ANALYSIS OF SOME PROTEINS OF EGG YOLK

Component	Vitellenin[1]	α-Vitellin[1]	β-Vitellin[1]	Phosvitin[2]
Alanine	0.54	0.63	0.61	1.56
Arginine	0.36	0.47	0.52	5.00
Aspartic acid	0.80	0.60	0.67	4.51
½ Cystine	0.04	0.12	0.13	0.00
Glutamic acid	0.80	0.82	0.86	4.71
Glycine	0.36	0.38	0.38	0.88
Hestidine	0.11	0.19	0.15	4.00
Isolencine	0.47	0.45	0.49	0.53
Leusine	0.79	0.70	0.73	0.76
Lysine	0.60	0.50	0.51	5.98
Methionine	0.15	0.19	0.21	0.28
Phenylalanine	0.32	0.27	0.25	0.76
Proline	0.26	0.43	0.44	0.87
Serine	0.51	0.54	0.51	30.9
Threonine	0.40	0.35	0.36	1.36
Tryptophan	0.13	0.13	0.13	0.41
Tyrosine	0.26	0.23	0.25	0.39
Valine	0.49	0.56	0.61	0.90
Nitrogen %	14.5	15.7	15.8	
Phosphorus %	0.13	0.53	0.30	9.6[3]
Sulfur %	0.87	1.19	1.30	—
Carbohydrate	—	—	—	3.93

[1] Cook *et al.* (1962). Amino acids in millimoles per gram dry protein.
[2] Allerton and Perlmann (1965). Amino acids as grams per 100 gm protein.
[3] Joubert and Cook (1958B).

(1960B). Both have the same lipid and nitrogen content, amino acid composition and molecular weight of 400,000. They differ in protein phosphorus content, electrophoretic mobility, adsorption on hydroxyapatite, ultracentrifugal behavior in alkaline media and in solubility. The difference in phosphorus content was reported by Burley and Cook (1961) as 0.50% for α- and 0.27% for β-lipovitellin. Sugano (1958) reported the phosphorus contents of α- and β-lipovitellins to be 0.7 and 1.6, respectively, with the protein in α-vitellin at 0.6% and the protein phosphorus in β-vitellin being 0.3%. Both lipovitellins gave positive Molisch reactions indicating the presence of a carbohydrate. About 87% of the phosphorus in lipovitellins was in the phospholipid fraction (Sugano 1957).

Cook *et al.* (1962) indicated that the two vitellins did differ in their content of histidine. This difference is not indicated in amino acid composition data given in Table 4.6. The lipid fraction of the lipovitellins is made up of 61% phospholipids, 35% triglycerides and 4% cholesterol according to Martin *et al.* (1963). Vandegaer *et al.* (1956) reported a lipid level of 18.2% for lipovitellins and a molecular weight of 40,000 by light scattering method of 33,000 hydrodynamic measurements. The amino acid composition of the vitellins is listed in Table 4.6.

Freezing and thawing of egg yolks resulted in changes in the electrophoretic patterns of the yolk lipoproteins according to Powrie *et al.* (1963). For this reason, most work has been done with fresh egg yolks. Radomski and Cook (1964A) were able to get clear separation of lipovitellins from phosvitin by gradient elution on Dowex-1 columns. β-lipovitellin was eluted as a homogeneous mass, but α-lipovitellin was heterogeneous. Using TEAE-cellulose had some advantages over Dowex-1 for gradient elution columns according to Radomski and Cook (1964B). Neelin and Cook (1961) attempted to differentiate the three lipoproteins of egg yolk by identification of terminal amino acids but were not successful. The major N terminal amino acids were arginine and lysine. The C terminal residue of vitellenin is glutamic acid.

Evans *et al.* (1974) reported that the lipovitellins in egg yolk did not change during 6 months of egg storage. They found 23% α- and 77% β-lipovitellin in the yolk they were working with. Burley and Kushner (1963) tested the reaction of a bacterial phosphidase on lipoproteins. They found no effect on lipovitellins unless chloroform were present. Lipovitellenin was rapidly hydrolyzed by the enzyme whether chloroform were present or not. Francis (1952) reported an antigenic response for vitellin.

The dissociation of lipovitellins has received attention. Kratohvil *et al.* (1962) reported that both lipovitellins dissociate in acid as well as alkaline solutions. Fifty percent dissociation of α-lipovitellin occurred at pH 10.6 and of β-lipovitellin at pH 7.8. Both dissociated to the same extent at pH 2.0. Burley and Cook (1962A) reported 50% dissociation for α- at pH 10.5 and β-lipovitellin at pH 7.8. Both aggregated irreversibly at pH 11.0 and above. The sulfhydryl groups in lipovitellin are not involved in the dissociation of α- and β-lipovitellin according to Burley and Cook (1962A). Cook and Wallace (1965) found that with β-lipovitellin dissociation was detectable at pH 5.8 and increased as the pH rose. The molecular weight of the dissociated subgroup was 1/2 that of the associated form.

Lipid-free Proteins of Egg Yolk

These proteins include phosvitin and the three livetins, according to Cook (1961). The livetins may contain from traces to 0.5% lipids. According to Parkinson (1966) the phosvitin contains 12% nitrogen and 10% phosphorus and less than 0.19% sulfur, while the livetins contain 15.3% nitrogen, a trace of phosphorus and 1.8% sulfur.

Phosvitin was analyzed by Mecham and Olcott (1949). They reported 10% phosphorus, 11.9% nitrogen, 0.7% amino nitrogen and less than 0.1% sulfur. Their amino acid listing was incomplete, but percentage values reported were similar to the more complete listing of amino acids by Allerton and Perlmann (1965). The latter values are included in Table 4.6. The two unusual characteristics of phosvitin are the high phosphorus and high serine content.

As indicated in Table 4.5, phosvitin makes up 5–6% of the yolk proteins. The molecular weight was reported by Wallace *et al.* (1966) to be about 48,000. Joubert and Cook (1958B) listed the molecular weight as 30,000 for phosvitin. In low concentrations of magnesium sulfate phosvitin forms a complex with a particle weight of about 1,400,000 and with calcium salts a particle of about 750,000 molecular weight.

Phosvitin was resolved into two fractions reported to be identical by Mok *et al.* (1961). Connelly and Taborsky (1961) also reported two fractions. They reported differences in metal content, amino acid composition and chemical stability at alkaline pH.

Electrophoretic studies with phosvitin have revealed that the phosvitin in fish roe is similar to egg yolk phosvitin (Mano and Lipmann 1966). Williams and Sanger (1959) expressed some surprise that conventional pyrophosphate bonds were not found in phosvitin with its high phosphorus content.

Phosvitin in egg yolk binds the approximately 3 mg of iron to prevent its migration into the egg white according to Greengard *et al.*

(1964). Phosvitin promotes a rapid oxidation of ferrous iron to ferric which it binds strongly (Taborsky 1963). Rearrangement of phosvitin at alkaline pH only occurs in the presence of metallic ions. Shantz and Dawson (1974) used electrophoresis to study egg yolk and found that all phosvitin bands stained positive for iron.

The livetins as now known were originally identified by Kay and Marshall (1928) who described "livetin" and gave a partial amino acid analysis. They indicated that from 20–25% of yolk proteins were livetin. Three principal fractions of livetin were found by Shepard and Hottle (1949). The molecular weights of α- and β-livetins were reported by Martin *et al.* (1957) as being about 70,000 and 48,000, respectively. The molecular weight of γ-livetin is 150,000 according to Martin and Cook (1958).

The livetins were found to be similar to some serum proteins. Williams (1962A) suggests that egg yolk γ-livetin and serum γ-globulin are similar. Mok and Common (1964) using serum antigens identified α-livetin with antigen a, and γ-livetin with antigen f. Serum albumin and α-livetin are likely the same protein and β-livetin is very similar to vitellomucoid, a glycoprotein according to Williams (1962A). The composition of the livetins is given in Table 4.7.

Minor Components of Egg Yolk Which Contain Nitrogen

A flavoprotein with a molecular weight of 38,000 was isolated from egg yolk by Ostrawski *et al.* (1962). One molecule of the protein binds one riboflavin. The protein suppresses the fluorescence of riboflavin until completely saturated with the vitamin. They reported that the flavoprotein also contained firmly bound cobalamine, vitamin B-12, that could be split off only by autoclaving at high temperature. The vitamin B-12 binder may be a contaminant of the flavoprotein according to these authors.

Lineweaver *et al.* (1948) reported several enzymes in egg yolk. Tributyrinase was found at a level of 0.4 units per kilogram of dry

TABLE 4.7

COMPOSITION OF LIVETINS

	α-Livetin	β-Livetin	γ-Livetin
Nitrogen	14.3	14.3	15.6
Hexose	—	7.0	2.6
Hexosamine	—	—	1.8

Source: Powrie (1973).

yolk. The tributyrinase was associated with the livetin proteins. Other esterases were also present. An alkaline phosphotase, 0.05 units per kilogram of dry weight, was found in the yolk. Peptidases were also present. An amylase was found in the livetin fraction of the egg yolk.

Free amino acids were reported in fresh egg yolk by Lea and Rhodes (1953) and by Ducay *et al.* (1960). All of the amino acids reported in egg yolk proteins except cystine were found in the free form.

Functional Properties of Egg Yolk

The functional property most often associated with egg yolk is emulsifying capacity. According to Harkins and Zollman (1926) an interfacial tension below 10 dynes per centimeter is indicative of excellent emulsifiability; with 1 dyne per centimeter or less emulsification is spontaneous. Vincent *et al.* (1966) reported interfacial tensions in egg yolk of about 5 dynes per centimeter even when diluted so that yolk made up only 1% of the mixture. Livetin acts to lower the interfacial tension. Chang *et al.* (1972) studied mayonnaise by electron microscopy. They postulated that the speckled layers observed were made up of coalesced low density lipoproteins of egg yolk and microparticles from the yolk granules. The low density lipoproteins have an intrinsic property of forming aggregates according to Kumar and Mahadevan (1970).

The gelling properties of egg yolk were studied by Mahadevan *et al.* (1969) and Kumar and Mahadevan (1970). Cleavage or alterations of lipoprotein bonds resulted in the formation of gels. Three or four of the proteins, hexoses, hexosamines and sialic acid are involved in gel formation.

The role of egg yolk macromolecules in cake structure was investigated by Kamat *et al.* (1973). The integrity of the egg yolk molecules is an essential prerequisite toward retaining cake quality of Madeira cakes and fatless sponge cakes. The earlier report of Joslin and Proctor (1954) that loss of foam volume in whipping tests with reconstituted whole egg powders was traced to the breakdown of the natural fat emulsion of fresh eggs might explain these results of Kamat *et al.* (1973). In their Madeira cakes, the lipovitellenin is the major determinant of cake quality. In the fatless sponge cakes, the lipovitellenin assists aeration and foaming while the lipovitellins inhibited aeration but helps to retain air whipped in by other egg constituents.

The nutritive value of egg yolk is high. Egg yolk is 100% digested whether raw or cooked according to Mauron (1972). Egg yolk proteins are not as good a source of the sulfur amino acids as egg whites, but the two major edible parts of the egg combined yield an amino acid mixture still used as a reference for protein quality.

A nonfood use for egg yolk was suggested by Nichols *et al.* (1954) in that egg yolk lipoproteins could be used in an assay for heparin.

BIBLIOGRAPHY

ABRAHAM, E. P. 1939. Some properties of egg white lysozyme. Biochem. J. *33*, 622-630.

AISEN, P., LANG, G., and WOODWORTH, R. C. 1973. Spectroscopic evidence for a difference between the iron-binding sites of conalbumin. J. Biol. Chem. *248*, 649-653.

ALDERTON, G., and FEVOLD, H. L. 1945. Preparation of the egg yolk lipoprotein, lipovitellin. Arch. Biochem. *8*, 415-419.

ALDERTON, G., WARD, W. H., and FEVOLD, H. L. 1945. Isolation of lysozyme from egg white. J. Biol. Chem. *157*, 43-58.

ALDERTON, G., WARD, W. H., and FEVOLD, H. L. 1946. Identification of the bacteria-inhibiting, iron-binding protein of egg white as conalbumin. Arch. Biochem. *11*, 9-13.

ALLERTON, S. E., and PERLMANN, G. E. 1965. Chemical characterization of the phosphoprotein phosvitin. J. Biol. Chem. *240*, 3892-3898.

AUGUSTYNIAK, J., MARTIN, W. G., and COOK, W. H. 1964. Characterization of lipovitellenin components and their relation to low density lipoprotein structure. Biochem. Biophys. Acta. *84*, 721-728.

AZARI, P., and BAUGH, R. F. 1967. A simple and rapid procedure for preparation of large quantities of pure ovotransferrin. Arch. Biochem. Biophys. *118*, 138-144.

AZARI, P. R., and FEENEY, R. E. 1958. Resistance of metal complexes of conalbumin and transferrin to proteolysis and thermal denaturation. J. Biol. Chem. *232*, 293-302.

AZARI, P. R., and FEENEY, R. E. 1961. The resistance of conalbumin and its iron complex to physical and chemical treatments. Arch. Biochem. Biophys. *92*, 44-52.

BAKER, C. M. A. 1960. The genetic basis of egg quality. Brit. Poultry Sci. *1*, 3-16.

BAKER, C. M. A. 1967. Molecular genetics of avian proteins. VII. Chemical and genetic polymorphism of conalbumin and transferrin in a number of species. Comp. Biochem. Physiol. *20*, 949-973.

BAKER, C. M. A. 1968A. Molecular genetics of avian proteins. IX. Interspecific and intraspecific variation of egg white proteins of the genus *Gallus*. Genetics *58*, 211-226.

BAKER, C. M. A. 1968B. The proteins of egg white. *In* Egg Quality, A Study of the Hen's Egg, T. C. Carter (Editor). Oliver and Boyd, Edinburgh, Scotland.

BAKER, C. M. A., and MANWELL, C. 1962. Molecular genetics of avian proteins. I. The egg white proteins of the domestic fowl. Brit. Poultry Sci. *3*, 161-174.

BAKER, R. C., HARTSELL, S. E., and STADELMAN, W. J. 1959. Lysozyme studies on chicken egg chalazae. Food Res. *24*, 529-538.

BAKER, R. C., and STADELMAN, W. J. 1958. Chicken egg chalazae. Strain and individual hen variations and their relation to internal quality. Poultry Sci. *37*, 558-564.

BALDWIN, R. E. 1973. Functional properties in foods. *In* Egg Science and Technology, W. J. Stadelman, and O. J. Cotterill (Editors). Avi Publishing Co., Westport, Conn.

BALDWIN, R. E., MATTER, J. C., UPCHURCH, R., and BREIDENSTEIN, D. M. 1967. Effects of microwaves on egg white. I. Characteristics of coagulation. J. Food Sci. *32*, 305-309.

BALL, H. R., and COTTERILL, O. J., 1970. Catalase activity of egg white. Poultry Sci. *49*, 1366.

BARMORE, M. A. 1934. The influence of chemical and physical factors on egg white foam. Colo. Agri. Expt. Sta. Tech. Bull. *9*.

BEELEY, J. G., and JEVONS, F. R. 1961. Heterogeneity of ovomucoid. Biochem. Biophys. Acta. *101*, 133-135.

BELLAIRS, R. 1961. The structure of the yolk of the hen's egg as studied by electron microscopy. I. The yolk of the unincubated egg. J. Biophys. Biochem. Cytology *11*, 207-225.

BERNARDI, G., and COOK, W. H. 1960A. An electrophoretic and ultracentrifugal study on the proteins of the high density fraction of egg yolk. Biochem. Biophys. Acta. *44*, 86-96.

BERNARDI, G., and COOK, W. H. 1960B. Separation and characterization of the two high density lipoproteins of egg yolk, α- and β-lipovitellin. Biochem. Biophys. Acta. *44*, 96-105.

BEYCHOK, S., and WARNER, R. C. 1959. Denaturation and electrophoretic behavior of lysozyme. J. Am. Chem. Soc. *81*, 1892-1897.

BIER, M. *et al.* 1953. Investigations on proteins and polymers. X. Composition and fractionation of ovomucoid. Arch. Biochem. Biophys. *47*, 465-473.

BLAKE, C. C. F. *et al.* 1967. On the conformation of the hen egg white lysozyme molecule. Proc. Roy. Soc. London, Ser. B *167*, 365-377.

BROOKS, J., and HALE, H. P. 1959. The mechanical properties of the thick white of the hen's egg. Biochem. Biophys. Acta. *32*, 237-250.

BROOKS, J., and HALE, H. P. 1961. The mechanical properties of the thick white of the hen's egg. II. The relation between rigidity and composition. Biochem. Biophys. Acta. *46*, 289-301.

BURLEY, R. W., and COOK, W. H. 1961. Isolation and composition of avian egg yolk granules and their constituent α- and β-lipovitellins. Can. J. Biochem. Physiol. *39*, 1295-1307.

BURLEY, R. W., and COOK, W. H. 1962A. The dissociation of α- and β-lipovitellin in aqueous solutions. Part I. Effect of pH, temperature, and other factors. Can. J. Biochem. Physiol. *40*, 363-372.

BURLEY, R. W., and COOK, W. H. 1962B. The dissociation of α- and β-lipovitellin in aqueous solution. Part II. Influence of protein phosphate groups, sulfhydryl groups, and related factors. Can. J. Biochem. Physiol. *40*, 373-379.

BURLEY, R. W., and KUSHNER, D. J. 1963. The action of *Clostridium perfringens* phosphotidase on the lipovitellins and other egg yolk constituents. Can. J. Biochem. Physiol. *41*, 409-416.

BUTTS, J. N., and CUNNINGHAM, F. E. 1972. Effect of dietary protein on selected properties of the egg. Poultry Sci. *51*, 1726-1734.

CANFIELD, R. E. 1963A. Peptides derived from tryptic digestion of egg white lysozyme. J. Biol. Chem. *238*, 2691-2697.

CANFIELD, R. E. 1963B. The amino acid sequence of egg white lysozyme. J. Biol. Chem. *238*, 2698-2707.

CANFIELD, R. E., and ANFINSEN, C. B. 1963A. Chromatography of pepsin and chymotrypsin digests of egg white lysozyme on phosphocellulose. J. Biol. Chem. *238*, 2684-2690.

CANFIELD, R. E., and ANFINSEN, C. B. 1963B. Nonuniform labeling of egg white lysozyme. Biochemistry *2*, 1073-1078.

CANFIELD, R. E., and LIN, A. K. 1965. The disulfide bonds of egg white lysozyme (Muramidase). J. Biol. Chem. *240*, 1997-2002.

CANN, J. R. 1949. Electrophoretic analysis of ovalbumin. J. Am. Chem. Soc. *71*, 907-909.

CHANG, P. K. 1973. Effect of polyphosphates on the functional properties of spray dried egg albumin. J. Food Sci. *38*, 239-241.

CHANG, P., POWRIE, W. D., and FENNEMA, O. 1970. Disc gel electrophoresis of proteins in native and heat-treated albumin, yolk and ultracentrifuged whole egg. J. Food Sci. *35*, 774-778.

CHANG, C. M., POWRIE, W. D., and FENNEMA, O. 1972. Electron microscopy of mayonnaise. Can. Inst. Food Sci. Technol. J. *5*, 134-137.

CHARGAFF, E. 1942. A study of lipoproteins. J. Biol. Chem. *142*, 491-512.

CHUNG, R. A., and STADELMAN, W. J. 1965. A study of variations in the structure of the hen's egg. Brit. Poultry Sci. *6*, 277-282.

CLARK, J. R., OSUGA, D. T., and FEENEY, R. E. 1963. Comparison of avian egg white conalbumins. J. Biol. Chem. *238*, 3621-3631.

COCHRANE, D., and ANNAU, E. 1962. Electrophoretic differences in the egg white proteins of the hen. Can. J. Biochem. Physiol. *40*, 1335-1341.

CONNELLY, C., and TABORSKY, G. 1961. Chromatographic fractionation of phosvitin. J. Biol. Chem. *236*, 1364-1368.

COOK, W. H. 1961. Proteins of the hen's egg yolk. Nature *190*, 1173-1176.

COOK, W. H. 1968. Macromolecular components of egg yolk. *In* Egg Quality, A Study of the Hen's Egg, T. C. Carter (Editor). Oliver and Boyd, Edinburgh, Scotland.

COOK, W. H., BURLEY, R. W., MARTIN, W. G., and HOPKINS, J. W. 1962. Amino acid compositions of the egg yolk lipoproteins and a statistical comparison of their amino acid ratios. Biochem. Biophys. Acta. *60*, 98-103.

COOK, W. H., and MARTIN, W. G. 1962. Composition and properties of soluble lipoproteins in relation to structure. Can. J. Biochem. Physiol. *40*, 1273-1285.

COOK, W. H., and WALLACE, R. A. 1965. Dissociation and subunit size of β-lipovitellin. Can. J. Biochem. *43*, 661-670.

COTTERILL, O. J., and WINTER, A. R. 1954. Egg white lysozyme. I. Relative lysozyme activity in fresh eggs having low and high interior quality. Poultry Sci. *33*, 607-670.

COTTERILL, O. J., and WINTER, A. R. 1955. Egg white lysozyme. 3. The effect of pH on the lysozyme-ovomucin interaction. Poultry Sci. *34*, 679-686.

CUNNINGHAM, F. E., and COTTERILL, O. J. 1962. Factors affecting the alkaline coagulation of egg white. Poultry Sci. *41*, 1453-1461.

CUNNINGHAM, F. E., COTTERILL, O. J., and FUNK, E. M. 1960. The effect of season and age of bird. 2. On the chemical composition of egg white. Poultry Sci. *39*, 300-308.

CUNNINGHAM, F. E., and LINEWEAVER, H. 1965. Stabilization of egg white proteins to pasteurizing temperatures above 60°C. Food Technol. *19*, 1442-1447.

CUNNINGHAM, F. E., and LINEWEAVER, H. 1967. Inactivation of lysozyme by native ovalbumin. Poultry Sci. *46*, 1471-1477.

DAM, R. 1971. In vitro studies on the lysozyme-ovomucin complex. Poultry Sci. *50*, 1824-1831.

DUCAY, E. D., KLINE, L., and MANDELES, S. 1960. Free amino acid content of infertile chicken eggs. Poultry Sci. *39*, 831-835.

EAKIN, R. E., SNELL, E. E., and WILLIAMS, R. J. 1940. A constituent of raw egg white capable of inactivating biotin in vitro. J. Biol. Chem. *136*, 801-802.

EAKIN, R. E., SNELL, E. E., and WILLIAMS, R. J. 1941. The concentration and assay of avidin, the injury-producing protein in raw egg white. J. Biol. Chem. *140*, 533-543.

EVANS, R. J., BAUER, D. H., and FLEGAL, C. J. 1974. Lipovitellins in fresh and stored shell eggs. Poultry Sci. *53*, 745-750.

FEENEY, R. E. *et al.* 1963A. A genetically varying minor protein constituent of chicken egg white. J. Biol. Chem. *238*, 1732-1736.

FEENEY, R. E. *et al.* 1960A. The comparative biochemistry of avian egg white proteins. J. Biol. Chem. *235*, 2307-2311.

FEENEY, R. E., DUCAY, E. D., SILVA, R. B., and MacDONNELL, L. R. 1952.

Chemistry of shell egg deterioration: The egg white proteins. Poultry Sci. *31*, 639-647.

FEENEY, R. E., RHODES, M. B., and ANDERSON, J. S. 1960B. The distribution and role of sialic acid in chicken egg white. J. Biol. Chem. *235*, 2633-2637.

FEENEY, R. E., SILVA, R. B., and MacDONNELL, L. R. 1951. Chemistry of shell egg deterioration: The deterioration of separated components. Poultry Sci. *30*, 645-650.

FEENEY, R. E., STEVENS, F. C., and OSUGA, D. T. 1963B. The specificities of chicken ovomucoid and ovoinhibitor. J. Biol. Chem. *238*, 1415-1418.

FERNANDEZ-DIEZ, M. J., OSUGA, D. T., and FEENEY, R. E. 1964. The sulfhydryls of avian ovalbumins, bovine lactoglobulin and bovine serum albumin. Arch. Biochem. Biophys. *107*, 449-458.

FEVOLD, H. L. 1951. Egg proteins. *In* Advances in Protein Chemistry, Vol. 6, M. L. Anson, and K. Bailey (Editors). Academic Press, New York.

FEVOLD, H. L., and LAUSTEN, A. 1946. Isolation of a new lipoprotein, lipovitellin, from egg yolk. Arch. Biochem. *11*, 1-7.

FORSYTHE, R. H., and BERGQUIST, D. H. 1951. The effect of physical treatments on some properties of egg white. Poultry Sci. *30*, 302-311.

FORSYTHE, R. H., and FOSTER, J. F. 1949. Note on the electrophoretic composition of egg white. Arch. Biochem. *20*, 161-163.

FORSYTHE, R. H., and FOSTER, J. F. 1950. Egg white proteins. II. An ethanol fractionation scheme. J. Biol. Chem. *184*, 385-392.

FRAENKEL-CONRAT, H. 1950. Comparison of the iron-binding activities of conalbumin and of hydroxylamidoproteins. Arch. Biochem. *28*, 452-463.

FRAENKEL-CONRAT, H., and FEENEY, R. E. 1950. The metal-binding activity of conalbumin. Arch. Biochem. *29*, 101-113.

FRAENKEL-CONRAT, H., and PORTER, R. R. 1952. The terminal amino groups of conalbumin, ovomucoid and avidin. Biochem. Biophys. Acta. *9*, 557-562.

FRAENKEL-CONRAT, H., SNELL, N. S., and DUCAY, E. D. 1952A. Avidin. I. Isolation and characterization of the protein and nucleic acid. Arch. Biochem. Biophys. *39*, 80-96.

FRAENKEL-CONRAT, H., SNELL, N. S. and DUCAY, E. D. 1952B. Avidin. II. Composition and mode of action of avidin A. Arch. Biochem. Biophys. *39*, 97-107.

FRANCIS, G. E. 1952. The immunological and serological properties of phosvitin. Biochem. J. *51*, 715-720.

FREDERICQ, E., and DEUTSCH, H. F. 1949. Studies on ovomucoid. J. Biol. Chem. *181*, 499-510.

GAFFIELD, W., VITELLO, L., and TOMIMATSU, Y. 1966. Optical rotary dispersion of egg proteins. II. Environment sensitive side chain chromophores in conalbumin. Biochem. Biophys. Res. Comm. *25*, 35-42.

GALYEAN, R. D., and COTTERILL, O. J. 1973. Chromatographic and electrophoretic studies of liquid and spray dried egg white. Poultry Sci. *52*, 2030.

GALYEAN, R. D., COTTERILL, O. J., and CUNNINGHAM, F. E. 1972. Yolk inhibition of lysozyme activity in egg white. Poultry Sci. *51*, 1346-1353.

GANDHI, S. K., KRISHNA, S., SCHULTZ, J. R., BOUGHEY, F. W., and FORSYTHE, R. L. 1968. Chemical modification of egg white with 3,3-dimethylglutanic anhydride. J. Food Sci. *33*, 163-169.

GARIBALDI, J. A., DONOVAN, J. W., DAVIS, J. G., and CIMINO, S. L. 1968. Heat denaturation of the ovomucin-lysozyme electrostatic complex—a source of damage to the whipping properties of pasteurized egg white. J. Food Sci. *33*, 514-524.

GREEN, N. M. 1963. Avidin. 3. The nature of the biotin-binding site. Biochem. J. *89*, 599-609.

GREENGARD, O., SENTENAC, A., and MENDELSOHN, N. 1964. Phosvitin, the iron carrier of egg yolk. Biochem. Biophys. Acta. *90*, 406-407.

GRUNDEN, L. P., VADEHRA, D. V., and BAKER, R. C. 1974. Effects of proteolytic enzymes on functionality of chicken egg albumin. J. Food Sci. *39*, 841-843.

HARKINS, W. D., and ZOLLMAN, H. 1926. Interfacial tension and emulsification. I. The effects of bases, salts, and acids upon interfacial tension between aqueous sodium oleate solutions and benzene. II. Extremely small interfacial tensions produced by solutes. J. Am. Chem. Soc. *48*, 69-80.

HAWTHORNE, J. R. 1950. The action of egg white lysozyme on ovomucoid and ovomucin. Biochem. Biophys. Acta. *6*, 28-35.

HEKTOEN, L., and COLE, A. G. 1928. The proteins of egg white. The proteins in egg white and their relationship to the blood proteins of the domestic fowl as determined by the precipitin reaction. J. Infectious Diseases *42*, 1-24.

HESTER, E. E., and PERSONIUS, C. J. 1949. Factors affecting the beading and leakage of soft meringues. Food Technol. *3*, 236-240.

JOLLÈS, P. 1967. Relationship between chemical structure and biological activity of hen egg-white lysozyme and lysozymes of different species. Proc. Roy. Soc. London, Ser. B *167*, 350-364.

JOLLÈS, J., JAUREGUI-ADELL, J., BERNIER, I., and JOLLÈS, P. 1963. La structure chimique lu lysozyme de blanc d'oeuf de poule: Etude detaillée. Biochem. Biophys. Acta. *78*, 668-689. (French)

JOSLIN, R. P., and PROCTOR, B. E. 1954. Some factors affecting the whipping characteristics of dried whole egg powder. Food Technol. *9*, 150-154.

JOUBERT, F. J., and COOK, W. H. 1958A. Separation and characterization of lipovitellin from hen egg yolk. Can. J. Biochem. Physiol. *36*, 389-398.

JOUBERT, F. J., and COOK, W. H. 1958B. Preparation and characterization of phosvitin from hen egg yolk. Can. J. Biochem. Physiol. *36*, 399-408.

JUTISZ, M., KAMINSKI, M., and LEGAULT-DÈMANE, J. 1957. Purification of ovomucoid by zone electrophoresis. Biochem. Biophys. Acta. *23*, 173-180. (French)

KAMAT, V. B., LAWRENCE, G. A., HART, C. J., and YOELL, R. 1973. Contribution of egg yolk lipoproteins to cake structure. J. Sci. Food Agri. *24*, 77-78.

KAMINSKI, M. 1954. Imunochemical studies and electrophoretics of globulins of hen's egg white. Characterization and attempted isolation. Biochem. Biophys. Acta. *13*, 216-223. (French)

KAY, H. D., and MARSHALL, P. G. 1928. The second protein (livetin) of egg yolks. Biochem. J. *22*, 1264-1269.

KLINE, L., SONODA, T. T., and HANSON, H. L. 1954. Comparisons of the quality and stability of whole egg powders desugared by the yeast and enzyme methods. Food Technol. *9*, 343-349.

KLOTZ, I. M. and WALKER, F. M. 1948. Complexes of lysozyme. Arch. Biochem. *18*, 319-325.

KETTERER, B. 1965. Ovoglycoprotein, a protein in hen's egg white. Biochem. J. *96*:372-376.

KRATOHVIL, J. P., MARTIN, W. G., and COOK, W. H. 1962. Molecular weight and dissociation behavior of α- and β-lipovitellin in acid solvents. Can. J. Biochem. Physiol. *40*, 877-883.

KUMAR, S. A., and MAHADEVAN, S. 1970. Physicochemical studies on the gelation of hen's egg yolk. Delipidation of yolk plasma by treatment with phospholipase-C and extraction with solvents. J. Agr. Food Chem. *18*, 666-670.

LANNI, F. *et al.* 1949. The egg white inhibitor of influenza virus hemagglutination. I. Preparation and properties of semipurified inhibitor. J. Biol. Chem. *179*, 1275-1287.

LEA, C. H., and RHODES, D. N. 1953. Phospholipins. Partition chromatography of egg yolk phospholipins on cellulose. Biochem. J. *54*, 467-469.

LEE, Y. C., and MONTGOMERY, R. 1961. The carbohydrate of ovalbumin. Arch. Biochem. Biophys. *95*, 263-270.

LEE, Y. C., and MONTGOMERY, R. 1962. Glycoproteins from ovalbumin: The structure of the peptide chain. Arch. Biochem. Biophys. *97*, 9-17.

LEPESCHKIN, W. W. 1922. The heat-coagulation of proteins. Biochem. J. *16*, 678-701.

LEWIS, J. C., SNELL, N. S., HIRSCHMANN, D. J., and FRAENKEL-CONRAT, H. 1950. Amino acid composition of egg proteins. J. Biol. Chem. *186*, 23-35.

LINDERSTROM-LANG, K., and OTTESEN, M. 1947. A new protein from ovalbumin. Nature *159*, 807-808.

LINEWEAVER, H., and CUNNINGHAM, F. E. 1966. Process for pasteurizing egg white. U.S. Pat. 3,251,697.

LINEWEAVER, H., CUNNINGHAM, F. E., GARIBALDI, J. A., and IJICHI, K. 1967. Heat stability of egg white proteins under minimal conditions that kill salmonellae. U.S. Dept. Agr.-ARS *74-39*.

LINEWEAVER, H. H., MORRIS, J., KLINE, L., and BEAN, R. S. 1948. Enzymes in fresh hen eggs. Arch. Biochem. *16*, 443-472.

LINEWEAVER, H., and MURRAY, C. W. 1947. Identification of the trypsin inhibitor of egg white with ovomucoid. J. Biol. Chem. *171*, 565-581.

LONGSWORTH, L. G., CANNAN, R. K., and MacINNES, D. A. 1940. An electrophoretic study of the proteins of egg white. J. Am. Chem. Soc. *62*, 2580-2590.

LUSH, I. E. 1961. Genetic polymorphism in the egg albumen proteins of the domestic fowl. Nature *189*, 981-984.

LUSH, I. E., and CONCHIE, J. 1966. Glycosidases in the egg albumen of the hen, the turkey and the Japanese quail. Biochem. Biophys. Acta. *130*, 81-86.

MacDONNELL, L. R., *et al.* 1955. The functional properties of egg white proteins. Food Technol. *9*, 49-53.

MacDONNELL, L. R., SILVA, R. B., and FEENEY, R. E. 1951. The sulfhydryl groups of ovalbumin. Arch. Biochem. Biophys. *32*, 288-299.

MAHADEVAN, S., SATYANARAYANA, T., and KUMAR, S. A. 1969. Physicochemical studies on the gelation of hen's egg yolk. Separation of gelling protein components from yolk plasma. J. Agr. Food Chem. *17*, 767-771.

MANDELES, S. 1960. Use of DEAE-cellulose in the separation of proteins from egg white and other biological materials. J. Chromatography *3*, 256-264.

MANO, Y., and LIPMANN, F. 1966. Characteristics of phosphoproteins (phosvitins) from a variety of fish roes. J. Biol. Chem. *241*, 3822-3833.

MARSHALL, M. E., and DEUTSCH, H. F. 1951. Distribution of egg white proteins in chicken blood serum and egg yolk. J. Biol. Chem. *189*, 1-9.

MARTIN, W. G., AUGUSTYNIAK, J., and COOK, W. H. 1964. Fractionation and characterization of the low density lipoproteins of hen's egg yolk. Biochem. Biophys. Acta. *84*, 714-720.

MARTIN, W. G., and COOK, W. H. 1958. Preparation and molecular weight of α-livetin from egg yolk. Can. J. Biochem. Physiol. *36*, 153-160.

MARTIN, W. G., TATTRIE, N. H., and COOK, W. H. 1963. Lipid extraction and distribution studies of egg yolk lipoproteins. Can. J. Biochem. Physiol. *41*, 657-666.

MARTIN, W. G., VANDEGAER, J. E., and COOK, W. H. 1957. Fractionation of livetin and the molecular weights of the α- and β-components. Can. J. Biochem. Physiol. *35*, 241-250.

MATSUSHIMA, K. 1958. An undescribed trypsin inhibitor in egg white. Science *137*, 1178-1179.

MAURON, J. 1972. Influence of industrial and household handling on food protein quality *In* Protein and Amino Acid Functions, E. J. Bigwood (Editor). Pergamon Press, New York.

MAY, K. N. 1959. The effect of certain factors on moisture, protein, threonine and other components of hens' eggs. Ph.D. Thesis, Purdue University, West Lafayette, Indiana.

MAY, K. N., and STADELMAN, W. J. 1960. Some factors affecting components of eggs from adult hens. Poultry Sci. *39*, 560-565.

McCORMICK, D. B. 1965. Specific purification of avidin by column chromatography on brotin cellulose. Anal. Biochem, *13*, 194-198.

McCREADY, S. T., and ROLAND, D. A. 1972. Electrophoretic characteristics of egg proteins from hens fed a calcium deficient diet. Poultry Sci. *51*, 1834.

McCULLY, K. A., MOK, C. C., and COMMON, R. H. 1962. Paper electrophoresis characterization of proteins and lipoproteins of hen's egg yolk. Can. J. Biochem. Physiol. *40*, 937-952.

McKENZIE, H. A., SMITH, M. B., and WAKE, R. G. 1963. The denaturation of proteins. I. Sedimentation, diffusion, optical rotation, viscosity and gelation in urea solutions of ovalbumin and bovine serum albumin. Biochem. Biophys. Acta. *69*, 222-239.

McNALLY, E. 1933. Relative amount of mucin the thick and in thin egg white. Proc. Soc. Exptl. Biol. Med. *30*, 1254-1255.

MECHAM, D. K., and OLCOTT, H. S. 1949. Phosvitin, the principal phosphoprotein of egg yolk. J. Am. Chem. Soc. *71*, 3670-3679.

MELAMED, M. D. 1966. Ovomucoid. *In* Glycoproteins, Their Composition, Structure and Function, A. Gottschalk (Editor). Elsevier Publishing Co., Amsterdam, the Netherlands.

MELAMED, M. D. 1967. Electrophoretic properties of ovomucoid. Biochem. J. *103*, 805-810.

MELAMED, M. D., and GREEN, N. M. 1963. Avidin. 2. Purification and composition. Biochem. J. *89*, 591-599.

MELNICK, D., and OSER, B. L. 1949. The influence of heat-processing on the functional and nutritive properties of proteins. Food Technol. *3*, 57-71.

MILLER, H. T., and FEENEY, R. E. 1966. The physical and chemical properties of an immunologically cross-reacting protein from avian egg whites. Biochemistry *5*, 952-958.

MOK, C. C., and COMMON, R. H., 1964. Studies on the livetins of hen's egg yolk. II. Immunoelectrophoretic identification of livetins with serum proteins. Can. J. Biochem. Physiol. *42*, 1119-1131.

MOK, C. C., MARTIN, W. G., and COMMON, R. H. 1961. A comparison of phosvitins prepared from hen's serum and from hen's egg yolk. Can. J. Biochem. Physiol. *39*, 109-117.

MONTGOMERY, R. 1970. Glycoproteins. *In* Advances in Carbohydrate Chemistry and Biochemistry M. L. Wolfrom, R. S. Tipson, and D. Horton (Editors). Academic Press, New York.

MONTGOMERY, R., LEE, Y. C., and WU, Y. C. 1965. Glycopeptides from ovalbumin. Preparation, properties, and partial hydrolysis of the asparginal carbohydrate. Biochemistry *4*, 566-577.

MONTGOMERY, R., and WU, Y. C. 1963. The carbohydrate-protein linkage in ovomucoid. Biochem. Biophys. Res. Commun. *11*, 249-254.

MULLER, H. G. 1961. Weissenberg effect in the thick white of the hen's eggs. Nature *189*, 213-214.

NEELIN, J. M., and COOK, W. H. 1961. Terminal amino acids of egg yolk lipoproteins. Can. J. Biochem. Physiol. *39*, 1075-1084.

NEUBERGER, A., and MARSHALL, R. D. 1966. Hen's egg albumin. *In* Glycoproteins, Their Composition, Structure and Function, A. Gottschalk (Editor). Elsevier Publishing Co., Amsterdam, The Netherlands.

NEUBERGER, A., and PAPKOFF, H. 1963. Carbohydrates in protein. 7. The nature of the carbohydrate in ovomucoid. Biochem. J. *87*, 581-585.

NICHOLS, A. V., RUBIN, L., and LINDGREN, R. T. 1954. Interaction of heparin active factor and egg yolk lipoprotein. Proc. Soc. Exptl. Biol. Med. *85*, 352-355.

NOWAK, C. M., POWRIE, W. D. and FENNEMA, O. 1966. Interaction of low density lipoprotein (LDL) from yolk plasma and methyl orange. J. Food Sci. *31*, 812-818.

OGDEN, A. L., MORTON, J. R., GILMOUR, D. G., and McDERMID, E. M. 1962. Inherited variants in the transferrin and conalbumins of the chicken. Nature *195*, 1026-1028.

ORR, M. L., and WATT, B. K. 1957. Amino acid content of foods. U.S. Dept. Agr.-ARS, Home Economics Res. Rept. *4*.

OSTRAWSKI, W., SKARZYNSKI, B., and ZAK, Z. 1962. Isolation and properties of flavoprotein from the egg yolk. Biochem. Biophys. Acta. *59*, 515-517.

PARKINSON, T. L. 1966. The chemical composition of eggs. J. Sci. Food Agr. *17*, 101-111.

PENNINGTON, D., SNELL, E. E., and EAKIN, R. E. 1942. Crystalline avidin. J. Am. Chem. Soc. *64*, 469.

POWRIE, W. D. 1973. Chemistry of eggs and egg products. *In* Egg Science and Technology, W. J. Stadelman, and O. J. Cotterill (Editors). Avi Publishing Co., Westport, Conn.

POWRIE, W. D., LITTLE, H., and LOPEZ, A. 1963. Gelation of egg yolk. J. Food Sci. *28*, 38-46.

RADOMSKI, M. W., and COOK, W. H. 1964A. Fractionation and dissociation of the avian lipovitellins and their interaction with phosvitin. Can. J. Biochem. *42*, 395-406.

RADOMSKI, M. W., and COOK, W. H. 1964B. Chromatographic separation of phosvitin, alpha and beta lipovitellin of egg yolk granules on TEAE cellulose. Can. J. Biochem. *42*, 1203-1215.

RHODES, M. B., AZARI, P. R., and FEENEY, R. E. 1958. Analysis, fractionation, and purification of egg white proteins with cellulose-cation exchanged. J. Biol. Chem. *230*, 399-408.

RHODES, M. B., BENNETT, N., and FEENEY, R. E. 1959. The flavoprotein-apoprotein system of egg whites. J. Biol. Chem. *234*, 2054-2060.

RHODES, M. B., BENNETT, N., and FEENEY, R. E. 1960. The trypsin and chymotrypsin inhibitors from avian egg whites. J. Biol. Chem. *235*, 1686-1693.

ROBINSON, D. S., and MONSEY, J. B. 1964. Reduction of ovomucin by mercaptoethanol. Biochem. Biophys. Acta. *83*, 368-370.

ROBINSON, D. S., and MONSEY, J. B. 1971. Studies on the composition of egg-white ovomucin. Biochem. J. *121*, 537-547.

ROMANOFF, A. L., and ROMANOFF, A. J. 1949. The Avian Egg. John Wiley & Sons, New York.

SAARI, A., POWRIE, W. D., and FENNEMA, O. 1964. Isolation and characterization of low density lipoproteins in native egg yolk plasma. J. Food Sci. *29*, 307-315.

SAUTER, E. A., HARNS, V., STADELMAN, W. J., and McLAREN, B. A. 1954. Effect of oil treating shell eggs on their functional properties after storage. Food Technol. *9*, 82-85.

SAUTER, E. A., and MONTOURE, J. E. 1972. The relationship of lysozyme content of egg white to volume and stability of foams. J. Food Sci. *37*, 918-920.

SAUTER, E. A., and PETERSON, C. F. 1969. Quality characteristics of eggs as influenced by lysozyme content. Poultry Sci. *48*, 1867.

SAUTER, E. A., and PETERSON, C. F. 1972. Quality characteristics of eggs from hens indexed for lysozyme content of egg white. Poultry Sci. *51*, 957-960.

SAUTER, E. A., PETERSON, C. F., and STEELE, E. E. 1970. The relationship of the lysozyme fraction in thick egg white to fertility and hatchability of eggs. Poultry Sci. *49*, 987-991.

SAUTER, E. A., PETERSON, C. F., WIESE, A. C. 1966. The effect of lysozyme content of thick egg white on the date of quality deterioration of shell eggs. Poultry Sci. *45*, 1121-1122.

SCHJEIDE, O. A., and URIST, M. R. 1960. Proteins induced in plasma by oestrogens. Nature *188*, 291-294.

SCHMIDT, C. L. A. 1938. The Chemistry of the Amino Acids and Proteins. Charles C. Thomas, Springfield, Ill.

SCHMIDT, G., BESSMAN, M. J., HICKEY, M. D., and THANNHAUSER, S. J. 1956. The concentrations of some soluble constituents of egg yolk in its soluble phase. J. Biol. Chem. *223*, 1027-1031.

SEIDEMAN, W. E., COTTERILL, O. J., and FUNK, E. M. 1963. Factors affecting heat coagulation of egg white. Poultry Sci. *42*, 406-417.

SHANTZ, R. C., and DAWSON, L. E. 1974. Electrophoretic examination of native and phosvitin fraction of avian egg yolk. Poultry Sci. *53*, 969-974.

SHENSTONE, F. S. 1968. The gross composition, chemistry and physicochemical basis of organization of the yolk and white. *In* Egg Quality, A Study of the Hen's Egg, T. C. Carter (Editor). Oliver and Boyd, Edinburgh, Scotland.

SHEPARD, C. C., and HOTTLE, G. A. 1949. Studies on the composition of the livetin fraction of the yolk of hen's eggs with the use of electrophoretic analysis. J. Biol. Chem. *179*, 349-357.

SLOSBERG, H. M., HANSON, H. L., STEWART, G. F., and LOWE, B. 1948. Factors influencing the effects of heat treatment on the leavening power of egg white. Poultry Sci. *27*, 294-301.

SMITH, M. B. 1964. Studies on ovalbumin. 1. Denaturation by heat, and the heterogeneity of ovalbumin. Australian J. Biol. Sci. *17*, 261-270.

SMITH, M. B., and BACK, J. F. 1965. Studies on ovalbumin. II. The formation and properties of δ-ovalbumin, a more stable form of ovalbumin. Australian J. Biol. Sci. *18*,, 365-377.

SMITH, A. H., WILSON, W. O., and BROWN, J. G. 1954. Composition of eggs from individual hens maintained under controlled environments. Poultry Sci. *33*, 898-908.

STEVENS, F. 1961. Starch gel electrophoresis of hen egg white, oviduct white, yolk, ova and serum proteins. Nature *192*, 972.

STEVENS, F. C., and FEENEY, R. E. 1963. Chemical modification of avian ovomucoids. Biochemistry *2*, 1346-1352.

ST. JOHN, J. L., and FLOR, I. H. 1931. A study of whipping and coagulation of eggs of varying quality. Poultry Sci. *10*, 71-82.

STRATIL, A. 1967. The effect of iron addition to avian egg white on the behavior of conalbumin fractions in starch gel electrophoresis. Comp. Biochem. Physiol. *22*, 227-233.

SUGANO, H. 1957. Studies on egg yolk proteins. II. Electrophoresis studies on phosvitin, lipovitellin and lipovitellenin. J. Biochem. *44*, 205-215.

SUGANO, H. 1958. Studies on egg yolk proteins. III. Preparation and properties of two major lipoproteins of the egg yolk, and lipovitellin J. Biochem. *45*, 393-401.

SUGANO, H., and WATANABE, I. 1961. Isolations and some properties of native lipoproteins from egg yolk. J. Biochem. *50*, 473-480.

TABORSKY, G. 1963. Interaction between phosvitin and iron and its effect on a rearrangement of phosvitin structure. Biochemistry *2*, 266-271.

TALLAN, H. H., and STEIN, W. H. 1951. Studies on lysozyme. J. Am. Chem. Soc. *73*, 2976-2977.

TALLAN, H. H., and STEIN, W. H. 1953. Chromatographic studies on lysozyme. J. Biol. Chem. *200*, 507-514.

TOMIMATSU, Y., CLARY, J. J., and BARTULOVICK, J. J. 1966. Physical characterization of ovoinhibitor, a trypsin and chymotrypsin inhibitor from chicken egg whites. Arch. Biochem. Biophys. *115*, 536-544.

TUNG, M. A., RICHARDS, J. F., and MORRISON, B. C. 1970. Rheology of fresh, aged and gamma irradiated egg white. J. Food Sci. *35*, 872-874.

TURNER, K. J., and COOK, W. H. 1958. Molecular weight and physical properties of a lipoprotein from the floating fraction of egg yolk. Can. J. Biochem. Physiol. *36*, 937-949.

VANDEGAER, J. E., REICHMANN, M. E., and COOK, W. H. 1956. Preparation and molecular weight of lipovitellin from egg yolk. Arch. Biochem. Biophys. *62*, 328-337.

VINCENT, R., POWRIE, W. D., and FENNEMA, O. 1966. Surface activity of yolk, plasma and dispersions of yolk fractions. J. Food Sci. *31*, 643-648.

WALLACE, R. A., JARED, D. W., and EISEN, A. Z. 1966. A general method for isolation and purification of phosvitin from vertebrate eggs. Can. J. Biochem. *44*, 1647-1655.

WARNER, R. C. 1954. Egg proteins. *In* The Proteins, 1st Edition, Vol. II, Part A, H. Neurath, and K. Bailey (Editors). Academic Press, New York.

WARNER, R. C., and WEBER, I. 1951. The preparation of crystalline conalbumin. J. Biol. Chem. *191*, 173-180.

WARNER, R. C., and WEBER, I. 1953. The metal combining properties of conalbumin. J. Am. Chem. Soc. *75*, 5094-5101.

WEINMAN, E. O. 1956. Lipid-protein complex in egg yolk. Fed. Proc. *15*, 381.

WENN, R. V., and WILLIAMS, J. 1968. The isoelectric fractionation of hen's egg ovotransferrin. Biochem. J. *108*, 69-74.

WILCOX, F. H., and COLE, R. K. 1954. Studies on lysozyme concentration in the egg white of the domestic fowl. Poultry Sci. *33*, 392-397.

WILLIAMS, J. 1962A. Serum proteins and the livetins of hen's egg yolks. Biochem. J. *83*, 346-355.

WILLIAMS, J. 1962B. A comparison of conalbumin and transferrin in the domestic fowl. Biochem. J. *83*, 355-364.

WILLIAMS, J., and SANGER, F. 1959. The grouping of serine phosphate residues in phosvitin and casein. Biochem. Biophys. Acta. *33*, 294-296.

WINDLE, J. J., WIERSEMA, A. K., CLARK, J. R., and FEENEY, R. E. 1963. Investigation of the iron and copper complexes of avian conalbumins and human transferrins by electron paramagnetic resonance. Biochemistry *2*, 1341-1345.

WINZOR, D. J., and CREETH, J. M. 1962. Physiochemical studies of ovalbumin. 3. The sulfhydryl and disulfide contents of ovalbumin and an iodine modified derivative. Biochem. J. *83*, 559-566.

WISHNIA, A., and WARNER, R. C. 1961. The kinetics of denaturation of conalbumin. J. Am. Chem. Soc. *83*, 2065-2071.

YOUNG, L. L., and GARDNER, F. A. 1972. Preparation of egg white ovomucin by gel filtration. J. Food Sci. *37*, 8-11.

CHAPTER 5

Herbert W. Ockerman

Meat Proteins

The casual chemistry student would define colloids as a jar full of colorful suspended particles that tend to settle very slowly. The more astute student would realize that in colloidal chemistry the particles and the suspending media could be in the solid, liquid or gaseous state. It would usually require either a colloidal chemist or a biochemist to rank muscle tissue very high on his list of colloidal chemical examples. The reason for this lack of association stems from the fact that in the early days of colloidal chemistry most of the information dealing with muscle was obtained by classical chemical and biochemical means and very few muscle characteristics were determined using colloidal techniques. The boundaries of a complex field such as colloidal chemistry can rarely be defined exactly, but even with a rough approximation it should become quite obvious that muscle fits into this branch of chemistry. To learn the muscles' secrets, more and more use is being made of colloidal techniques such as ultracentrifugation, X-ray analysis, streaming birefringence, diffusion, light scattering, viscosity measurements, electrophoresis and electron microscopy.

Colloids as defined by Jirgensons and Straumanis (1954) are substances consisting of a homogeneous medium and particles dispersed therein. Dean (1948) described a colloidal system as one in which an appreciable fraction of the molecules lies at the surface or interface between the two states. Jirgensons (1958) stated that the particles of a colloidal system are invisible with an ordinary microscope but can be seen with an electron microscope. He also noted that colloid particles will pass through the pores of ordinary filters but are retained by semi-permeable membranes.

Dawning of Colloidal Chemistry

The English scientist Thomas Graham in the 19th century observed wide differences in the spreading and filtering rate of various substances in pure liquids. Salt and sugar solutions were mobile, but glue spread very slowly. This led Graham to a classification of substance into slow spreading glue-like colloids ("colla" in Greek means glue) and fast spreading crystalloids.

Colloidal chemistry was established as a branch of chemistry in the early 1900's when the German chemist, Wolfgang Ostwald, established a Society of Colloid Chemists and a "Colloidal Chemical Magazine" was initiated.

Cell as a Colloidal System

Dean (1948) interpreted colloidal science as a link between chemistry and the biological sciences and stated that the animal kingdom is wholly colloidal, reporting that proteins represent the largest and most diverse group of biological colloids. Muscle tissue is defined by Jirgensons and Straumanis (1954) as a gel (two components, one solid and the other liquid) possessing some proteins which are colloids. Jirgensons (1958) later stated that in animal tissue the macromolecules of proteins are joined in complex higher order systems and that the muscles are composed of several kinds of large protein molecules. These protein molecules resemble strings or short rods that are connected end to end or overlapped in a staggered or pleated configuration. Briskey and Fukazawa (1971) described muscle as a delicately balanced protein suspension in a dilute salt solution.

Colloidal Classification

The properties and behavior of a colloidal system are largely dependent upon the size and shape of the dispersed material which may be small particles or large molecules.

Like most scientific boundaries the maximum and minimum size of a colloidal particle is open to debate and certainly colloidal properties do not start and end at an absolute dimension. Table 5.1 shows the approximate sizes that may be used as guideposts and also some of the muscle molecules that fall within these ranges.

Shape is also an important property of a colloidal particle. This is particularly true in biological colloidal systems where many long thin particles or films are encountered. Often both length and width are outside the classical colloidal size range, with the length being longer and the width being shorter than that stated in Table 5.1. For this reason muscle components are often classified according to their shape as follows: (1) globular (sphero-) protein: hemoglobin, G-actin, myogen, and (2) fibrous (linear) proteins: myosin, F-actin, tropomyosin, collagen, elastin, keratin.

Structural configuration often has a strong influence on functional properties of proteins. For example, the long thin particles or films have good physical properties (tensile strength and elasticity), resist flow and set easily.

Other properties besides shape, such as amino acid composition, solubility, and bonding, are also used to subdivide colloidal systems

TABLE 5.1

SIZE OF PARTICLES IN A DISPERSE SYSTEM (A SUBSTANCE DISPERSED IN A HOMOGENOUS MEDIUM)

Angstroms	mm	Average No. of Atoms	Avg. Mol Wt of Organic Atoms	Analysis Equipment Size Parameters	Description	Particles or Molecules
10^7 Å	10^0					
10^6 Å	10^{-1}					Sand
10^5 Å	10^{-2}					
10^4 Å	10^{-3}	10^9	10^{10}	Size of ordinary filter paper pores	Coarse colloids (show colloidal properties)	*Bacillus coli* large fat droplets, foams, emulsions
10^3 Å	10^{-4}			Lower limits of of light micro-scope	Colloids	myosin medium fat droplets, gelatin, myogen
10^2 Å	10^{-5}					small fat droplets, large protein molecules, hemoglobin, actin, common organic molecule
10^1 Å	10^{-6}	10^3	10^4	Lower limits of electron micro-scope	Micromolecules Small molecules Atoms	polypeptides, vitamins, sugars, amino acids, inorganic molecules, most atoms, oxygen
10^0 Å	10^{-7}					

and these are outlined in Fig. 5.1. In every case, some muscle cell components fit into both classifications; however, the particles in major abundance fit into the classification on the left-hand side of Fig. 5.1.

Cell Construction

Colloidal chemists are vitally interested in the construction of the colloidal system because it often influences the system's activity. This is of major importance in muscle tissue since a number of the physical and chemical properties of the cell are modified by the colloidal construction of this unit. Meyerhof (1944) stated that muscle contraction in itself is essentially a problem of colloid chemistry or at least has many aspects which are colloidal in nature. Muscle structure is altered, moderately to severely, in changing muscle into human or animal food and an understanding of the initial construction as well as these alterations are indispensable to the meat scientist.

MUSCLE COMPOSITION

The composition of muscle is influenced by age, sex, specie, genetic makeup and state of nutrition of the animal from which it was obtained. Within the animal, muscle work load and anatomical location of the muscle also influence the ratio of components. In the post-mortem tissue such things as continuing enzymatic reactions and moisture evaporation will alter the tissue's composition. An analysis of an average lean muscle tissue from a well-fed animal will yield the following approximate values.

19.5% Nitrogen fraction
- 18% Protein
- 1.5% Nonprotein nitrogen

3.0% Fat
75.0% Water (inversely related to fat content)
1.0% Glycogen
1.5% Ash

The protein fraction is composed of amino acids that have the following general structure:

Carboxyl group

R-CH-C-O-H (C=O)

Amino group: N with H H

$$R{-}CH(NH_2){-}C(=O){-}O{-}H$$

Amino acids as well as products constructed from them (polypeptides and proteins) are ampholytes in solution which means that the molecules can carry both a positive (amino group) and a negative charge at the same time. This is extremely important for 3 dimensional bonding and attracting or repelling other types of molecules. Amino acids commonly found in natural proteins are listed in Table 5.2.

All amino acids with the exception of glycine possess one or more

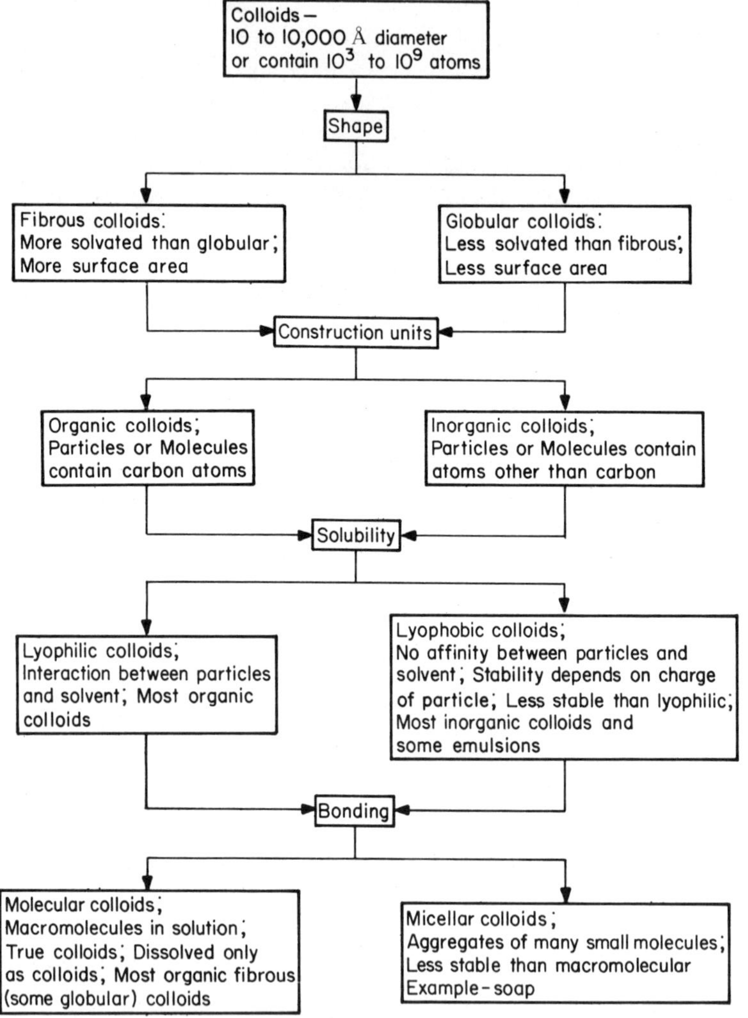

FIG. 5.1. CLASSIFICATION OF COLLOIDS INTO SUBCATEGORIES

TABLE 5.2

AMINO ACIDS COMMONLY FOUND IN NATURAL PROTEINS

Classification	Amino Acid	Required Amino Acid for Life and Growth in Man	Abbreviation
Aliphatic			
Monoaminomonocarboxylic	glycine		gly
	alanine		ala
	valine	*	val
	leucine	*	leu
	isoleucine	*	ile
	serine		ser
	threonine	*	thr
Sulfur containing	cysteine		CySH
	cystine		CyS-SCy
	methionine	*	met
Monoaminodicarboxylic	aspartic		asp
	glutamic		glu
Amides of monoaminodicarboxylic	asparagine		asn
	glutamine		gln
Basic	lysine	*	lys
	hydroxylysine		hyl
	arginine	* (growth only)	arg
	histidine	* (growth only)	his
Aromatic	phenylalanine	*	phe
	tyrosine		tyr
Heterocyclic	tryptophan	*	try
	proline		pro
	hydroxyproline		hyp

Source: Ockerman (1974).

asymmetric carbons which make them optically active. This gives the amino acids the ability to rotate polarized light which is useful in their study.

The nonprotein nitrogen fraction of muscle is composed of nitrogen compounds other than protein and includes: creatine, free amino acids and nucleotides.

Most of the fat (lipid) in the cells is in the triglyceride form which has the following general formula.

```
     H
     |
H-C-O-C-R'
     |    ||
     |    O
H-C-O-C-R''
     |    ||
     |    O
H-C-O-C-R'''
     |    ||
     |    O
     H
```

This structure is composed of three fatty acids connected by an ester linkage to a glycerol molecule. Generally, the fatty acids on the glycerol are different and the triglyceride is called a mixed triglyceride. The fatty acids commonly attached to the glyceride in animal fat are shown in Table 5.3.

Lipids vary quantitatively and qualitatively with the dietary level, maturity, species, muscle and fiber types.

The largest (75%), and often overlooked, component of the cell is water, the quantity of which is inversely related to the lipid content of the muscle. Water is a very polar compound with the following configuration.

$$+ \begin{matrix} H \diagdown \\ \\ H \diagup \end{matrix} O \; -$$

The water is found both in and between the cells of muscle being electrostatically and chemically bound as well as physically entrapped in the cell. The ratio of free (water removed by physical pressure) to bound (water remaining after meat is subjected to physical pressure) water is largely dependent on water-protein and protein-protein interrelationships. The polar structure of water has a hydrophilic relationship with protein based on molecular attraction by electrical charges of both signs and by polar groups. The protein-protein interrelationship determines the size of the spaces that can contain water in the protein network. The proteins of the myofi-

TABLE 5.3

STRAIGHT CHAIN FATTY ACIDS

Common Name	Formula	Systematic Name
Saturated fatty acids		
Butyric	C_3H_7COOH	butanoic
Caproic	$C_5H_{11}COOH$	hexanoic
Caprylic	$C_7H_{15}COOH$	octanoic
Capric	$C_9H_{19}COOH$	decanoic
Lauric	$C_{11}H_{23}COOH$	dodecanoic
Myristic	$C_{13}H_{27}COOH$	tetradecanoic
Palmitic	$C_{15}H_{31}COOH$	hexadecanoic
Stearic	$C_{17}H_{35}COOH$	octadecanoic
Arachidic	$C_{19}H_{39}COOH$	eicosanoic
Unsaturated fatty acids		
Palmitoleic	$C_{15}H_{29}COOH$	9-hexadecenoic
Oleic	$C_{17}H_{33}COOH$	9-octadecenoic
Linoleic	$C_{17}H_{31}COOH$	9,12-octadecadienoic
Linolenic	$C_{17}H_{29}COOH$	9,12,15-octadecatrienoic
Arachidonic	$C_{19}H_{31}COOH$	5,8,11,14-eicosatetraenoic

Source: Ockerman (1974).

brils (contractile thread-like elements) play a major role in water retention within the muscle (Briskey and Fukazawa 1971).

The glycogen component of the muscle is composed of polymers of D-glucopyranose connected in a bush-like branching system. The linkage (Merkel 1971) is α-1, 4 and α-1, 6 and can be represented as follows.

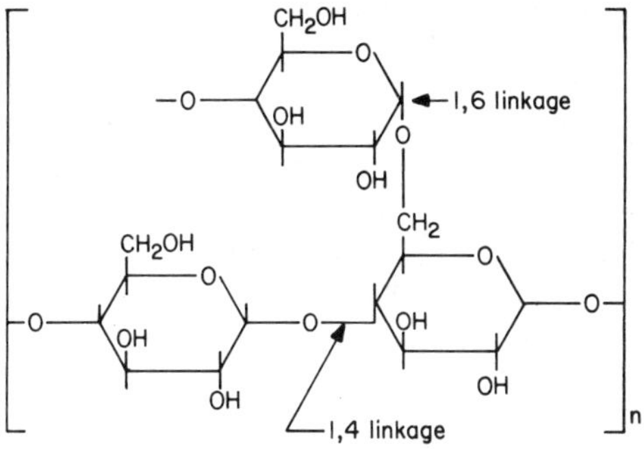

The carbohydrate content of muscle is often increased during the production of sausage through the inclusion of such items as sugar, corn syrup products, starches and milk products.

The inorganic component of muscle tissue is analyzed by incineration and referred to as ash. These inorganic elements exist in combination with each other and often in combination with the organic molecules. The approximate percentages found in lean meat are shown in Table 5.4. Since salt is added (2–5%) during the curing and sausage manufacturing process the quantity of sodium and chloride found in cured tissue is much higher than the naturally occurring quantities.

The main thrust of this chapter will concentrate upon the colloidal

TABLE 5.4

INORGANIC ELEMENTS FOUND IN LEAN TISSUE

Element	% in Lean Meat
Sulfur	0.187–0.230
Potassium	0.184–0.415
Phosphorus	0.131–0.343
Sodium	0.066–0.168[1]
Chlorine	0.040–0.076[1]
Magnesium	0.018–0.033
Calcium	0.008–0.020
Iron	0.001–0.005
Zinc	0.002–0.006

[1] Large increase when tissue is salt cured.

aspects of meat proteins. This should not suggest that the other components are not important from a physiological or, for that matter, from a colloidal standpoint. However, due to space limitations, boundaries have to be drawn at some point and some of the other components will be described in other chapters.

MUSCLE PROTEIN MICRO STRUCTURE

Amino Acid Linkage

Micromolecular substances such as amino acids (Table 5.2) and simple polypeptides are too small to be considered colloids, but since they are the building blocks of proteins it is important to understand their linkages. The amino acids are linked through peptide bonds to form the protein chains. A short general protein structure (tripeptide, constructed from three amino acids) would have the following configuration.

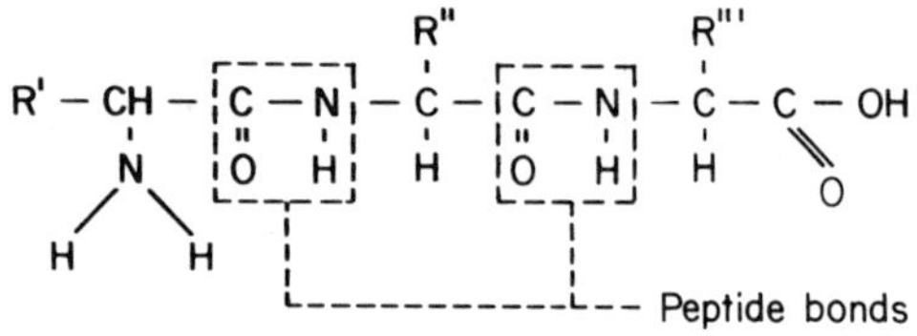

```
                          R"               R'''
            ┌─────┐       |     ┌─────┐    |
R' — CH — │C — N│ — C — │C — N│ — C — C — OH
       |    │‖    |│   |    │‖    |│   |    ‖
       N    │O    H│   H    │O    H│   H    O
      / \   └──┬──┘         └──┬──┘
     H   H     |               |
               └───────────────┴── Peptide bonds
```

Most native protein molecules occur in the form of a helix (with one or more strands) but may occur in some circumstances as a random coil, or as a beta or pleated sheet.

Many proteins have been separated from muscle by various extraction procedures. The most important contractile proteins as currently defined are those listed in Table 5.5. References are provided to assist the reader interested in details of their discovery. Figures 5.2, 5.3, and 5.4 diagram the gross and molecular structure of a muscle, highly magnify the muscle cells, and diagrammatically illustrate the muscle's sub-cellular components.

The construction of the major proteins found in muscle and their combinations will be described in subsequent sections. The average percentage of these proteins found in skeletal muscle tissue can be found in Table 5.6.

Myosin

Myosin is a myofibril protein that can be extracted from muscle with a strong salt solution such as 0.5 *M* KCl. Under the electron microscope myosin resembles a rod from 1400–2300 Å long, has a variable diameter which averages 20 Å and a thickened end with a

TABLE 5.5

PROTEINS OF MYOFIBRILS

Name of Protein	Description	References
Actomyosin	Combination of actin and myosin	
Myosin	Major structure of thick filaments	Szent-Györgyi (1941–1943)
Actin	Major structure of thin filaments	Kuhne (1863) Halliburton (1887), Straub (1942), Finck (1968)
Native tropomyosin	Tropomyosin and troponin	Ebashi and Ebashi (1964), Endo *et al.* (1966)
Tropomyosin	In thin filament and Z line	Baily (1946, 1948)
Troponin	Calcium receptive part of thin filament	Ebashi and Kodama (1965, 1966 A, 1966B) Ebashi *et al.*, 1968
α-actinin	Resembles actin in amino acid composition; exists in Z-band and M line	Ebashi and Ebashi (1965), Nonomura (1967)
M-substance	Part of α-actinin and possibly related to 10-S[1] α-actinin	Masaki *et al.* (1968)
6 S[1] -component	Part of α-actinin	Masaki *et al.* (1967)
β-actinin	Regulates length of F-actin	Maruyama (1965 A, 1965B)

[1] Sedimentation coefficient.

TABLE 5.6

PERCENTAGE OF MYOFIBRIL PROTEINS

Protein	% of Total Myofibrillar Protein
Actomyosin	
Myosin	50–54
Actin	21–27
Tropomyosin	11–15
Troponin	0.8–1.2
α-actinin	4.4–6.6
β-actinin	2.2–3.3

diameter from 30–60 Å (Zobel and Carlson 1963; Huxley 1963B; Rice *et al.* 1966). See Fig. 5.2 and 5.5. Its molecular weight has been calculated at 470,000–850,000 (Hamm 1970; Briskey and Fukazawa 1971) and this extracted molecule has high viscosities, a sedimentation constant of 6.4 and double refractions to flow. The myosin molecule is amphoteric (capable of acting either as an acid or as a base) and is stiffened by the relatively high charge density on its surface. Its shape, however, may be slightly altered in living muscle tissue.

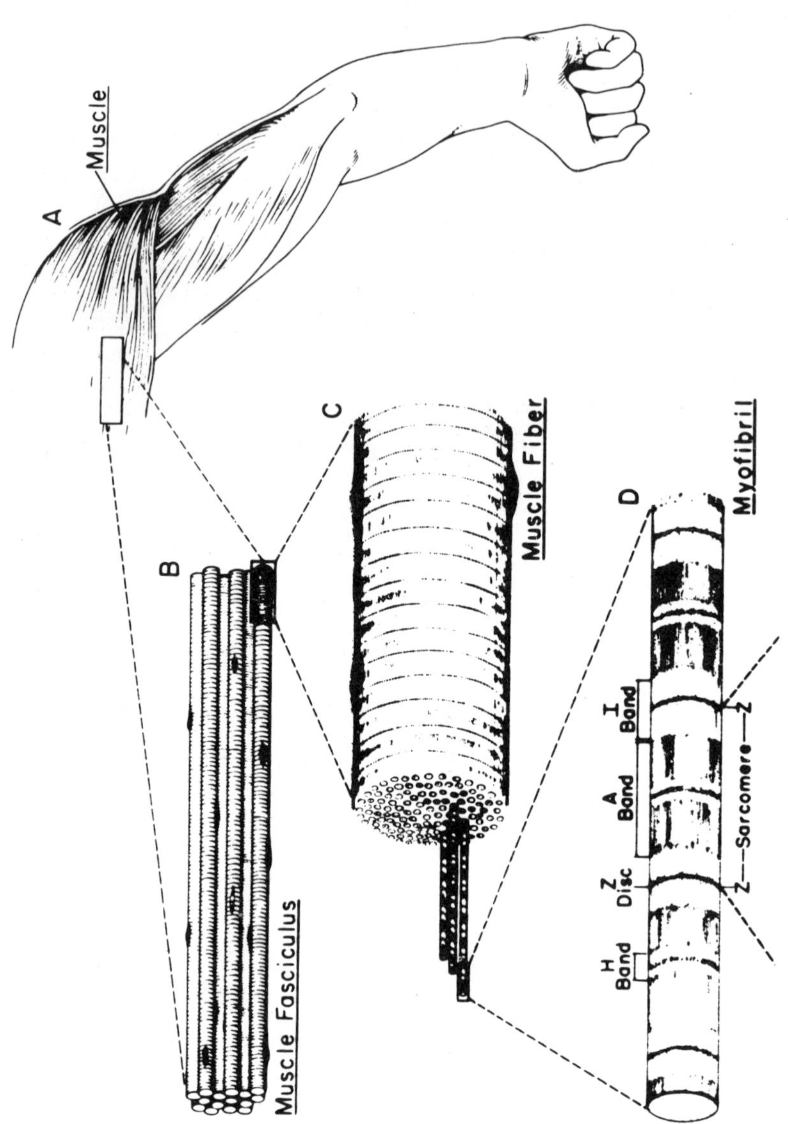
A
Muscle
B
Muscle Fasciculus
C
Muscle Fiber
D
Myofibril
H Band
Z Disc
A Band
I Band
Z
Sarcomere
Z

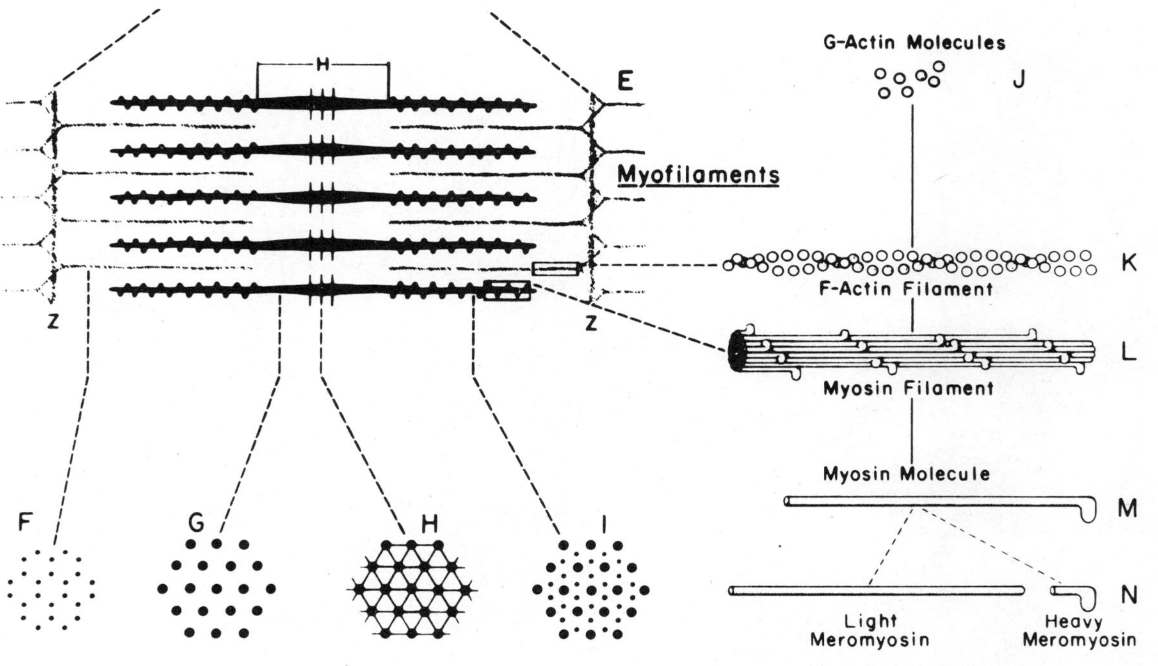

From Bloom and Fawcett (1968)

FIG. 5.2. DIAGRAM OF THE ORGANIZATION OF SKELETAL MUSCLE FROM THE GROSS TO THE MOLECULAR LEVEL. F, G, H, and I ARE CROSS SECTIONS AT THE LEVEL INDICATED

Drawings by S. C. Keene, W. B. Saunders Co.

Courtesy of H. E. Huxley, Cambridge University

FIG. 5.3. PHOTOMICROGRAPH OF A PORTION OF EIGHT MUSCLE MYOFIBRILS. DIRECTION OF FIBER IS TOP LEFT TO BOTTOM RIGHT

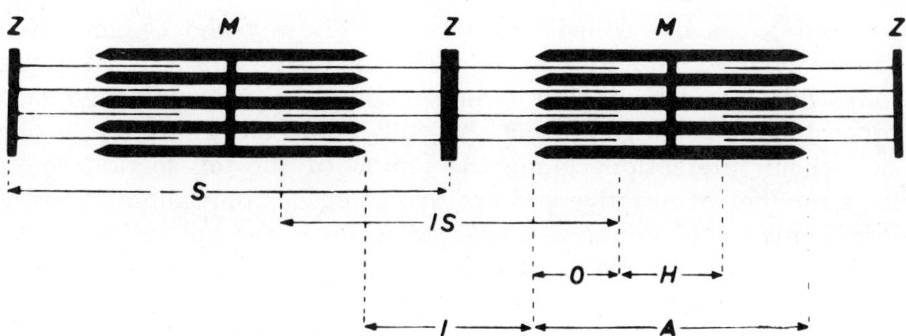

FIG. 5.4. DIAGRAMMATIC REPRESENTATION OF THE ARRANGEMENT OF FILAMENTS IN A VERTEBRAE MUSCLE

S—Sarcomere (from Z-line to Z-line). O—Overlap zone. IS—length of thin filaments. (F—Actin molecule). A—length of thick filament (myosin molecule). I—Distance between thick filaments (contains only thin filaments but is not their total length).
Reprinted from Contractile Proteins and Muscle by Laki, p. 7, courtesy of Mercel Dekker, Inc.

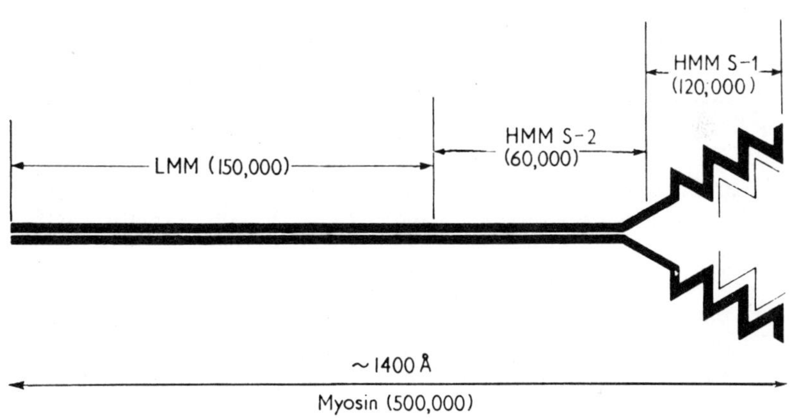

From Lowey et al. (1969)

FIG. 5.5. SCHEMATIC REPRESENTATION OF MYOSIN MOLECULE

Reprinted with permission of Academic Press.

Myosin is constructed of amino acids joined by peptide bonds into polypeptide chains. The molecule has a head, neck and tail section and the head section is the thickened end which can be observed in the electron microscope. The head contains little or no helical conformation and a secondary peptide in addition to the primary peptide; it is rather globular in structure.

The tail section (1100-1300 × 20 Å) is composed of two polypeptide helical coils that are, in turn, coiled around each other and referred to as the coiled coil structure (Lowey and Cohen 1962; Lowey *et al.* 1966; Rice *et al.* 1966; Bendall 1969). Its structure looks like two screen door springs moderately twisted around each other but not so tightly that the coils interlock. There are strong side chain interactions along the length of the tail section and a large number of negative and positive charges. Approximately every third amino acid contains a charged group in the tail region of the molecule (Bendall 1969). The amino acid proline, when used in polypeptide construction, causes a kink in the chain; however, this amino acid is seldom encountered in the tail section of the molecule (Bendall 1969). All of these factors led to a highly helical structure and a fairly stiff (foundation for contraction) section for this region of the myosin molecule.

Between the head and tail section is the neck area which some investigators believe is composed of two single, nonhelical polypeptide chains. Regardless of its structure there is fairly general agreement that the neck area has great flexibility which allows for movement of the head during contraction.

The myosin molecule is double headed (Slayter and Lowey 1967) with each head having a molecular weight of 120,000–180,000 (Mueller and Perry 1962) and a dimension of 200 × 40 Å (Bendall 1969). The head region of this protein has a high proline content, has little or no helical shape and is often described as globular in structure. There are two current theories (Perry 1967) dealing with the construction of the head.

(1) The peptide chain of the tail when it reaches the head folds back and forth in a randomized arrangement creating the globular head.

(2) Independent shorter chains (some researchers believe two in number) of a globular nature are attached to the main chain at this point.

The head portion of the myosin molecule contains the ATPase enzyme activity and actin combining sites and strongly binds both magnesium and calcium ions. This is the portion of the molecule that forms the bridges between the myosin and actin filaments (Huxley 1957). The enzymatic activity of myosin is activated by calcium ions and inhibited by magnesium ions.

A great deal has been learned about the myosin molecule because it can be chopped (Fig. 5.5) at two locations by specific enzymes. Trypsin, chymotrypsin or subtilisin will sever approximately 2/3 of the tail section from the rest of the molecule and this roughly splits the molecule in 1/2 (Gergely *et al.* 1955; Mihalyi and Harrington 1959; Middlebrook 1959). This section of the tail portion is called light meromyosin (LMM), has a molecular weight from 96,000–150,000 and a dimension of 600–900 Å × 20 Å (Huxley 1963A; Rice *et al.* 1966; Hamm 1970). Like the total tail previously described, the LMM portion is a highly coiled coil helix low in phenylalanine and is soluble at high ionic strength (Lowey and Cohen 1962; Lowey *et al.* 1966). The remainder of the molecule following the severing procedure is called heavy meromyosin (HMM) and is composed of the head (200 Å), neck and part of the tail (400 Å) of the myosin molecule (Szent-Gyorgyi 1960; Lowey *et al.* 1966). It has a molecular weight of 232,000–380,000 and is 750 × 60 Å in size. The HMM has all the enzymatic properties of myosin (Szent-Gyorgyi 1960; Lowey *et al.* 1966; Hamm 1970). It is soluble at low ionic strength of KCl and has a structure of both globular and helical configuration.

Papain guillotines the heads of the myosin molecule from the rest of the structure. This decapitation severs the globular heads from the helical tail and facilitates study of the head section. From research it has been learned that each head contains two types of SH groups. One of these groups on each head catalyzes ATPase activity and participates in forming actomyosin with actin. This hydrolysis reaction, catalyzed by myosin ATPase, cleaves the terminal phosphate bond in ATP (adenosine triphosphate—a high energy molecule) and transfers part of the released energy into mechanical work called contraction (Mommaerts 1966). This ATPase activity is stimulated by Ca^{++}.

$$\text{ATP} + \text{H}_2\text{O} \xrightarrow[\text{Ca}^{++}]{\text{myosin ATPase}} \text{ADP} + \text{Pi} + \text{energy}$$

Thick Filaments

Most researchers agree that the thick filaments are primarily composed of an aggregation of large linear polymers of myosin molecules. These take on something of a cigar-like shape with projections on each end positioned in a spiral fashion (Fig. 5.2).

The thick filaments are constructed by linking two myosin molecules tail to tail, thus resulting in a polymer which has heads on both ends and a bare midriff. There is some speculation that a yet unidentified protein may be used as a cement for the tail center sections and thus account for the slight bulging found in the center of this structure. Other myosin molecules with highly charged tails, which hold the structure together with electrostatic charges, are then added to this structure. The tails have a staggered, overlapping configuration, which also aids in binding the filaments together. There are 12-20 parallel myosin molecules in a cross section of the thick filament (Holtzer *et al.* 1961). Each filament is approximately 12 molecules long. This leaves the heads protruding in orderly rows on each end of the filament (Fig. 5.2). The myosin heads are arranged in pairs with one head positioned on each side of the filament. Each pair of heads is rotated 120° from its neighbor and yields 6 rows of myosin heads on a filament. Longitudinal spacing (Bendall 1969) per pair of heads is at intervals of 143 Å which yields 429 Å spacing per row. The head and neck region is flexible and movable, extending approximately 135 Å from the center of the filament. There are from 30-36 protruding heads in each of the 6 rows yielding a total of 180-216 heads per thick filament. The myosin molecules are arranged with opposite polarities on each end of the thick filaments since their heads are pointed in opposite directions. Since the heads move in unison this filament resembles a racing shell (Fig. 5.2) with rowers placed in each end (none in the center) facing the center of the shell (Murray and Weber 1974). Upon contraction the myosin heads or oars are moved by the crew in an attempt to propel the shell in opposite directions. This moves the attached (to the oars) thin filaments on both ends of the thick filament toward the center of the thick filament and results in a shortening or contraction of the muscle.

The overall thick filaments are 15,000 Å in length and 100-120 Å in diameter (Huxley 1953A, 1957, 1963A). The myosin heads are protruding on each end leaving a bare center section 1,500-2,000 Å in length (Bendall 1969). The middle of this center section is slightly swollen where myosin tails join from opposite directions.

Actin

G-actin (globular actin) is composed of a polypeptide chain which has a high proline content and a large number of bulky nonpolar side chains. The high proline content puts a kink in the polypeptide chain (Bendall 1969); the nonpolar chains are tucked inside the molecule; and there is a low surface charge on the exterior which yields a molecule with a very compact structure, a low helical content and a spherical shape (Fig. 5.2). This bead shaped (globular-spherical) molecule has a molecular weight of 41,770-57,000, a diameter of 55 Å and contains 374-450 residues (Mommaerts 1952; Szent-Gyorgyi 1966; Reese and Young 1967; Bendall 1969; Hamm 1970; Elzinga and Collins 1973). G-actin has a sedimentation constant of 3.3 (Harper 1972).

ATP is bound to G-actin and this is converted to ADP (adenosine diphosphate) when the G-actin beads are strung together to yield F-actin (fibrous actin) (Straub and Feuer 1950; Bendall 1969).

$$\text{n (G-actin-ATP)} \xrightleftharpoons{\text{Polymerizes (-S-S-: bond formation)}} \underbrace{(\text{G-actin-ADP})_n}_{\downarrow \\ = \text{F-actin}} + \text{Pi}$$

Kasai (1969) reported that activation enthalpy (heat content per unit mass) for polymerization was 20-25 kcal per mole and the activation enthalpy for depolymerization was approximately 10 kcal per mole.

The polymerization of G-actin units to yield F-actin changes the structure from a globular to a linear aggregate (Fig. 5.2). There are 400-600 monomers (tip to tip through the Z line) per filament and two filaments are intertwined to yield F-actin (Bendall 1969). This resembles twisting two strings of beads together. There are 13-15 globular sub-units per turn of the helix and the crossover points are 340-420 Å (Hanson and Lowy 1965; Hanson 1967; Huxley and Brown 1967; Millmann *et al.* 1967). This yields a filament with an overall diameter of 60-80 Å and Vibert *et al.* (1972) reported that actin can exist in two distinct configurations—a resting and an activated state.

Tropomyosin

Tropomyosin is a short thin (450 × 20 Å) molecule with a molecular weight of approximately 70,000-130,000 (Tsao *et al.* 1951; Holtzer *et al.* 1965; Woods 1967) and a sedimentation constant of 2.6 (Harper 1972). The individual molecule resembles the tail of a myosin molecule because it is 95% helical and is constructed of two α-helical coils that are intertwined (Kominz *et al.* 1954; Lowey *et al.* 1966). Unlike the head of the myosin molecule, however, there is no ATPase activity in this structure (Bailey 1948). Tropomyosin molecules attach end to end (40-60 of them) yielding a long thin fiber (Murray and Weber 1974) that is located in the thin filaments and the Z-line.

Troponin

Troponin is an oblong globular protein with a molecular weight of 50,000-85,000 (Ebashi *et al.* 1967) and a sedimentation constant of 3.5 (Harper 1972). Its high proline content can account for the lack of a coiled structure. Troponin binds Ca^{++} and confers Ca sensitivity to the actin myosin complex (Ebashi *et al.* 1967; Fuchs and Briggs

1968). Troponin may also play a cementing role between tropomyosin and actin (Drabikowski *et al.* 1968). There are 40–60 troponin molecules in a thin filament (Murray and Weber 1974) as shown in Fig. 5.6.

α- and β-Actinin

There are additional minor protein components of the thin filaments. α-actinin is involved in cross-linking of the protein and β-actinin (mol wt 60,000) is associated with troponin. The α-actinin accelerates the gelatinization and polyermization of F-actin and is also located in the Z-band which is illustrated in Figures 5.4, 5.6, 5.7, and 5.8 (Briskey *et al.* 1967; Masaki *et al.* 1967). α-actinin is often divided (Nonomura 1967) into component parts as follows.

Separated Into	Sedimentation Coefficient
Light component (mol wt 160,000)	6 S
Intermediate component	10 S
Heavy component	25 S

Thin Filaments

Thin filaments are primarily composed of F-actin, tropomyosin, troponin and α- and β-actinin. The total filament has a molecular weight of several million (Hamm 1970).

The backbone of the thin filaments is the twisted beaded structure of actin (Fig. 5.2). Lying in the actin grooves is the long thin filament of tropomyosin. Straddling the thin tropomyosin filament, also in the actin groove, is the globular protein troponin (Fig. 5.6). The troponin is very near one end of the tropomyosin molecule which gives one troponin molecule per tropomyosin. One tropomyosin molecule extends over 7 G-actin beads. This also gives a ratio of 1 troponin/7 G-actin beads (Murray and Weber 1974). This calculates to 2 molecules of troponin and 2 molecules of tropomyosin/pitch of 400 Å on the actin coil.

The thin filaments are 60–80 Å in diameter and 20,000–22,000 Å in length (both sides of the Z-line) (Huxley and Hanson 1957; Huxley 1963A). The thin helical filaments are linked together at the Z-disk through a tropomyosin-like molecule which then runs in either direction in the grooves of the actin superhelices (Huxley 1953B; Bendall 1969). The G-actin, even though globular in shape, has one polarity (direction of connection with actin) on one side of the Z-disk and an opposite polarity on the other side. This gives the G-actin an arrowhead appearance (pointing away from Z-disk) when it combines with myosin in contrast to a strict spherical shape when

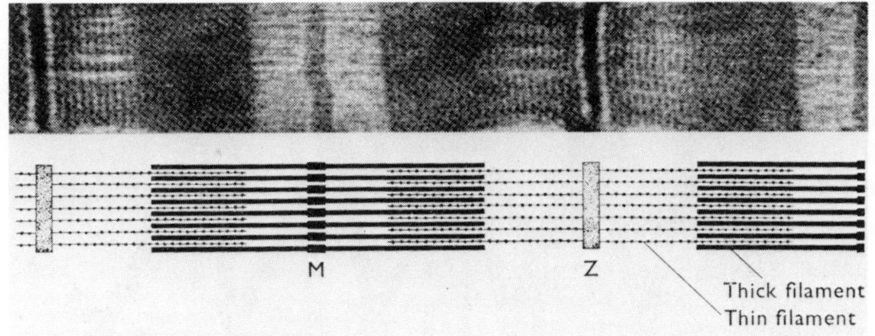

From Ebashi, Endo and Ohtsuki (1969)

FIG. 5.6. ANTI-TROPONIN STAINING ON A CHICKEN SKELETAL MUSCLE FIBER

Notice appearance of regular striation along the whole thin filament region; 24 striations can be counted on either side of the Z-band. This indicates that troponin molecules are distributed along each thin filament with a period of about 400 Å as shown schematically in the lower part of the figure.

viewed under the electron microscope (Huxley 1963A). If the thin filaments are visualized as extending through the Z-disk then this change of polarity on each end occurs in both the thick and thin filaments.

Z-disk

The construction of the flat protein structure of the Z-disk is still somewhat debatable. The thin filaments are attached to it (or pass through it, but with a change of orientation) by a tropomyosin-like molecule. Actinin has also been isolated from the Z-disk and appears to induce a crosslink formation between strands of F-actin. The Z-disk is estimated to be 800 Å wide. An electron microscope view of this structure yields a tetragonal array instead of the hexagonal one expected. It is suggested that the thin filament splits into strands as it approaches the Z-disk and that the disk has a somewhat knitted appearance as shown in Fig. 5.7.

Muscle Fiber Arrangement

The muscle fiber is 200,000–800,000 Å in diameter and seldom longer than 12 cm although a few have been isolated as long as 47 cm (Bendall 1969).

The thick and thin filaments are alternately aligned in transverse register which yields cross-striations of the fiber as shown in Fig. 5.2, 5.8 and 5.9. This banding is caused by optical dense areas called

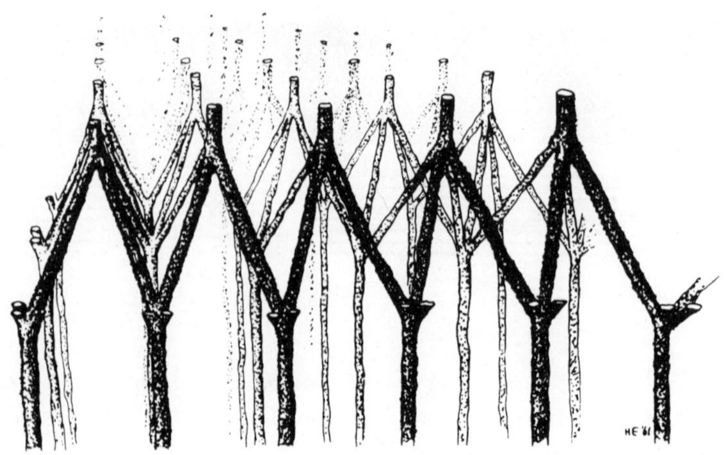

From Knappeis and Carlsen (1962)

FIG. 5.7. ILLUSTRATION OF ATTACHMENT OF THIN FILAMENTS TO THE Z-DISC

Drawing by E. Elias, The Rockefeller Institute Press.

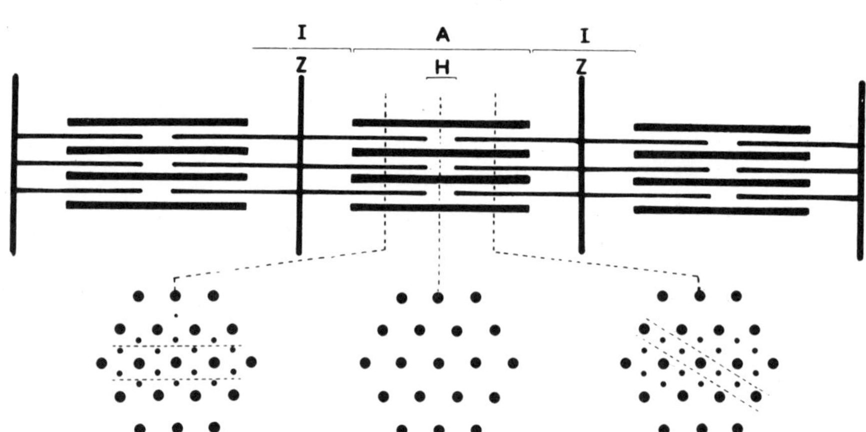

From Huxley and Hansen (1960)

FIG. 5.8. DIAGRAM ILLUSTRATING THE THICK AND THIN FILAMENTS IN A LONGITUDINAL MYOFIBRIL THREE SARCOMERES IN LENGTH

Vertical dotted lines indicate the location of three transverse sections taken through the A-zone. Dotted lines in the transverse sections indicate how an electron micrograph can show different arrangements of thick and thin filaments depending on how the section is taken.

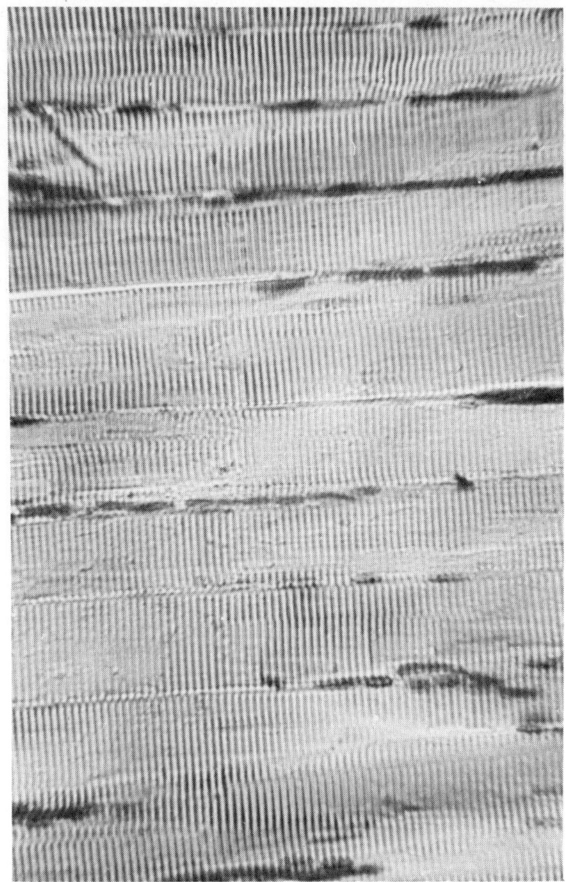

FIG. 5.9. LIGHT PHOTOMICROGRAPH SHOWING LIGHT AND DARK BANDS ACROSS A MUSCLE

anisotropic (A) and less dense bands called isotropic (I). In the resting state the A band is 15,000–16,000 Å in length and the I band is approximately 8,000–10,000 Å (Huxley 1958; Huxley and Hanson 1960). The A band maintains its length during contraction but the I band becomes shorter. The I band is composed of thin filaments arranged in a hexagonal array and connected in the center by the Z line as shown by "F" in Fig. 5.2. The distance from one Z line to the next is called the sarcomere length and the sarcomere is the functional unit of the muscle fiber. In the resting state it is approximately 26,000 Å in length and shortens 20% or more on contraction (Huxley 1965). The A band is composed of thick filaments for its entire length and overlapping ends of the thin filaments are present in the ends of this zone. This gives the A band a darker appearance on the ends (both thick and thin filaments) and a lighter appearance in the center (only thick filaments). This light area in the center of the A band is known as the H zone. The H zone is approximately 5,000 Å in the resting state. The ratio of these areas in the A band varies considerably with contraction. In fact the H zone may disappear and a darker zone appear in the center of the A band on super contraction. On stretching, the H zone can occupy most of the A band as the thin filaments almost pull out of the thick filaments. In the resting cell in the center of the H zone there is a small darker area known as the M band and is caused by the thickened area in the center of the thick filaments. In the muscle fiber the ratio of actin to myosin on a molar basis is 4:1 and on a weight basis 1:2 (Hanson and Huxley 1960; Hasselbach 1964; Huxley and Brown 1967; Perry 1967; Bendall 1969). In the overlapping area of the A band a cross section would show 6 thin filaments around each thick filament and 3 thick filaments around each thin filament. The thick filaments are approximately 450 Å apart and the distance between thick and thin filaments is approximately 250 Å. This distance increases with contraction as the muscle cell swells. The myosin heads of the thick filaments form the bridges between the thick filaments and G-actin monomers of the thin filaments. There are 6 bridges in a helical fashion per 430 Å and these are in pairs and the next pair is rotated 120° from the preceding pair.

Bendall (1966, 1969) reported there are approximately 1,000-2,000 fibrils (minute filaments from which the fiber is constructed and presumably represents the contractile elements) of 10,000-20,000 Å diameter in a fiber (cell) of 500,000 Å diameter. There is no membrane component around the myofibrils (muscle fibrils) and they remain structurally intact because they are insoluble at the ionic strength of the cell (Briskey and Fukazawa 1971). If the thick and thin filaments are extracted from the cell and placed in a proper environment they will combine in a fashion very similar to that found in the intact cell except the filaments lose their three dimensional orientation. Arnaud *et al.* (1970) suggested a structure for actomyosin in solution which is constructed of F-actin (mol wt of 4×10^6) on which 28 molecules of A-myosin (mol wt of 5×10^5) are attached. This would yield a cylindrical molecule 5800 Å long and 1160 Å in diameter with a mol wt of $18 \pm 4 \times 10^6$. D'Haese and Komnick (1972A, 1972B) indicated that isolated actomyosin in the uncontracted state consists of actin filaments which are loaded with myosin molecules in the arrowhead pattern and interconnected by a few myosin bridges, whereas contracted threads are made up of both actin and myosin filaments.

ADDITIONAL COMPONENTS OF THE FIBER

Sarcoplasmic Reticulum

The exterior of the muscle fiber acts as a semi-permeable membrane and is called the plasmalemma (sarcolemma). It is triple layered and 100 Å thick (Paul 1965). The sarcolemma maintains the fiber's osmotic pressure and conveys the action potential along the fiber from its origin at the motor end plates (Hodgkin 1965; Katz 1966). This membrane is in turn connected to the sarcoplasmic reticulum.

The myofibrils are surrounded by tubules and vesicles known as the sarcoplasmic reticulum (Fig. 5.10). This net is divided into two systems (T and L) depending on interconnection and direction of orientation. The T system's thin (400 Å diameter) tubules (transverse or intermediate system) are connected to the plasmalemma (sarcolemma) and continue inward, enwrapping each fibril at the A-I junction and are continuous from one fiber to the next across the muscle. The T channels are not only connected to the fiber sheath (sarcolemma) but are part of the cell membrane itself, enfolded deep into the interior of the fiber (Porter and Franzini-Armstrong 1965). This system develops rapidly during early animal growth (Schiaffino and Margreth 1969). The L system's tubules (longitudinal or sarcotubular system) are parallel to the myofibril and almost totally enwrap the fiber.

Where the L system meets the T system at the A-I junction the L system swells on each side of the T system and these swellings are called terminal cisternae. These two terminal cisternae (expanded parts of L system) and the T tubule between them are known as the triad (T system) and they are located at the junction of the L and T system on the A-I border (in some animals on the Z line, as shown in Fig. 5.10.).

These tubules which delineate the fibers are used as a transmission system, from the sarcolemma, for nerve impulses necessary to trigger contraction (Franzini-Armstrong 1963). The sarcoplasmic reticulum has the ability to bind and release Ca^{++} in an active process that is dependent on ATP utilization (Cassens and Cooper 1971), pH, and bound divalent cations (Duggan and Martonosi 1970). The primary area of calcium storage and the calcium pump is located in the terminal cisternae.

The sarcoplasmic reticulum has been extracted from muscle and

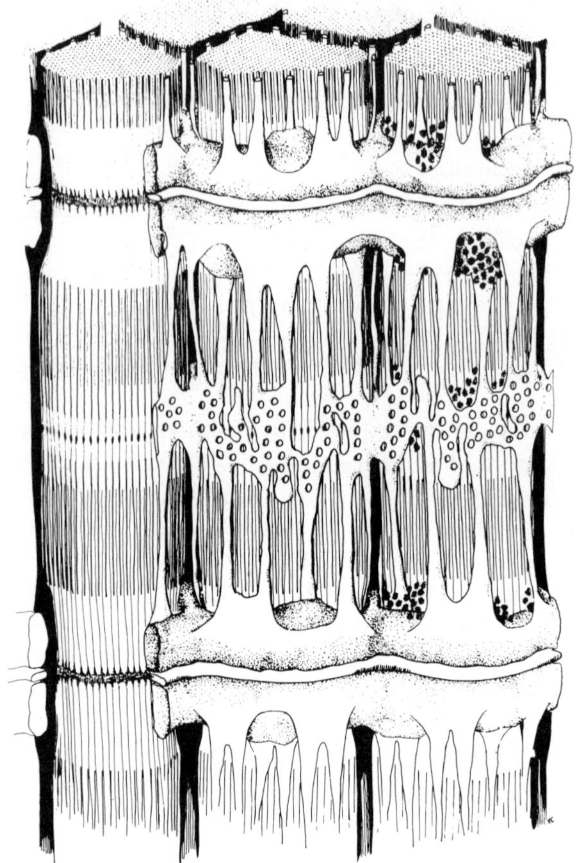

From Peachey (1965)

FIG. 5.10. THREE-DIMENSIONAL RECONSTRUCTION OF THE SARCOPLASMIC RETICULUM ASSOCIATED WITH SEVERAL MYOFIBRILS

Reprinted with permission of Rockefeller Institute Press.

examined in detail. Martonosi (1968) reported that the biological membrane of this structure is a bimolecular leaflet of phospholipids coated on both sides with proteins. The presence of particles (30–40 Å) with physiological unknown roles on the surface of the sarcoplasmic reticulum has also been noted (Martonosi 1968; Deamer and Baskin 1969). Solubility studies of the sarcoplasmic reticulum have yielded material of 125,000-160,000 and 100,000, 80,000 and 62,000 daltons in molecular weight (McFarland and Inesi 1970; Ikemoto *et al.* 1971). Protein from the sarcoplasmic reticulum can be separated into a large number of bands. The band has been lo-

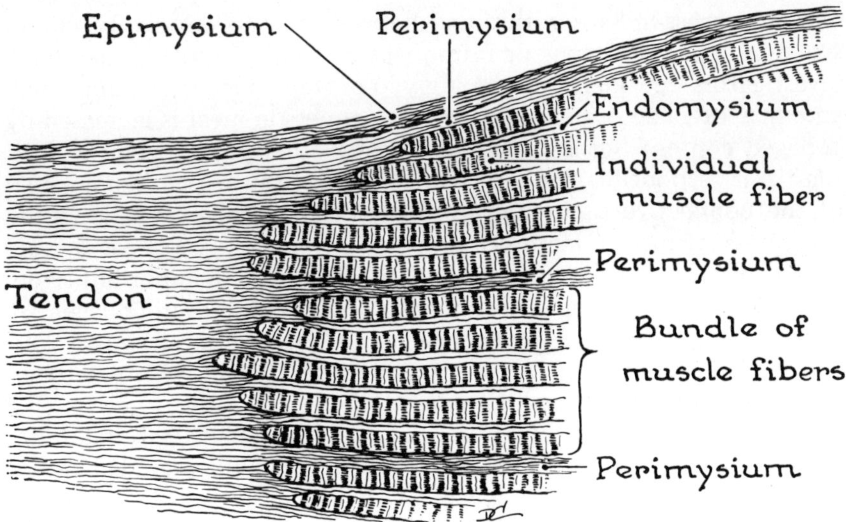

From Ham (1969)

FIG. 5.11. DIAGRAM OF A LONGITUDINAL SECTION OF MUSCLE, SHOWING HOW THE MUSCLES TERMINATE IN TENDONS

The connective tissue of the epimysium, the perimysium and the endomysium is continuous with the connective tissue of the tendon. The sarcolemma covering of each muscle fiber also is adherent to the connective tissue of the tendon.

cated that contains the ATPase enzyme implicated in the calcium transport system (Martonosi 1969).

Connective Tissue

Connective tissue is a fibrous protein often referred to as the stromal fraction and it is used to form and support other tissue and to regulate and control the extent of contraction. It is the structural matrix of the skin, cartilage, bone, tendons (Fig. 5.11), vessels and the covering of most of the animal body tissue including muscles and muscle sub-units. It is composed of two major types of tissue known as collagen and elastin.

Collagen is the major component of connective tissue and is screw-shaped in configuration with three (two similar and one more basic) helices wound together (Hamm 1970) which are, in turn, cross-linked in a staggered configuration by the following structure.

$$\underset{\text{N}}{\overset{-\text{C}-}{}}\!\!\!\!\!\!\;\text{C-(CH}_2)_3\text{-}\underset{\text{H}}{\underset{|}{\text{C}}}\text{=N-(CH}_2)_3\text{-C}\overset{\text{-C-}}{\underset{\text{N}}{}}$$

Covalent cross-linkage within and between the helices can be extensive as found in tendons or infrequent as found in soft tissue. Some investigators consider collagen a glyco-protein since it contains small amounts of galactose and glucose. Toughness in meat is increased by collagen content but is also influenced by other protein components. Elastin is responsible for less than 3% (except in *M. semitendinosus*) of the connective tissue and probably contributes very little to the toughness of meat (Hamm 1970). It is beyond the scope of this report to review the current state of the art for these important structural tissues but rather recent reviews were reported on collagen by Gross (1961), Piez (1966) and Veis (1970); and on elastin by Partridge (1966); and on connective tissue alterations by Herring *et al.* (1967) and Carmichael and Lawrie (1967).

Connective tissue is composed of macromolecules of indefinite length. Collagen (white connective tissue) can be degraded rather easily.

$$\text{Collagen} \xrightarrow{\text{boiling}} \text{gelatin}$$

$$\text{Bones and connective tissue} \xrightarrow[\text{more degraded}]{\text{prolonged boiling}} \text{impure material called glue}$$

On the other hand elastin (yellow connective tissue) is almost non-destructible even when attacked by acids.

The connective tissue is held together by a ground substance which acts as a cementing matrix. It is a firm gel of linear biocolloids that form a very fine meshwork.

The amino acid composition of connective tissue is different than other muscle components and this is used for quantitative determination. This difference in amino acid composition makes collagen one of the few digestible animal proteins that does not have a fairly well-balanced amino acid profile (Table 5.7).

Smith *et al.* (1972) divided the elastin fraction into two components, one of which is soluble. The ratio of amino acids is alike in both types and the soluble type (mol wt 74,000) is thought to be a precursor of the insoluble type. As the elastin changes from the soluble type to the insoluble type there is an increase in the cross-linkage.

Sarcoplasm

Any space in the muscle not taken up by contractible proteins is filled with an intercellular fluid called sarcoplasm which surrounds and bathes the myofibrils. Recent reviews by Scopes (1970) and Kastenschmidt (1970) described this fluid and soluble protein in de-

TABLE 5.7

AMINO ACID CONTENT OF COLLAGEN AND ELASTIN

	Molecular Weight	High Content of	Low Content of[1]
Collagen	300,000	glycine proline hydroxyproline	sulfur amino acids methionine tryptophan
Elastin	>1,000,000	glycine alanine valine proline	hydroxyproline

[1] Low content of methionine and tryptophan is primarily responsible for the low food value of collagen.

tail. These water-soluble proteins comprise about 20–25% of the total muscle protein. Sarcoplasm contains the following major constituents.

(1) Myogen (an early term used to designate soluble sarcoplasmic proteins) includes glycolytic enzymes as well as other soluble cell components; molecular weight range from 30,000–100,000.
(2) Myoglobin is a tissue pigment that has a globular structure (40 × 35 × 25 Å), a chain that is 70% α-helical, and a heme residue; molecular weight 16,890–18,000.
(3) Peptides.
(4) Amino acids.
(5) Glycogen.
(6) Glycolytic enzymes.
(7) ATP.
(8) Phosphocreatine.
(9) Inorganic electrolytes.

Mitochondria

Mitochondria (Fig. 5.12) are organelles, approximately 20,000 Å long, in the sarcoplasm and the number varies greatly from muscle to muscle (Bendall 1969). In active muscles the mitochondria are very profuse and regularly spaced whereas in inactive muscles these components are fewer in number and irregularly spaced. Mitochondria always lie outside the contractile fibrils distributed between the myofibrils in longitudinal rows or along horizontal planes at I-bands, Z-disks or A-I junctions (Bennett 1960). These organelles often straddle the Z-disk, practically surrounding the myofibril.

The mitochondria are referred to as the powerhouse of the cell (Ham 1965) and supply energy for many energy-requiring activities

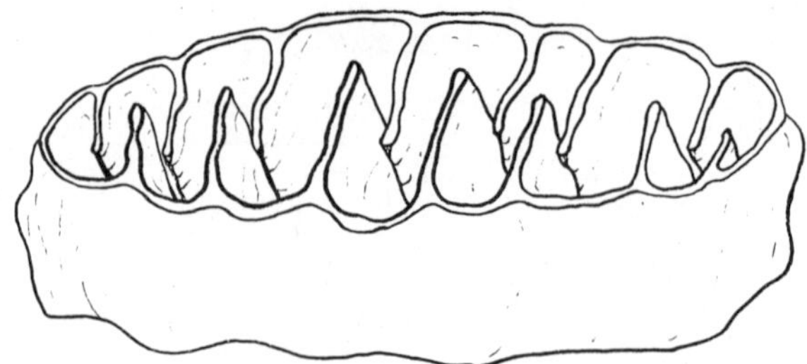

FIG. 5.12. A THREE-DIMENSIONAL DIAGRAM OF A MITOCHONDRION SHOWING THE SHELF-LIKE MITOCHONDRIAL CRESTS

of cells, including that necessary for the contractile units of muscle. The mitochondria are oval to cylindrical in shape and vary in size. This structure has two membrane walls and the inner membrane has many folds called cristae that project like shelves into the mitochondria. The interior is filled with a dense fluid that has no observable structure. These small bodies utilize oxygen and contain (Southard and Hultin 1969; Harmon 1964; Ham 1965) the following enzymes or enzymatic processes.

(1) The electron transport system of enzymes.
(2) Fatty acid activating enzymes.
(3) Linkage of electron transport system with citric acid cycle.
(4) A number of citric acid cycle enzymes.
(5) Provides energy from oxidative phosphorylation.

$$(\text{ADP} + \text{Pi} \xrightarrow{\text{energy}} \text{ATP})$$

The energy rich ATP is made available to the sarcoplasm outside the mitochondria where it supplies energy for many of the metabolic processes.

Ribosomes

Ribosomes are spherical objects that appear granular and located in the sarcoplasm and in the nucleus. These granules are rich in RNA and are the site where messenger RNA becomes attached to transfer RNA. Ribosomes are also the site where amino acids are linked to form proteins. The number of ribosomes have been used as an index to the protein synthesis activity of the cell. The ribosomes may be located free in the cytoplasmic matrix or attached to membrane vesicles of the endoplasmic reticulum.

Nuclei

In contrast to most animal cells the muscle fiber is multinucleated. A fiber several millimeters in length will contain several hundred nuclei. These nuclei are distributed throughout the length of the fiber but are more numerous and irregularly spaced at the ends. The nuclei are oval in form, 80,000-100,000 Å in length (Briskey and Fukazawa 1971) and are located just under the sarcolemma or fiber membrane (Fig. 5.13). The nuclear matrix contains (1) chromatin—DNA (chromasomal substances) which determines the specificity of the cell and metabolic activity, and (2) nucleolus—RNA which functions in protein manufacturing, and synthesis and nucleic acid metabolism.

MUSCLE FIBER TYPES

Casual observation of skeletal muscles even within the same animal will reveal that these muscles are not homogenous in type. Some of the most obvious differences may be found in comparing light and dark poultry meat but even in other species closer examination will reveal that muscles are different. Further investigation will show that areas within some muscles are different than adjacent ones in the same muscle. This type of variability is caused by the construction of muscles from two types of fibers. These types of fibers are most commonly referred to as red and white fiber, although other names (dark-light, slow-fast) have often been used. Many researchers have also given intermediate classification between these two extremes. Detailed examination will reveal that some of the color difference can be attributed to myoglobin and cytochrome content. The myoglobin content of two species is shown in Table 5.8. Muscle fibers have been classified and a very complete review of fiber typing can be found in the report by Goldspink (1970). Not only are muscles different in myoglobin content but also in physical, biochemical (Beatty and Bocek 1970) and metabolic characteristics as may be seen in Table 5.9.

By examining the very significant difference in the structure and function of these fiber types it is obvious that the number and distribution of fiber types plays a major role in the animal's metabolism. In the red fibers, respiration is the principal source of energy for rephosphorylation via oxidative phosphorylation in the mitochondria

$$(\text{ADP} + \text{Pi} \xrightarrow{\text{rephosphorylation}} \text{ATP}).$$

In the white fiber, glycolysis supplies the energy for rephosphorylation of ADP. Fiber types are also of extreme importance to the uti-

FIG. 5.13. LIGHT MICROSCOPE VIEW OF A LONGITUDINAL SECTION OF MUSCLE FIBERS. THE PHOTOMICROGRAPH SHOWS A PORTION OF THE DISTRIBUTION OF NUCLEI IN THE CELL

lization and conversion of these muscles into food. Such factors as color, emulsifying capacity, water-holding capacity, pH, oxidation, and tenderness, as well as many more food-related parameters are influenced by muscle fiber type.

CONTRACTION

Now that the major fiber components have been discussed, the next question becomes—how do they work in concert to achieve contraction. Muscle shortening is an extremely complex series of re-

TABLE 5.8

PERCENTAGE OF MYOGLOBIN CONTENT PER WET WEIGHT OF THE MUSCLE

	Horse (%)	Pig (%)
Longissimus muscle (white)	0.465	0.280
Psoas muscle (red)	0.705	0.437

Source: Lawrie (1950).

TABLE 5.9

COMPARISON OF RED AND WHITE FIBERS AND MUSCLES

Red Fiber (or Muscle)	White Fiber (or Muscle)
Sudan black B positive (lipid stain)	Sudan black B negative (lipid stain)
Contains more myoglobin (and more iron)	Contains less myoglobin (and less iron)
Fiber—smaller in diameter	Fiber—larger in diameter
Fiber—arrangement checkerboard—most species	Fiber—arrangement checkerboard—most species
Fiber—in clumps that are surrounded by white fibers (pig)	In the periphery of the bundle (pig)
More mitochondria	Fewer mitochondria
Fiber—rich in sarcoplasm	Fiber—less sarcoplasm
Wide Z-line	Narrow Z-line
Surrounded by many capillaries (higher capillary to fiber ratio)	Surrounded by fewer capillaries (lower capillary to fiber ratio)
Less connective tissue (collagen)	More connective tissue (collagen)
Slow-twitch contraction	Fast-twitch contraction
More blood flow per unit mass	Less blood flow per unit mass
Contains less glycogen	Contains more glycogen
Less soluble protein (sarcoplasmic)	More soluble protein (sarcoplasmic)
More lipids	Less lipids
Less saturated lipids	More saturated lipids
Greater lipase activity	Less lipase activity
Lower ATPase activity	Higher ATPase activity
Rate and total Ca^{++} uptake lower	Rate and total Ca^{++} uptake higher
Less creatine phosphate	More creatine phosphate
Less ATP	More ATP
More AMP	Less AMP
More RNA	Less RNA
More rapid protein turnover rate	Less rapid protein turnover rate
Aerobic metabolism (oxygen required)	Anaerobic metabolism (no oxygen required)
High oxidative enzyme (citric acid cycle) activity	High glycolytic enzyme activity
Muscle	Muscle
Psoas	Longissimus
Soleus (nearly homogenous-red fibers)	Brachioradialis (superficial portion—75% white fibers)
Trapezius	Gastrocnemius
Sartorius (60–70% red fibers)	Adductor magnus
Pig Semitendinosus (axial)	Pig Semitendinosis (peripheral)

Source: Author's summary of review by Cassens and Cooper (1971).

actions that occur exceedingly fast and, at the present time, some of the steps have not been elucidated. For these reasons an outline approach will be used to present this information (Winton and Bayliss 1962; Bendall 1969).

(1) A signal is generated in the brain and a nerve impulse is carried through the spinal cord to the nerve cell in the ventral horn of the spinal cord.
(2) From this cell a nerve fiber transmits the signal. This transmission through the nerve is as follows (A–D).
  (A) The resting nerve cell is polarized by maintaining a negative internal charge and a positive external charge (Hodgkin 1965). This charge differential is maintained by a resting membrane that is more permeable to K^+ and Cl^- than it is to Na^+ and this yields (Bendall 1969) (a) and (b).
    (a) High internal concentration of K^+ and protein ions.
    (b) High external concentration of Na^+ and Cl^- ions.
  (B) The nerve signal is an electrical impulse that adds a positive charge to the interior of the cell. This causes the nerve membrane to become depolarized.
  (C) Local depolarization changes the permeability of the membrane and it locally becomes more permeable to Na^+ and less permeable to K^+ which in turn generates a new local potential difference.
  (D) This potential difference causes current to flow down the nerve carried by an inward flow of Na^+ ions.
(3) This nerve fiber travels through a nerve trunk until it reaches the muscle.
(4) At the muscle level the nerve fiber branches and terminates in a nerve terminal called a motor end plate.
(5) This motor end plate is connected to a single muscle fiber. At the motor end plate the impulse encounters its first membrane discontinuity.
(6) A recent theory is that the action potential circumvents this barrier by releasing acetylcholine from the motor end plates which in turn depolarizes the muscle membrane (Katz 1966). This propagated depolarized wave is conducted on the muscle membrane very similarly to the way it was conducted in the nerve fiber by local depolarization which changes permeability, which in turn generates a potential difference that flows along the membrane.
(7) This action potential is transmitted inwardly by the sarcoplasmic reticulum transverse invaginations which are con-

nected to (or are a part of) the membrane. Signals passing through these tubules toward the A-I junction have the following effect on the conducting membranes (A–D).

(A) Changes in semi-permeability occur.

(B) Small number of Na^+ penetrate the membrane.

(C) Small number of K^+ escape the membrane.

(a) A Na^+/K^+ pump is necessary to restore resting concentrations after the action potential passes.

(b) This pump is powered by ATP energy.

(D) An Na^+ depolarization wave is generated.

(8) As the action potential arrives at the triad system another gap in connection is encountered. Current data are not clear as to how the potential gets around this barrier but naturally acetylcholine is again suspected. Regardless of the last hurdle the action potential does arrive at the triad system and the permeability of the terminal cisternae is altered.

(9) Ca^{++} are released from the terminal cisternae (Jobsis and O'Connor 1966) of the triads (Fig. 5.10) and longitudinal transmission is in the form of Ca^{++} and possibly Mg^{++}.

(10) Ca^{++} diffuse to active enzyme sites on myosin and actin (Fig. 5.2 and 5.4).

(A) Two calcium ions are needed per enzyme site on myosin.

(B) Ca^{++} is bound to troponin which is held at the actin site by tropomyosin. This releases Mg^{++} to activate ATPase on the myosin head.

(C) If a Ca^{++} concentration of 1×10^{-7} *M* in the fibrils is maintained, the muscle is in a relaxed state. On release of Ca^{++} from the sarcoplasmic reticulum to a fibril concentration of 1×10^{-6} *M*, contraction is initiated (Duggan and Martonosi 1970).

(11) Myosin ATPase on the myosin head splits ATP to release energy (Podolsky and Morales 1958; Nanninga and Mommaerts 1960; Alberty 1968; Bendall 1969).

$$\text{Myosin(M)} + \text{ATP} \rightleftarrows \text{M} - \text{ATP} \; \underset{-\Delta F = 6-10 \text{ Kcal/mole}}{\overset{H_2O}{\rightleftarrows}} \; \text{M}\begin{matrix} / \text{ADP} \\ \backslash \text{Pi} \end{matrix} \rightleftarrows \text{M} + \text{ADP} + \text{Pi}$$

(12) This release of Ca^{++} and ATP chemical energy initiates movement of the myosin head. Theories on these movements are as follows in (A) through (C).

(A) Swivel action of myosin heads.

(B) Sweeping (or rowing) motion of myosin heads.

(C) Miniature contractions of the head.

But regardless of the nature of the myosin head movement, chemical energy is converted to mechanical energy.

(13) The myosin heads link with the actin and pull or push the thin filaments toward the center of the sarcomere. One connection gives a relative movement of 100 Å or 1% of the muscle length (Murray and Weber 1974). The heads then disconnect and reattach, and the operation continues and the fiber becomes shorter. Contraction results from the sliding of the actin filaments and the myosin filaments together (Fig. 5.2 and 5.4).

(14) Contraction can reduce muscle length to approximately 40% of its resting length.

(A) The thick and thin filaments do not alter their length but slide past one another.
(B) Distance between Z-disks shorten.
(C) Z-disk moves toward the A-band.
(D) Narrowing of the I-band.
(E) A-band maintains its length.
(F) H-zone narrows and on super contraction becomes a darkened zone due to overlapping of the thin filaments in the center of the A-band.

(15) Since liquid is virtually noncompressible the volume of the muscle remains relatively constant with contraction. Therefore, a muscle must bulge as it gets shorter. If you plot myosin to myosin distance ("G" in Fig. 5.2—directly proportional to radius in the following formula) versus the square root of some measure of fiber length (e.g., sarcomere length), a straight line will be obtained (Elliot *et al.* 1963; Millmann *et al.* 1967). The following formula indicates this mathematical relationship.

$$r = \sqrt{V/\pi L}$$

where: V = volume—constant
π = 3.14—constant
L = fiber length
r = radius (directly proportional to myosin—myosin cross-sectional distance)

Relaxation

(1) Action potential stops.
(2) Na/K pump re-establishes polarization.
(3) Sarcoplasmic reticulum re-accumulates Ca^{++} (Ca^{++} pump).

(4) Calcium concentration is lowered by the calcium pump to 10^{-7} M. Calcium returns to terminal cisternae.
(5) Troponin re-establishes suppression of Mg^{+2}-activated ATPase.
(6) ATP is formed (Weber 1970).

$$\text{Phosphocreatine} + \text{ADP} \xrightarrow[\text{phosphokinase}]{\text{creatine}} \text{Creatine} + \text{ATP}$$

(energy reserve-storage of high energy bond)

(A) One ATP is required to break one cross linkage.
(B) ATP needed to power Ca^{++} pump.
(C) ATP saturates binding sites on myosin.

(7) Myosin dissociates from actin.
(8) Filaments slide to relaxed state.
(9) Resting actin and myosins are separated by repulsive electrostatic forces.

MUSCLE PROTEIN MACRO STRUCTURE

Skeletal Muscle

The muscle fiber (cell) sarcolemma (nonconnective tissue covering) is held in position by a network of connective tissue referred to as the endomysium (Fig. 5.14, 5.15, 5.16, and 5.17). The sarcolemma (plasmalemma) is triple-layered and approximately 100 Å thick (Bendall 1969). Blood capillaries may be found between the fibers. Many cells (20–50 fibers each) are assembled into a group known as a muscle bundle (or fascicle) (Bendall 1969). This bundle is covered by a thin (20,000–30,000 Å) white glistening connective tissue called the perimysium. The cross-sectional size of the muscle bundle is referred to in the meat trade as texture. Muscles that are capable of fine movement generally have small bundles and consequently fine texture. Muscles that are capable of great power have larger bundles and coarser texture. Marbling (fat within the muscle), nerve bundles and blood vessels are located between the muscle bundles. The bundles are grouped together to form an intact muscle and they are covered with a thicker, white, glistening connective tissue called the epimysium. Regardless of the skeletal muscle function, or size, this same general pattern is followed.

Smooth Muscle

Smooth, involuntary, nonstriated muscle is the simplest form of the three major muscle types. This type is found in the walls of hollow ducts and in the skin. The fiber is spindle-shaped with a single,

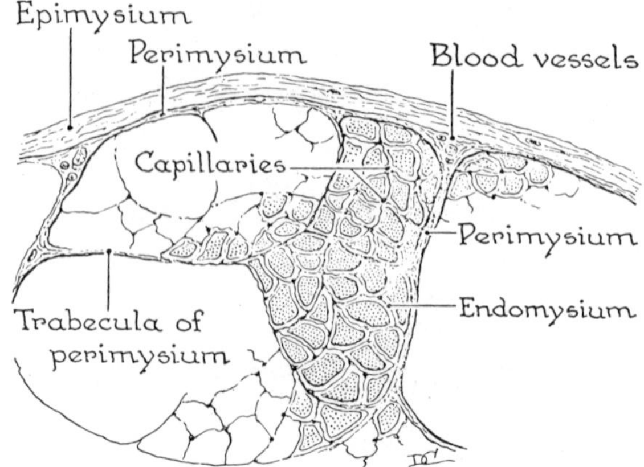

From Ham (1969)

FIG. 5.14. DIAGRAM OF A CROSS SECTION OF MUSCLE

This shows the connective tissue epimysium that surrounds the muscle. The partitions of perimysium that divide the muscle into bundles and the delicate connective tissue endomysium that extends between the individual fibers in the bundle and carries the capillaries which may also be seen.

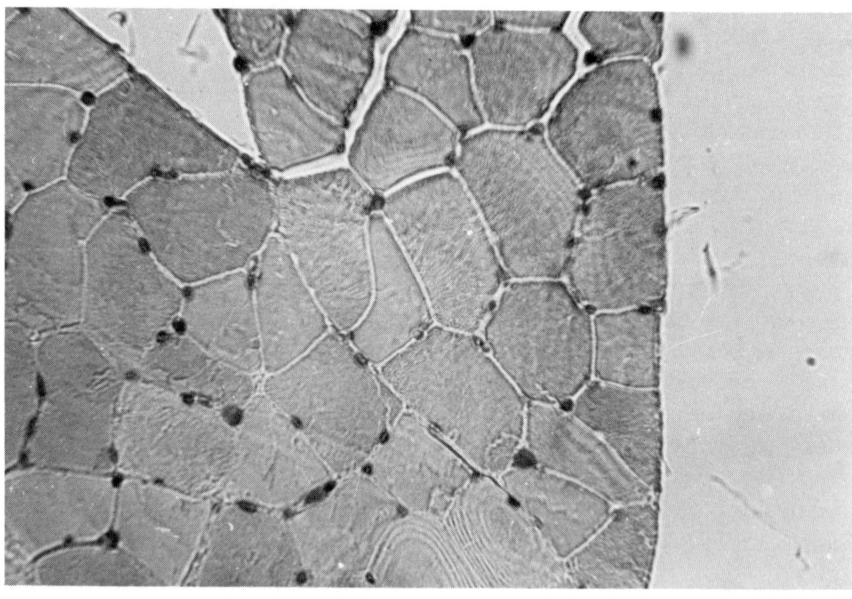

FIG. 5.15. LIGHT MICROSCOPE CROSS SECTIONAL VIEW OF MUSCLE FIBERS SHOWING ARRANGEMENT, SHAPE AND CONNECTIVE TISSUES

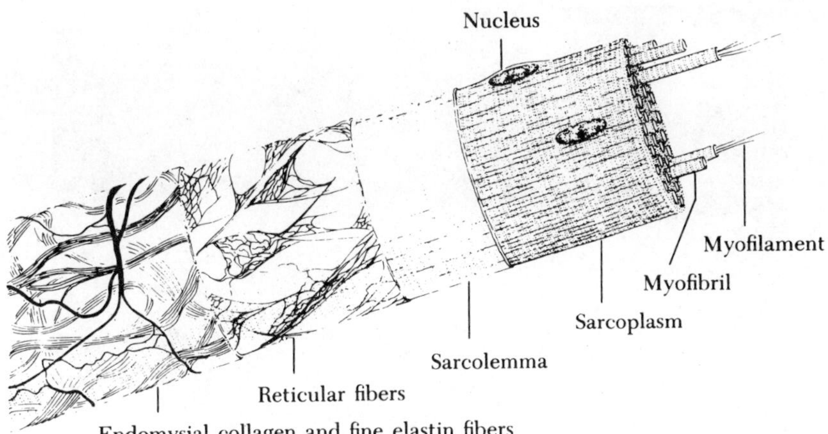

FIG. 5.16. DIAGRAMMATIC LONGITUDINAL SKETCH OF A MUSCLE FIBER
From the Science of Meat and Meat Products, Second Edition, by J. F. Price and B. S. Schweigert. W. H. Freeman and Co. © 1971.

centrally located nucleus. The fibers collect together in a bundle with the thick middle region overlapping the thin ends of a neighboring fiber. In a hollow duct the smooth muscle has two orientations: the inside layer is circular in shape and runs around the tube and the outside smooth muscle layer is longitudinal and runs parallel with the duct. This type of muscle contraction is slow, sustained and resistant to fatigue.

Cardiac Muscle

The cardiac muscle is an involuntary, striated muscle that is found in the heart and in the root of large vessels close to the heart. Cardiac looks much like skeletal muscle but functions very similarly to smooth muscle. This muscle has a central location for the nuclei within the fiber, and has a branching, meshwork structure. The fiber contains some dark staining cross-fiber markings called intercalated disks, which aid tremendously in identification.

SOLUBILITY OF PROTEIN COMPONENTS

A great deal of the knowledge concerning the structure and biochemical behavior of muscle has been obtained by extracting the individual components and studying them independently. Using this technique the influence of altered environments could also be evaluated. Components could then be recombined to study their additive effect. By taking advantage of their individual solubilities the muscle

From Parrish and Busch (1971)

FIG. 5.17. AN ELECTRON MICROGRAPH (X 41,731) OF A PORTION OF 3 MUSCLE MYOFIBRILS IS SHOWN

is often divided (Ockerman 1974) into the fractions described in the following paragraphs.

Sarcoplasmic Protein

The sarcoplasmic protein can be solubilized in a dilute salt (KCl) solution (approximately 0.1 *M*) and comprises about 6% of the muscle. These proteins are in the fluid that surrounds and penetrates the myofibrils and other organelles in the muscle fiber. Sarcoplasmic protein is composed of the following sub-units: (1) *cytoplasmic supernatant* (4% of muscle)—soluble glycolytic enzymes; (2) *nuclear fraction*—protein, ribonucleic acid (RNA), deoxyribonucleic acid (DNA), lipoproteins; and (3) *microsomial fraction*—(A) proteins—myoglobin (0.2-0.4% of muscle—tissue pigment), hemoglobin (0.4% of muscle—blood pigment); (B) organelles—mitochondria—structure involved in cell respiration, lysosomes—contains hydrolytic enzymes, sarcoplasmic reticulum—tubular membrane structure.

Myofibrillar Protein

The myofibrillar proteins are not extracted in dilute salt solutions (0.1 *M*) but are soluble in concentrated salt (KCl) solutions (approximately 0.6 *M*) and comprise about 10% of the wet weight of the muscle. These proteins are the protein constituents of the myofibrils which give rigidity to the cell and are mechanically responsible for the conversion of chemical energy into mechanical energy. The myofibrillar proteins may be divided into the following fractions.

(1) Myosin (5.8% of muscle)—thick filaments, cross-bridges between myosin and G-actin, ATPase.
   - (A) Light meromyosin (LMM) (4% of muscle)—tail of thick filament.
   - (B) Heavy meromyosin (HMM) (2% of muscle)—head, neck and part of tail of thick filament, ATPase.

(2) M-Protein ($<$ 0.1% of muscle)—middle of H-zone and believed to connect the tails of the myosin.

(3) Actin (2.5% of muscle).
   - (A) F-actin (filamentous aggregates)—primary structural unit of the thin filament.
   - (B) G-actin (globular)—building blocks of F-actin of the thin filaments.

(4) Tropomyosin (0.5-1.5% of muscle)—attaches actin to Z-line and runs the length of the thin filament in the F-actin grooves.

(5) Troponin (1% of muscle)—protein located periodically in the F-actin grooves, astride the tropomyosin, Ca^{++} receptive part of thin filament.

(6) α-actinin (0.2% of muscle)—located in Z-disk and M-band area.

(7) β-actinin ($<$ 0.1% of muscle)—occurs in thin filament and is believed to keep interactions of strands to a minimum and also believed to regulate length of F-actin.

(8) Actomyosin—interaction of actin and myosin, formed and broken in contraction and relaxation.

Connective Tissue Protein

Connective tissue proteins often referred to as stroma proteins are insoluble in concentrated salt solutions at cold temperatures and comprise 2% of the muscle. These tissues are the major supporting tissues of the body, enclose structures, cover the body, and connect muscle, organs and bones. These connective tissue structures are a very significant part of the muscle since no fiber exists without part of its structure being in contact with this structural tissue. Connective tissue proteins are found in the following types of tissue.

(1) Structure.
  Sarcolemma—elastic sheath that surrounds the muscle fiber and is composed of some collagen as well as lipoproteins.
  Perimysium—connective tissue covering of muscle bundle.
  Epimysium—connective tissue covering of the muscle.
  Reticulum fibers—fibrous network around the cells that holds the tissues together.
  Mitochondria—filamentous or granular organelles of the cytoplasm that manufacture ATP to power muscle activity.
  The mitochondria are composed of lipoproteins, enzymes of the Krebs cycle, the electron transport system, including flavin nucleotides and cytochromes.

(2) Connective tissues are of the following types (protein).
  (A) Collagen—white connective tissue that can be converted to gelatin upon heating.
  (B) Elastin—yellow connective tissue that is almost nondestructible.
  (C) Reticulin—connective tissue that resists conversion to gelatin upon moist heating.

Nonprotein Nitrogen

Nonprotein nitrogen fraction is salt water soluble and is not precipitated by trichloroacetic acid which would precipitate myofibril and sarcoplasmic proteins. This fraction as the name indicates is composed of nitrogen containing compounds other than protein and comprises 1.5% of the muscle. Examples found in muscle are as follows:

(1) Creatine—the phosphorylated form stores high energy phosphate.
(2) Free amino acids—unconnected sub-units of protein.
(3) Nucleotides—basic unit of nucleic acid, ribonucleic acid (RNA) and deoxyribonucleic acid (DNA) consisting of a base, sugar

and phosphate unit. Nucleotides include (1) inosine monophosphate (IMP), (2) adenosine triphosphate (ATP), (3) adenosine diphosphate (ADP), and (4) adenosine monophosphate (AMP).

(4) Carnosine—a dipeptide.

CHANGES IN MUSCLE STRUCTURE AS IT IS CONVERTED INTO FOOD

Changes During Rigor

The most obvious physical change that occurs after death is the stiffening of the muscles. The actin and myosin are locked together, the sarcomere length decreases and the whole system becomes inextensible. Any attempt to stretch the rigor muscle would result in a fracture at the I-band. Two chemical factors that accompany rigor and are at least partially responsible for it are a decrease in pH and the exhaustion of ATP.

The pH decreases from 7.2 to approximately 5.5 during rigor because glycogen in the muscle at the time of death is converted to lactic acid through the anaerobic glycolytic system (Fig. 5.18). This process cannot continue on through the citric acid cycle as it would in the live tissue since the citric acid cycle requires oxygen that is no longer available after death. Animal species and genetics both influence ultimate pH and rate of pH decline. Since the production of lactic acid is dependent on the supply of the raw material, glycogen, the quantity of this material at the time of death can influence ulti-

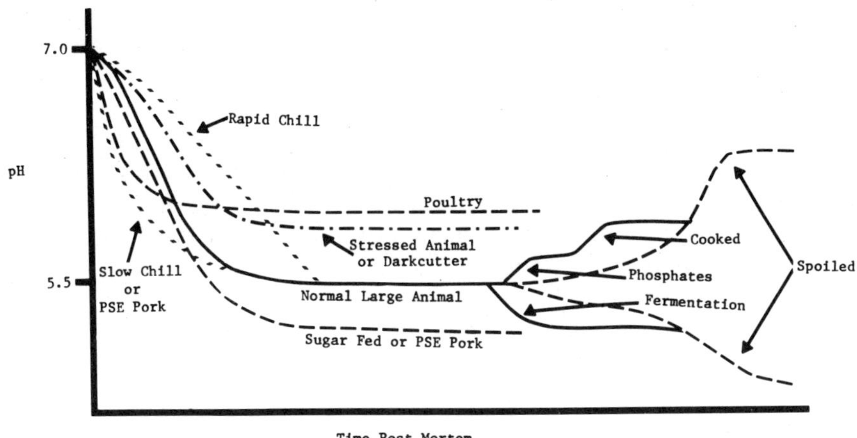

FIG. 5.18. PH ALTERATIONS PLOTTED AGAINST THE TIME POST-MORTEM FOR A VARIETY OF TISSUES AND CONDITIONS

mate pH. Such things as stress prior to slaughter can deplete the glycogen supply whereas feeding may increase the muscle content of glycogen. Temperature of the carcass post-mortem can alter the rate of pH decline. Like most chemical reactions the warmer the environment the faster the reaction and in this case the faster the pH decline. With a rapidly chilled carcass the reverse is true.

ATP continues to be utilized after death and continues to be rephosporylated until the reserve supply of creatin phosphate is exhausted. At this point the quantity of ATP decreases and the cross-links between actin and myosin filaments become rigid.

The rate of decline and ultimate pH have an undesirable influence on two conditions that develop post-mortem. Dark cutting beef is associated with a high ultimate pH. This is a genetic as well as a pre-slaughter stress associated problem. Pale-soft-exudative (PSE) pork is the other end of the spectrum and is currently believed to be associated with high temperatures (> 30° C) and low pH approximately 45 min post-mortem which leads to a reduction in extractability of sarcoplasmic and myofibrillar proteins (Briskey *et al.* 1966). Again, this is a genetic and stress problem and also there seems to be a correlation with stress susceptible (death caused by stress) strains of hogs.

Tenderness is drastically altered during rigor. Initially the muscle is tender but becomes less tender as rigor progresses. Aging accompanied by natural proteolytic enzymatic degradation causes the muscle sarcomere length to extend (Gotthard *et al.* 1966) and to become more tender. This is frequently accompanied by a fracture in the area of the Z-line, probably by hydrolysis of tropomyosin (Penny 1968). Rigor (actin and myosin connecting) may occur when the muscle is in any state of contraction or relaxation and consequently the muscle may have a variety of sarcomere lengths (length shortens slightly during rigor). Generally, the longer the sarcomere length, the more tender the muscle. It has been suggested that muscles be stretched prior to rigor development and maintained in this position to improve tenderness (Locker 1959, 1960; Herring *et al.* 1965A, 1965B). This leads to some additional tenderization but it has been learned that it is more important to prevent shortening than it is to assure maximum stretch. McCrae *et al.* (1971) reported that most muscles toughened when skeletal attachments were severed before chilling. A dramatic increase in tenderness can be achieved by microwave cooking immediately after death (Streitel 1974).

The degree of shortening of the sarcomere length (and consequently tenderness alteration) during rigor development is influenced by environmental temperature. As temperature is lowered to 15°C the degree of shortening decreases, and reaches a minimum at 15°C. Lowering the temperature further (below 15°C) causes the degree of shortening to increase once more (Hamm 1970). This means from a tenderness standpoint if a carcass is rapidly reduced in temperature below 15°C that the muscle will toughen and this is referred to as cold shortening (Locker and Hagyard 1963; Marsh and Leet 1966). Bacterial growth can also have a major influence on muscle protein degradation with prolonged storage (Ockerman *et al.* 1974; Ockerman *et al.* 1972; Hasegawa *et al.* 1970). The influence of bacteria seems to follow two routes: the first is degradation caused by bacterial growth and the second is the influence of altered pH produced by these microorganisms. Since specific microorganisms have the ability to alter the pH in either direction the organism involved will determine the changes that occur in the tissue. These changes can alter many parameters important to food utilization and are reviewed by Ockerman *et al.* (1972).

Changes During Freezing

Speed of freezing is quite important to the tissue alteration. Slow freezing begins by ice crystal formation in the extracellular spaces, concentration of soluble material in the frozen extracellular fluid, migration of water from muscle cells, dehydration and denaturation of cell protein, and large ice crystal formation in extracellular area causing a rupture of cell membranes. On the other hand, rapid freezing (fast transition through the 0 to -5°C zone) produces numerous small ice crystals located uniformly throughout the muscle, less translocation of water and less protein denaturation (Hamm 1970). If tissue is frozen pre-rigor and then thawed unrestrained, the muscle will shorten in a violent manner (thaw rigor) to 20–25% of its original length probably due to calcium release from a damaged sarcoplasmic reticulum (Perry 1950; Marsh and Thompson 1957, 1958). Frozen storage increases denaturation of protein that can be observed by a reduction in protein solubility.

Changes During Cooking

Changes in muscle protein during cooking are usually rather drastic. As the muscle is heated, the tissue shrinks and juice is lost due to changes in myofibrillar protein (Tyszkiewicz *et al.* 1966). During heating, color changes also occur, and loss of muscle enzyme activity is encountered due to denaturation of sarcoplasmic proteins. The solubility of myofibrils and sarcoplasmic proteins is also reduced during cooking (Cohen 1966; Paul *et al.* 1966). Physical and chemical changes in tissue (Hamm 1966; Hofmann 1966; Lawrie 1968; Thompson *et al.* 1968; Field *et al.* 1970; Hamm 1970; Schmidt and Parrish 1971) that occur during heating are outlined as follows.

(1) 20–30° C—No change in colloidal-chemical properties, solubility, or ion binding; reduction in myosin ATPase at 30° C.
(2) 30–50° C—Myofibrillar protein—unfolding of peptide chains, formation of cross-linkages, sarcomere shortened, and reduction in water holding capacity; sarcoplasmic protein, portions are denatured.
(3) 50–55° C—Myofibrillar protein—rearrangement of myofibril, stable cross bridges are formed; sarcoplasmic protein, denaturation continues.
(4) 55–80° C—Myofibril and globular protein are coagulated causing disintegration; collagen shrinks; loss of M-line.
(5) Above 80° C—Formation of disulfide bonds due to oxidation—Maillard reaction occurs (browning of meat).
(6) Sterilization temperature—Some amino acids destroyed.

Changes During Curing

Water binding is of major importance in all tissues, but this is particularly true with cured meat. Myosin, actin, pH, ions that can alter charges, and dissolved salt in the sarcoplasma, all have a marked influence on water binding capacity of meat (Hamm 1970). During rigor this water-holding capacity is drastically reduced. The addition of sodium chloride to muscle tissue increases the water-holding capacity at post-mortem pH by changing the charge on the myofibrillar proteins, causing adjacent peptide chains to repel each other due to electrostatic charges thus causing the tissue to swell and bind more water (Ockerman 1973). Phosphates also increase water-holding capacity by raising muscle pH, breaking myosin bridges, binding Ca^{++} and shifting the isoelectric point to a lower pH value for some proteins (Ockerman 1973). The increased ionic strength due to additives, and consequently the increased protein solubility, is taken advantage of in the new tumbling procedures used for cured tissue.

Changes During Comminution

Changes during comminution include physical rupture as well as the alterations caused by additives discussed in the previous section. The emulsion theory is that protein is extracted from the tissue and encapsulation of fat occurs forming an emulsion. Hegarty *et al.* (1963) reported that myosin is a much better emulsifier than actin for sausage production and Hamm (1970) found the emulsifying capacity series to be G-actin-myosin-actomyosin-sarcoplasmic proteins-F-actin. The sarcoplasmic proteins may take on increased importance in practice because they are more readily dissolved. Preblending is an attempt to increase extraction time and to produce a firmer emulsion. Photomicrographs of a model system incorporating excess quantities of fat show this encapsulated structure (Plimpton *et al.* 1972). More recent work on a conventional type of product would suggest that fat is distributed as single fat cells and clusters of fat cells suspended in a meat matrix in a dispersion rather than an emulsion form (Van den Oord and Visser 1973).

SUMMARY

Muscle is a very highly organized colloidal system that is composed of cells that have thick and thin filaments arranged in register so that they can slide together and the whole unit shortened for contraction. This complex system has its own energy manufacturing unit and is controlled by a complicated system of electrical and chemical impulses. This system is not only an architectural and functional masterpiece in life but upon slaughter the tissue can be utilized as a very excellent dietary protein source. These tissues can also be manipulated to produce a wide variety of useful dietary items.

BIBLIOGRAPHY

ALBERTY, R. A. 1968. Equilibrium hydrolysis of ATP and the binding constants of MgATP, CaATP, MgADP, and CaADP. J. Biol. Chem. *243*, 1337–1343.

ARNAUD, Y., LEGER, J. J., GRUDA, J., and LEROUX, L. 1970. To the purpose of actomyosin and a few of its physicochemical properties. European J. Bioch. *16*, 336–341. (French)

BAILEY, K. 1946. Tropomyosin: A new asymmetric protein component of muscle. Nature *157*, 368–369.

BAILEY, K. 1948. Tropomyosin: A new asymmetric protein component of the muscle fiber. Biochem. J. *43*, 271–279.

BEATTY, C. H., and BOCEK, R. M. 1970. Biochemistry of red and white muscle. *In* Physiology and Biochemistry of Muscle as Food, 2nd Edition, E. J. Briskey, R. G. Cassens, and B. B. Marsh (Editors). Univ. of Wisconsin Press, Madison.

BENDALL, J. R. 1966. Muscle as a contractile machine. *In* Physiology and Biochemistry of Muscle as a Food, E. J. Briskey, R. G. Cassens, and J. C. Trautman (Editors). Univ. of Wisconsin Press, Madison.

BENDALL, J. R. 1969. Muscles, Molecules and Movement. American Elsevier Publishing Co., New York.

BENNETT, H. S. 1960. The structure of striated muscle as seen by the electron microscope. *In* Structure and Function of Muscle, G. H. Bourne (Editor). Academic Press, New York.

BLOOM, W. and FAWCETT, D. W. 1968. A Textbook of Histology. W. B. Saunders Co., Philadelphia, Pa.

BRISKEY, E. J. and Fukazawa, T. 1971. Myofibrillar proteins of skeletal muscle. *In* Advances in Food Research, Vol. 19, C. O. Chichester, E. M. Mrak, and G. F. Stewart (Editors). Academic Press, New York.

BRISKEY, E. J. *et al.* 1966. Biochemical aspects of post-mortem changes in porcine muscle. J. Agr. Food Chem. *14*, 201-207.

BRISKEY, E. J., SERAYDARIAN, K., and MOMMAERTS, W. F. H. M. 1967. The modification of actomyosin by α-actinin. II. The interaction between α-actinin and actin. Biochim. Biophys. Acta. *133*, 424-434.

CARMICHAEL, D. J., and LAWRIE, R. A. 1967. Bovine collagen. I. Changes in collagen solubility with animal age. J. Food Technol. *2*, 299-311.

CASSENS, R. G., and COOPER, C. C. 1971. Red and white muscle. Advances in Food Research, Vol. 19, C. O. Chichester, E. M. Mrak, and G. F. Stewart (Editors). Academic Press, New York.

COHEN, E. M. 1966. Protein changes related to ham processing temperatures. I. Effect of time-temperature on amount and composition of soluble proteins. J. Food Sci. *31*, 746-750.

DEAMER, D. W., and BASKIN, R. J. 1969. Ultrastructure of sarcoplasmic reticulum preparations. J. Cell Biol. *42*, 296-307.

DEAN, R. B. 1948. Modern Colloids. D. Van Nostrand Co., New York.

DeROBERTIS, E. D. P., NOWINSKI, W. W., and SAEZ, F. A. 1970. Cell Biology, 5th Edition, W. B. Saunders Co., Philadelphia, Pa.

D'HAESE, J. and KOMNICK, H. 1972A. Fine structure and contraction of isolated muscle actomyosin. I. Evidence for a sliding mechanism by means of oligomeric myosin. Z. Zellforsch. Mikroskop. Anat. *134*, 411-426.

D'HAESE, J., and KOMNICK, H. 1972B. Fine structure and contraction of isolated muscle actomyosin. II. Formation of myosin filaments and their effect on contraction. Z. Zellforsch. Mikroskop. Anat. *134*, 427-434.

DRABIKOWSKI, W., KOMINZ, D. R., and MARUYAMA, K. 1968. Effect of troponin on the reversibility of tropomyosin binding to F-actin. J. Biochem. (Tokyo) *63*, 802-804.

DUGGAN, P. F., and MARTONOSI, A. 1970. Sarcoplasmic reticulum. 9. Permeability of sarcoplasmic reticulum membranes. J. Gen. Physiol. *56*, 147.

EBASHI, S., and EBASHI, F. 1964. A new protein component participating in superprecipitation of myosin B. J. Biochem. (Tokyo) *55*, 604-613.

EBASHI, S., and EBASHI, F. 1965. α-Actinin, a new structural protein from striated muscle. I. Preparation and action on actomyosin-ATP interaction. J. Biochem. *58*, 7-12.

EBASHI, S., EBASHI, F., and KODAMA, A. 1967. Troponin as a Ca^{++} receptive protein in the contractile system. J. Biochem. (Tokyo) *62*, 137-140.

EBASHI, S., ENDO, M., and OHTSUKI, I. 1969. Control of muscle contraction, Quart. Rev. Biophys. *2*, 351-384.

EBASHI, S., and KODAMA, A. 1965. A new protein factor promoting aggregation of tropomyosin. J. Biochem. (Tokyo) *58*, 107-108.

EBASHI, S., and KODAMA, A. 1966A. Native tropomyosin-like action of troponin on trypsin-treated myosin B (Actomyosin). J. Biochem. (Tokyo) *60*, 733-744.

EBASHI, S., and KODAMA, A. 1966B. Interaction of troponin with F-actin in the presence of tropomyosin. J. Biochem. (Tokyo) *59*, 425.

EBASHI, S., KODAMA, A., and EBASHI, F. 1968. Troponin. I. Preparation and physiological function. J. Biochem. (Tokyo) *64*, 465-526.

ELLIOTT, G. F., LOWY, J., and WORTHINGTON, C. R. 1963. X-ray diffraction studies of resting, living muscle. J. Molec. Biol. *6*, 295-305.

ELZINGA, M., and COLLINS, J. H. 1973. Complete amino acid sequence of actin. 166th National Meeting of American Chemical Society, Chicago, Ill.

ENDO, M. *et al.* 1966. Localization of native tropomyosin in relation to striation patterns. Jap. J. Biochem. *60*, 605-608.

FIELD, R. A., PEARSON, A. M., and SCHWEIGERT, B. S. 1970. Hydrothermal shrinkage of bovine collagen. J. Animal Sci. *30*, 712-716.

FINCK, H. 1968. On discovery of actin. Science *160*, 332.

FRANZINI-ARMSTRONG, C. 1963. Sarcolemmal invaginations and the T-system in skeletal muscle fibers. J. Cell Biol. *19*, 24A.

FUCHS, F., and BRIGGS, F. N. 1968. The site of calcium binding in relation to the activation of myofibrillar contraction. J. Gen. Physiol. *51*, 655-676.

GERGELY, J., GOUVEA, M. A., and KARIBIAN, D. 1955. Fragmentation of myosin by chymotrypsin, J. Biol. Chem. *212*, 165-177.

GOLDSPINK, G. 1970. Morphological Adaptation Due to Growth and Activity. In Physiology and Biochemistry of Muscle As a Food, 2nd Edition, E. J. Briskey, R. G. Cassens, and B. B. Marsh. Univ. of Wisconsin Press, Madison.

GOTTHARD, R. M., MULLENS, A. M., BOULWARE, R. S., and HANSARD, S. L. 1966. Histological studies of post-mortem changes in sarcomere length as related to bovine meat tenderness. J. Food Sci. *31*, 825-828.

GROSS, J. 1961. Collagen. Sci. Am. *204*, No. 5, 120-130.

HALLIBURTON, W. D. 1887. On muscle-plasma. J. Physiol. *8*, 133-202.

HAM, A. W. 1965. Histology. J. B. Lippincott Co., Philadelphia, Pa.

HAM, A. W. 1969. Histology, 6th Edition, J. B. Lippincott Co., Philadelphia, Pa.

HAMM, R. 1966. Heating of muscle systems. *In* The Physiology and Biochemistry of Muscle as a Food. E. J. Briskey, R. G. Cassens, and J. C. Trautman. Univ. of Wisconsin Press, Madison.

HAMM, R. 1970. Properties of meat protein. *In* Proteins as Human Food, R. A. Lawrie (Editor). Avi Publishing Co., Westport, Conn.

HANSON, J. 1967. Axial period of actin filaments. Nature (London) *213*, 353-358.

HANSON, J., and HUXLEY, H. E. 1960. The molecular basis of contraction in cross-striated muscles. *In* Structure and Function of Muscle, G. H. Bourne (Editor). Academic Press, New York.

HANSON, J. and LOWY, J. 1965. Molecular basis of contractility in muscle. Brit. Med. Bull. *21*, 264-271.

HARMON, J. W. 1964. The ultrastructure of the muscle cell. *In* Disorders of Voluntary Muscle, J. N. Walton (Editor). Little, Brown, Boston, Mass.

HARPER, W. J. 1972. Meat Proteins. Food Science and Nutrition 830.02. Ohio State Univ., Columbus.

HASEGAWA, T., PEARSON, A. M., PRICE, J. F., and LECHOWICH, R. V. 1970. Action of bacterial growth on sarcoplasmic and urea-soluble proteins from muscle. 1. Effects of *Clostridium perfringens, Salmonella enteritidis, Achromobacter liquefaciens, Streptococcus faecalis* and *Kurthia zopfii.* Appl. Microbiol. *20*, 117-122.

HASSELBACH, W. 1964. Relaxing factor and the relaxation of muscle. Progr. Biophys. Mol. Biol. *14*, 169-222.

HEGARTY, G. R., BRATZLER, L. J., and PEARSON, A. M. 1963. Studies on the emulsifying properties of some intercellular beef muscle proteins. J. Food Sci. *28*, 663-668.

HERRING, H. K., CASSENS, R. G., and BRISKEY, E. J. 1965A. Sarcomere length of free and restrained bovine muscles at low temperature as related to tenderness. J. Sci. Food Agr. *16*, 379-384.

HERRING, H. K., CASSENS, R. G., and BRISKEY, E. J. 1965B. Further studies on bovine muscle tenderness as influenced by carcass position, sarcomere length and fiber diameter. J. Food Sci. *30*, 1049-1054.

HERRING, H. K., CASSENS, R. G., and BRISKEY, E. J. 1967. Tenderness and associated characteristics of stretched and contracted bovine muscle. J. Food Sci. *32*, 317-323.

HODGKIN, A. L. 1965. Conduction of the Nervous Impulse. Liverpool Univ. Press, Liverpool, England.

HOFMANN, K. 1966. Über den nährwert des fleisches und seine veränderung beim erhitzen. Fleischwirts *46*, 1121-1124.

HOLTZER, A., CLARK, R., and LOWEY, S. 1965. The conformation of native and denatured tropomyosin B. Biochemistry *4*, 2401-2411.

HOLTZER, A., LOWEY, S., and SCHUSTER, T. M. 1961. Myosin. *In* Molecular Basis of Neoplasia (Papers presented at the 15th Annual Symposium on Fundamental Cancer Research). Univ. of Texas Press, Austin.

HUXLEY, H. E. 1953A. Electron microscope studies of the organization of the filaments in striated muscle. Biochim. Biophys. Acta. *12*, 387-394.
HUXLEY, H. E. 1953B. X-ray analysis and the problem of muscle. Proc. Roy. Soc., Sec. B *141*, 59-62.
HUXLEY, H. E. 1957. The double array of filaments in cross-striated muscle. J. Biophys. Biochem. Cytol. *3*, 631-647.
HUXLEY, H. E. 1958. The contraction of muscle. Sci. Am. *199*, No. 19, 66-82.
HUXLEY, H. E. 1963A. Electron microscope studies on the structure of material and synthetic protein filaments from striated muscle. J. Mol. Biol. 7, 281-308.
HUXLEY, H. E. 1963B. Electron microscope studies on the organization of filaments in striated muscle. Biochim. Biophys. Acta. *12*, 387-394.
HUXLEY, H. E. 1965. The mechanism of muscular contraction. Sci. Am. *213*, No. 6:18-27.
HUXLEY, H. E., and BROWN, W. 1967. X-ray diffraction studies of muscle. J. Mol. Biol. *30*, 383-434.
HUXLEY, H. E., and HANSON, J. 1957. Quantitative studies on the structure of cross-striated myofibrils. I. Investigations by interference microscope. Biochim. Biophys. Acta. *23*, 229-249.
HUXLEY, H. E., and HANSON, J. 1960. The molecular basis of contraction in cross-striated muscles. *In* Structure and Function of Muscle, G. H. Bourne (Editor). Academic Press, New York.
IKEMOTO, N., BHATNAGAR, G. M., and GERGELY, J. 1971. Fractionation of solubilized sarcoplasmic reticulum. Biochem. Biophys. Res. Commun. *44*, 1510-1517.
JIRGENSONS, B. 1958. Organic Colloids. American Elsevier Publishing Co., New York.
JIRGENSONS, B., and STRAUMANIS, M. E. 1954. A Short Textbook of Colloid Chemistry. John Wiley & Sons, New York.
JOBSIS, F. F., and O'CONNOR, M. J. 1966. Calcium release and reabsorption in the sartorius muscle of the toad. Biochem. Biophys. Res. Commun. *25*, 246-252.
KASAI, M. 1969. Thermodynamical aspect of G-F transformation of actin. Biochim. Biophys. Acta. *180*, 399-409.
KASTENSCHMIDT, L. L. 1970. Metabolism of muscle as a food. *In* Physiology and Biochemistry of Muscle as a Food, 2nd Edition, E. J. Briskey, R. G. Cassens, and B. B. Marsh (Editors). Univ. of Wisconsin Press, Madison.
KATZ, B. 1966. Nerve, Muscle and Synapse. McGraw-Hill and Co., New York.
KNAPPEIS, G. G., and CARLSEN, F. 1962. The ultrastructure of the Z-disc in skeletal muscle. J. Cell Biol. *13*, 323-335.
KOMINZ, D. R., HOUGH, A., SYMONDS, P., and LAKI, K. 1954. The aminoacid composition of actin, myosin, tropomyosin, and the meromyosins. Arch. Biochem. Biophys. *50*, 148-159.
KUHNE, W. 1863. A living nematode observed in a living muscle fiber. Arch. Pathol. Anat. Physiol. Virchow's *26*, 222-224. (German)
LAKI, K. 1971. Contractile Proteins and Muscle. Marcel Dekker, New York.
LAWRIE, R. A. 1950. Some observations on factors affecting myoglobin concentrations in muscle. J. Agr. Sci. *40*, 356-366.
LAWRIE, R. A. 1966. Meat Science. Pergamon Press, Oxford.
LAWRIE, R. A. 1968. Chemical changes in meat due to processing—a review. J. Sci. Food Agr. *19*, 233-240.
LOCKER, R. H. 1959. Striation patterns of ox muscle in rigor mortis. J. Biophys. Biochem. Cytol. *6*, 419-422.
LOCKER, R. H. 1960. Degree of muscular contraction as a factor in tenderness of beef. Food Res. *25*, 304-307.
LOCKER, R. H., and HAGYARD, C. J. 1963. A cold shortening effect in beef muscle. J. Sci. Food Agr. *14*, 787-792.

LOWEY, S., and COHEN, C. 1962. Studies on the structure of myosin. J. Mol. Biol. *4*, 293-308.
LOWEY, S., GOLDSTEIN, L., and LUCK, S. 1966. Isolation and characterization of a helical sub-unit from heavy meromyosin. Biochem. Z. *345*, 248-254.
LOWEY, S., SLAYTER, H. S., WEEDS, A. G., and BAKER, H. 1969. Substructure of myosin molecule. J. Mol. Biol. *42*, 1-30.
MARSH, B. B., and LEET, N. G. 1966. Studies in meat tenderness. III. The effects of cold shortening on tenderness. J. Food Sci. *31*, 450-459.
MARSH, B. B., and THOMPSON, J. F. 1957. Thaw rigor and the delta state of muscle. Biochim. Biophys. Acta. *24*, 427-428.
MARSH, B. B., and THOMPSON, J. F. 1958. Rigor mortis and thaw rigor in lamb. J. Sci. Food Agr. *9*, 417-424.
MARTONOSI, A. 1968. Sarcoplasmic reticulum. V. The structure of sarcoplasmic reticulum membranes. Biochim. Biophys. Acta. *150*, 694-704.
MARTONOSI, A. 1969. Protein composition of sarcoplasmic reticulum membranes. Biochem. Biophys. Res. Commun. *36*, 1039-1044.
MARUYAMA, K. 1965A. A new protein-factor hindering network formation of F-actin in solution. Biochem. Biophys. Acta. *94*, 208-225.
MARUYAMA, K. 1965B. Some physico-chemical properties of β-actinin, actin-factor, isolated from striated muscle. Biochim. Biophys. Acta. *102*, 542-548.
MASAKI, T., ENDO, M., and EBASHI, S. 1967. Localization of 6 S component of α-actinin at Z-band. J. Biochem. (Tokyo) *62*, 630-632.
MASAKI, T., TAKAITI, O., and EBASHI, S. 1968. M-substance, a new protein constituting the M-line of myofibrils. J. Biochem. (Tokyo) *64*, 909-910.
McCRAE, S. E., SECCOMBE, C. G., MARSH, B. B., and CARSE, W. A. 1971. Studies of meat tenderness. 9. Tenderness of various lamb muscles in relation to their skeletal restraint and delay before freezing. J. Food Sci. *36*, 566-570.
McFARLAND, B. H., and INESI, G. 1970. Studies of solubilized sarcoplasmic reticulum. Biochem. Biophys. Res. Commun. *41*, 239-243.
MERKEL, R. A. 1971. Carbohydrates. *In* The Science of Meat and Meat Products, J. F. Price, and B. S. Schweigert (Editors). W. H. Freeman and Co., San Francisco.
MEYERHOF, O. 1944. Physical changes of muscle related to activity. Colloid Chem. *5*, A883-900.
MIDDLEBROOK, W. R. 1959. The sub-unit structure of myosin. Abstr. Intern. Congr. Biochem. *4*, 84.
MIHALYI, E., and HARRINGTON, W. F. 1959. Studies of the tryptic digestion of myosin. Biochem. Biophys. Acta. *36*, 447-466.
MILLMANN, B. M., ELLIOTT, G. F., and LOWY, J. 1967. X-ray diffraction studies. Nature (London) *213*, 356-358.
MOMMAERTS, W. F. H. M. 1952. The molecular transformation of actin. I. Globular actin. J. Biol. Chem. *198*, 445-457.
MOMMAERTS, W. F. H. M. 1966. Molecular alterations in myofibrillar proteins. *In* Physiology and Biochemistry of Muscle as a Food, E. J. Briskey, R. G. Cassens, and J. C. Trautman (Editors). Univ. of Wisconsin Press, Madison.
MUELLER, H., and PERRY, S. V. 1962. The degradation of meromyosin by trypsin. Biochem. J. *85*, 431-439.
MURRAY, J. M., and WEBER, A. 1974. The cooperative action of muscle proteins. Sci. Am. *230*, No. 2, 58-71.
NANNINGA, L. B., and MOMMAERTS, W. F. H. M. 1960. Formation of an enzyme-substrate complex between myosin and ATP. Proc. Natl. Acad. Sci. *46*, 1166-1173.
NONOMURA, Y. 1967. A study on the physico-chemical properties of α-actinin. J. Biochem. (Tokyo) *61*, 796-802.
OCKERMAN, H. W. 1973. Chemistry of Meat Tissue. Animal Science Dept. Ohio State Univ., Columbus.

OCKERMAN, H. W. 1974. Chemistry of muscle and major organs. *In* Meat Hygiene, J. A. Libby (Editor). Lea and Febiger, Philadelphia.
OCKERMAN, H. W., CAHILL, V. R., and PROCTOR, G. O. 1974. Evaluation of bovine muscle inoculated post-mortem with *Pseudomonas putrefaciens, Bacillus subtilis*, or *Leuconostoc mesenteroides.* IV. International Congress of Food Science and Technology, Madrid.
OCKERMAN, H. W. *et al.* 1972. Germfree muscle techniques as a tool for studying microbiological roles in sausage production. 18th European Meat Research Conference, Guelph, Canada.
PARRISH, F. C., and BUSCH, W. A. 1971. Extent and role of proteolysis in post-mortem muscle. *In* Reciprocal Meat Conference Proceedings. National Live Stock and Meat Board, Chicago.
PARTRIDGE, S. M. 1966. Elastin. *In* Physiology and Biochemistry of Muscle as a Food, E. J. Briskey, R. G. Cassens, and J. C. Trautman (Editors). Univ. of Wisconsin Press, Madison.
PAUL, J. 1965. Cell Biology. Heinemann and Co., London.
PAUL, P. C., BUCHTER, L., and WIERENGA, A. 1966. Solubility of rabbit muscle proteins after various time-temperature treatments. J. Agr. Food Chem. *14*, 490-492.
PEACHEY, L. D. 1965. The sarcoplasmic reticulum and transverse tubules of the frogs sartorius. J. Cell Biol. *25*, 209-230.
PENNY, I. F. 1968. Effect of ageing on the properties of myofibrils of rabbit muscle. J. Sci. Food Agr. *19*, 518-524.
PERRY, S. V. 1950. Studies on the rigor resulting from the thawing of frozen frog sartorius muscle. J. Gen. Physiol. *33*, 563-577.
PERRY, S. V. 1967. The structure and interactions of myosin. Progr. Biophys. Mol. Biol. *17*, 325-381.
PIEZ, K. A. 1966. Collagen. *In* Physiology and Biochemistry of Muscle as a Food. E. J. Briskey, R. G. Cassens, and J. C. Trautman (Editors). Univ. of Wisconsin Press, Madison.
PLIMPTON, R. F. *et al.* 1972. Emulsion characteristics and processing feasibility of sausage products incorporating unconventional protein sources. 18th European Meat Research Conference, Guelph, Canada.
PODOLSKY, R. J., and MORALES, M. F. 1958. The enthalpy change of ATP hydrolysis. J. Biol. Chem. *218*, 945-959.
PORTER, K. R., and FRANZINI-ARMSTRONG, C. 1965. The sarcoplasmic reticulum. Sci. Am. *212*, No. 3, 72-80.
PRICE, J. F., and SCHWEIGERT, B. S. 1971. The Science of Meat and Meat Products. W. H. Freeman and Co., San Francisco.
REESE, M. K., and YOUNG, M. 1967. Studies on the isolation and molecular properties of globular actin. Evidence for a single polypeptide chain structure. J. Biol. Chem. *242*, 4449-4458.
RICE, R. V., BRADY, A. C., DEPUE, R. H., and KELLY, R. E. 1966. Morphology of individual macromolecules and their ordered aggregates by electron microscope. Biochem. Z. *345*, 370-393.
SCHIAFFINO, S., and MARGRETH, A. 1969. Coordinated development of sarcoplasmic reticulum and T system during postnatal differentiation of rat skeletal muscle. J. Cell Biol. *41*, 855-875.
SCHMIDT, J. G., and PARRISH, F. C. 1971. Molecular properties of post-mortem muscle. 10. Effect of internal temperature and carcass maturity on structure of bovine Longissimus. J. Food Sci. *36*, 110-119.
SCOPES, R. K. 1970. Characterization and study of sarcoplasmic proteins. *In* Physiology and Biochemistry of Muscle as a Food, 2nd Edition, E. J. Briskey, R. G. Cassens, and B. B. Marsh (Editors). Univ. of Wisconsin Press, Madison.
SLAYTER, H. S., and LOWEY, S. 1967. Substructure of the myosin molecule as visualized by electron microscopy. Proc. Natl. Acad. Sci. *58*, No. 4, 1611-1618.

SMITH, D. W., BROWN, D. M., and CARNES, W. H. 1972. Preparation and properties of salt-soluble elastin. J. Biol. Chem. *247*, 2427-2432.
SOUTHARD, J. H., and HULTIN, H. O. 1969. Glycolytic activity of chicken breast muscle mitochondria. J. Food Sci. *34*, 622-624.
STRAUB, F. B. 1942. G- and F-actin and effect of ATP. Studies Inst. Med. Chem. Univ. Szeged. *2*, 3.
STRAUB, F. B., and FEUER, G. 1950. Adenosinetriphosphate, the functional group of actin. Biochim. Biophys. Acta. *4*, 455-470.
STREITEL, R. H. 1974. Maintenance of the Inherent Tenderness of Beef by the Alteration of Rigor Mortis. M. S. Thesis, Ohio State University, Columbus.
SZENT-GYÖRGYI, A. G. 1941-1943. Studies Inst. Med. Chem., Univ. Szeged *1*, 4.
SZENT-GYÖRGYI, A. G. 1960. Proteins in the myofibril. *In* Structure and Function of Muscle, Vol. II, G. H. Bourne (Editor). Academic Press, New York.
SZENT-GYÖRGYI, A. G. 1966. Nature of actin-myosin complex and contraction. *In* Physiology and Biochemistry of Muscle as a Food, E. J. Briskey, R. G. Cassens, and J. C. Trautman (Editors). Univ. of Wisconsin Press, Madison.
THOMPSON, G. B., DAVIDSON, W. D., MONTGOMERY, M. W. and ANGLEMIER, A. F. 1968. Alterations of bovine sarcoplasmic protein as influenced by high temperature aging. J. Food Sci. *33*, 68-72.
TSAO, T. C., BAILEY, K., and ADAIR, G. S. 1951. The size, shape, and aggregation of tropomyosin particles. Biochem. J. *49*, 27-36.
TYSZKIEWICZ, S., TYSZKIEWICZ, I., and DUKALWSKA, M. 1966. Annual Report of the Meat Research Institute *3*, No. 1, 29. (Polish)
VAN DEN OORD, A. H. A., and VISSER, P. R. 1973. State and distribution of fat in finely comminuted meat products. Die Fleischwirtschaft *53*, No. 10, 1427-1432. (German)
VEIS, A. 1970. Collagen. *In* Physiology and Biochemistry of Muscle as a Food, 2nd Edition, E. J. Briskey, R. G. Cassens, and B. B. Marsh (Editors). Univ. of Wisconsin Press, Madison.
VIBERT, P. J., HASELGROVE, J. C., LOWY, J., and POULSEN, F. R. 1972. Muscle-structural changes in actin-containing filaments. Nature—New Biol. *236*, 182-183.
WEBER, A. 1970. Relaxation: interactions of ions and proteins. In Physiology and Biochemistry of Muscle as a Food, 2nd Edition, E. J. Briskey, R. G. Cassens, and B. B. Marsh (Editors). Univ. of Wisconsin Press, Madison.
WINTON, R. F., and BAYLISS, L. E. 1962. Nerve Neuromuscular and Synaptic Transmission, and the Central Nervous System. *In* Human Physiology, 5th Edition, Churchill Ltd., London.
WOODS, E. F. 1967. Molecular weight and sub-unit structure of tropomyosin B. J. Biol. Chem. *242*, 2859-2876.
ZOBEL, C. R., and CARLSON, F. D. 1963. An electron microscope investigation of myosin and some of its aggregate. J. Mol. Biol. 7, 78-89.

CHAPTER 6

Arthur F. Novak
Ramu M. Rao
Durward A. Smith

Fish Proteins

INTRODUCTION

A thorough knowledge of the chemical and physical properties of fish is of paramount importance in evaluating its use as a safe and nutritious food. Some of the more important physical properties of fish proteins are related to their colloidal properties and thus have great importance in both the physical structure of the fish and the foods of fish origin.

To achieve an understanding of fish as a food product, we must examine the composition of its edible portion. While in some fishery products the whole fish may be used, most human foods utilize only the "edible portion" or the skin- and bone-free fillet of fish.

Of this edible portion, water is the major component, amounting to as much as 80% of the total weight. The water content varies only slightly and is inversely related to the oil content, so that the sum of the oil and water is very close to 80% of the total muscle weight. The water of fish flesh exhibits a number of particular traits which are directly related to both colloidal and chemical forces. Moisture in fish tissue is held tightly by these forces so that flesh subjected to high pressures does not release much water. This retention of water can be measured by centrifuging the fish and determining the amount of fluid which separates after applying a stipulated centrifugal force for a stated period of time. The water retentivity of fish flesh is greatest in freshly caught, untreated fish. Cold storage and freezing both reduce the retentivity.

Protein, the component which ranks second to water in quantity, ranges from 6-28% of the total weight, but usually comprises approximately 18-20% of the total weight, and is by far the most important component of fish used as food. These proteins can be broken down to α-amino acids by treatment with acids or enzymes. The assortment of amino acids, their various orders, and their spacial arrangements determine the functions which the protein may perform. In fish, the skin and fins, the working parts of the muscle, the enzymes and hormones, the blood and muscle pigments, the liver and kidney cells and the linings of the intestinal tract are mostly or entirely protein in nature.

The other major constituent of fish is the body oils which differ from plant and animal fats in two ways.

(1) Animal and plant fats very seldom have fatty acids with more than 18 carbon atoms. In fish the chain length up to 1/3 of the fatty acids exceeds 18 carbon atoms. Most of the long-chain fatty acids consist of C_{20} and C_{22} fatty acids.
(2) The fatty acids of fish oils contain more double bonds than do plant or vegetable oils, with the C_{20} fatty acids often being hexenes.

The deposit fats of fish consist primarily of triglycerides while the lipids associated with the cells of the flesh occur in nontriglyceride forms such as phospholipids. These lipids, along with the various inorganic components shown in Tables 6.1 and 6.2 and minor components such as glycogen and urea, are all associated in the colloidal suspension known as the sarcoplasma or cell fluid. A detailed examination of these colloidal components must be undertaken to appreciate the properties which fish muscle protein possesses.

Many species of fish have muscle tissue which is higher in protein, lipids, and water than domestic animal flesh. In most fatty fish the fat content increases in spring and summer with correspondent de-

TABLE 6.1

VITAMIN CONTENT OF EDIBLE FLESH OF FISH

Vitamin	Units	Content per 100 gm		
		Average	Usual Range	
			Low	High
Vitamin A[1]	μg	25.0	10.0	100
B Vitamins				
Thiamine	μg	50.0	10.0	100
Riboflavin	μg	120.0	40.0	700
Nicotinic acid	mg	3.0	0.5	12
Vitamin B-12	μg	1.0	0.1	15
Pantothenic acid	mg	0.5	0.1	1
Pyridoxine	μg	500.0	50.0	1,000
Biotin	μg	5.0	0.001	8
Folic acid	μg	80.0	71.0	87
Vitamin C	mg	3.0	1.0	20
Vitamin D[1]	μg	15.0	6.0	30
Vitamin E[1]	μg	12.0	4.0	35

[1] Fish of medium or high oil content.

TABLE 6.2

INORGANIC ELEMENTS IN THE EDIBLE FLESH OF FISH

Mineral	Average Content mg %
Potassium	300
Chloride	200
Phosphorous	200
Sulfur	200
Sodium	63
Magnesium	25
Calcium[1]	15
Iron	1.5
Manganese	1
Zinc	1
Fluorine	0.5
Arsenic	0.4
Copper	0.1
Iodine	0.1

[1] Calcium values for canned fish with bone included is approximately 200 mg %.

creases in water content, while the nitrogen values remain nearly constant. In many species the water content of the muscle accounts for 3/4 of the muscle mass (75–80%). The solid matter consists mainly of nitrogenous compounds (17–24%), and lipids (0–20%) (McCallum *et al.* 1969). Proximate analysis of whole beach-spawning capellin shows a sharp decline in fat content, a slight increase in moisture and a small decrease in protein during winter. The same results were observed in yellowtail muscle over an 11-month period between May and March (Shimizu *et al.* 1973).

Frontier-Abou (1969) examined the muscle of 743 fish from 31 edible species for the determination of water, lipid and protein content. The average values obtained were: water 77.46%, lipids 0.83% and protein 20.34%. Intraspecies variability was often found to be greater than interspecies differences.

Red lateral muscles of certain fish have been shown to be rich in the chromoproteins, cytochrome and myoglobin (Umemura 1951; Huys 1954; Matsurra and Hashimoto 1955; Rossi-Fanelli and Antonini 1955). The red lateral muscles have been shown to be especially rich in cytochrome-c, which differs spectroscopically from that of horse heart muscle (Matsurra and Hashimoto 1955). The crystalline cytochrome-c isolated from the heart muscle of tuna and bonito, however, appears to be nearly identical to the crystalline isolate from beef heart muscle (Hagihara *et al.* 1957).

The hemoglobin content of fish apparently depends upon various environmental and physiological conditions. The blood hemoglobin content of migratory fish has been shown to vary from as low as 2.0 gm per 100 ml to a high of 11.8 gm per 100 ml (Gelineo and Gelineo 1955; Brull and Cnypers 1954). Saito (1954) reported the blood hemoglobin content of migratory fish to be greater than that of less active fish. Hemoglobin exhibits species specific chemical variations and can be classified starch-gel electrophoresis (Tsuyuki and Roberts 1966; Yamanaka *et al.* 1965, 1967; Amano *et al.* 1971; Yoshiyasu and Humoto 1972). Yoshiyasu and Humoto (1972) used this method to confirm that all *Salvelinus* fishes in Japan except *ashorokoma* can be classified as one species on the basis of the geographically continuous variation in the number of plyoric caecums and gill rakers.

Protamines are a group of proteins which were first isolated from fish sperm. The distinctive features of these proteins are their size and amino acid content. Protamines are among the smallest proteins, having molecular weights in the neighborhood of 5,000. The protamines are exceedingly rich in the basic amino acid arginine, which may account for from 70-80% of the total amino acid content. There are no amino acids containing sulfur in the protamines. Their high basicity (pH ranges from 10-12) accounts for their ready formation of salts and their tendency to associate with acidic proteins and nucleic acids.

The connective tissue constitutes a large portion of the protein in fish. The connective tissue which has been most extensively studied is collagen, the major constituent of cartilage and the other connective tissues. Collagen is characterized by a high content of glycine, proline, and hydroxyproline, but is distinctly lacking in amino acids containing sulfur. The collagens are the only proteins known to contain significant quantities of hydroxyproline. They are solubilized in the presence of dilute acidic buffers of pH 3-4, and are reprecipitable at neutral pH. Prolonged boiling in water converts collagens to soluble proteins known as gelatins (Bear 1952).

FUNCTIONAL PROPERTIES

Fish proteins are divided into classes based upon their solubilities or on the nonprotein components which in some instances are integral parts of the substances. Water-soluble proteins are called albumins, and the salt-soluble proteins are known as globulins. Albumins comprise 10-20% of the proteins of fish muscle, while 70-90% are globulins. Thus nearly all of fish muscle is soluble if the proper combination of salts and water is used. The insoluble proteins, keratin and collagen, are characterized by their conversion to gelatins and glues in hot water. An appreciable part of the structure of fish is collagen. This includes the structural components of tendons, cartilage, skin and the eye. The skeletal systems of elasmobranchs are also collagens, as are numerous types of specialized proteins. Fish eggs contain lipoproteins in which the protein is combined with fat-soluble components. The heme-proteins, hemoglobin in the blood and myoglobin of the muscle, contain small amounts of iron. Other specialized proteins contain various metal and phosphorous groups.

Fish muscle is approximately 16% nitrogen by weight. This total nitrogen is divisible into two fractions, (1) nitrogenous extractives, and (2) proteins (Shewan 1951). Nitrogenous extractives, or nonprotein nitrogen, have been reported at between 9–14% of the total nitrogen in flatfishes, between 14–18% in the herring group, and between 34–38% in the elasmobranchs. The nonprotein nitrogen consists of: creatin, an energy storehouse in the muscle which is also found in amphibians and mammals; carnosine and ansarine which are short-chain peptides sometimes found in fish muscle but normally absent (Shewan 1951); and trimethylamine oxide, bitain and ammonia which are common in fish muscle but not found in mammalian muscle. Trimethylamine oxide is primarily responsible for the high nonprotein nitrogen content of elasmobranchs, where it functions as an osmo-regulator.

The colloidal suspension within the muscle cell is a dynamic system with fluctuations in the composition of fish which seem to be due largely to the replacement of a portion of the muscle water by lipids and proteins. The composition of the free amino acid pool in muscle of freshly killed North Atlantic cod has been reported by Danberges *et al.* (1968), showing the total amount of free amino acids to comprise both the intracellular plasma pools, and the amino acids in the form of small peptides. A total of 19 free amino acids have been reported in fish muscle. Histidines were found to be the major component in the free amino acid pool, with more than 90% of the free amino acid pool represented by histidine, taurine, glycine, lysine and alanine. The free amino acid pool, however, constitutes only 2.4% of the total amino acids present in fish protein. Protein extractability decreases as the free amino acid content increases in cod muscle which is aged on ice (Anderson and Steinberg 1964). This illustrates one of the problems encountered in accurately determining the chemical composition of fish. Problems such as post-mortem changes, seasonal changes and geographic variability make the usability of much of the early work done on the chemical composition of fish questionable.

Structurally, fish muscle contains two main groups of proteins, (1) the soluble proteins of the sarcoplasm or intercellular fluid, and (2) the structural proteins of the myofibrils. The main components of structural protein are actomyosin, topomyosin, myosin and actin. Their solubilization is influenced by many factors including (1) the nature of the tissue, (2) the physiological state of the fish, (3) the degree of subdivision, (4) the ionic strength of the solution, (5) the pH, and (6) the duration of the extraction. The minimum ionic strength necessary to bring these proteins into solution at neutral pH ranges between 0.3–0.4 depending upon species. Actomyosin is more easily extracted from red muscles than from white due to the higher adenosine triphosphate content of red muscles (Crepax 1952; Lawrie 1953).

The extractability of the protein of frozen stored cod muscle in cold, neutral 5% sodium chloride solution can be used as a measure of cold storage protein denaturation. Such denaturation is defined as "a change in the protein, such that it is no longer soluble or extractable by salt solutions under conditions in which the native protein is soluble or extractable" (Cowei and Mackie 1968). It is of interest to use such a test to compare the characteristics of fish muscle with those of other animals such as the rabbit. Hasselbach and Webster (1953) showed that myosin can be quantitatively extracted from coarsely minced rabbit muscle in a buffer at pH 6.0–6.4 with an ionic strength of 0.6 *M*. Actin was brought into solution by further mincing the muscle in a Waring blender. Actomyosin was found to go into solution only at pH values of 7.0 and above. This contrasts with fish extracts obtained at a pH of 6 which exhibit a strong flow birefringence and thus contain actomyosin. The long F-actin filaments of the muscle fibers are apparently already cut into shorter fragments at slightly acid pH permitting solubilization, while the selective extraction of myosin does not appear to be possible. From this we can see that the stabilities of fish and rabbit muscles are quite different in the low pH ranges. When the pH is adjusted to 7.5 the minimal ionic strength necessary to solubilize fish proteins is no longer lower than that necessary for the solubilization of animal protein (Hamoir 1955).

Fish muscle is less stable in the frozen state than is mammalian muscle. It loses its tenderness and becomes tough while a simultaneous denaturation of actomyosin occurs. This general instability does not seem to be attributable to the lower content of stroma or insoluble protein in the fish muscle, but may be due to such things as drip loss and loss of free amino acids.

Changes in proteins and free amino acids seriously affect the organoleptic quality of the flesh as was reported in pieces of carp fillets during storage at -3°C, -8°C, and -38°C. The most serious changes were observed at -3°C. From this it follows that the increase in maximum storage time obtainable by chilling the catch in iced holds at -3°C is of questionable advantage when compared to normal icing. The amount of protein denaturation found during storage at a few degrees below 0°C seems to depend partially on the initial freezing temperature (Partmann 1968). Objective tests for fish freshness commonly include pH, trimethylene concentration and hypoxanthene concentration for iced fish; while deterioration, pH, extractable protein, cell fragility, and color ratio are used as criteria for fish held in frozen storage (Connell and Howgate 1969).

Most of these storage changes can be tied to a breakdown of cell protein. Protein extractability decreases as free fatty acid content increases in cod muscle aged in ice. Increases in free fatty acids indicate a breakdown of the fish oils which is an indication of decreasing freshness (Anderson and Steinberg 1964). Ultracentrifugal patterns of protein extracted from aging muscle showed increasing polydispersity, while phase contrast microscopy showed that the inextractable material contained muscle fragments consisting of bundles of myofibrils, some being as wide as the full muscle fiber. These results tend to indicate that in aging muscle an interaction of contractile protein with free fatty acids occurs which results in the formation of a cross-linked network within the muscle fiber causing resistance to fragmentation and protein extractability. The smaller loss observed during aging in ice is in part caused by dissociation occurring during extraction. When muscle is aged in the frozen state, the reaction between contractile protein and free fatty acids increases and the complexes formed are stabilized. Work with protein extractability by Umemoto and Kanna (1969) indicates that there is little protein denaturation even after long periods of cold storage. The solubility of myosins soluble in 0.6 *M* potassium chloride is greater at rigor than at the pre- or post-rigor stages, while the amount of sarcoplasmic protein extractable with 0.005 *M* potassium chloride has been found to decrease during storage. Such salting out curves of myosins show that peaks of actomyosin or myosin and tropomyosin appear at any stage of rigor while the actin peak appears only in rigor and post-rigor mortis (Maruyama and Suzuki 1968).

Actomyosin is a compound formed between actin and myosin during contraction of the muscle. The investigations considering fish actomyosin were first oriented towards the study of the denaturation of actomyosin. Fish actomyosin easily becomes insoluble in salt solutions with its instability increased by successive precipitations. This change in solubility has been used by several investigators to study the denaturation of actomyosin. The denaturation caused by freezing actomyosin gels *in vitro* has been compared with the denaturation caused by freezing muscle *in situ*, indicating that the influence of the speed of freezing and duration of cold storage is the same in both cases, but that the denaturation occurs more slowly in muscle. Snow (1950), using pure haddock actomyosin, concluded that precipitated protein is denatured much more quickly by freezing than is dissolved protein. The extent of denaturation parallels the freezing time, with more rapid freezing causing less denaturation than slow freezing. Denaturation from solution is accelerated at pH values lower than 6.5 or higher than 7.8. Correlations have been drawn between the denaturation of structural protein and the oxidation of lipids or fatty acids (King *et al.* 1962; Anderson and Steinberg 1964). The role of tocopherol on the quality of fish muscle has been examined with specific reference to the effect of α-tocopherol on the solubility of actomyosin (Ikeda and Taguchi 1966, 1967; Ikeda *et al.* 1966). By the addition of α-tocopherol, the increase in the thiobarbituric acid value was effectively diminished, while the insolubilization of actomyosin was only slightly prevented. On the other hand, the stability of actomyosin itself in the absence of fatty acid was markedly improved.

The proteins of a muscle cell are numerous and comprise those important in contraction, the function of the nucleus and in enzymatic reactions of the cell. The protein myosin is one of these proteins and is necessary both for contraction and as an enzyme for ATP. Together with actin, myosin forms the contractile protein actomyosin.

For some years the existence of myosin in fish was questionable due to the inability to isolate and identify it as a myosin (Hamoir 1949, 1951, 1955). Then Connell (1954) described a method for the isolation of cod myosin, whereby coarsely minced muscle was extracted with a 0.35 *M* phosphate potassium chloride solution at pH 6.5 containing 0.01 *M* pyrophosphate. The myosin was then separated by water dialysis and the precipitate was washed and reprecipitated under the same conditions. The resulting clear solution was then purified by electrophoresis and was shown to have nearly the same intrinsic viscosity and isoelectric point as rabbit myosin. When combined with cod F-actin it produced an artificial actomyosin, having an intrinsic viscosity of about 1/2 that of the natural actomyosins. Ultracentrifugal studies showed the preparation to differ widely in size and shape from rabbit myosin. The existence of true myosin in fish muscle, however, has now been unequivocally established by ultracentrifugation of carp actomyosin in the presence of ATP (Hamoir 1951, 1955). Under these conditions the actomyosin peak disappears completely and a new one which sediments more slowly becomes visible. The sedimentation constant is 6.55.

Another major structural protein is actin. The study of the electrophoretic behavior of actin presents a special problem in view of its well-known ability to occur in two forms: (1) the globular form existing in salt-free solution (G-actin) and (2) the fibrillar form, which is formed by the polymerization of G-actin with salts (F-actin).

The freezing point of fish flesh is also related to the colloidally and chemically bound water. The water in fish does not freeze at 0°C, but rather it begins to freeze at approximately -1°C. As the temperature of the fish is lowered the amount frozen at successively lower temperatures increases. At -30°C about 90% of the water is frozen, and finally at -34°C essentially all of the water is frozen. Although the serum contains various salts, urea and free amino acids which depress the freezing point, the macromolecular glycoproteins appear to be the biological anti-freeze most responsible for the freezing point depression.

Glycogen, which functions as a nutrient, is an important component of the colloidal structure of living fish muscle, and is present in remarkable quantities (Dill 1921; MacLeod and Simpson 1927; Partmann 1965). Historically the glycogen content of fish muscle has been considered to be lower than that of mammalian muscle. However, many species of fish have a muscle glycogen content which compares favorably with that of mammals. Since the procedures used to capture fish almost invariably involves excessive struggling, the glycogen content of fish flesh reaching the market is usually very low. The products of post-mortem degradation of glycogen, however, are present and undoubtedly contribute to both the flavor and the texture of the fish. The glycogen content represents between 0–0.85% of the fresh weight of fish (Amano *et al.* 1971, 1953). A study of glycogen reduction related to the conditions of catching and handling showed the glycogen content to be 0.6% when the motor nerve is severed immediately after catching, 0.34% after beheading, and 0.11% when the fish dies after struggling.

The carbohydrate content has a decided effect on salt solubility of the protein. The salt solubility of freshly prepared myofibrillar protein ranges from 70–80%. Freeze drying immediately reduces this solubility to about 15%, indicating severe denaturation of the proteins. Yasui and Hashimoyo (1966), however, showed that sugars can exert a protective effect on the native properties of myosin even after freezing and freeze drying. Love (1962) reported that related compounds such as glycerol also protect fish protein against denaturation during frozen storage. The salt solubility of the protein is almost directly related to the sugar content. Glucose exerted slightly more protection than sucrose at lower concentration. The effect of fructose is much different from glucose and sucrose. At concentrations from 0-15%, fructose exerts practically no effect on the solubility of the protein.

Freeze-dried samples of myofibrillar protein were white in color and remained organoleptically stable for about 7-14 days of storage at room temperatures in air-permeable containers. Longer storage periods resulted in the development of off odors and flavors. These flavors were typified as rancid or "fishy." Vacuum packing aided in maintaining the original organoleptic integrity of the product by retarding lipid alteration.

The enzymes rank among the most important of the proteins in the colloidal system of fish muscle. Crystalline pepsins have been prepared from the stomachs of salmon (Norris and Elam 1940) and three species of tuna (Norris and Mathies 1953), and were found to differ from hog or beef pepsins in substrate specificity, crystal structure and amino acid composition. The presence of a cathepsin has been demonstrated in the stomachs of certain fresh water teleost fish (Buchs 1953), while the gastric mucosa of elasmobranch fish possesses both a cathepsin and a pepsin with optimum pH activity of 2.0 and 4.5, respectively.

Most spoilage in fish is due to post-mortem proteolyic activity. Such spoilage is studied by inhibiting bacterial growth in the muscle during storage by treating the muscle with chemical preservatives such as toluene, or by prior sterilization of the tissues with ionizing irradiation. Slow proteolysis is observed in nearly every case, as judged by increases in tyrosine, ammonia or amino nitrogen. An investigation in which extracts of tissue homogenates were used for determining cathepsin (pH 4.0) and neutral protease (pH 7.6) activity showed no neutral protease activity in the muscles and organs of several widely different fishes, while the muscles and more especially the liver and spleen possessed a much more marked cathepsin activity than did similar mammalian tissues.

Optimum conditions of autolysis of various fish muscles in the presence of a toluene-chloroform mixture have been determined. Manita *et al.* (1969) investigated autolysis under aseptic conditions without the addition of any antiseptic. The autolysis of mackerel muscle homogenate was shown to proceed readily at 45° C, and pH 3, and at 60-65° C in a slightly alkaline pH range. Antiseptics such as toluene, chloroform and thymol did not affect the optimum pH (3.5) of autolysis at 45° C, but considerably lowered the rate of autolysis. These antiseptics inhibited the brown discoloration and fat oxidation in the mackerel muscle homogenate during the autolysis.

Arginase reportedly occurs in red lateral muscle of tuna (Matsuura

et al. 1953) and in dogfish, but not in skate muscles. Amino acid decarboxylases have been variously reported as being absent and present (Simidu *et al.* 1953) in fish muscles and livers. Carbonic anhydrase content varies with the season and decreases markedly when fish struggle prior to capture. The quantity present never exceeds 850 mg per 100 gm (Tarr 1950; MacLeod and Simpson 1932; Sharp 1934). Post-mortem glycogenolysis results in the formation of lactic acid, and proceeds more rapidly at 15° C than at 0° C (Odense *et al.* 1966). In the temperature range of 0° C to -10° C the maximum has been found to occur at about -3° C (Sharp 1935). The glycogen content of pike was found to remain relatively high even after seven days of ice storage. This contrasts with the findings for several other species including the white sucker, where the muscle glycogen is nearly completely degraded in 3–4 days. Higher glycogen content in the posterior portion of pike muscle as compared with that of the anterior portion may partially explain the apparently high glycogen content in this fish after several days of storage (Manohar 1970). Mechanisms for the breakdown and synthesis of glycogen in the tissues of marine animals have been discussed by Nagayama (1966). Free reducing sugar (MacPherson 1932), identified as glucose, occurs in fish muscles in concentrations which range from 1–74 mg per 100 gm (Jones 1955; Tarr 1954). There has been some confusion regarding the origin and fate of glucose in post-mortem, because certain fish frozen in liquid nitrogen have a muscle glucose content that is negligible initially, but which increases markedly after a few days at 0° C (Tarr 1954). Normally glucose is present in cod muscles at the time of slaughter and decreases very slowly at 0° C due to bacterial action (Jones 1955). In fish liver both lactic acid and reducing sugar accumulates post-morten (Sharp 1935). The enzymatic steps which are involved in these degradations remain to be determined.

Reports have been published on the solubility and recovery characteristics of the isolates of whole raw fish and frozen fish at various pH's, salt concentrations and temperatures. Two major protein fractions of fish muscle (myofibrillar and sarcoplasmic) were separated and some of the physical, chemical and organoleptic properties of each were determined before and after drying, and the use of some proteolytic enzymes were evaluated as an aid in modifying the physical and chemical characteristics of the separated protein fractions. Basically, the protein components of fish after evisceration, skinning and deboning consist of myofibrillar and sarcoplasmic proteins in a ratio of about 70% myofibrillar to 30% sarcoplasmic (Dassow *et al.* 1970). Muscle tissue also contains highly unsaturated lipids, phospholipids and considerable amounts of nitrogenous compounds

(NPN) including peptides, amino acids, free amines, betaines, trimethylamine oxide and urea. After three extractions, approximately 95% of the sarcoplasmic protein is removed from the tissue. Estimation of protein removal was made by subtracting the amount of protein found in the extracting solution from the original total protein content of the comminuted tissue. The small amount of protein found in the extracting solution after the fourth extraction was probably due to some solubilization of actomyosin. The amount of non-protein constituents that was removed after each extraction followed the same pattern as that found for sarcoplasmic proteins.

The total lipid content of the tissue was reduced by 70% after three extractions. The first extraction removed 40%, the second 20% and the third 10%. A fourth extraction did not significantly reduce the lipid content. An increase in the ratio of phosphorus to total residual lipids after the first extraction of the myofibrillar fraction shows that the lipids removed were mostly triglycerides. The subsequent decrease in phosphorus content of the lipids which remained on the myofibrillar protein corresponded with increases in the phosphorus content of the lipids associated with the sarcoplasmic fraction, showing that such extractions progressively remove increasing amounts of phospholipids. However, after three extractions, the phospholipid content of the lipids remaining on the myofibrillar fraction showed only a slight decrease and the gross composition of the lipids remaining in this fraction was nearly the same as that originally present in the tissue.

Myofibrillar protein has low water solubility near neutrality (Fig. 6.1) but is highly soluble, forming very viscous solutions at pH's about 10.0 and below 3.5. There is a significant decrease in solubility at pH values below 4, with the isoelectric point shifting from 5-4.5. The dried proteins have a comparatively higher solubility between pH 4.5 and 8.0 and a lower solubility at pH values ranging from 9-11. Myofibrillar protein isolates produced by extracting with hot polar solvents such as isopropanol, have drastically reduced functional properties. For example, freshly prepared freeze-dried myofibrillar protein will emulsify approximately 185 gm oil per gram of protein. Extracting the myofibrillar protein with IPA (at 20°C) prior to freeze drying reduces the emulsifying capacity to 136 gm oil per gram of protein. Extractions carried out at 50°C reduce the emulsifying capacity to less than 90 gm oil per gram of protein and at 70°C all emulsifying capacity is destroyed. In addition, extractions at temperatures lower than 50°C do not effectively remove all of the lipids particularly the phospholipids, resulting in organoleptic deterioration after about one month of storage at room temperatures. Extraction

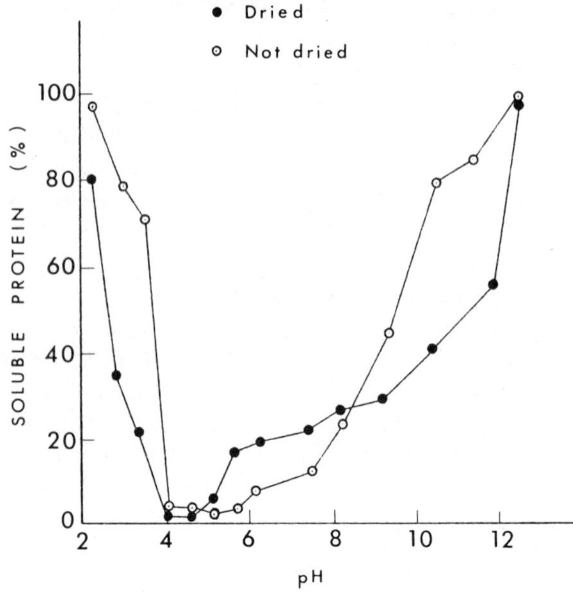

FIG. 6.1. SOLUBILITY OF MYOFIBRILLAR PROTEIN BEFORE AND AFTER FREEZE DRYING AT pH VALUES RANGING FROM 2-12

of the dry myofibrillar preparations with nonpolar solvents such as hexane (either hot or cold) did not remove sufficient phospholipids to ensure against organoleptic deterioration during storage. Subsequent work showed that, if myofibrillar protein were modified by partial hydrolysis with a proteolytic enzyme and recovered as a phosphate complex, the lipids could be removed by extracting the wet complex with IPA, which resulted in small losses of functional properties.

The nucleotides play a major role in the properties of fish muscle after harvest. The changes in NAD are related indirectly to the development of rigor mortis. Numerous acid-soluble nucleotides such as ATP, ADP, AMP, IMP, guanosine, and uridine have been detected in fish muscle. The degradation process of nucleotides in fish muscle is commonly considered to be the following.

ATP------- ADP------- AMP------- IMP------ Inosine------Hypoxanthine

The degradation time changes according to the storage temperature, the kind of fish, the kind of muscle, and the condition of the fish before slaughter. Some attempts have been made using enzyme inhibitors such as polymetaphosphate and fluoride to prevent the degradation of IMP and prolong the keeping time of a product of acceptable flavor.

Interactions And Analytical Procedures

The colloidal properties of fish protein involve many factors such as the fish used, whether whole or partial, the protein extracted, method of extraction and degree of denaturation.

Measurement of the colloidal properties of proteins involves physical and chemical methods developed especially for colloids. Osmotic pressure, one of the simplest of the methods, uses the fact that solvent molecules permeate membranes while proteins molecules cannot. Under perfect conditions, a protein solution that is 1 *M* in concentration exerts an osmotic pressure equal to 22.4 atmospheres at 0°C. From this measurement, the molecular weight can be determined. Since pressure is very high, dilute solutions may be used to protect the membrane.

Ultracentrifugation is another important analytical tool. It increases the gravitational force and causes proteins to settle at various rates proportional to their respective molecular weights. Several of the components may also be detected using this method.

Fractionation with solvents, particularly at low temperatures, is more suitable for producing proteins for consumption. Addition of alcohol to the colloidal system lowers the dielectric constant, increasing the electrical interaction and reducing solubilities. Proper adjustment of the temperature, pH, protein and alcohol with low salt concentration, makes separation of constituent proteins more selective than the older salting-out procedure.

Dialysis removes smaller crystalloidal substances from proteins in the colloidal state. This method removes salts and leaves proteins in solution.

Muscle protein classification has been achieved by quantitatively identifying several fractions. This was first done successfully with rabbit muscle where four fractions were identified (Weber and Meyer 1933): (a) the water-soluble albumin fraction, (b) the globulin fraction which is extractable at low ionic strength ($\leqslant 0.15\ M$), (c) the myosin fraction which is extractable at high ionic strength ($\leqslant 0.5\ M$), and (d) the stroma, a residue which is insoluble in dilute solutions of hydrochloric acid (Weber 1934). The albumin fraction has since been subdivided into a number of fractions. One of the most notable of these fractions is myoalbumin which comprises a small amount of the muscle albumin.

The myosin fraction is a structural protein which is present in the myofibril along with actin and tropomyosin (Perry 1953; Bailey 1948). Myosin has been shown to exist as a complex mixture of enzymes (Mommaerts and Parrish 1951), while the plural of the term has been used for mixtures of muscle globulins isolated at high ionic strength and precipitated by dilution (Hamoir 1951). Quantitative fractionation of fish muscle protein has been employed by numerous authors (Reay and Kuchel 1937). The comparative data from these analysis is shown in Table 6.1.

The influence of the physiological state of the muscle upon protein extraction has been investigated in rabbit and tortoise muscle by Dubuisson (1950, 1953). The extractability was found to be maximal in normal resting, relaxed muscle, and decreased in stimulated or contracted muscle. Such an extensive study seems impossible with fish muscle because of the inexcitability which occurs shortly after exeresis. Thus the changes occurring during rigor mortis are the only physiological changes which have been investigated. The amount of actomyosin which can be extracted decreases at the onset of rigor mortis, then increases again at the resolution of rigor. Extractability is also influenced by the degree of subdivision of the tissue. When cod muscle is comminuted for 3–5 min, a quantitative extraction of the structural proteins can be readily obtained without the influence of salt as is normally necessary for extraction. The influence of pH on the extractability of cod muscles has been investigated by blending the muscle for 3–5 min in a large volume of 5% sodium chloride (0.85 *M*) adjusted to various pH's with buffer solutions. The solubilization of the protein was found to increase sharply between pH 5 and pH 6. The extractability curve obtained did not depend upon the species of fish or the method of subdivision of the tissue, however the position was slightly influenced by the ionic strength.

The components of proteins are isolated by various fractionation methods such as electrophoresis and ultracentrifugation after the proteins have been separated into fractions by their various extractabilities. There are two general categories: (1) those proteins extractable by salts of low ionic strength ($\leqslant$ 0.15 *M*) and (2) those proteins extracted by high ionic strength ($\geqslant$ 0.15 *M*).

The minimal salt concentration necessary to solubilize carp muscle at a pH of 7.5 was found to be 0.45, while the ionic strength required for the solubilization of ling cod was 0.3 *M* (Connell 1953). This illustrates that variations in the colloidal properties of the muscle do occur and are attributable to such factors as species, age and muscle state.

The fish muscle proteins extractable by salt solutions of low ionic strength include the myogens, the myoalbumins and the globulin X.

This fraction normally accounts for about 21–22% of the total muscle protein of Teleost fish (Hamoir 1955), and is divided into the three main groups (myoalbumins, myogins, and globulin X by electrophoresis. Data from electrophoresis, however must be carefully examined since very small amounts of contaminating protein can result in apparent electrophoretic heterogeneity (Hamoir 1955, 1957; Henrotte 1952). Another important group of proteins is the chromoproteins, which constitutes a large portion of the sarcoplasm.

Good fractionation of protamine may be achieved by means of countercurrent distribution. The best results are obtained by using a 15% solution of sodium acetate in water as the lower stage, and a 5% solution of lauric acid in *n*-butanol as the upper phase. Such analysis has shown that clupeine, salmine and iridine are not homogeneous compounds (Felix 1955). Electrophoretic analysis of protamines was attempted using various buffer solutions in a Tiselius apparatus. This analysis showed clupeine and salmine to be electrophoretically homogeneous with their respective mobilities dependent upon the buffers used. The phenomenon was probably due to the combination of protamines with the anions of the buffer, particularly the phosphate ion. One molecule of salmine is thought to be capable of combination with four moles of phosphate ($HP0_4$). Later attempts at electrophoretic fractionation of protamines proved to be more successful. A Tiselius apparatus with 0.1 *M* phosphate buffer was used. Results were greatly improved by the addition of secondary butanol. The best mixture was found to be one of 18 parts of 86% ethanol, 6 parts trichloroacetic acid, and 1 part of secondary butanol. This mixture was found to separate clupeine, salmine, iridine, foninine, truttine and sturine into three distinct spots on the chromotography paper when stained with amino-shwarz indicator but not when stained with ninhydrin. The addition of sodium oxalate to the origin of the chromatogram effects a sharper separation of the three spots. Fractionation by chromotography and by solubilities of picrates and sulfates of the protamines have been attempted with varying degrees of success.

As would be expected, species-specific differences have been found in collagens. X-ray diffraction showed differences in the teleost collagen, ichthyotoidin, and the elasmobranch collagen, elastoidin. The comparison of codfish skin collagen and bovine collagens showed fish collagen to be far higher in hydroxyproline and proline. Elastoidin consists of at least two components, one a typical collagen and the other a noncollagenous protein rich in tyrosine (Kubota and Kimura 1967A, 1967B). The weight ratio of the collagen to noncollagenous protein components is eight to two (Kimura and Kubota 1967). The two may be separated only by severe treatment which may cause cleavage of covalent linkages.

Fresh collagenous tissues contain appreciable amounts of soluble collagen which can be extracted by neutral salt, dilute alkali and dilute acid. This soluble collagen is assumed to be a precursor of collagen. Kimura and Kubota (1966) showed that a small amount of soluble protein, rich in tyrosine and cystine but poor in hydroxyproline, can be extracted from elastoidin fiber by cold dilute acid. This protein is identical with the insoluble protein component of elastoidin fiber in its amino acid composition. Little is known about the interrelationship between collagen and noncollagenous protein which contributes to the stability of the fiber (Kimura and Kubota 1966). The amino acid composition of elastoidin fiber when compared to this noncollagenous protein was high in tyrosine, cystine and histidine, but poor in hydroxyproline, hydroxylysine and glycine (Kimura and Kubota 1968, 1969). The amino acid composition of soluble elastidin is shown in Table 6.2. Part of the hydroxylysine in vertebral collagen is O-glycosidically linked to glucosylgalactose and galactose through its O-hydroxy group (Spiro 1969). There are several other connective tissues in fish including scales which consist of 56% mineral matter, 21% ichthylepedin and 23% gelatin (Solomons 1955). A pseudokeratin, ichthyokeratin, has been reported in salmon eggs (Young and Smith 1956) and the eye lens.

The proteins extractable by cold neutral salt solutions of high ionic strength include myosin, actin and tropomyosin. Salt (e.g. sodium chloride, potassium chloride, lithium chloride) solutions of 0.5–1.0 *M* and pH of 7.0-7.5 have proven most satisfactory (Hamoir 1955; Dyer 1954) for extraction.

Superprecipitation of actomyosin of aged fish muscle was examined by Partmann (1963) who reported that when aged muscles of beef or fish were homogenized in 0.16 *M* potassium chloride solution with up to 10 *M* ATP, all of the fiber fragments contracted. Suzuki *et al.* (1967) reported that actomyosin of aged fish muscle showed superprecipitation when (2.0×10) *M* ATP was added to the gel. These experiments indicate that structural proteins do not change appreciably during post-mortem aging. This phenomenon of superprecipitation of actomyosin depends on the concentration of calcium, ATP, magnesium and upon the presence of a relaxing factor.

When compared to rabbit, actomyosin in the fish protein shows similarity from an electrochemical point of view. The maximum acid and base-binding capacities (Hamoir 1951), the solubilities, and the electrophoretic mobilities are very close for rabbit and fish actomyosin. The properties related to the size and shape of the particles differ widely, however. Data from viscosities taken at gradient 0 and of a

TABLE 6.3

CONTENT AND PARTITION OF PROTEIN NITROGEN IN RABBIT AND FISH MUSCLE

Scientific Name	Protein N as % of the Fresh Weight	Albumins	Globulins Soluble at 0.15	Structural Proteins	Stroma
Lepus cuniculus					
L. White muscles	2.54	22	22	39	17
Red muscles	2.25	17	17	39	27
White muscles	—	10	9	63	18
Gadus aeglefinus L.	—	30	—	67	3
Torpedo marmorata Risso Ray and *Torpedo ocellata* Rafinesque	2.35–2.83	26	—	64	10
Gadus aeglefinus L.	2.54	15	—	—	3
Gadus morrhua Day	2.50	15	—	—	3
Pleuronectes Microcephalus Day	2.71	16	—	—	4
Raja batis L.	2.73	—	—	—	8
Scyllium canicula Day	2.71	17	—	—	10
Gadus morrhua Day	2.50	13	8	76	3

sedimentation constant set at concentration 0 show that the differences are due to the structure of the solution which is much more stable in the case of fish actomyosin (Tsao and Bailey 1953). From the point of view of extractability, however, the behavior of fish actomyosin is not due to a difference in properties of solubility of the actomyosin, but to the secondary properties of these particles within the myofibril.

Comparative biochemical studies on highly purified myosins from dorsal muscle of carp (*Cyprinus carpio*) and tilapia (*Tilapia mossambica*), and the skeletal muscle of rabbit were performed by Takashi (1973). The following biochemical properties were investigated: chromatographic profile on DEAE-sephadex column, effects of Ca^{++}, Mg^{++}, EDTA and PCMB on myosin ATPase activity, sulfhydryl group content, and thermostability of Ca^{++} -ATPase activity. Most of these tests showed remarkable similarity between rabbit and fish myosin. The most striking difference between fish and rabbit myosins was observed in the thermostability of Ca^{++} -ATPase activity. Inactivation rates of the myosins from carp, tilapia and rabbit were shown to be 6.1×10^{-5} sec^{-1}, 18.3×10^{-5} sec^{-1} and 5.0×10^{-5} sec^{-1} respectively at pH 7 and 30°C.

Purification of myosin and actomyosin from carp muscle by the

TABLE 6.4

AMINO ACID COMPOSITION OF SOLUBLE ELASTOIDIN

Amino Acid	Residues per 1,000 Residues
Hydroxyproline	35.0
Aspartic acid	77.5
Threonine	29.6
Serine	53.0
Glutamic acid	75.7
Proline	91.0
Glycine	214.0
Alanine	89.6
½-CyS	8.2
Valine	39.2
Methionine	11.3
Isoleucine	20.8
Leucine	30.8
Tyrosine	85.8
Phenylalanine	13.0
Tryptophan	trace
Hydroxylysine	4.9
Lysine	28.1
Histidine	34.0
Arginine	58.3
Tyrosine derivatives	present
Amide ($-NH_2$)	(50.6)[1]

[1] Non-amino acid residues.

dilution method has been investigated by Takashi (1972). Fish myosin has been believed to denature at a faster rate than rabbit muscle. This was tested by measuring ATPase activity, ATP sensitivity, and the activities of the contaminating enzymes, myokinase and adenylic deaminase, at each step of the dilution method. The crude preparations of myosin usually contained myokinase and adenylic deaminase. Repeated precipitation resulted in virtually complete removal of myokinase, but had little effect on adenylic deaminase. In this experiment specific ATPase activity of myosin preparations was shown to slightly decrease during repeated precipitation by dilution, while actomyosin preparations retained constant activity. These results suggest that myosin does denature slightly during repeated precipitation by dilution. Further chromatographic studies showed carp myosin to be mainly monomeric.

Purified fish actin was first described by Straub (1942). Numerous reports on isolation have followed (Mommaerts and Parrish 1951; Tsao and Bailey 1953). One of the most successful methods has used acetone to dry the muscle (Barany *et al.* 1957; Ueda *et al.* 1968), and is based on the principle that actin exists either in a free state or in a conjugated state with myosin (actomyosin) with alkaline-earth

TABLE 6.5

ACTIN CONTENT IN FISH MUSCLE

Species	Actin (mg N/gm of Meat)[1]
Star-spotted shark	1.13
Mackerel	1.27
White croaker	1.35
Truelizard fish	3.51
Swordfish	2.13
Yellow-tail	1.97
Perch	1.66
Shotted halibut	0.36
Sardine	0.63
Rock-fish	1.10
Sea-robin	1.13

[1] Extracted by grinding with water for 1 hr.

metals associated in the connection between actin and myosin. If this is true, actin should be easily separated by removing the metals from the complex. In this process the muscle is acetone dried then extracted with water or a salt solution by blending. Actin extracts are then separated by centrifuging. The components separated by this method are studied by salting out and electrophoresis. The components thus separated are F-actin and tropomyosin. Comparison of the actin content of various species has shown no differences among the species. The amount of actin extractable from the various kinds of fish are seen in Table 6.5.

Another of the globulins is tropomyosin. Studies of the tropomyosin component of the muscle fibril have generally followed modifications of the isolation procedure of Bailey (1948) in which the sulfhydril groups of the protein are protected by the presence of dithiothreitol. One problem with the Bailey method in fish muscle is that it apparently includes significant amounts of nucleotides (Dingle 1959).

Hamoir (1951) applied a salt extraction procedure, without using organic solvents, to carp muscle. By diluting at neutrality and at acid pH levels, two fractions containing tropomyosin were thus isolated. The fraction isolated from neutral solution contained 12% RNA and was termed nucleotropomyosin, while the fraction isolated from acid solution had a lower nucleic acid content and resembled the tropomyosin isolated by the Bailey procedure. Low nucleotide tropomyosin has also been isolated with salt extraction of the heat stable fraction of protein (Odense *et al.* 1969). Tropomyosin obtained by this method is thus often referred to as "heat-treated tropomyosin."

Odense *et al.* (1969) compared the protein extracted from cod muscle by a simple procedure involving magnesium chloride and elevated temperatures with tropomyosin isolated by the classical method of Bailey. The tropomyosin derived from heat treatment of cod muscle contains much lower amounts of associated nucleotide material. The amino acid analysis of tropomyosin is given in Table 6.6, and can be compared with the amino acid content of the other muscle components as shown in Table 6.7.

APPLICATION OF COLLOIDAL PROPERTIES

The nutritional needs of a significant segment of the world population are undersatisfied. In most cases, the most critical nutritional insufficiency is a lack of high quality protein. The lack of an economic source of animal protein has long prevented any great improvement in this situation. The major underutilized source of animal protein is in the world's oceans. The historical problem encountered in using fish as a dietary supplement has been providing it in an economical manner to the people who most need it. The goal of many food researchers thus has been to develop an economical method of producing a fish protein concentrate which retains the functional properties of the native protein. Three general methods of producing fish protein concentrates have been developed. In the first, chemical methods use solvent extractions to remove water and lipids from the raw material yielding final products which are bland-tasting, light-colored, protein powders. Chemical methods are used mostly in the western hemisphere. To be used for extraction of raw fish or fish meal, the solvent should have the following qualities.

(1) Non-toxic
(2) Available in pure form
(3) Safe and easy to handle
(4) Low in price
(5) Free from fire or explosion hazard
(6) Efficient in removal of water, lipids, and odor-bearing compounds, without dissolving or reacting with the proteinaceous components
(7) Of low boiling point to allow for inexpensive desolventization of the concentrate, and to avoid the necessity of heating the extracted material to excessive temperature
(8) Suitable for solvent recovery and deodorization
(9) Inert toward equipment construction materials

Secondly, biological methods (enzymatic or fermentation) are equally important. This is the oldest method of producing fish concen-

TABLE 6.6

AMINO ACID ANALYSIS OF TROPOMYOSIN
(RESULTS ARE GIVEN IN MOLE PERCENTAGES)

Amino Acid	Bailey Process	Heat Treated
Lys	13.6	13.6
His	0.7	1.1
Arg	4.9	5.9
Asp	12.4	14.9
Thr	3.3	2.9
Ser	4.0	3.7
Glu	24.1	23.7
Pro	0.2	1.2
Gly	2.0	2.8
Ala	11.8	10.4
Cys	0.6	0.2
Val	3.1	3.2
Met	2.2	0.9
Ile	3.4	3.4
Leu	11.5	9.9
Tyr	1.8	1.5
Val	0.4	0.9
Trp	0.0	0.0

Source: Odense *et al.* (1969).

TABLE 6.7

AMINO ACID CONTENT OF FISH PROTEINS

Name[1]	Symbol	Approximate amount in gm/100 gm			
		Flesh	Myosin[2]	Actin[2]	Collagen
Alanine	Ala	7.1	6.5	5.4	10.4
Arginine	Arg	6.9	6.7	7.4	9.1
Aspartic Acid	Asp	11.2	11.5	9.7	7.5
Cystine	Cys	1.4	0.9	1.4	0.9
Glutamic Acid	Glu	16.9	21.7	13.3	11.3
Glycine	Gly	5.1	3.4	5.0	28.2
Histidine	His	3.6	2.1	3.3	1.2
Hydroxyproline	—	0.0	0.0	0.0	9.0
Hydroxylysine	—	0.0	0.0	0.0	1.0
Isoleucine[1]	Ileu	5.0	4.6	7.7	1.7
Leucine[1]	Leu	9.2	9.4	6.6	3.2
Lysine[1]	Lys	10.6	10.6	6.5	3.7
Methionine[1]	Met	2.7	3.0	4.1	2.0
Phenylalanine[1]	Phe	4.7	3.9	4.6	2.0
Proline	Pro	4.4	3.5	6.0	12.4
Serine	Ser	5.8	4.9	5.9	7.9
Threonine[1]	Thr	5.5	4.3	6.9	0.6
Tryptophan[1]	Try	1.4	0.8	1.6	0.0
Tyrosine	Tyr	4.1	2.7	6.0	0.6
Valine[1]	Val	5.8	5.3	5.9	2.3

[1] Amino acids essential to human nutrition. Histidine is essential for laboratory and stock animals. Arginine can be synthesized but only slowly. Cystine may replace part of the methionine requirement, and tyrosine may replace part of the phenylalanine requirement.

[2] Myosin and actin account for 50% and 20% respectively of the total protein of white muscle. They combine to form actomyosin during muscle contraction.

trate, and is still used in the Far East for producing fish sauces and pastes. The biological methods generally result in protein breakdown, which allows subsequent separation of water and lipids by simple physical means. The fish tissue can be broken down by the introduction of either living proteolytic microorganisms such as yeasts, bacteria and molds, or by the use of isolated proteolytic enzymes. The resulting hydrolysates are then separated from the undigested sludge by filtration and centrifugation. Biologically produced concentrates contain protein cleavage products, which characterize the flavor of the finished product making it distinct from the solvent-extracted products, which are usually bland in taste.

Finally, relatively sophisticated physical techniques have been developed for the removal of water and oil from fish. Passage of electric discharges through a raw fish slurry makes the separation of oils and water from the water-soluble constituents and solids much easier. H. Doevenspeck, a German engineer, developed the procedure which involves a periodic electrical discharge from capacitors through the slurry, followed by centrifugal separation of the components. A likely explanation of the phenomenon is that the cells rupture, with consequent release of their contents.

Another physical method has been designed to remove moisture from fish. Here the raw fish is finely divided and dispersed in a nonvolatile, immiscible liquid which can act as a heat transfer medium. The dispersion is then heated in a vessel and the pressure is suddenly reduced over the surface of the dispersion containing the heated fish particles. The moisture present is removed by flash evaporation, while the nonvolatile heat transfer medium retains the dehydrated fish particles. Heating before vacuum application can be done under an atmosphere of nitrogen to minimize oxidation. The dehydrated fish particles are eventually separated from the heat transfer liquid by filtration or centrifugation and the particles can then be extracted with solvent if desired.

The goal of the FPC research has been to produce a functional protein isolate, utilizing the only inexpensive source of high quality protein available. Tannenbaum *et al.* (1970) have reported the use of FPC as a starting material for the preparation of protein isolates. In that work, FPC was partially hydrolyzed with sodium hydroxide and recovered as an isoelectric protein. An alkaline hydrolysis process is also being investigated. Although partial alkaline hydrolysis may prove useful in the preparation of functional protein isolates, this procedure can lower the nutritional quality through the racemization (Tannenbaum *et al.* 1970) and the possible conversion of cysteine to lanthionine (Lindley and Philips 1945). In addition the economics of using a once processed product as a raw material source might lessen the economic advantage that unprocessed, raw fish would enjoy. This latter factor, of course, cannot be fully evaluated unless functional and organoleptic properties of the isolates are fully ascertained. For example, the price of sodium caseinate varies widely depending on its functional properties.

ACKNOWLEDGEMENT

This chapter is a contribution of the Louisiana State University Department of Food Science Sea Grant Program.

BIBLIOGRAPHY

AMANO, H., HASHIMOTO, K., and MATSUURA, F. 1971. Starch-gel electrophoresis of hemoglobin of "Funa." Bull. Japan. Soc. Sci. Fisheries *37*, No. 1, 48-54.

AMANO, K., BITO, M., and KWABATA, T. 1953. Handling effect upon biochemical change in the fish muscle immediately after catch. I. Difference of glycolysis in the frigate mackerel killed by various methods. Bull. Japan. Soc. Sci. Fisheries *19*, 487-498.

ANDERSON, M. L., and STEINBERG, M. A. 1964. Effect of lipid content on protein-sodium linolenate interaction in fish muscle homogenates. J. Food Sci. *29*, 327-330.

BAILEY, K. 1948. Tropomyosin: A new asymmetric protein component of the muscle fibril. Biochem. J. *43*, 271.

BARANY, M., BARANY, K., and GUBA, F. 1957. Preparation of actin without extraction of myosin. Nature *179*, 818-819.

BEAR, R. S. 1952. The structure of collagen fibrils. *In* Advances in Protein Chemistry, Vol. 7, M. L. Anson, K. Bailey, and J. T. Edsall (Editors). Academic Press, New York.

BRULL, L., and CNYPERS, Y. 1954. Blood perfusion of the kidney of *Lophius piscatorius* L. VI. Influence of perfusion pressure on urine volume. J. Marine Biol. Assoc. United Kingdom *33*, No. 3, 733-738.

BUCHS, S. 1953. Fundamental observations on the existence, extraction and activation of gastric cathepsin. Enzymologia *16*, No. 4, 193-214.

CONNELL, J. J. 1953. Studies on the proteins of fish skeletal muscles. 2. Electrophoretic analysis of codling extracts of low ionic strength. Biochem. J. *54*, 119-126.

CONNELL, J. J. 1954. Studies on the proteins of fish skeletal muscles. 3. Cod myosin and cod actin. Biochem. J. *58*, 360-367.

CONNELL, J. J., and HOWGATE, P. F. 1969. Sensory and objective measurements of quality of frozen stored haddock of different initial freshness. J. Sci. Food Agr. *20*, No. 8, 469-476.

COWIE, W. P., and MACKIE, I. M. 1968. Examination of protein extractability method for determining cold-storage protein denaturation in cod. J. Sci. Food Agr. *19*, 696-698.

CREPAX, P. 1952. Electrophoretic properties of extracts of muscles possessing different morphological properties. Biochim. Biophys. Acta. *9*, 385-398.

DAMBERGES, N., ODENSE, P., and GUILBAULT, R. 1968. Changes in free amino acids in skeletal muscle of cod *Gadus morhua* under conditions simulating gillnet fishing. J. Fisheries Res. Board Can. *25*, No. 5, 935-942.

DASSOW, J. A., PATASHNIK, M., and KOURY, B. J. 1970. Characteristics of

Pacific hake, *Merluccius productus*, that affect its suitability for food. U.S. Fisheries Wildlife Serv. Circ. *332*, 127-136.

DILL, D. B. 1921. A chemical study of certain Pacific coast fishes. J. Biol. Chem. *48*, 73-82.

DINGLE, J. R. 1959. Proteins in fish muscle. XV. Preparation of actin from cod muscle with potassium iodide. J. Fisheries Res. Board Can. *16*, 243-246.

DINGLE, J. R., and DYER, W. J. 1955. Note on albumin protein fractions in a sturgeon. J. Fisheries Res. Board Can. *12*, 646-648.

DUBUISSON, M. 1950. Muscle activity and muscle proteins. Biol. Rev. Cambridge Phil. Soc. *25*, 46-72.

DUBUISSON, M. 1953. Structural proteins extractable from turtle muscle and their alterations during contraction. Bull. Classe Sci. Acad. Roy. Belg. *39*, 35-41.

DYER, W. J. 1954. Main problems of protein denaturation. Proc. Symp. Cured Frozen Fish Technol., Swed. Inst. Food Preserv. Res., Goteborg, Publ. *100* (1953), Art. VII, 9.

FELIX, K. 1955. Protamines, nucleoprotamines and nuclei. Am. Scientist *43*, 431-449.

FRONTIER-ABOU, D. 1969. Variations in the total composition of the muscles in 3 species of *Carangideae.* Ann. Nutr. Aliment. *23*, No. 6, 313-334.

GELINEO, A., and GELINEO, S. 1955. Concentration of hemoglobin in the blood of some fresh water fishes. Compt. Rend. Soc. Biol. *149*, 1411.

HAGIHARA, B., MATSUBARA, H., HORIO, T., and OKUNUKI, K. 1957. Crystallization of cytochrome-c from fish. Nature *179*, 249-251.

HAMOIR, G. 1949. Comparative study of the myosins of fish and rabbit. Bull. Soc. Chim. Biol. *31*, 118-122.

HAMOIR, G. 1951. Fish tropomosin and fish nucleotropomyosin. Biochem. J. *48*, 146-151.

HAMOIR, G. 1955. Contribution to the study of the muscle proteins of fish. Research on the striated muscle of the carp. Arch. Intern. Physiol. Biochim. *63*, Part 4, Suppl. 152.

HAMOIR, G. 1957. Isolation and properties of a low molecular weight protein of carp myogen. Biochem. J. *66*, 29.

HASSELBACH, W., and WEBER, H. H. 1953. The influence of the MB-factor on the contraction of the fiber model. Biochim. Biophys. Acta. *11*, No. 1, 160-161.

HENROTTE, J. G. 1952. A crystaline constituent from myogen of carp muscles. Nature *169*, 968-969.

HUYS, J. V. 1954. Isolation and crystalization of the myoglobin of thon. Arch. Intern. Physiol. *62*, 296-297.

IKEDA, S., SATO, M., and TAGUCHI, T. 1966. Biochemical studies of L-ascorbic acid in aquatic animals. VI. Effect of α-tocopherol on L-gulonolactone dehydrogenase activity in carp hepatopancreas. Bull. Japan. Soc. Sci. Fisheries *32*, No. 7, 585-589.

IKEDA, S., and TAGUCHI, T. 1966. Improved assay method and levels of vitamin E in fish tissues. Bull. Japan. Soc. Sci. Fisheries *32*, No. 4, 346-351.

IKEDA, S., and TAGUCHI, T. 1967. Protective effect of α-tocherol on the solubility of actomyosin from yellowtail muscle. Bull. Japan. Soc. Sci. Fisheries *33*, No. 6, 567-571.

JONES, N. R. 1955. Free amino acids of fish. 1-methylhistidine and β-alamine liberation by skeletal muscle anserinase of codling (*Gadus callarias*). Biochem. J. *60*, 81-87.

KIMURA, S., and KUBOTA, M. 1966. Studies on elastoidin. I. Some chemical and physical properties of elastoidin and its components. J. Biochem. *60*, No. 6, 615-621.

KIMURA, S., and KUBOTA, M. 1967. Elastoidin. II. Dialyzable peptides released by pepsin digestion of elastoidin. Bull. Japan. Soc. Sci. Fisheries *32*, No. 5, 430-437.

KIMURA, S., and KUBOTA, M. 1968. Studies on elastoidin. III. Protein components of soluble elastoidin. Bull. Japan. Soc. Sci. Fisheries *34*, No. 6, 535-540.

KIMURA, S., and KUBOTA, M. 1969. Tyrosine derivatives in a structural protein elastoidin. J. Biochem. *65*, No. 1, 141-143.

KING, F. J., ANDERSON, M. L., and STEINBERG, M. A. 1962. Reaction of cod actomyosin with linoleic and linolenic acids. J. Food Sci. *27*, 363-366.

KUBOTA, M., and KIMURA, S. 1967A. Collagen in aquatic animals. III. Physiochemical properties of the pepsin-solubilized collagen of dolphin hide. Bull. Japan. Soc. Sci. Fisheries *33*, No. 3, 217-223.

KUBOTA, M., and KIMURA, S. 1967B. Skill collagen of the great blue shark. Bull. Japan. Soc. Sci. Fisheries *33*, No. 4, 338-344.

LAWRIE, R. A. 1953. Activity of the cytochrome system in muscle and its relation to myoglobin. Biochem. J. *55*, 298-305.

LINDLEY, H., and PHILIPS, H. 1945. The identity of the lanthionine forming and bisulfite-reactive fractions of the combined cystine of wool. Biochem. J. *39*, 17-23.

LOVE, R. M. 1962. New factors involved in the denaturation of frozen cod muscle protein. J. Food Sci. *27*, 544.

MacCALLUM, W. A., *et al.* 1969. Newfoundland capelin. Proximate composition. J. Fisheries Res. Board Can. *26*, No. 8, 2027-2035.

MacLEOD, J. J. R., and SIMPSON, N. W. 1927. The immediate post-mortem changes in fish muscle. Contrib. Can. Biol. Fisheries *3*, 439-456.

MacPHERSON, N. L. 1932. Studies on the behavior of the carbohydrates and lactic acid of the muscle of haddock (*Gadus aeglefinus*) after death. Biochem. J. *26*, 80-87.

MANITA, H., KOIZUMI, C., and NONAKA, J. 1969. Aspectic autolysis of mackerel muscle. Bull. Japan. Soc. Sci. Fisheries *35*, No. 10, 1027-1033.

MANOHAR, S. V. 1970. Postmortem glycolytic and other biochemical changes of white sucker (*Catostomus commersoni*) and northern pike (*Esox lucius*) at 0° C. J. Fisheries Res. Board Can. *27*, No. 11, 1997-2002.

MARUYAMA, Y., and SUZUKI, T. 1968. Post-mortem change of horse mackerel muscle proteins. Bull. Japan. Soc. Sci. Fisheries *34*, No. 5, 415-419.

MATSUURA, F., BABA, H., and MORI, T. 1953. Chemical studies on the red muscle (chiai) of fishes. I. Occurrence of arginase in the red muscle of fishes. Bull. Japan. Soc. Sci. Fisheries *19*, No. 8, 893-898.

MATSUURA, F., and HASHIMOTO, K. 1954. Chemical studies on the red muscle (chiai) of fishes. II. Determinations of the content of hemoglobin, myoglobin and cytochrome-c in the muscle of fishes. Bull. Japan. Soc. Sci. Fisheries *20*, No. 4, 308-312.

MOMMAERTS, W. F. H. M., and PARRISH, R. G. 1951. Myosin. I. Preparation and criteria of purity. J. Biol. Chem. *188*, 545-552.

NAGAYAMA, F. 1966. Phosphorylases. Decomposition and synthesis of glycogen in living organisms. Bull. Japan. Soc. Sci. Fisheries *32*, No. 2, 188-195.

NORRIS, E. R., and ELAM, D. W. 1940. Preparation and properties of crystalline salmon pepsin. J. Biol. Chem. *134*, 443-454.

NORRIS, E. R., and MATHIES, J. C. 1953. Preparation, properties and crystallization of tuna pepsin. J. Biol. Chem. *204*, 673-680.

ODENSE, P. H., ALLEN, T. M., and LEUNG, T. C. 1966. Multiple forms of lactate dehydrogenase and aspartate aminotransferase in herring. (*Clupea harengus harengus L.*) Can. J. Biochem. *44*, No. 10, 1319-1326.

ODENSE, P. H., LEUNG, T. C., GREEN, W. A., and DINGLE, J. R. 1969. The isolation of cod muscle tropomyosin by heat treatment. Biochem. Biophys. Acta. *188*, No. 1, 124-131.

PARTMANN, W. 1963. Post-mortem changes in chilled and frozen muscle. J. Food Sci. *28*, 15-27.

PARTMANN, W. 1965. Changes in proteins, nucleotides and carbohydrates during rigor mortis. *In* The Technology of Fish Utilization, R. Kreuzer (Editor). Fishing News, London.

PARTMANN, W. 1968. Quantitative measurement of the contractility of muscle fiber fragments with addition of adenosine-triphosphate. Z. Lebensmittel-Unters Forsch *137*, No. 2, 74-78.

PERRY, S. V. 1953. The protein components of the isolated myofibrils. Biochem. J. *55*, 114-122.

REAY, G. A., and KUCHEL, C. C. 1937. The proteins of fish. Gt. Brit. Dept. Sci. Ind. Res. Rept. Food Invest. Board 1936, 93-95.

ROSSI-FANELLI, A., and ANTONINI, E. 1955. Purification and crystallization of the myoglobin of salt water fish. Arch. Biochem. Biophys. *58*, No. 2, 498-500.

SAITO, K. 1954. Biochemical studies on fish blood. Bull. Japan. Soc. Sci. Fisheries. *19*, No. 12, 1134-1143.

SHARP, J. G. 1934. Post-mortem breakdown of glycogen and accumulation of lactic acid in fish muscle. Proc. Roy. Soc. (London) Ser. B. *114*, No. 790, 506-512.

SHARP, J. G. 1935. Glycogenolysis in fish liver at low temperatures. Biochem. J. *29*, No. 4, 854-859.

SHEWAN, J. M. 1951. The chemistry and metabolism of nitrogenous extractives in fish. Biochem. Soc. Symp. (Cambridge, Engl.) No. *6*, 2847.

SHIMIZU, W., KURORAWA, Y., and IKEDA, S. 1953. XVII. Imidazole compounds in fish muscles, with special reference to taste of red-muscle fishes. 2. Bull. Res. Inst. Food Sci., Kyoto U. No. 12, 40-48; Chem. Abstr. *48*, 13121.

SHIMIZU, Y., TADA, M., and ENDO, K. 1973. Seasonal variations in chemical constituents of yellow-tail muscle. I. Water, lipid and crude protein. Bull. Japan. Soc. Sci. Fisheries *39*, No. 9, 993-999.

SIMPSON, W. W., and MacLEOD, J. J. R. 1926. Post-mortem changes in the free sugar, glycogen, phosphates and lactic acid in mammalian muscle. Trans. Roy. Soc. Can. *20* (Sect. V), 371-375.

SNOW, J. M. 1950. Proteins in fish muscle. III. Determination of myosin by freezing. J. Fisheries Res. Board Can. *7*, 599-607.

SOLOMONS, C. 1955. Protein in fish scale. S. African J. Med. Sci. *20*, 27-28.

SPIRO, R. G. 1969. Characterization and quantitative determination of the hydroxylysine-linked carbohydrate units of several collagens. J. Biol. Chem. *244*, No. 3, 603-612.

STRAUB, F. B. 1942. Actin. Studies Inst. Med. Chem. Univ. Szeged. *2*, 3-15.

SUZUKI, A., NAKAZATO, M., and FUJIMAKI, M. 1967. Studies on proteolysis in stored muscle. I. Changes in nonprotein nitrogenous compounds of rabbit muscle during storage. Agr. Biol. Chem. *31*, No. 8, 953-957.

TAKASHI, R. 1972. Studies on muscular proteins of fish. VII. Chromatography of carp myosin on diethylaminoethyl-sephadex A-50. Bull. Japan. Soc. Sci. Fisheries. *38*, No. 2, 126-132.

TAKASHI, R. 1973. Studies on muscular proteins of fish. VIII. Comparative studies on the biochemical properties of highly purified myosins from fish dorsal and rabbit skeletal muscle. Bull. Japan. Soc. Sci. Fisheries. *39*, No. 2, 197-205.

TANNENBAUM, S. R., AHERN, M., and BATES, R. P. 1970. Solubilization of fish protein concentrate. I. An alkaline process. Food Technol. *24*, No. 5, 604-607.

TARR, H. L. A. 1950. The acid-soluble phosphorus compounds of fish skeletal muscle. J. Fisheries Res. Board Can. *7*, 608-612.

TARR, H. L. A. 1954. Microbiological deterioration of fish post-mortem, its detection and control. Bacteriol. Res. *18*, 1-15.

TSAO, T. C., and BAILEY, K. 1953. The extraction, purification and some chemical properties of actin. Biochim. Biophys. Acta. *11*, No. 1, 102-113.

TSUYUKI, H., and ROBERTS, E. 1966. Interspecies relationships within the genus *Oncorhynchus* based on biochemical systematics. Res. Board Can. J. *23*, No. 1, 101-107.

UEDA, T., SHIMIZU, U., and SIMIDU, W. 1968. Species differences in fish actomyosin. III. The velocity and mechanism of heat-denaturing reaction. Bull. Japan. Soc. Sci. Fisheries. *34*, No. 4, 351-356.

UMEMOTO, S., and KANNA, K. 1969. Studies on gel-filtration of fish muscle protein. II. Gel-filtration of sarcoplasmic and myofibrillar proteins on sepharose 213. Bull. Japan. Soc. Sci. Fisheries *35*, No. 6, 555-558.

UMEMURA, K. 1951. Respiratory enzymes of tiai (fish red muscle). Nagoya J. Med. Sci. *14*, 81-85.

WEBER, H. H. 1934. The fine structure and mechanical properties of the myosin thread. Arch. Ges. Physiol. *235*, 205-233.

WEBER, H. H., and MEYER, K. 1933. Colloidal behavior of muscle proteins. V. Quantitative relationships between muscle proteins and its significance for the structure of striated rabbit muscle. Biochem. Z. *266*, 137-152.

YAMANAKA, H., YAMAGUCHI, K., HASHIMOTO, K., and MATSUURA, F. 1967. Starch-gel electrophoresis of fish hemoglobins. III. Salmonoid Fishes. Bull. Japan. Soc. Sci. Fisheries *33*, No. 3, 195-203.

YAMANAKA, H., YAMAGUCHI, K., and MATSUURA, F. 1965. Starch gel electrophoresis of fish hemoglobins. I. Usefulness of cyanmethemoglobin for the electrophoresis. Bull. Japan. Soc. Sci. Fisheries *31*, No. 10, 827-832.

YASUI, T., and HASHIOMOYO, Y. 1966. Effect of freeze-drying on denaturation of myosin from rabbit skeletal muscle. J. Food Sci. *31*, 295-299.

YOSHIYASU, K., and HUMOTO, Y. 1972. Starch-gel electrophoresis of hemoglobins of freshwater salmonoid fishes in southwest Japan. I. Genus: *Salvelinus* (char). Bull. Japan. Soc. Sci. Fisheries *38*, No. 7, 779-788.

YOUNG, E. G., and SMITH, D. G. 1956. The amino acids in the ichthulokeratin of salmon eggs. J. Biol. Chem. *219*, No. 1, 161-164.

CHAPTER 7

Chifa F. Lin

Interaction of Sulfated Polysaccharides with Proteins

INTRODUCTION

The physical stability of fluid type foods is a determining factor in judging the quality as well as the shelf-life of the products. Instability of the products is often caused by the incompatibility of the components of the fluid or by the unwarranted interaction between components. Some of the means to improve the stability of the fluid type foods include heat treatment, mechanical treatment and the use of additives. Heat treatment, such as forewarming, is used in evaporated milk processing to enhance the stability of the products. Mechanical treatment by homogenization is applied to fluid foods of the emulsion type to prevent fat separation. However, it is not always possible to create a fully stable product by manipulation of processing steps and in such instances it may be necessary to resort to the use of stabilizing agents. The use of such additives is known as a beneficial practice in the manufacturing of evaporated milk, chocolate milk, ice cream and other formulated fluid type beverages.

Most of the stabilizers used in the food industry are derived either from plants or marine algae. With the seasonal variation, age difference and different local environments of these plant species, it is not surprising to find that the stabilizer extracts perform differently from time to time. Moreover, in a product consisting of proteins, lipids, carbohydrates and electrolytes, the interaction between various components needs to be well balanced so that a stable system can be achieved.

The interaction between various components has been a subject for extensive review by many investigators; for example, Waugh (1954) has treated protein-protein interaction with emphasis on association, aggregation and denaturation. Bettelheim-Jevons (1958) has reviewed complex formation between proteins and aminopolysaccharides and mucopolysaccharides. Polysaccharide-lipoprotein interactions have been described in detail with reference to molecular complex formation between some chemically sulfated polysaccharides and lipoproteins in a paper by Cornwell and Kruger (1961). The effect of cations on protein stability has been dealt with in several reviews (Gurd and

Wilcox 1956; McMeekin and Polis 1949; Putnam 1948; Steinhardt and Beychok 1964). The particular binding and effects of calcium on casein and serum proteins have been reviewed by Greenberg (1944). The present chapter deals with the specific protein-sulfated polysaccharide interactions which occur in complex food systems.

IONIC INTERACTION BETWEEN SULFATED POLYSACCHARIDES AND PROTEINS

For ease of discussion, the term sulfated polysaccharide will be limited to agar-agar, porphyran, furcellaran, fucoidan and carrageenan. As shown in Table 7.1 these polysaccharides contain various amounts of ester sulfate. Structurally these polysaccharides are similar, especially carrageenan and furcellaran. Protein reactivity is exhibited by all sulfated polysaccharides. Inasmuch as a sulfated polysaccharide is a negatively charged polymer over a wide range of pH, it is capable of forming complexes with a positively charged substrate such as a protein molecule. The mechanism by which this interaction between the sulfated polysaccharide and the protein takes place can be visualized as shown in Fig. 7.1.

Above the isolectric point of the protein (Fig. 7.1, I) polyvalent metal ions in solution act as bridges between the negatively charged carboxyl groups on the protein and the negatively charged ester sulfates of the polysaccharide. At a pH below the isolectric point of the protein (Fig. 7.1, IV) similar electrostatic interactions take place between the ester sulfate of the polysaccharide and the protonated amine groups on the protein. As shown in Fig. 7.1 intermediate or transitional degrees of association are to be expected (II and III).

Among the sulfated polysaccharides mentioned earlier, one of the most widely used is carrageenan. The major use of carrageenan in the food industry is in dairy and its related products. Carrageenan was found to interact with the casein fraction both at the pH of normal

TABLE 7.1

SULFATE CONTENT[1] OF SEVERAL SULFATED POLYSACCHARIDES

	%
Agar-agar	1–3
Porphyran	3–18
Furcellaran	16–20
Fucoidan	18–38
k-carrageenan	25–30
i-carrageenan	28–35
λ-carrageenan	32–39

[1] Expressed as % SO_4.

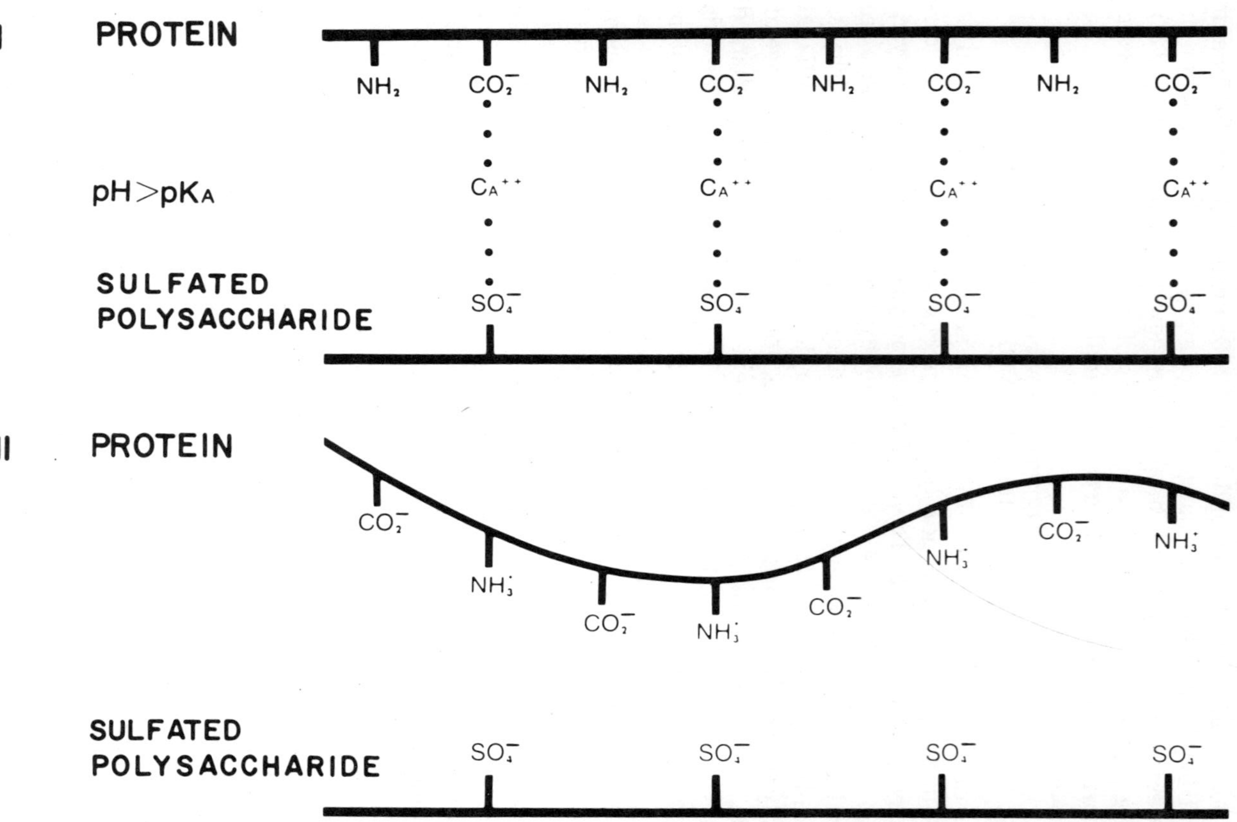
I
PROTEIN
NH2
CO2−
CA++
pH>pKA
SULFATED
POLYSACCHARIDE
SO4−
II
PROTEIN
CO2−
NH3+
SULFATED
POLYSACCHARIDE
SO4−

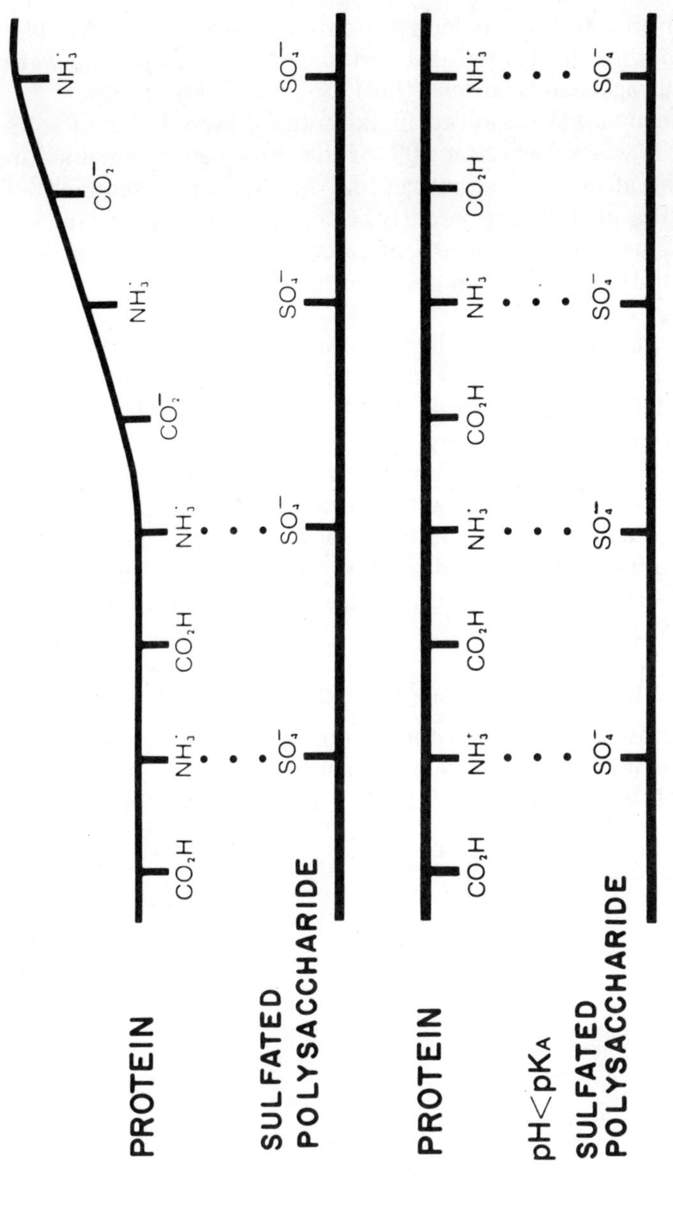

FIG. 7.1. IONIC INTERACTION BETWEEN SULFATED POLYSACCHARIDES AND PROTEINS

From Stanley (1972)

milk (pH 6.7) and at the isoelectric point of casein (pH 4.6). The acid whey and the ultracentrifugal serum from these systems were both void of carrageenan (Hansen 1966) (See Fig. 7.2 and 7.3).

When carrageenan was dissolved in skimmilk dialyzed free of soluble milk salts, it was shown that 60% of the carrageenan remained in the supernatant after ultracentrifugation, but still 40% sedimented with some of the protein. Apparently, calcium ions are important in this reaction, since upon addition of calcium chloride to these systems all the soluble carrageenan sedimented with the casein (Hansen 1966). (See Fig. 7.4 and 7.5).

Carrageenans interact with β-lactoglobulin at pH 4.0 or below. When the pH of the solution was raised to 5 or above, no interaction was observed (Hildago and Hansen 1966). Since at pH 4.0 or below, β-lactoglobulin is positively charged, it is evident that the type of interaction involved here is a direct ionic interaction between the negatively charged carrageenan and the positively charged protein. Therefore, these experimental findings clearly indicate that the general interaction between the sulfated polysaccharides and protein described earlier is not a hypothetical situation, but rather it occurs in complex food systems.

STABILIZATION OF CASEIN BY CARRAGEENAN

One of the most interesting properties of carrageenan is its ability to stabilize calcium-sensitive casein against precipitation by calcium ions. As discussed earlier, the stability of the casein system is depen-

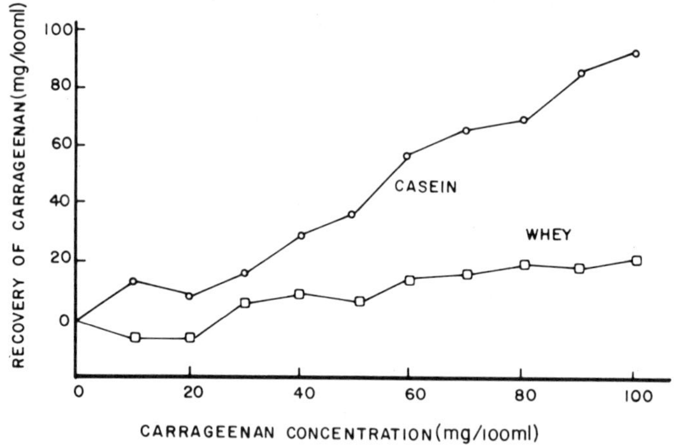

From Hansen (1966)

FIG. 7.2. THE DISTRIBUTION OF CARRAGEENAN IN SKIMMILK AT pH 4.6

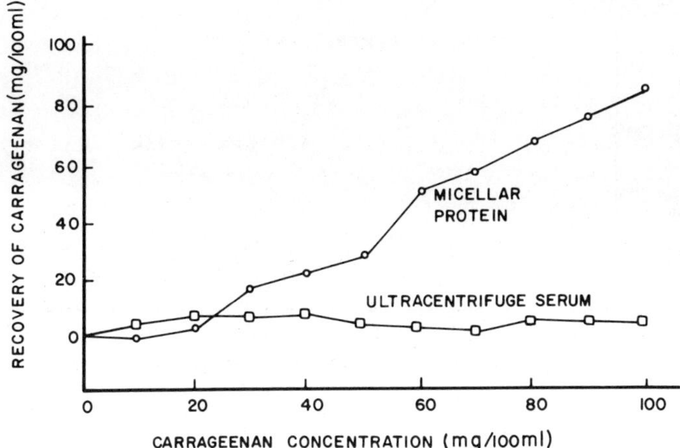

From Hansen (1966)

FIG. 7.3. THE DISTRIBUTION OF CARRAGEENAN IN SKIMMILK AT pH 6.7

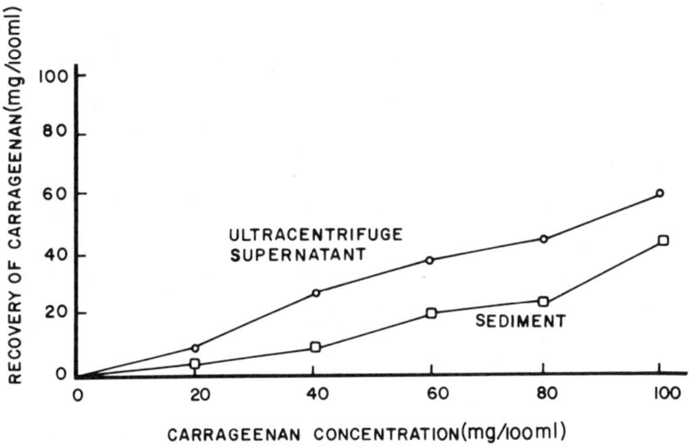

From Hansen (1966)

FIG. 7.4. DISTRIBUTION OF CARRAGEENAN IN DIALIZED SKIMMILK AT pH 7

dent upon the protective action of κ-casein. Carrageenan and furcellaran were found to be the only naturally occurring polysaccharides to possess this property of κ-casein (Lin and Hansen 1970). None of the stabilizers represented in the neutral group, in the carboxylated group, or in the mixed carboxylated-sulfated group provided any comparable protective effect against the precipitation of the protein by calcium (Table 7.2).

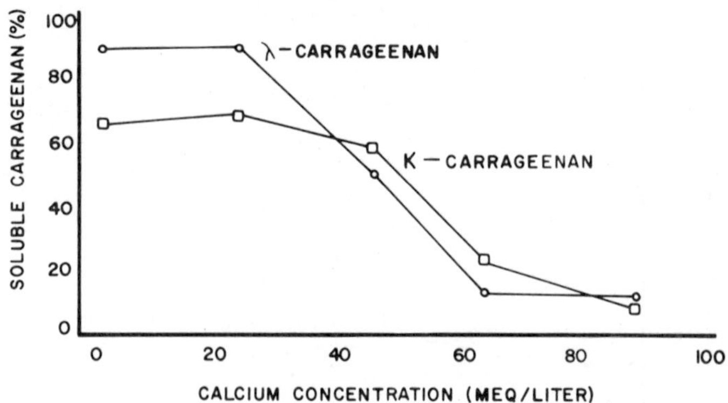

From Hansen (1966)

FIG. 7.5. EFFECT OF CALCIUM ION CONCENTRATION ON THE DISTRIBUTION OF 0.1% CARRAGEENAN IN DIALIZED SKIMMILK AT pH 7

Ultrastructure of Carrageenan α_s-casein Complex

A comparison of the effectiveness of carrageenan extracts obtained from selected botanical species of seaweeds (Fig. 7.6) revealed a marked difference among the samples. Analysis of the chemical properties of these samples (Table 7.3) showed that there was a direct relationship between their stabilizing strength and their 3,6-anhydrogalactose (3,6-AG) content and also a possible correlation with their potassium sensitivity. The stabilizing strength was dependent neither on the total sulfate content nor on the viscosity values of the extracts.

The corresponding infrared spectra obtained from dry films of the carrageenan samples are presented in Fig. 7.7. Such spectra have frequently been used to identify the position of sulfate groups in the pyranose rings. The most effective carrageenans (Fig. 7.7A) were those which contained the typical features of κ- or ι-carrageenan, namely (1) a peak at 930 cm^{-1} due to 3,6-AG, (2) a strong peak at 850 cm^{-1} due to the sulfate on the C-4 position, and (3) the virtual absence of any absorbance at 820 cm^{-1} from primary sulfate in the C-6 position. In contrast, the relatively ineffective fractions (Fig. 7.7B) all displayed the λ-carrageenan characteristics of very little 3,6-AG, little or no C-4 sulfate, but they contained a rather broad sulfate band in the 830 cm^{-1} region reflecting the presence of sulfate in both the C-2 and the C-6 positions.

The absence of 6-sulfate and the presence of 3,6-anhydride in κ- and ι-carrageenan are known (Rees 1969) to favor the formation of

TABLE 7.2

STABILIZATION OF α_s-CASEIN (0.15%) BY SOME HYDROCOLLOIDS AT pH 6.7 (HYDROCOLLOID/α_s-CASEIN = 1/4)

Groups	Glycosidic Linkages	Stabilized α_s-casein, %
Neutral		
Guar gum	β-1,4 (branch α-1,6)	0
Locust bean gum	β-1,4 (branch α-1,6)	0–0.1
Agarose	β-1,4 and β-1,3	0–1.5
Carboxylated		
CMC	β-1,4	0–0.1
Algin	β-1,4	0
Pectin	α-1,4	0–2.8
Gum arabic	β-1,3 and α-1,6	0–1.3
Hyaluronic acid	β-1,3 and β-1,4	0
Mixed carboxylated and sulfated		
Heparin	α-1,3 and α-1,4	0–6.0
Chondroitin sulfate A (4-sulfate)	β-1,3 and β-1,4	0–10.8
Chondroitin sulfate C (6-sulfate	β-1,3 and β-1,4	0–1.5
Chondroitin sulfate D	β-1,3 and β-1,4	0–0.5
Sulfated		
Sulfated cellulose	β-1,4	6–15.4
Fucoidan	α-1,2 and β-1,4	0
Furcellaran	α-1,3 and β-1,4	40–50
λ-Carrageenan	α-1,3 and β-1,4	40.0–50.0
i-Carrageenan	α-1,3 and β-1,4	92.0–100
k-Carrageenan	α-1,3 and β-1,4	90.0–100

Source: Lin (1970).

a helical conformation. X-ray diffraction (Anderson *et al.* 1969) and optical rotary dispersion studies (Rees *et al.* 1969) revealed that κ- and ι-carrageenan exist as random coils in water solution at temperatures above 50°C. As the temperature drops, the randomness disappears and a double helix structure is formed. (Fig. 7.8, Gel I).

Further temperature decreases result in intermolecular interactions and additional double helix formation. These bundles of double helices serve as the backbone of the carrageenan gel structure. The loose end on both sides of the double helices becomes the junction for interactions with other bundles of carrageenan double helices, creating the three-dimensional gel network as shown in Gel II of Fig. 7.8.

Electron microscopical study (Chakraborty and Hansen 1971) of the ultrastructure of carrageenan/α_s-casein system provided evidence of a similar structural network. α_s-casein was found to exist as electron dense aggregates (1,000–1,500 Å diameter) entrapped by interconnecting strands of relatively electron-transparent κ-carrageenan.

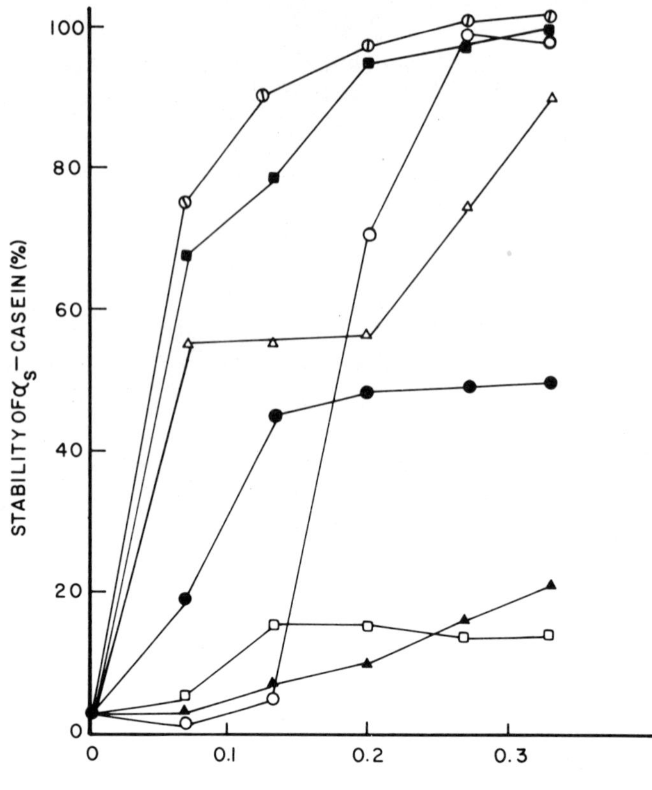

From Lin and Hansen (1970)

FIG. 7.6. STABILIZATION OF α_s-CASEIN (0.15%) BY CARRAGEENANS AT pH 6.7

○ = κ-carrageenans from *C. crispus*; ■ = unfractionated carrageenan extract from *G. stellata*; ○ = ι-carrageenan from *E. spinosum*; △ = κ-carrageenan from *G. pistillata*; ● = λ-carrageenan from *C. crispus*; □ = λ-carrageenan from *P. rotundus.* Calcium ion concentration is 0.01 *M.*

The α_s-casein aggregates were distributed as discrete particles in this three-dimensional network with very little or no protein in evidence in the polysaccharide strands between the particles. A diagram is given in Fig. 7.9 to illustrate the postulated involvement of the double helix in the stabilization mechanism. The fundamental concept of this model is that the interconnecting bridges are bundles of κ-carrageenan double helices which possess little or no protein reactivity as the result of self-aggregation. The loose ends of the double helices serve as junction zones to connect other strands of polysac-

TABLE 7.3

STABILIZING ABILITY OF SEVEN CARRAGEENAN SAMPLES IN COMPARISON WITH THEIR ANALYTICAL DATA

Stabilizing Strength in Descending Order	(α)D	Potassium Sensitivity	η_{inh}	Galactose (%)	SO_3Na, %	3,6-AG, %	Molar Ratio Gal:SO_3Na:3,6-AG
C. crispus (κ fraction)	50	+++	11.6	28.6	29.6	22.9	1:1.63:0.90
G. stellata (unfractionated)	55	+	14.4	27.7	33.8	19.7	1:1.23:0.60
E. spinosum (ι fraction)	35	+-	7.2	30.3	37.8	19.0	1:1.97:0.71
G. pistillata (κ fraction)	88	++	10.5	35.0	35.0	12.8	1:1.57:0.41
C. crispus (λ fraction)	38	-	14.4	39.4	34.6	4.7	1:1.38:0.13
G. pistillata (λ fraction)	83	-	20.6	43.6	41.1	4.2	1:1.48:0.11
P. rotundus (unfractionated)	23	-	5.1	25.4	36.3	2.8	1:2.25:0.12

Source: Black *et al.* (1965).

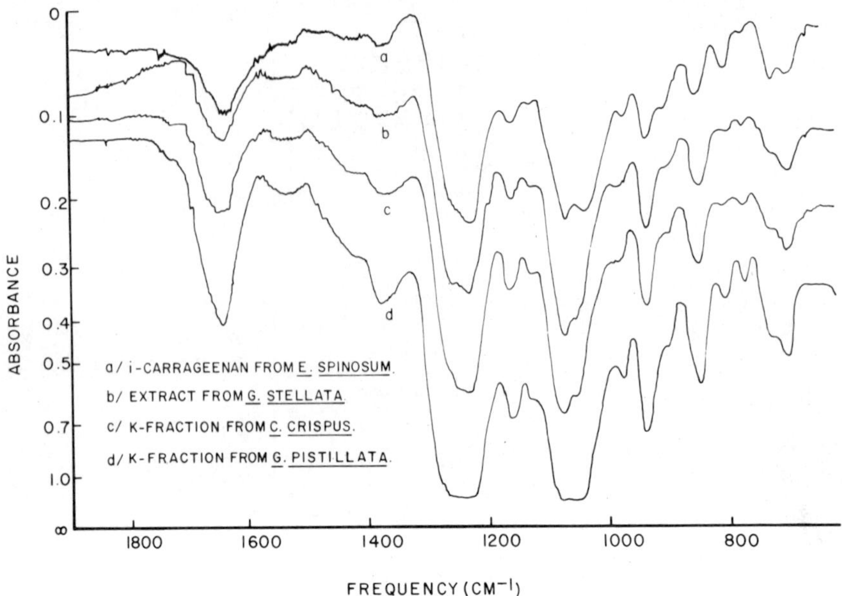

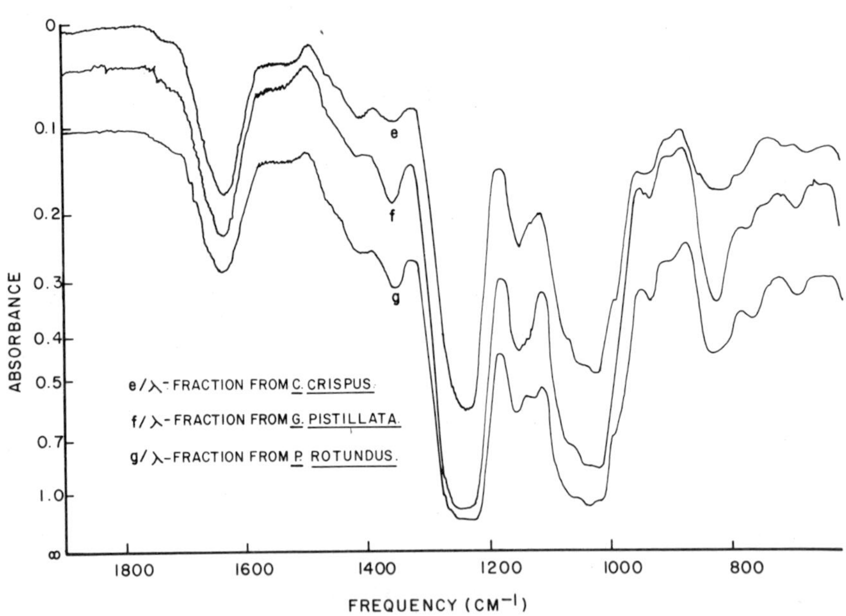

From Lin and Hansen (1970)

FIG. 7.7. INFRARED SPECTRA OF VARIOUS CARRAGEENANS

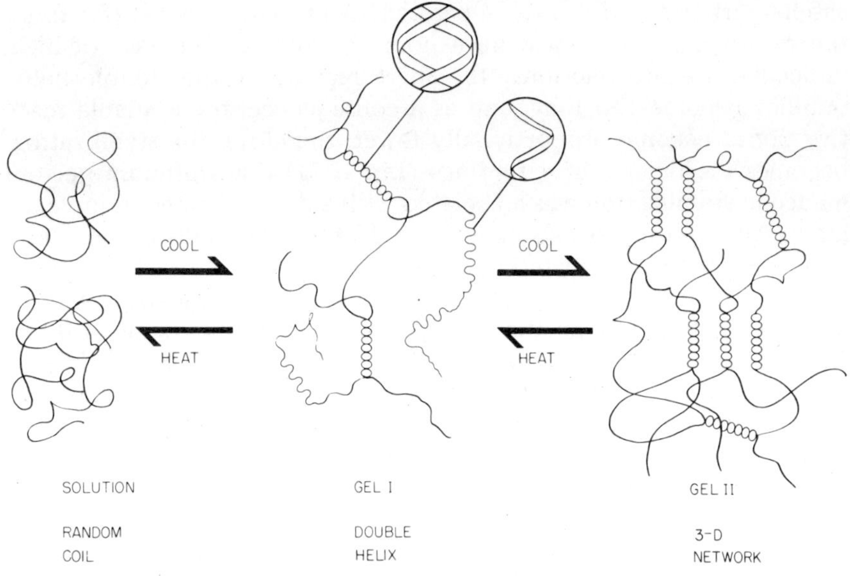

Anderson et al. (1969)

FIG. 7.8. MECHANISM OF CARRAGEENAN GEL FORMATION

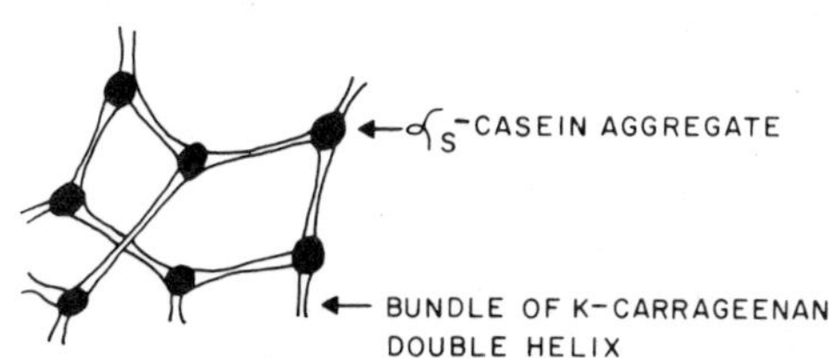

From Chakraborty and Hansen (1971)

FIG. 7.9. ULTRASTRUCTURE OF κ-CARRAGEENAN/α_s-CASEIN COMPLEX

charides to form the three-dimensional network. It also serves to "trap" α_s-casein micelles and keep them apart, so that further aggregation of protein under the influence of calcium will be prevented.

Factors Affecting the Stability of Carrageenan/α_s-casein Complex

One critical factor in this proposed model is the length of the interconnecting carrageenan double helices. On the assumption that physical separation of α_s-casein by κ-carrageenan is the major criterion for a stable complex, low molecular weight fractions will have "short arms" (double helix bundles) which will not be able to keep the α_s-

casein particles sufficiently separated to prevent aggregation; thus, the resulting complex will have poor stability. In the case of high molecular weight fractions, the inert regions of the double helix bundles become too long, and as a consequence, the available reactive zones become proportionally fewer; therefore, the stabilization becomes inefficient. The findings (Lin 1971) that optimum performance in stabilization was associated with a definite range of molecular weight of the carrageenans from 100,000–300,000 can be best explained by the postulated model.

A drastic decrease in stabilizing ability of κ-carrageenan was revealed (Lin 1971) when the pH was lowered, and below pH 4.0 the carrageenans were completely ineffective (Fig. 7.10). Since double helix formation is highly dependent on pH, it is to be expected that a shift of pH value would affect the stabilizing ability of carrageenans.

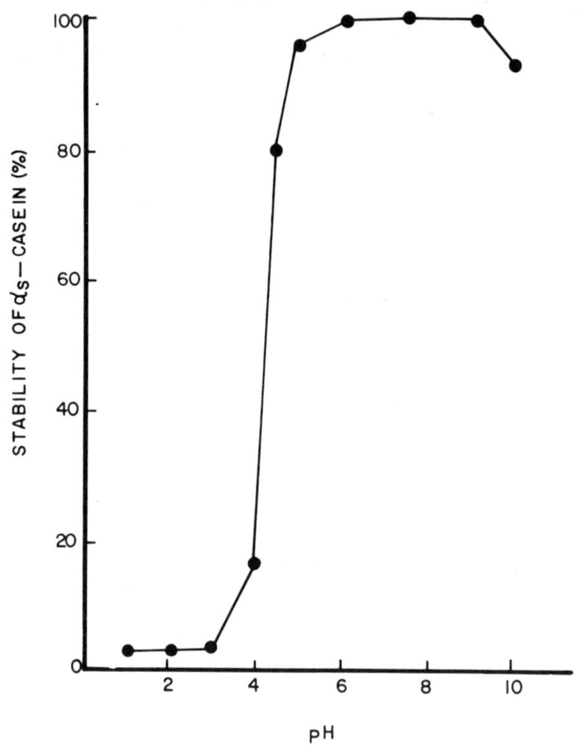

From Lin (1971)

FIG. 7.10. EFFECT OF pH ON THE STABILIZING ABILITY OF κ-CARRAGEENAN. PROTEIN CONCENTRATION, 0.15%; κ-CARRAGEENAN, 0.0375%; AND $CaCl_2$, 0.01 *M*

Furthermore, below ph 4.0, α_s-casein is positively charged; therefore, a nonspecific direct ionic interaction may take place between the negatively charged carrageenan and the positively charged protein resulting in a complete precipitation of protein.

Temperature is another important factor affecting the double helix formation. In general, heat does not favor hydrogen bonding (Lehninger 1970). Therefore, the double helices start to unwind with increasing temperature and eventually become random coils. In an optical rotary dispersion study (Rees *et al.* 1969), the optical rotation of the carrageenan solution was found to decrease when the temperature was raised. This evidence clearly suggests the unwinding of the double helices. The fact (Fig. 7.11) that the protective ability of κ-carrageenan was impaired by increasing temperature would, therefore, also be consistent with the possible involvement of a double helix conformation in the stabilization mechanism.

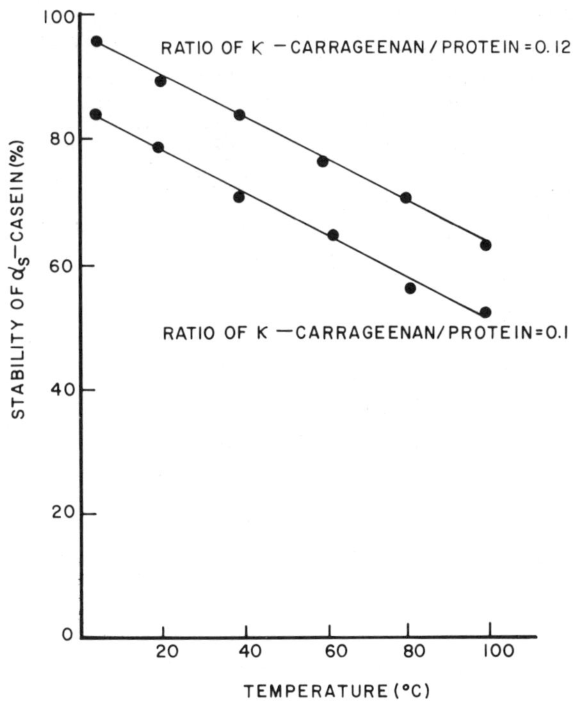

From Lin (1971)

FIG. 7.11. EFFECT OF TEMPERATURE ON THE STABILIZING ABILITY OF κ-CARRAGEENAN. PROTEIN CONCENTRATION, 0.15; pH 7.2; $CaCl_2$, 0.01 *M.* SOLUTIONS WERE HEATED AT THE RESPECTIVE TEMPERATURES FOR 15 MIN

In trying to explain the ability of carrageenan to stabilize α_s-casein, the possibility that the stability is achieved through binding of calcium by the carrageenan should not be overlooked, because calcium ions which were bound to the negatively charged molecules would no longer be available to precipitate α_s-casein. However, evidence (Lin 1971) (Fig. 7.12) indicated that the addition of calcium ion up to 8 *mM* to a solution containing 0.015% κ-carrageenan (approximately 0.2 *mM* of ester sulfate) and 0.15% α_s-casein showed only a slight decrease in the stability of the complex. Further increase of calcium ion up to 80 *mM* did not affect the stability of the complex at all. These results clearly indicated that calcium binding did not play an important role in the stabilization of α_s-casein.

Free boundary electrophoresis (Hansen 1968) and sedimentation velocity data (Lin 1971) suggested that there was no complex formation in the absence of calcium ions. This raised the possibility of a direct ionic interaction between the negatively charged groups of α_s-casein and the ester sulfate groups of carrageenan through the mediation of the calcium ion. However, this possibility was discounted by the evidence (Lin 1971) that the stability of the complex was independent of the ionic strength of the solution.

The major role of calcium ions is postulated here to be to neu-

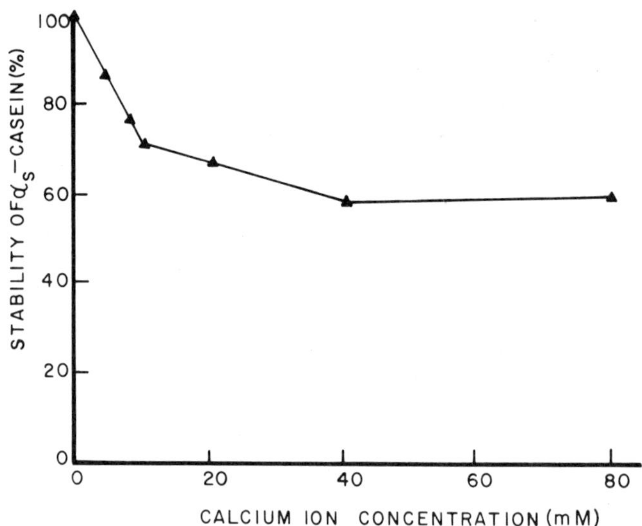

From Lin (1971)

FIG. 7.12. EFFECT OF CALCIUM ION CONCENTRATION ON THE STABILIZING ABILITY OF κ-CARRAGEENAN. α_s-CASEIN, 0.15%; κ-CARRAGEENAN, 0.015%; pH 7.2

tralize charge groups of ester sulfate so that the charge repulsion forces are eliminated, permitting double helix formation. This effect may be fairly specific because X-ray diffraction studies (Rees 1969) indicated that double helix formation for κ-and ι-carrageenan was favored in the presence of cations such as K^+, NH_4^+, Cs^+ and Rb^+ which cause gelation of carrageenan solutions. Although Ca^{++} was not mentioned in the literature, this divalent ion has a potent gelling effect on both κ-carrageenan and ι-carrageenan. Therefore, it is very likely that the presence of Ca^{++} will promote double helix formation.

Figure 7.13 (Lin 1971) shows the plot of α_s-casein stability against protein concentration at the constant weight ratio of carrageenan to protein of 0.2. A rapid decrease of the stability of the complex was

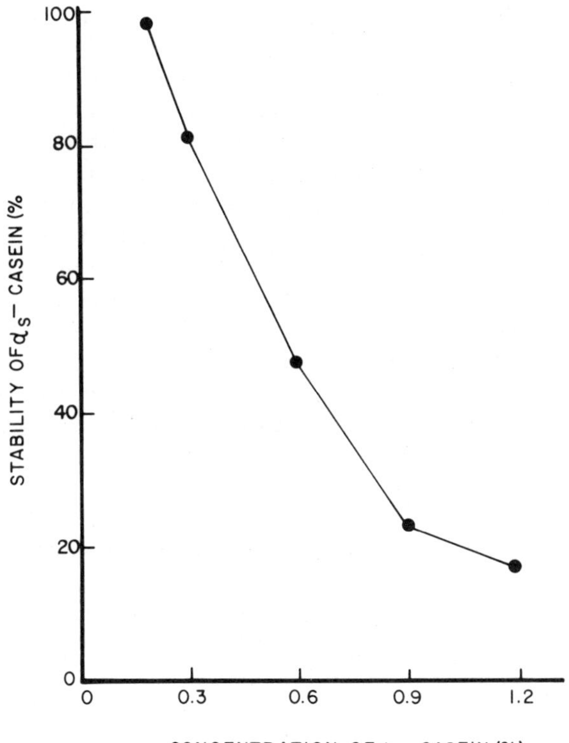

From Lin (1971)

FIG. 7.13. EFFECT OF α_s-CASEIN CONCENTRATION ON THE STABILIZING ABILITY OF κ-CARRAGEENAN. CARRAGEENAN/α_s, 0.2; pH 7.2; $CaCl_2$, 0.01*M*

observed when the protein concentration was increased. At the highest concentration, only 20% of the protein remained soluble. This observation is consistent with the assumption that the physical separation of α_s-casein is the major criterion in the stabilization of α_s-casein by carrageenans. At higher concentration of carrageenan and α_s-casein, the chance of collision between the entrapped α_s-casein increases. As a result of collision, the stability of the complex decreases.

The relative ineffectiveness of λ-carrageenan in stabilizing α_s-casein against calcium precipitation is largely due to the presence of the 6-sulfate on the 1,4-linked galactose unit. When a 1,4-linked unit is 6-sulfated, it tends to exist in the favored C1 chair conformation. As can be seen in Fig. 7.14, this introduces a kink into the polymer chain. The formation of the helical conformation is interrupted whenever a kink appears. Because of the difficulty in forming the helical structure, it is understandable that the λ-carrageenan is relatively ineffective in stabilizing α_s-casein as compared with κ-carrageenan.

Alkaline modification of λ-carrageenan removes the 6-sulfate and results in the formation of a 3,6-anhydride. As a result of these changes, the 1,4-linked pyranose unit changes to the 1C chair conformation with resultant removal of the kink (Mueller and Rees 1967). Thus, from a conformational standpoint, the alkaline modification is essentially a chain straightening process leading to a greater regularity in the carrageenan polymer. From the proposed model (Fig. 7.9), that helical conformation is the most important requirement for carrageenans in stabilizing α_s-casein, it is to be expected that the alkaline modification will greatly enhance the effectiveness of λ-carrageenan to stabilize α_s-casein (Fig. 7.15). However, the effectiveness of the alkaline-modified λ-carrageenan (also called θ-carrageenan) still does not fully match that of κ-carrageenan, especially at lower carrageenan concentration. This is because 70% of the 1,3-linked units in θ-carrageenan are sulfated at the C-2 position. An examination of molecular models shows that the presence of this 2-sulfate group will prevent the formation of the double helix. Because the single helix conformation (θ-carrageenan) will be less rigid than that of the double helix (κ-carrageenan), it is obvious that the effectiveness of θ-carrageenan cannot fully match that of κ-carrageenan.

Stabilization of Other Proteins by Carrageenans

In addition to the ability to stabilize α_s-casein, carrageenan can also stabilize β-casein (Lin and Hansen 1970), para-casein (O'Loughlin and Hansen 1973), and soy protein, peanut protein, cottonseed protein and cocoanut protein (Chakraborty and Randolph 1972) against

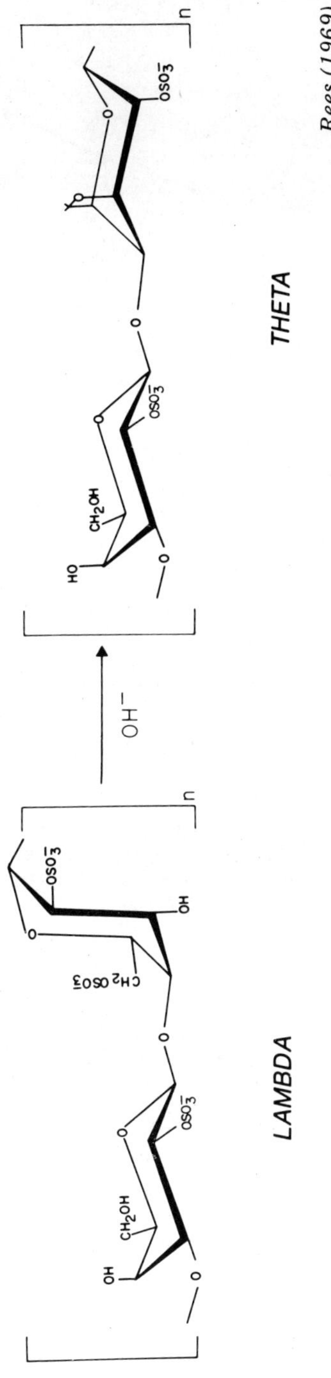

FIG. 7.14. ALKALINE MODIFICATION OF λ-CARRAGEENAN

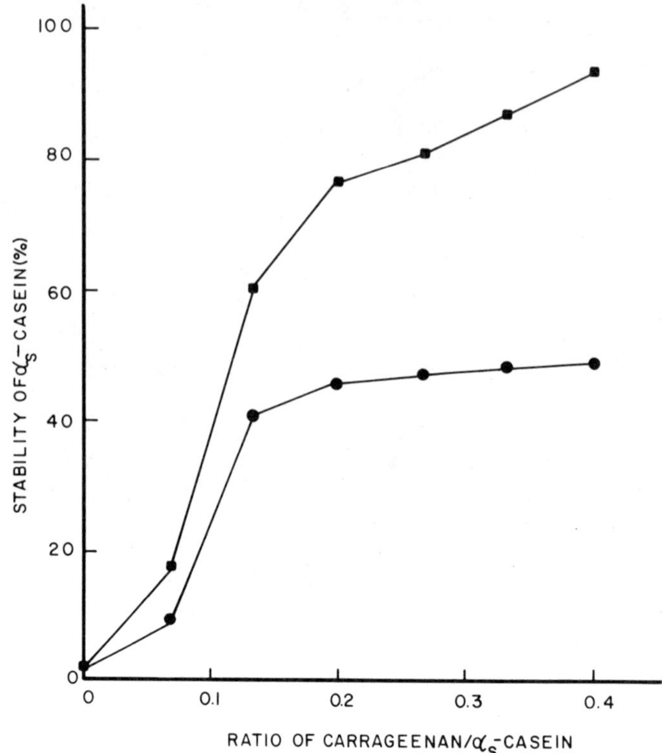

From Lin and Hansen (1966)

FIG. 7.15. EFFECT OF ALKALINE TREATMENT ON STABILIZING ABILITY OF λ-CARRAGEENAN FROM *C. CRISPUS.* ■ = ALKALINE-TREATED λ-CARRAGEENAN; ● = ORIGINAL λ-CARRAGEENAN

precipitation with calcium. The mechanism involved in stabilizing these proteins is presumably the same as that of the carrageenan/α_s-casein system.

In summary, the phenomenon of casein micelle stabilization described here is believed to account for the effectiveness of carrageenans as stabilizers for evaporated milk and infant formulas. It may also account for the cocoa-suspending property of carrageenan in the chocolate milk.

STABILIZATION OF α_s-CASEIN BY SULFATED GALACTOMANNAN

Proposed Model for Sulfated Locust Bean Gum/α_s-casein Complex

The stabilization of calcium-sensitive casein by protective colloids other than native κ-casein was first reported (Lin and Hansen 1970)

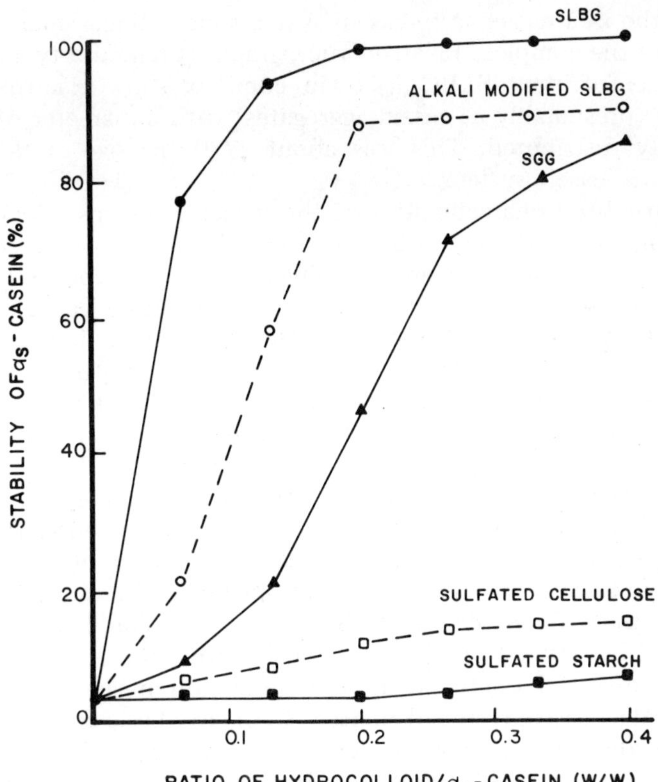

From Lin (1971)

FIG. 7.16. STABILIZATION OF α_s-CASEIN BY SELECTED CHEMICALLY SULFATED POLYSACCHARIDES. PROTEIN CONCENTRATION, 0.15%; pH 7.2; $CaCl_2$, 0.01 *M*.

to be limited to the carrageenan type polysaccharides. However, it was revealed later that some synthetically sulfated polysaccharides also possessed this ability.

As shown in Fig. 7.16 (Lin and Hansen N.D.), sulfated cellulose and sulfated starch were totally ineffective. Sulfated guar gum gave a 85% stabilizing ability at the highest ratio (0.4) of hydrocolloid/protein tested, however, its effectiveness was poor at the low concentration of hydrocolloid. In contrast, sulfated locust bean gum (SLBG) showed a remarkably high performance (85%) even at the ratio of hydrocolloid/protein of 0.067. This performance at low concentrations was even better than that of κ-carrageenan from *Chondrus crispus.* Complete stabilization was achieved at a ratio of SLBG/α_s-casein of 0.2. The turbidity of the resulting solution was less than that of the κ-carrageenan/α_s-casein systems indicating a smaller particle size of the complex. Electron micrographs (Chakraborty and Hansen 1971) of the SLBG/α_s-casein complex show electron-dense particles (presumably α_s-casein aggregates) with a diameter of 100Å uniformly distributed. This was about 1/10 the size of the κ-carrageenan/α_s-casein system.

The structural characteristics of the galactomannans effective for stabilizing casein are quite different from those of the carrageenans. They consist mainly of mannose joined by β-1,4 glycosidic linkages with galactose units branching along the mannose chain. Recent studies (Courtois and LeDizet 1970; Dea *et al.* 1972) indicated that the backbone of the locust bean gum can be divided into two regions: the "smooth regions," which are the unsubstituted part and the "hairy regions," which carry grouped galactose substitutes. The "smooth regions" may convert from the random coil to a regular conformation which is extended and ribbon-like, if they can be stabilized in this form by binding to κ-carrageenan or furcellaran.

In the absence of knowledge about the distribution of ester sulfate groups on SLBG, one would have to speculate that sulfate groups would exist both in the "smooth regions" and the "hairy regions" on the SLBG. Based on the available information, a model is proposed here (Fig. 7.17) to explain the interaction mechanism between SLBG and α_s-casein (Lin and Hansen N.D.). The basic concept of this model is that the ester sulfate groups in the "smooth regions" of SLBG would form intramolecular linkages with adjacent subunits through the mediation of calcium ions; this would assist in reducing electrostatic tensions and result in forming an "inert area" which is void of protein reactivity. The branched subunits, galactose sul-

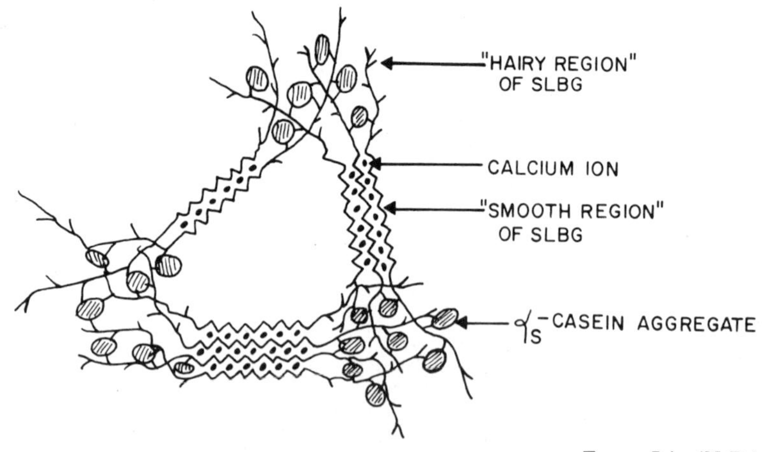

From Lin (N.D.)

FIG. 7.17. ULTRASTRUCTURE OF SLBG/α_s-CASEIN COMPLEX

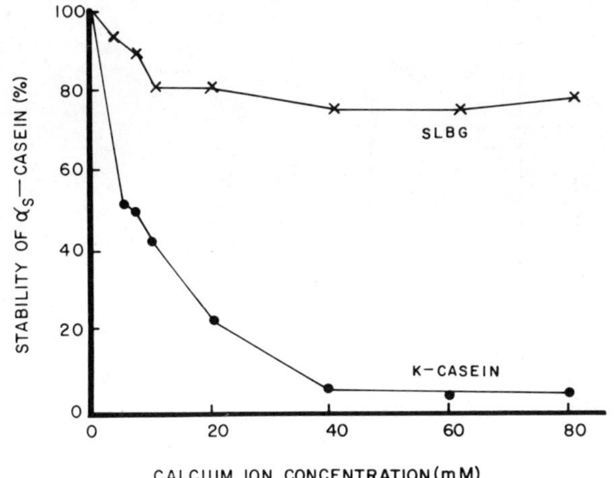

From Lin (N.D.)

FIG. 7.18. EFFECT OF CALCIUM CONCENTRATION ON THE STABILIZING ABILITY OF SLBG AND κ-CASEIN. SLBG, 0.015%; α_S-CASEIN, 0.15%; pH 7.0

fates, would then serve as junction zones to link other strands of SLBG to form a three-dimensional network. Small particles of α_s-casein aggregates (100 Å) participate in forming these junction zones.

The resulting network would be somewhat rigid because of the numerous junctions. This model shares the same principle as the one postulated for the carrageenan-stabilized systems, namely that the conformational arrangement of the polysaccharides is the controlling factor for creating segments with a sufficiently rigid structure to resist collapse.

The mode of action for the SLBG system explains why the turbidity of the stabilized protein system is less than for κ-casein or κ-carrageenan: there are many more protein-reactive centers in this complex in which the aggregates of α_s-casein are trapped and, therefore, the individual protein particles are smaller.

Factors Affecting the Stability of Sulfated Locust Bean Gum/α_s-casein Complex[1]

As in the case of carrageenan, the action of sulfated galactomannan does not depend on a simple calcium binding effect by the ester sulfate on the polysaccharide. The protective ability of sulfated locust bean gum was not impaired when the calcium ion concentration increased to 80 *mM* (Fig. 7.18). This is in contrast to the native protec-

[1] Based on Lin 1971; Lin and Hansen N.D.

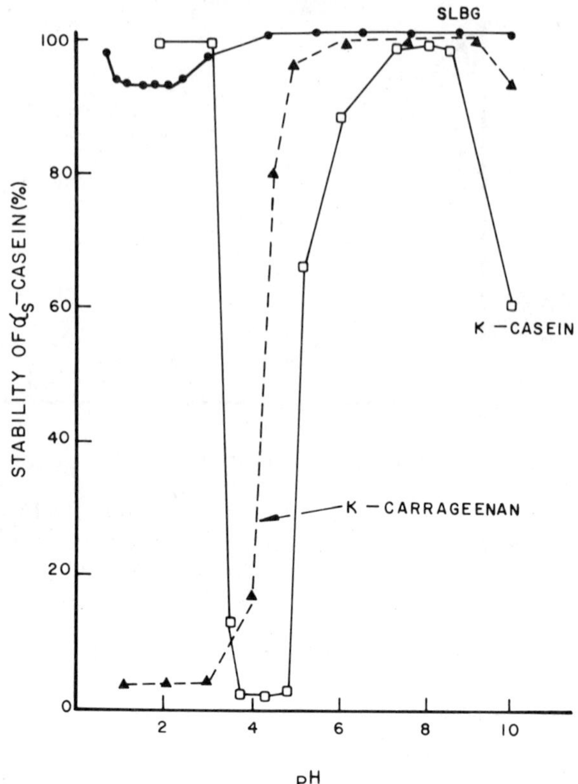

From Lin (1971)

FIG. 7.19. EFFECT OF pH ON THE STABILIZING ABILITY OF SLBG. CURVES OF κ-CASEIN AND κ-CARRAGEENAN ARE INCLUDED FOR COMPARISON. PROTEIN CONCENTRATION, 0.15%; SLBG CONCENTRATION, 0.0375%; $CaCl_2$, 0.01 *M*.

tive colloid, κ-casein, which completely lost its stabilizing ability when the calcium ion concentration increased beyond 40 *mM*. pH exerts little effect (Fig. 7.19) on the stability of the SLBG/α_s-casein complex. This was a property totally different from that of the carrageenan and of κ-casein. This evidence is an indication that the detailed mode of action of SLBG is different from carrageenan and that the conformation of the polysaccharides may be a controlling factor in the complex formation. It also provides evidence to preclude the possibility that the direct ionic-type interaction between the negatively charged SLBG and the positively charged groups of α_s-casein is involved in the complex formation. This is because at the pH corresponding to the pKa of ester sulfate of 2.0, the loss of complex stability was negligible.

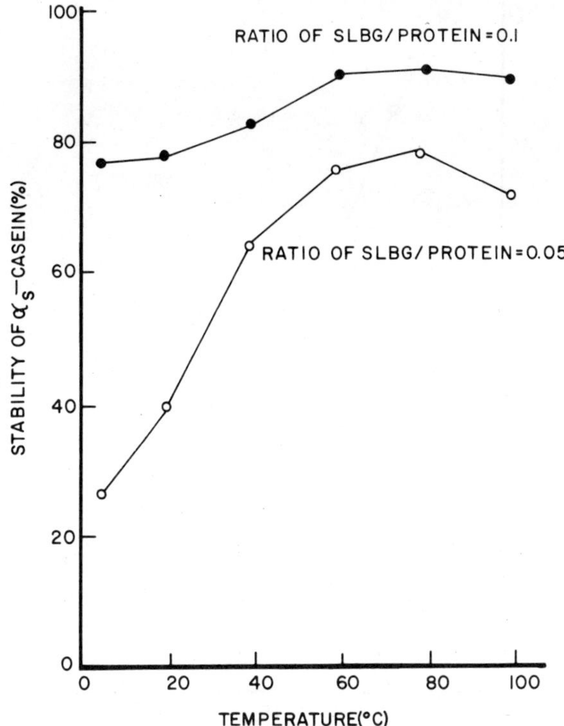

From Lin (1971)

FIG. 7.20. EFFECT OF TEMPERATURE ON THE STABILIZING ABILITY OF SLBG. PROTEIN CONCENTRATION, 0.15%; pH 7.2; $CaCl_2$, 0.01 *M.* SOLUTIONS WERE HEATED AT THE RESPECTIVE TEMPERATURES FOR 15 MIN

Figure 7.20 shows the effect of the temperature of heating at constant heating time on the stabilizing ability of SLBG. It is apparent that maximum effectiveness of SLBG was within the range of 60–80° C and that outside this range the stability decreased. These results are in marked contrast to the findings for κ-carrageenan in which case the effectiveness decreased linearly with temperature. Therefore, there is a possibility that hydrophobic bonding may be involved in the SLBG stabilization. For carrageenans, however, this situation is different.

It is also possible that the segments of the polysaccharide strands may be hydrogen-bonded to form a closely-packed aggregate with little opportunity for forming an extended three-dimensional network. One might picture this model as a wet chicken feather in which the details of a network are not prominent. On drying, the feather opens up and exposes the individual branches. The closed

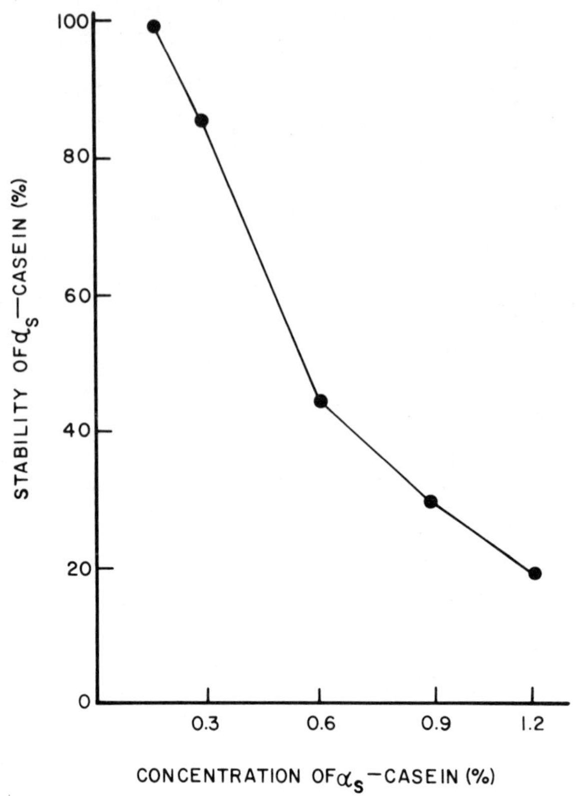

From Lin (1971)

FIG. 7.21. EFFECT OF α_s-CASEIN CONCENTRATION ON THE STABILIZING ABILITY OF SLBG. RATION OF SLBG/α_s-CASEIN, 0.2; pH 7.2; $CaCl_2$, 0.01 *M.*

hydrogen-bonded structure apparently is less conducive for the formation of the required three-dimensional network.

The removal of the 6-sulfate group by alkaline modification lowers the stabilizing ability of SLBG (Fig. 7.16). This observation is consistent with the proposed model. Since the branch galactose 6-sulfate is involved in the junction zones to form the three-dimensional network, a removal of the sulfate group and replacement with 3,6-anhydrogalactose would certainly hinder its ability to attain this stabilized conformation.

A linear decrease in stability was observed (Fig. 7.21) when the concentration of SLBG and α_s-casein was increased. These results were identical to the response of κ-carrageenan discussed previously (Fig. 7.13). This suggests that physical separation of α_s-casein aggregates is the major requirement for a stable complex system. This requirement is the same for both the κ-carrageenan and the SLBG models. The fact that both types of complexes responded in an identical manner to a change in concentration supports the validity of the assumption. When the complex concentration increases, the chance for a collapse of the entrapped protein particles becomes greater, and as a result, the stability of the complex decreased.

CONCLUSION

The application of gums or hydrocolloids in the food industry has a long history. However, most of the food gums are used predominantly for their viscosity building properties and their gelling properties. Little is known about the possible chemical interaction occurring between gums and the food components.

Because of a lack of adequate fundamental knowledge, the use of certain stabilizers in a particular product relies to a large extent upon trial and error. A better understanding of how stabilizers function in food products is needed and will be extremely helpful in designing new foods with greater predictive success.

BIBLIOGRAPHY

ANDERSON, N. S. *et al.* 1969. X-ray diffraction studies of polysaccharides: double helix models for κ- and ι-carrageenans. J. Mol. Biol. *45*, 85–89.

BETTELHEIM-JEVONS, F. R. 1958. Protein-carbohydrate complexes. *In* Advances in Protein Chemistry, Vol. 13, C. B. Anfinsen, M. L. Anson, and J. T. Edsall (Editors). Academic Press, New York.

BLACK, W. A. P., BLAKEMORE, W. R., COLQUHOUN, J. A., and DEWAR, E. T. 1965. The evaluation of some red marine algae as a source of carrageenan and of its κ- and λ-components. J. Sci. Food Agr. *16*, 573–585.

CHAKRABORTY, B. K., and HANSEN, P. M. T. 1971. Electron microscopy of protein/hydrocolloid interacting system, M27. J. Dairy Sci. *54*, 754.

CHAKRABORTY, B. K., and RANDOLPH, H. E. 1972. Stabilization of calcium-sensitive plant protein by κ-carrageenan. J. Dairy Sci. *37*, 719–721.

CORNWELL, D. G., and KRUGER, F. A. 1961. Molecular complexes in the isolation and characterization of plasma lipoproteins. J. Lipid Res. *2*, No. 2, 110–134.

COURTOIS, J. E., and LeDIZET, P. 1970. Galactomannans. VI. Action of some mannanases on galactomannan. Bull. Soc. Chim. Biol. *52*.

DEA, I. C. M. *et al.* 1972. Tertiary and quaternary structure in aqueous polysaccharide systems which model cell wall cohesion. Reversible changes in conformation and association of agarose, carrageenan and galactomannans. J. Mol. Biol. *68*, 153–172.

GREENBERG, D. M. 1944. The interaction between the alkali earth cations, particularly calcium, with proteins. *In* Advances in Protein Chemistry, Vol. 1, M. L. Anson, and J. T. Edsall (Editors). Academic Press, New York.

GURD, F. R. N., and WILCOX, P. E. 1956. Complex formation between metallic cations and proteins, peptides and amino acids. *In* Advances in Protein Chemistry, Vol. 11, C. B. Anfinsen, M. L. Anson, and J. T. Edsall (Editors). Academic Press, New York.

HANSEN, P. M. T. 1966. Distribution of carrageenan stabilizers in milk. J. Dairy Sci. *49*, 698.

HANSEN, P. M. T. 1968. Stabilization of α_s-casein by carrageenan. J. Dairy Sci. *51*, 192–195.

HILDAGO, J., and HANSEN, P. M. T. 1969. Interaction between food stabilizers and β-lactoglobulin. J. Agr. Food Chem. *17*, 1089–1092.

LEHNINGER, A. L. 1970. Biochemistry: The Molecular Basis of Cell Structure and Function. Worth Publishers, New York.

LIN, C. F. 1971. The casein stabilizing function of sulfated polysaccharides. Ph.D. Dissertation, Ohio State Univ., Columbus.

LIN, C. F., and HANSEN, P. M. T. 1970. Stabilization of casein micelles by carrageenan. Macromolecules *3*, 269–274.

LIN, C. F., and HANSEN, P. M. T. (N. D.) Stabilization of α_s-casein by sulfated locust bean gum. Manuscript in preparation.

McMEEKIN, T. L., and POLIS, B. D. 1949. Milk proteins. *In* Advances in Protein Chemistry, Vol. 5, M. L. Anson, and J. T. Edsall (Editors). Academic Press, New York.

MUELLER, G. P., and REES, D. A. 1968. Current structural views of red seaweed polysaccharides. *In* Transactions of the Drugs from the Sea Symposium, University of Rhode Island, 27–29 August 1967, H. D. Preudenthal (Editor). Marine Technology Society, Washington, D.C.

O'LOUGHLIN, K., and HANSEN, P. M. T. 1973. Stabilization of rennet-treated milk protein by carrageenan. J. Dairy Sci. *56*, 629.

PUTNAM, F. W. 1948. The interactions of proteins and synthetic detergents. *In* Advances in Protein Chemistry, Vol. 4, M. L. Anson, and J. T. Edsall (Editors). Academic Press, New York.

REES, D. A. 1969. Structure, conformation and mechanism in the formation of polysaccharide gels and networks. *In* Advances in Carbohydrate Chemistry and Biochemistry, Vol. 24, M. L. Wolfrom, T. S. Tipson, and D. Horton (Editors). Academic Press, New York.

REES, D. A. *et al.* 1969. Conformational analysis of polysaccharides. III. The relation between stereochemistry and properties of some natural polysaccharide sulfates. J. Polymer Sci., Part C. *28*, 261–276.

STANLEY, N. F. 1972. Properties of carrageenans as related to structure. Paper presented at the Maine Section of Am. Chem. Soc. Meeting on Oct. 20.

STEINHARDT, J., and BEYCHOK, S. 1964. Interaction of proteins with hydrogen ions and other small ions and molecules. *In* The Proteins, Vol. 2, H. Neurath (Editor). Academic Press, New York.

WAUGH, D. A. 1954. Protein-protein interaction. *In* Advances in Protein Chemistry, Vol. 9, M. L. Anson, K. Bailey, and J. T. Edsall (Editors). Academic Press, New York.

CHAPTER 8

Arthur L. Moirano

Sulfated Seaweed Polysaccharides

INTRODUCTION

In order to categorize the food colloids which may be extracted from the wide variety of available sea plants, it may be desirable to briefly review the classifications of marine algae. Algae are normally divided into four major classes based partly on their predominant photosynthetic pigments. Figure 8.1 shows these four classes in abbreviated outline.

The green and blue-green algae are commonly associated with fresh water, while the brown and red algae are found almost exclusively in marine habitats and are usually referred to as seaweeds. Of the four classes, the brown and red marine algae provide the algal polysaccharides of commerce. The brown seaweeds, often referred to as kelp and rockweed, are employed for the extraction of alginates, polysaccharides composed of mannuronic and guluronic acid units. The red seaweeds yield agar, carrageenan and furcellaran. Not only do all of these red seaweed polysaccharides contain galactose in one form or another as a building block, but all the galactose units in the structures that have been defined are alternately α-1,3; β-1,4 glycosidically linked. The amount of sulfate conveniently serves to broadly differentiate the various red seaweed polysaccharides. As stated in the Federal Register, Title 21, Section 121.1066, a red seaweed polysaccharide must have 20% or more of sulfate on a dry weight basis if it is to be classed as a carrageenan. The 1974 Food Chemicals Codex has lowered the minimum for sulfate to 18% on a dry weight basis (Natl. Acad. Sci. 1974). For many years, commercial agar was considered as a sulfated polysaccharide since typical analysis yielded from 3–4% sulfate. It is now known that many commercial agars may be separated into two fractions, i.e., an essentially sulfate-free nonionic polysaccharide known as agarose and a sulfated fraction broadly known as agaropectin. The latter has no commercial value at present and much of it is removed and discarded during commercial production of agar. Furcellaran has 8–19% sulfate.

It may thus be seen that only carrageenan and furcellaran represent the class of naturally sulfated food colloids of present commercial importance. In view of the many similarities in properties and struc-

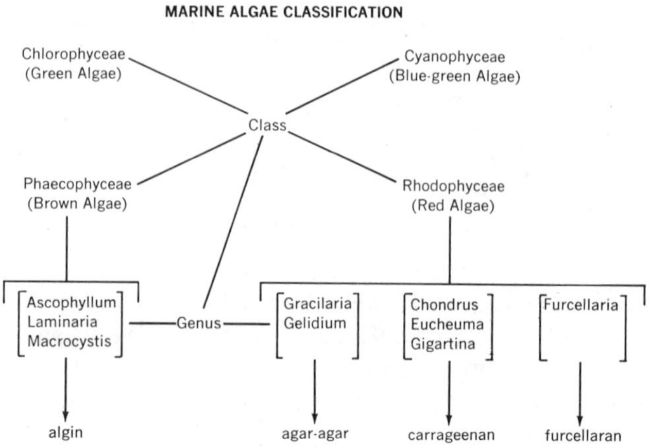

FIG. 8.1. ABBREVIATED MARINE ALGAE CLASSIFICATION CHART

ture between furcellaran and *kappa*-carrageenans, furcellaran will not be treated separately but rather discussed where appropriate in the following detailed discussion of carrageenan.

CARRAGEENAN STRUCTURES

When carrageenan was first produced commercially, circa 1937, the principal source was the red seaweed *Chondrus crispus* which is native to the Canadian Maritimes. Other than the fact that it contained galactose and ester sulfate, and was probably heterogeneous, very little was known about its structure. It was not until later that Smith and Cook (1953) were able to separate two fractions from *Chondrus crispus.* They took advantage of the fact that one of the fractions could be selectively precipitated by potassium ions and they designated this as *kappa*-carrageenan. The nonpotassium sensitive fraction was called *lambda*-carrageenan. Recent work indicates that these fractions do not occur in the same plant but rather in individual plants which commonly grow together (McCandless *et al.* 1973). The basic structures of *kappa* and *lambda* were determined by O'Neill (1955). Further characterization of these was done by Rees as well as the elucidation of a third type of carrageenan, i.e. *iota*, which is derived from the red seaweed *Eucheuma spinosum* (Rees 1969).

A number of red seaweeds are now used as sources of carrageenan. The carrageenans extracted from these weeds differ principally from one another in the amount of 3,6 anhydro-D-galactose (3,6-AG) they contain and the number and position of ester sulfate groups. Based on what is now known, it is no longer practical to define various native carrageenans as to weed source alone. Neither is it practical to try to define them strictly by physical properties such as solubility in potassium solutions. In the broad sense carrageenan may be defined as "that group of galactan polysaccharides extracted from red seaweeds of the *Gigartinaceae*, *Solieraceae* and *Phyllophoraceae* families that have ester sulfate content of 18% or more and are alternately α-1,3; β-1,4 glycosidically linked." While it is true that a spectrum of carrageenans exists (Pernas *et al.* 1967), it is nevertheless possible to distinguish a number of these which will serve as a basis for discussions of structure and properties. Figure 8.2 shows the basic repeating disaccharide structures which make up the various carrageenans. The B units represent the 1,3-linked galactosides whereas the A units represent the 1,4-linked galactosides. For most food applications, carrageenans having estimated molecular weights ($\bar{M}_n$) of 100,000–500,000 are generally employed.

Kappa-carrageenan is composed of alternating 1,3-linked galactose 4-sulfate and 1,4-linked 3,6 anhydro-D-galactose residues (Anderson *et al.* 1968A). An inferred precursor, *mu*-carrageenan, theoretically

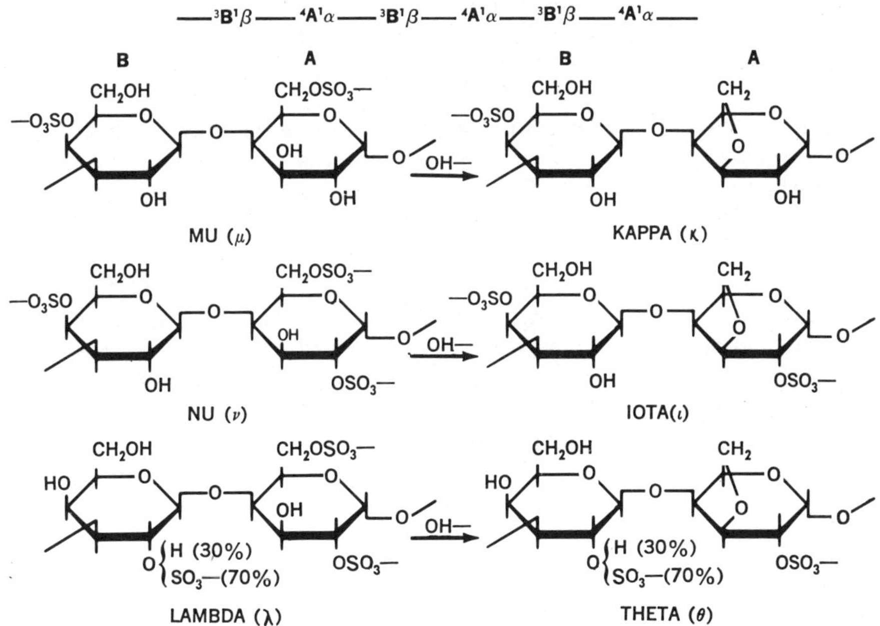

FIG. 8.2. REPEATING UNITS OF CARRAGEENANS

contains no anhydride groups but rather 6-sulfate in the 1,4-linked galactoside (Anderson *et al.* 1968B). An enzyme, given the name of "dekinkase" by Lawson and Rees (1970) is present in *kappa*-bearing sea plants and has the ability to remove this sulfate with concomitant ring closure to form a 3,6-anhydride. In nature, neither *mu* with no 3,6-AG nor limit *kappa* with theoretical 3,6-AG of about 35% exists. Rather, there are intermediates containing moderate amounts of 3,6-AG, such as the *kappa* from *Chondrus crispus* which has a 3,6-AG content of about 28%. These intermediate products exhibit greater potassium sensitivity than calcium sensitivity. In general, it is convenient to refer to those carrageenans which exhibit greater potassium sensitivity than calcium sensitivity as *kappa* types. As will be shown later, an increase in anhydride content from 28–35% results in increased potassium sensitivity and gelling ability.

When *kappa*-bearing seaplants are extracted under mild alkaline conditions, the anhydride content is unaltered. While there are a number of applications that require *kappa* with a moderate 3,6-AG content, it is frequently desirable to employ *kappa*-carrageenans having as high a level of 3,6-AG as possible in order to maximize certain properties. Stanley (1963) obtained a patent for increasing the amount of 3,6-AG in various carrageenans through the use of sufficiently alkaline extraction techniques which provides an S_N2 elimination of 6-sulfate from the 1,4-linked galactoside with concomitant ring closure to form the anhydride. Carrageenans with 3,6-AG very close to the theoretical limit may be prepared using this process.

Kappa-carrageenans from different red seaweeds, even after modification to develop maximum 3,6-AG, may still differ from one another in several respects. They may have somewhat different amounts of 4-sulfate on the 1,3-linked galactoside, for example. They may also differ from one another in that not all of the 1,4-linked galactosides have 6-sulfate. No 3,6-AG can be formed when 6-sulfate is not present. Still another difference is in the amount of 2-sulfate on the 1,4-linked galactoside. For example, the *kappa* from *Chondrus crispus* has 5–10% of the 3,6-AG sulfated in the 2-position whereas the *kappa* from *E. cottonii* has about 4–7% and that from *Hypnea* has none (Stancioff and Stanley 1969). In those cases where 2-sulfate is present, the properties of the *kappa* may change depending on whether the 2-sulfates are randomly distributed or in blocks. All of the aforementioned carrageenans are still conveniently classified as *kappas* because they all exhibit more potassium sensitivity than calcium sensitivity. Furcellaran does not contain any 2-sulfate on the 1,4-linked galactoside and has less 4-sulfate on the 1,3-linked galactoside than *kappa*-carrageenans.

Iota-carrageenan is composed of alternating 1,3-linked galactose 4-sulfate and 1,4-linked 3,6-anhydro-D-galactose-2-sulfate residues (Rees 1969). An inferred precursor *nu*-carrageenan theoretically contains no anhydride but rather 6-sulfate in the 1,4-linked galactoside (Stancioff and Stanley 1969). Neither *nu* nor limit *iota* exists in nature but rather intermediates with moderate amounts of 3,6-AG occur. The 3,6-anhydride may be increased close to the theoretical limit using the same techniques as for *kappa*. It may thus be seen that the major difference between fully anhydrized *kappa*- and *iota*-carrageenans is the amount of 2-sulfate on the 1,4-linked galactoside. As the 2-sulfate- or *iota*-moieties of fully anhydrized *kappas* increase to as high as 25–50%, the potassium sensitivity decreases as evidenced by marked weakening of gelling properties. When more than 80% of the 2-positions are sulfated, calcium sensitivity becomes predominant. When the carrageenan becomes more calcium sensitive than potassium sensitive, it is convenient to refer to it as an *iota*-carrageenan. As will be described in the next section, the nature of its gel properties compared to *kappa* also changes.

Lambda-carrageenan is primarily composed of alternating 1,3-linked galactose (about 70% contain 2-sulfate) and 1,4-linked galactose 6-sulfate residues. Note that the 1,3-linked galactose residues are not sulfated at C4 such as in *kappa* and *iota* but rather are about 70% sulfated at C2 (Dolan and Rees 1965). Theoretically there is no 3,6-AG present in *lambda*. This also holds true in practice, which suggests that there is no enzymatic pathway that converts the 1,4-linked units in *lambda* to the anhydride. If, however, *lambda* is extracted under sufficiently alkaline conditions, 6-sulfate will be eliminated from the 1,4-linked galactoside and 3,6-AG will be formed. The limit polysaccharide, where all of the 6-sulfate has been eliminated, has been designated as *theta*-carrageenan. Other than some loss in viscosity that takes place during the anhydrizing process, the properties of *theta* as compared to *lambda* are basically unchanged. As a result, *theta* is not intentionally produced commercially. In those cases where *kappa*-bearing weeds grow mixed with *lambda*-bearing weeds, the formation of *theta* is simply incidental when anhydrizing of the *kappa* takes place during processing.

An extremely useful tool for the characterization of the structures of red seaweed extractives is infrared analysis. The relation of IR spectra to physical properties and chemical composition has been discussed by Stancioff and Stanley (1969).

BASIC PROPERTIES OF CARRAGEENAN

In a previous section, it was pointed out that *kappa*-carrageenan is derived from *mu*, *iota* from *nu* and *theta* from *lambda*. Since only *kappa*, *iota* and *lambda* types are of any commercial significance, the following discussions will be limited primarily to these three. The principal basic properties of these food colloids are summarized in Table 8.1. In the case of *kappa*, the properties described are those of a carrageenan having ester sulfate of about 25% and 3,6-AG close to the theoretical maximum, since this is the most widely used type in food applications. Since *kappa*-carrageenans, dependent on the source, can differ from one another in the quantity of 3,6-AG and/or ester sulfate, distinctions will be made where necessary.

For the reasons given in the paragraphs under "Applications of Carrageenan," a differentiation between carrageenan behavior in aqueous and milk (dairy) systems will be made where appropriate.

Solubility

Water.—*Hot.*—all carrageenans and furcellaran are soluble in hot waters.

Cold.—Na^+ salts of *kappa* and *iota* are soluble in cold water while salts of other cations such as K^+ or Ca^{++} do not dissolve completely but exhibit limited to very high swelling depending upon such factors as type and level of cations present, particle density of the carrageenan, etc. *Lambda* is fully soluble in cold water regardless of the cations with which it is associated. *Iota* is particularly sensitive to calcium ions with which it forms highly thixotropic cold dispersions, thus lending itself very well as a suspending agent. Similar effects are obtained with potassium ions when used at sufficient concentration.

The question as to the degree of solubility of a product which swells on hydration often arises. In practice, if the viscosity of an aqueous carrageenan solution that has been heated and cooled is essentially the same as that of a cold dispersion, it is presumed that the product is cold water soluble. This measurement should be accompanied by visual observations to ensure that the cold mix is really a solution instead of a swollen dispersion, as evidenced by the smoothness.

Another question that frequently arises concerns the temperature of solubility. In the case of carrageenans which require heat for solution, the temperature required will depend on such factors as the concentration of carrageenan, cations associated with the carrageenan and cations that are a part of the system into which the carrageenan is to be incorporated. In the majority of food applications in

TABLE 8.1

SELECTED PROPERTIES OF CARRAGEENAN

	Kappa	Iota	Lambda
Solubility			
Hot water	Soluble above 70°C	Soluble above 70°C	Soluble
Cold water	Na^{++} salt soluble. From limited to high swelling of K^+, Ca^{++} and NH_4^+ salt	Na^+ salt soluble. Ca^{++} salt gives thixotropic dispersions	All salts soluble
Hot milk	Soluble	Soluble	Soluble
Cold milk	Insoluble	Insoluble	Disperses with thickening
Cold milk (TSPP added)	Thickens or gels	Thickens or gels	Increased thickening or gelling
Concentrated sugar solutions	Soluble hot	Difficulty soluble	Soluble hot
Concentrated salt solutions	Insoluble cold and hot	Soluble hot	Soluble hot
Water miscible solvents	See text		
Organic solvents	Insoluble	Insoluble	Insoluble
Gelation			
Effect of cations	Gels most strongly with K^+	Gels most strongly with Ca^{++}	Nongelling
Type of gel	Brittle with syneresis	Elastic with no syneresis	Nongelling
Locust bean gum effect	Synergistic	None	None
Stability			
Neutral and alkaline pH	Stable	Stable	Stable
Acid (pH 3.5)	Hydrolysis of solution, accelerated by heat. Gelled state stable.		Hydrolysis
Compatibility	Generally compatible with nonionics and anionics, but not with cationics.		

which they are used, solution of *kappa* and *iota* types is accomplished above 70° C.

Milk.—The unique ability of carrageenan to interact with the micellar casein of milk has led to the development of numerous applications which will be described later in this chapter. Lin and Hansen (1968, 1970) have attempted to clarify the behavior of carrageenan in milk. They demonstrated that carrageenan has the ability to effectively stabilize the calcium sensitive α_s- and β-casein fractions derived from milk. In natural milk, these fractions are stabilized by another casein fraction, *kappa*-casein. When *kappa*-casein is specifically removed by rennet enzyme, for example, the calcium sensitive protein fractions are free to react with calcium in the milk and coagulate. Carrageenan, of all the natural colloids examined, has the greatest ability to stabilize calcium sensitive fractions against precipitation and to exhibit appreciable micelle building properties. It was found that *kappa*-carrageenan was as effective as *kappa*-casein in accomplishing this.

While this work does not explain the mechanism of carrageenan behavior in actual milk applications, it nevertheless is a very important first step in its understanding.

Hot Milk.—All carrageenans and furcellaran are soluble in hot milk.

Cold Milk.—*Lambda*-carrageenan has the greatest ability to disperse in and thicken milk without the need of solubilizing salts. Its insensitivity to potassium and calcium ions, along with its high ester sulfate content, may account for this cold milk action. The higher the molecular weight of the *lambda*, the lower the concentration required for a given degree of thickening. Also, the state of dispersion and rate of hydration are much dependent on both the applied shear and particle size of the carrageenan. High shear and/or use of fine mesh *lambda* will serve to hasten the thickening of cold milk.

With regard to *kappa*- and *iota*-carrageenans, the higher the 3,6-AG content, and the lower the ester sulfate content, the more insoluble they are in cold milk. This is attributable in part to the increased sensitivity of these carrageenans to cations such as Ca^{++} and K^+ which are constituents of the milk. Furcellaran is also insoluble in cold milk. Even *kappa* and *iota* types which are practically insoluble in cold milk may effectively be used for thickening and gelling if tetrasodium pyrophosphate (TSPP) is used. This salt has the ability to form gels in cold milk and is commonly used in conjunction with pre-gelatinized starches for the preparation of cold set milk puddings. While its action on the carrageenan/milk protein system is not known, it is possible to obtain much firmer gels when carrageenan is present than could be obtained with phosphate alone.

Concentrated Sugar and Salt Solutions.—*Kappa*- and *lambda*-carrageenans are soluble in hot sucrose solutions as high as 65% concentration. *Iota*-carrageenan, however, is sparingly soluble under these conditions. On the other hand, *iota* and *lambda* solutions will tolerate high concentrations of strong electroytes, e.g. 20–25% of NaCl, while *kappa* will be salted out.

Water Miscible Solvents.—Water miscible solvents, e.g., alcohol, propylene glycol, glycerin, may be incorporated into carrageenan solutions. The concentration of solvent tolerated depends on the molecular weight of carrageenan, type of carrageenan and balance of cations as well as the method of incorporation of solvent. In general, those factors which tend to make the carrageenan more hydrophilic will increase the solvent tolerance. For example, the higher the ester sulfate content of the carrageenan and the lower its molecular weight, the higher the solvent tolerance. Also, the lower the concentration of cations present, the higher the tolerance.

Organic Solvents.—All carrageenan products are generally insoluble in organic solvents.

Gelation

Water Gels.—*Kappa*- and *iota*-carrageenans as well as furcellaran have the ability to form gels upon cooling of a hot solution. These gels are all thermally reversible, i.e., they melt on heating and gel again on cooling. According to Rees (1969) carrageenans which form aqueous gels do so because of double helix formation, as shown in Fig. 8.3. At temperatures above the melting point of the gel, thermal agitation overcomes the tendency to form helices and the polymer exists in solution as a random coil. On cooling, a three dimensional polymer network builds up in which double helices form the junction points of the polymer chains (Fig. 8.3, Gel I). Further cooling leads to aggregation of these junction points (Fig. 8.3, Gel II). The inhibitory effect of kinks in the polymer chain can be seen. Anderson (1968C) has shown that the presence of even one kink in 200 residues has a dramatic effect in lowering gel strength.

The effect of sulfation on gelling properties likewise can be explained sterically on the basis of the double helix secondary structure. Examination of molecular models shows that sulfate at C2 of the 1,3-linked units such as occurs in *lambda*, acts as a wedging group to prevent the double helix from forming. Sulfate at C2 on the 3,6-AG units, as occurs in *iota*, projects outward and does not sterically interfere with double helix formation. Sulfate at C4 on the 1,3-linked galactoside, as occurs in *kappa* and *iota*, similarly projects outward

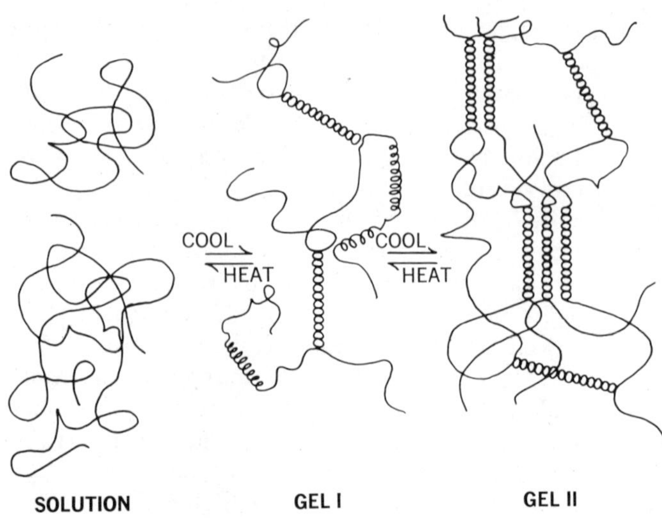

FIG. 8.3. GELLING MECHANISM OF CARRAGEENAN ACCORDING TO REES

and does not interfere with double helix formation. Sulfation at C6 of the 1,4-linked galactoside forms kinks in the chains which tend to inhibit double helix formation. This is understandable if we consider the effect of 3,6-anhydride on the conformation of the carrageenan chain. Where a 1,4-linked unit is 6-sulfated, it tends to exist in the Cl chair conformation, as do all the 1,3-linked units. As can be seen (Fig. 8.4) this introduces a kink into the polymer chain. Closure of the ring to form the 3,6-anhydride, constrains the 1,4 pyranose unit to the 1C form with resultant removal of the kink (Mueller and Rees 1967). The introduction of 3,6-anhydride is thus a chain straightening process leading to greater regularity in the polymer which results in enhanced gel potential due to increased capability of forming a double helix.

Thus, the more the idealized structures of *kappa* and *iota* can be approached by conversion to 3,6-AG, the higher the gelling potential. On the other hand, 3,6-AG formation in going from *lambda* to *theta* has no effect since gelation is still precluded by the 2-sulfate on the 1,3-linked galactoside. In order for gelation of *kappa* and *iota* to take place, it is essential that certain cations be present. These may be associated with the carrageenan or may be constituents of the system in which the carrageenan is used. In food applications, the three most commonly encountered cations are Na^+, K^+ and Ca^{++} with NH_4^+ being used to a lesser extent. Salts of these cations may already be a constituent of certain food products or it may be desirable to utilize

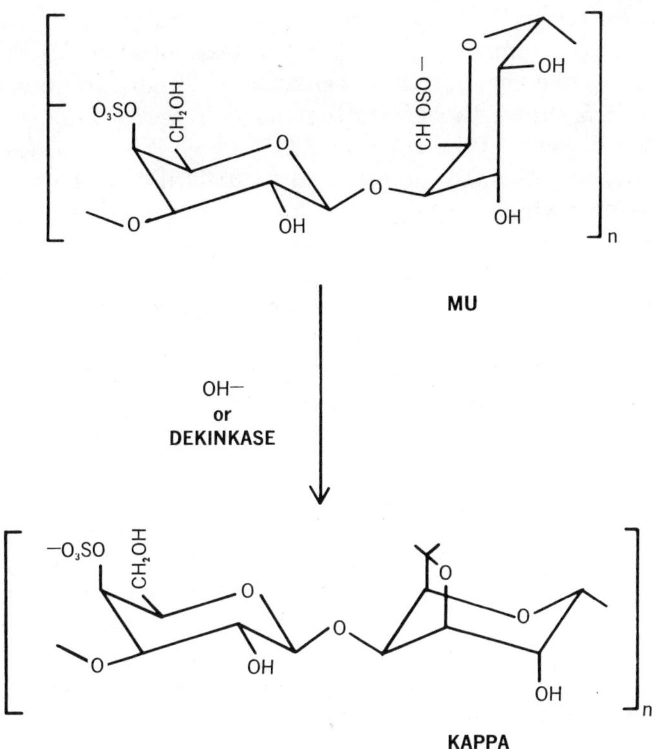

FIG. 8.4. CHANGE FROM C1 TO 1C CONFORMATION AS 3,6-AG IS PRODUCED

them in order to develop specific effects with carrageenan, as may be illustrated by the following.

High 3,6-AG *kappa*-carrageenans will not gel if in the Na^+ form. For food applications, K^+, Ca^{++} or $NH_4{}^+$ ions may be added to induce gel formation. Potassium ions produce the strongest gel, although the reasons for this are still not understood. Also, the higher the 3,6-AG content, the greater the enhancement of the gel strength by gelling cations. Pure potassium *kappa* produces a gel that is somewhat elastic. In practice, however, this product is not realized and the amount of Ca^{++} normally present in the carrageenan produces a very brittle gel.

Kappa-carrageenans, as commercially manufactured for gelling applications, produce brittle aqueous gels that are subject to syneresis. The latter may be defined as the loss of interstitial fluid from the interior of a gel due to shrinkage of the gel and expulsion of the fluid which then appears on the surface of the gel. Commercial *kappas* are generally employed with additional K^+ salts to take advantage of the enhanced gelling properties imparted by this cation. Furcellaran, like *kappa*, exhibits greatest sensitivity to potassium.

Figure 8.5 shows the effect that an increase in concentration of K^+ ions will have on the gel strength of a high 3,6-AG *kappa*. The gel strength is conveniently determined by a Marine Colloids Gel Tester, for example, which measures the force in grams required to drive a plunger of specified dimensions into a gel until it ruptures.

The judicious use of a specific cation may very well permit much lower utilization of a carrageenan product than would normally be required if this cation were not present. Care must be taken, however, to avoid excessive use which may cause undesirable syneresis development. Carrageenan manufacturers will often include the required cations in their products or make specific recommendations regarding further additions.

Iota-carrageenan, like *kappa*, does not form a double helix when it is in the sodium form. Although *iota* will gel with K^+ and NH_4^+ ions, it gels most strongly with Ca^{++} to form a very elastic gel that is not subject to syneresis. In appearance, this gel very closely resembles a gelatin gel. The much higher gelling temperature of *iota* gels, however, allows this product to be used in preparation of dessert gels that do not require refrigeration. To date, none of the carrageenan products have been used effectively for cold set water gel preparations.

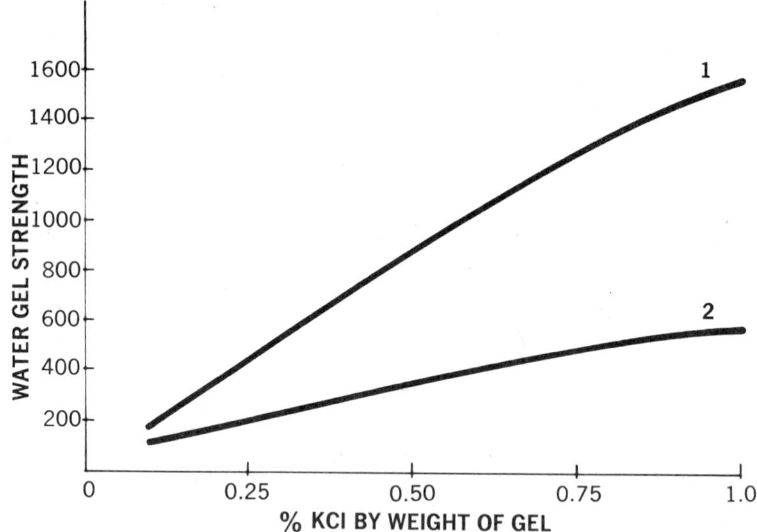

FIG. 8.5. EFFECT OF KCl ON GEL STRENGTH FOR 0.5% *KAPPA*-CARRAGEENAN BY WEIGHT OF SOLUTION (CURVE 2) AND 1.0% *KAPPA*-CARRAGEENAN (CURVE 1).

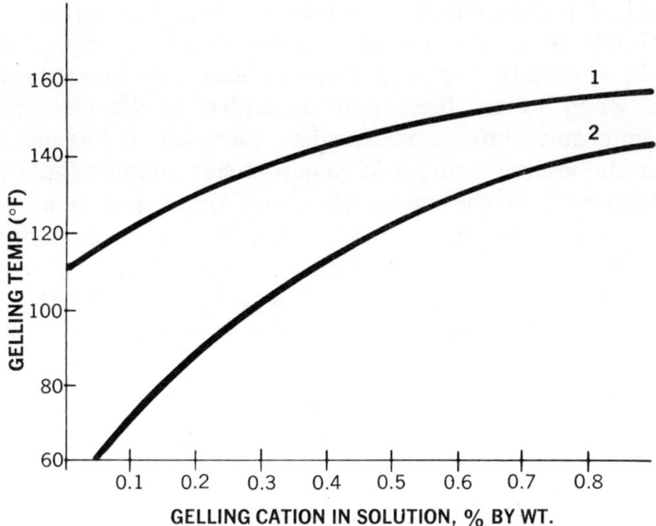

FIG. 8.6. GELLING TEMPERATURE AS A FUNCTION OF Ca^{++} AND K^{+} CATIONS CURVE 1—*IOTA* WITH Ca^{++}. CURVE 2—*KAPPA* WITH K^{+}. (GELLING TEMPERATURE AT ZERO ADDED CATIONS IS DEPENDENT ON CATIONS PRESENT IN CARRAGEENAN)

It was previously mentioned that carrageenan gels are thermally reversible in that they remelt upon heating and gel again on cooling. The gelling temperature of a specific type of carrageenan is relatively insensitive to absolute carrageenan concentration and is primarily a function of the concentration of gelling cation present. Figure 8.6 shows the effect of increasing concentration of gelling cations on the gelling temperature of various products. All carrageenan gels exhibit a higher remelt temperature than the setting temperature. In the case of *kappa*-carrageenans, it is from 10–15°C higher while with *iota*-carrageenans, it is from 4–5°C higher. Aqueous *kappa* and furcellaran gels do not normally exhibit excellent freeze/thaw properties. Considerable change in the gel texture may take place as well as substantial amounts of water release. *Iota* shows considerable stability to freeze/thaw and compares favorably to gelatin in this respect.

Synergism with Locust Bean Gum.—Two polysaccharides extensively used in the food industry are guar gum and locust bean gum. They belong to the galactomannan group in that they contain a mannose backbone structure with galactose side chains. The major difference between guar and locust bean is that the former has approximately 1 galactose side chain for every 2 mannose units whereas the ratio in locust bean gum is about 1:4. Locust bean gum exhibits a synergism with *kappa*-carrageenan, as evidenced by marked enhancement of the gel strength as well as a transformation of the gel from a brittle to a more elastic structure, also accompanied by reduction in syneresis. Figure 8.7 shows that gel strength markedly increases when locust bean gum is added to *kappa*-carrageenan whereas guar gum shows no synergistic behavior. It has been postulated that the greater number of galactose side chains in guar prevent proper alignment of the carrageenan and this galactomannan. Furcellaran is not as synergistic with locust bean gum as *kappa.* Also *kappas* from various sources exhibit different degrees of synergism with locust bean gum. Overall, the synergistic effect is most evident when the *kappa* is of a high 3,6-AG type with ester sulfate between 20–25%.

Iota, even of a high 3,6-AG type, exhibits no synergism with locust bean gum. Of course, the same is true of *lambda.*

Milk Gels.—All carrageenan products have the ability to form milk gels by cooling a solution of the carrageenan in hot milk. Even *lambda*, which does not gel in water regardless of the cations present, will form milk gels at levels of 0.2% by weight of the milk. Since *lambda* by itself is not sensitive to metallic ions, this gelation may be attributable to the formation of carrageenan-protein bonds, perhaps mediated by the Ca^{++} of the milk. Work done by Marine Colloids, Inc. demonstrated that the interaction of the carrageenan in neutral milk systems is with the casein rather than the albumin fractions.

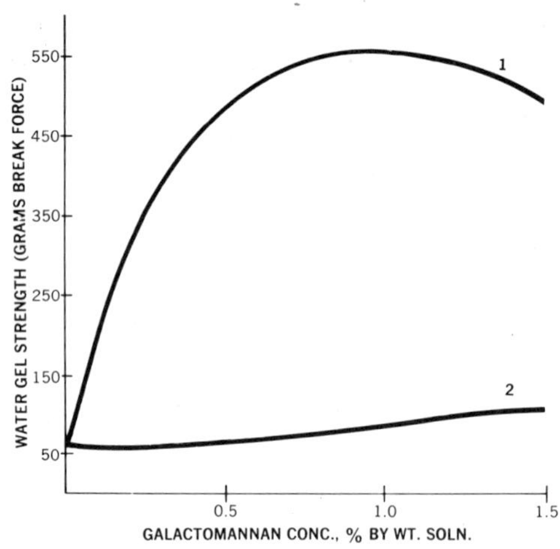

FIG. 8.7. EFFECT OF GALACTOMANNANS ON GEL STRENGTH OF A 0.50% *KAPPA*-CARRAGEENAN PLUS 0.25% KCl BY WEIGHT OF SOLUTION. CURVE 1 SHOWS SYNERGISTIC EFFECT WITH LOCUST BEAN GUM. CURVE 2 SHOWS NONSYNERGISTIC EFFECT WITH GUAR GUM

With *kappa* and *iota* there is, in addition to the carrageenan/casein bonding, a water-gelling effect from the cations present in the carrageenan as well as the cations, e.g., Ca^{++} and K^{+}, present in the milk. There appears to be a similar enhancement of gelling properties in milk as there is in water when appropriate cations are present and when the carrageenan concentration is sufficiently high. In effect, the same enhancement of gelling that takes place in water takes place in milk.

The presence of fat also influences the behavior of the carrageenan in milk, as discussed in the paragraphs under "Applications of Carrageenan."

Certain salts may be used to modify the type of gel obtained upon cooling of hot milk containing dissolved carrageenan. *Kappa* produces gels in milk which have the same brittle nature that they have in water. Moreover, these gels are very prone to syneresis when cut. Furcellaran milk gels are much smoother textured than carrageenan milk gels and less subject to syneresis. The addition of various salts such as orthophosphates, polyphosphates, carbonates, citrates, etc., which have the ability to chelate or precipitate Ca^{++} may be used to modify *kappa* milk gels so that they texturally approach furcellaran gels.

Iota-carrageenan, dissolved in hot milk and allowed to cool, does not produce the same syneresis-free gels that it does in water. If TSPP is included, however, syneresis is markedly reduced and the gel becomes more compliant. This is all the more interesting since TSPP dissolved in hot milk by itself does not produce a milk gel upon cooling. Similar, but not as dramatic, syneresis reductions of *iota* milk gels are achieved with polyphosphates, orthophosphates, citrates and carbonates. In cold milk systems, soft gels may be produced by *lambda* when used at sufficient concentration. *Theta* shows no difference compared to *lambda* as far as cold milk gelling capability is concerned, while moderate to high 3,6-AG *kappa*- and *iota*-carrageenans by themselves are virtually insoluble in cold milk. As previously mentioned, only TSPP appears to have the ability to allow *kappa* and *iota* to disperse in cold milk and set to a gel. This gel, incidentally, is much firmer than could be obtained with TSPP alone.

Viscosity

When carrageenan products are to be used for thickening purposes, it is frequently necessary that they be standardized for viscosity. Even in those cases, for example, where the carrageenan is to be used primarily for gelation, some viscosity will be contributed to the system before gelation takes place. It is important that the carrageenan provide a viscosity within predetermined limits. The same applies to those applications where carrageenans are used in combination with other thickening agents such as starches and gums.

The viscosity of a carrageenan solution is generally measured under conditions where any gelling tendency has been eliminated. When a hot carrageenan solution is cooled, the viscosity increases gradually until the temperature of gelation is reached. At this point, there is a sudden and very substantial increase in viscosity as the gel begins to set. When considering carrageenan flow properties, it is therefore necessary to distinguish between the sol and the gel state. The sol-gel transition depends on the type and concentration of cations present as was previously discussed. Viscosity measurements are therefore made at sufficiently high temperatures, e.g. 75°C, to avoid gelation. The carrageenan concentration employed is generally 1.5% by weight of the aqueous solution. In this way, both gelling and non-gelling types of carrageenan may be compared to each other as well as to other food colloids. Cold water soluble carrageenans are frequently dissolved at 1.0% concentration in room temperature water and the viscosity measured at 25°C.

In actual industry practice, viscosity measurements are often made with easily operated rotational viscometers such as the Brookfield. Commercial carrageenans are generally available in viscosities ranging from about 5-800 cps when measured at 1.5% at 75°C. Carrageenans having viscosities of less than 100 cps have flow properties very close to Newtonian. The degree of deviation from Newtonian flow increases with increasing concentration and molecular weight of the carrageenan. The solutions exhibit pseudoplastic flow properties and are sufficiently shear dependent that it is necessary to specify the shear rate used for making the viscometer measurement. This is recognized in industry and the specifying of spindle and speed is well accepted.

Carrageenan products intended for thickening applications show an exponential increase in viscosity as the concentration is increased. For example, a sodium *kappa* typically having a viscosity of 500 cps. in a 1% solution will give a viscosity of about 3500 cps. in a 2% solution. Very low molecular weight carrageenans, however, show a nearly linear increase in viscosity with concentration. A useful empirical relationship between concentration and viscosity was derived by Stoloff (1959).

Stability

Carrageenan products are most stable at neutral and alkaline pH, even at elevated temperatures. As the pH is lowered, hydrolysis of the glycosidic linkages takes place resulting in a loss of viscosity and gelling potential. The hydrolysis is greatly accelerated by heat when the pH is low. The rate of hydrolysis of a particular carrageenan is influenced by several factors. Cleavage occurs preferentially at the 1,3-glycosidic linkages and is accelerated by the presence of the strained ring system of the anhydride. Sulfation at C2 of the 1,4-linked units, as in *iota*, moderates the attack by acid to the extent that *iota* exhibits a degradation rate of about 1/2 that of *kappa*. Even *kappa*- and *iota*-carrageenans which do not contain high 3,6-AG exhibit sufficient hydrolysis so that they cannot normally be used for food applications at low pH, e.g. 3.0–4.0, when the primary function is that of thickening. The same holds true for *lambda*.

On the other hand, *kappa*- and *iota*-carrageenans may very effectively be used for gelling applications at low pH since they become fairly resistant to hydrolysis in the gelled state. Their secondary and tertiary structures may exert a shielding effect on the glycosidic bonds to minimize attack by acid. In order to keep hydrolysis at a minimum above the gelling temperature when the pH is low, HTST systems are recommended.

Compatibility

The various carrageenan products are generally compatible with the wide range of products such as starches, sugars, gums, etc., used in the food industry. Since carrageenans are anionic, they are compatible with other anionics as well as nonionics. They are, however, incompatible with most cationics. The reaction with gelatin and other proteins may be used as an example. At pH values below the isoelectric point of the protein, co-precipitation with carrageenan usually takes place. In certain applications, this may be utilized to advantage in removal of unwanted proteins. In other applications, this may not be desirable and can sometimes be avoided.

Although quaternary ammonium salts are not employed in the food industry, it should be mentioned that carrageenan is incompatible with these compounds to the point of quantitative precipitation in most cases.

COMMERCIAL PROCESSING OF CARRAGEENAN

After harvesting, the seaweeds are either sun-dried or mechanically dried, then baled and shipped to the manufacturing plant. There, the seaweeds are screened to remove sand and shells, and water washed to remove extraneous salts. The carrageenan is extracted from the seaweed with hot water containing an appropriate concentration of alkali. The latter as well as the temperature and time of cooking determines the amount of 3,6-AG that will be produced. In the case of *kappa* and *iota*, processing is usually done to develop as much 3,6-AG as possible whereas the opposite is true for *lambda*. The "paste," as it is often referred to, is then filtered with diatomaceous earth or perlite to separate cellulose and pigments. The filtrate containing the dissolved carrageenan at a concentration of about 1.5% is further concentrated by evaporation to about 2.5–3.0% in order to facilitate recovery of the carrageenan by either one of two processes currently in use today.

In one case, the filtrate is mixed with a sufficient quantity of isopropyl alcohol and the carrageenan separates as a fibrous precipitate. Extraneous salts, which are soluble in the aqueous alcohol, are conveniently removed at this point. The alcohol is separated from the fibrous precipitate or coagulum by screening and squeezing. The alcohol removed is distilled for reuse and the moist carrageenan coagulum is further purified by washing with fresh alcohol. Repeating the procedure for alcohol separation, the carrageenan coagulum is vacuum-dried and ground to a powdered form.

Carrageenan may also be recovered from the filtrate by another process. First, a small percentage of stripping agent such as a mono- or diglyceride is added to the filtrate. The mixture is then passed over heated rolls which remove the water by evaporation and leave a film of carrageenan. Because of stripping agent, which represents about 2% by weight of the carrageenan, the carrageenan is readily scraped off in sheets which are then ground.

Roll-dried products generally contain a higher processing salt concentration than alcohol products since the salts are not removed as in the alcohol process. On the other hand, roll-dried products are frequently more economical to manufacture and are usually used in milk applications where clarity of solution is not required and where the darker powder color as compared to alcohol products is not objectionable. Moreover, they generally exhibit greater dispersibility in water without the lumping tendency of alcohol precipitated products.

APPLICATIONS OF CARRAGEENAN

Carrageenan applications may be divided into two broad categories, i.e., those applications based on milk (dairy) and those based on water (nondairy). Because the unique interaction of carrageenans with the casein micelle in milk has resulted in the development of many important applications, it is convenient to group these and differentiate them from primarily water-based systems where the carrageenans are not complexed with protein. Some applications may involve the use of a protein such as sodium caseinate which is derived from milk. These are to be considered as nondairy applications since sodium caseinate in aqueous solution does not have the same micellar state of undenatured casein in milk and exhibits the typical carrageenan milk reactivity to a much lesser extent.

Tables 8.2 and 8.3 show the principal uses of carrageenan in milk and water-based systems. Each of these will be discussed as to its basic composition, the reasons for incorporating carrageenan, and the type of carrageenan best suited for the application. The task of selecting the proper carrageenan for a given application is frequently simplified by reviewing the basic properties which were discussed in detail previously.

MILK APPLICATIONS

Frozen Desserts

Ice Cream, Ice Milk.—The majority of frozen desserts manufactured today are stabilized to control ice crystal formation, and to prevent whey separation both during storage prior to freezing and at melt-down. In order to meet these requirements, several ingredients are needed. Commonly used gums to control ice crystal formation are locust bean gum, guar gum, carboxymethylcellulose and their combinations. These gums are generally used at levels from 0.1–0.2% by weight of the finished product. While they are very effective in controlling the texture of the ice cream, they do have a tendency to induce whey separation during storage of the mix prior to freezing as well as during melt-down. It is primarily as a whey prevention agent that carrageenan finds utility. Dependent upon the type of ice cream and fat content, i.e., 10–12% or 3–6%, the concentration of carrageenan used is from 0.01–0.03% by weight of the finished product. This is also frequently expressed as 100–300 parts per million (ppm). The most widely used carrageenan in high fat mixes is a strong milk-gelling *kappa* type with an ester sulfate content of about 25% and a high 3,6-anhydrogalactose (3,6-AG) content. Low fat mixes commonly employ weak milk-gelling *kappas* with high ester sulfate and moderate to high 3,6-AG. The reason that strong gelling types can be employed in high fat but not in low fat systems is due in part to the presence of the dispersed fat phase. The latter apparently tempers the carrageenan/milk protein complex, serving to interrupt it to some extent. In the case of low fat mixes, a high gel strength carrageenan could actually promote whey separation. While it is true that the weaker gelling *kappa* types described above could also be used in high fat products, the level required is sufficiently high to make them less economical compared to strong gelling types.

TABLE 8.2

TYPICAL MILK (DAIRY) APPLICATIONS OF CARRAGEENAN

Use	Function	Product	Approximate Use Level (%)
Frozen Desserts			
Ice cream, ice milk	Whey prevention Control meltdown	*Kappa*	0.010–0.030
Pasteurized Milk Products			
Chocolate, egg-nog, fruit-flavored	Suspension, bodying	*Kappa*	0.025–0.035
Fluid skimmilk	Bodying	*Kappa, iota*	0.025–0.035
Filled milk	Emulsion stabilization, bodying	*Kappa, iota*	0.025–0.035
Creaming mixture for cottage cheese	Cling	*Kappa*	0.020–0.035
Sterilized Milk Products			
Chocolate, etc.	Suspension, bodying	*Kappa*	0.010–0.035
Controlled calorie	Suspension, bodying	*Kappa*	0.010–0.035
Evaporated	Emulsion stabilization	*Kappa*	0.005–0.015
Infant formulations	Fat and protein stabilization	*Kappa*	0.020–0.040
Milk Gels			
Cooked flans or custards	Gelation	*Kappa, kappa + iota*	0.20–0.30
Cold prepared custards (with added TSPP)	Thickening, gelation	*Kappa, iota, lambda*	0.20–0.50
Pudding and pie fillings (starch base)			
Dry mix cooked with milk	Level starch gelatinization	*Kappa*	0.10–0.20
Read-to-eat	Syneresis control, bodying	*Iota*	0.10–0.20
Whipped Products			
Whipped cream	Stabilize overrun	*Lambda*	0.05–0.15
Aerosol whipped cream	Stabilize overrun, stabilize emulsion	*Kappa*	0.02–0.05
Cold Prepared Milks			
Instant Breakfast	Suspension, bodying	*Lambda*	0.10–0.20
Shakes	Suspension, bodying, stabilize overrun	*Lambda*	0.10–0.20
Acidified Milks			
Yogurt	Bodying, fruit suspension	*Kappa* + locust bean gum	0.20–0.50

TABLE 8.3

TYPICAL WATER APPLICATION OF CARRAGEENAN

Use	Function	Carrageenan Type	Approximate Use Level (%)
Dessert gels	Gelation	*Kappa* + *iota* *Kappa* + *iota* + locust bean gum	0.5–1.0
Low calories jellies	Gelation	*Kappa* + *iota* *Kappa* + galactomannans	0.5–1.0
Pet foods (canned)	Fat stabilization, thickening, suspending, gelation	*Kappa* + locust bean gum	0.5–1.0
Fish gels	Gelation	*Kappa* + locust bean gum *Kappa* + *iota*	0.5–1.0
Syrups	Suspension, bodying	*Kappa*, *lambda*	0.3–0.5
Fruit drink powders and frozen concentrates	Bodying	Sodium *kappa*, *lambda*	0.1–0.2
	Pulping effects	Potassium/calcium *kappa*	0.1–0.2
Relishes, pizza, barbecue sauces	Bodying	*Kappa*	0.2–0.5
Imitation milk	Bodying, fat stabilization	*Iota*, *lambda*	0.03–0.06
Imitation coffee creams	Emulsion stabilization	*Lambda*	0.1–0.2
Whipped tooppings (artificial)	Stabilize emulsion, overrun	*Kappa*, *iota*	0.1–0.3
Puddings (nondairy)	Emulsion stabilization	*Kappa*	0.1–0.3

Pasteurized Milks

Chocolate, Eggnog, Fruit-flavored.—For these applications we must consider both the method of preparation as well as the end product. In the category of pasteurized milks, the principal product that employs carrageenan is chocolate milk. The carrageenan functions to keep the cocoa in suspension and at the same time provides a thickening of the milk to create a rich mouthfeel. This effect of carrageenan may be extended to other milks such as eggnog or fruit-flavored types.

Milks of this type were once prepared by vat pasteurization methods but today high temperature/short time (HTST) techniques are more commonly employed. A typical chocolate milk contains 1% cocoa, 6% sugar, and from 0.025–0.035% carrageenan. Heating to pasteurization temperatures (e.g. 82°C) also serves to dissolve the carrageenan. The milk is then force-cooled under agitation using plate coolers, for example, and filled into containers at about 4°C.

If the milk is not agitated during cooling, a definite, undesirable gel structure from the carrageenan/casein interaction is visible. Moreover, in the case of chocolate milk, the cocoa all settles out before the carrageenan/casein structure develops at about 38°C. As a result of agitation during cooling, a finely dispersed broken gel structure results and the cocoa is suspended in this structure. The milk is free-flowing with no visible signs of gelation (livering), which is obtained only if an excessive amount of carrageenan is used. On the other hand, if either an insufficient amount or the wrong type of carrageenan is employed, the cocoa will settle out.

The majority of these pasteurized milks utilize from 0.025–0.035% of predominately *kappa*-carrageenans having ester sulfate of about 25% and high 3,6-AG close to 35%. *Iota* is functional but generally not as economical to use as *kappa*. *Lambda* is the least economical to employ in this system. This may in part be explained if we consider the fact that *lambda* is not sensitive to K^+ or Ca^{++} ions and does not form aqueous gels with these cations which are present in milk. As a result, its action in milk is primarily a binding through its ester sulfate groups to the casein. In the case of *kappa* and *iota* there is, in addition to casein binding, intermolecular carrageenan gel structure formation based on K^+ and Ca^{++} ions present in the carrageenan and in the milk.

Filled Milk and Skimmed Milk.—Filled milk may be described as a fluid skimmilk or reconstituted nonfat dry milk in which vegetable oils replace butterfat in whole or in part. Carrageenan functions to stabilize the emulsion as well as to impart richness of mouthfeel, particularly to those products containing relatively low fat. For example, when carrageenan is used, the richness of a 3.5% butterfat milk may be obtained with only 1% of fat.

Fluid skimmilk contains less than 0.5% of butterfat and suffers from the disadvantage of being watery in appearance and mouthfeel. Improved mouthfeel may be achieved through the use of carrageenan.

In both of these products, *iota-* or *kappa*-carrageenans of moderate 3,6-AG and ester sulfate of at least 25% are very effective at levels of 0.02-0.04% by weight of the finished product.

Creamed Cottage Cheese.—The creaming mixture used for cottage cheese must be stable without separation and it is also desirable that it have sufficient body to cling to the cottage cheese curds. Very frequently, moderate 3,6-AG *kappa*-carrageenans with ester sulfate of at least 25% are used at levels of 0.02-0.03% in combination with locust bean gum at 0.1-0.2% to provide these requirements.

Sterilized Milks

Chocolate, Eggnog, Fruit-flavored.—These products differ principally from the pasteurized types in that they are subjected to much higher heat in order to achieve sterilization. At one time it was customary to sterilize such products in stationary autoclaves, typically at temperatures of 116°C, for prescribed lengths of time. Slow heat transfer, resulting in relatively long sterilization times, yielded products with impaired flavor. Cooling was accomplished by moving the containers through cooling water without appreciable agitation of the contents. As a result, cocoa suspension with carrageenan was very difficult to accomplish. This is attributable to the fact that the carrageenan/casein complex does not "set up" until the temperature has been reduced to about 38°C. Above this temperature, constant agitation is required to prevent the cocoa from settling out. As a result, a process without agitation such as stationary autoclaving, allows almost all of the cocoa to settle out. Nevertheless, the presence of carrageenan still prevents the cocoa from packing solidly at the bottom of the container and it can easily be distributed by shaking on the part of the consumer.

Modern systems employ either rotary type autoclaves in which the contents are under agitation both during the heating and cooling period and HTST systems wherein the fluids are rapidly brought to sterilization temperatures of about 143°C using heat exchangers, held for a prescribed length of time (about 10 sec) and then cooled and aseptically filled at approximately room temperature into previously sterilized containers.

Factors which determine the proper type of carrageenan to be used are time and temperature of sterilization, and fat and solids content of the system. In general, the longer the heating time at elevated temperature, the more chance there is for carrageenan/protein destabilization to take place if high 3,6-AG, high gel strength *kappa*-carrageenans are used. Either moderate 3,6-AG *kappa*-carrageenans having ester sulfate of 25% or more or high 3,6-AG *iota*-carrageenans would then be recommended, usually at levels ranging from 0.01–0.035%. *Lambda*-carrageenan, while functional, is generally not as economical to employ as the others, for reasons previously given.

Controlled Calorie Milk Drinks.—These products differ from the conventional chocolate, eggnog, and fruit-flavored types in that they contain higher solids, usually in the form of additional milk solids, caseinates or soy protein. They are designed to provide a controlled calorie intake.

The products are conventionally prepared using rotary autoclaving or aseptic packing techniques. Carrageenan is used to prevent insolubles from settling out, to prevent fat separation and to impart richness of mouthfeel. A high 3,6-AG *kappa*-carrageenan having ester sulfate of at least 25% is most frequently used at levels of from 0.02–0.03% by weight of finished product.

Evaporated Milk.—The standards of identity for evaporated milk permit the use of carrageenan. Prior to the inclusion of carrageenan, the evaporated milk industry experienced many difficulties with fat separation. Evaporated milk normally contains about 10% fat. To minimize the effects of separation, it was found necessary to turn over the cases containing the cans of evaporated milk in storage in order to redistribute the separated fat. Phosphate salt additions helped to some extent but left much to be desired. As little as 0.005% (50 ppm) of a high 3,6-AG, *kappa*-carrageenan having ester sulfate of about 25% is required to effectively prevent separation.

As was previously mentioned, high fat systems permit the utilization of high strength carrageenans which would otherwise create destabilization in low fat systems. Both *lambda* and *iota* could be used but would require somewhat higher concentrations and would not be as economical to employ.

We previously mentioned the carrageenan/casein bond and intermolecular carrageenan bonds as contributing to the ability to suspend insoluble ingredients such as cocoa. In the case of evaporated milk, not only do these interactions take place, but there is another that must also be considered. The dispersed butterfat globules have a monomolecular layer of phospholipid (lecithin) covering the globules. Since lecithin contains basic amino groups with which the ester sulfate groups of carrageenan are quite reactive, it is very likely that carrageenan/phospholipid bonds are also formed. This may account in part for the efficiency of carrageenan.

Infant Formulations.—Carrageenan is required for fat and protein stabilization in these products, which are marketed in cans or bottles, the former being primarily sold in supermarkets and the latter used in hospitals. Concentrates are intended for dilution with water while the single-strength types are ready for consumption as is. In addition, certain infant formulations are based on soy protein and vegetable fats which substitute for milk protein and butterfat respectively. Infant formula products are most generally prepared using rotary sterilization techniques, and a *kappa* of moderate 3,6-AG content, with 25% or more ester sulfate is most widely employed. Usage level is normally from 0.02–0.04%. High 3,6-AG *kappa*-carrageenans invariably cause destabilization of the protein during sterilization.

Milk Gels

All of the milk applications that have been discussed require a carrageenan with high ester sulfate in the range of 25% or higher. It is interesting to note that *kappa*-carrageenans having ester sulfate near the 20% level do not function to prevent whey separation in ice cream, suspend cocoa in chocolate milk, prevent fat separation in evaporated milk or stabilize protein in infant formulations. Neither can furcellaran, which has even lower ester sulfate, function in the aforementioned systems.

For the preparation of milk gels, the requirement of high ester sulfate is not nearly so critical. As a matter of fact, excellent milk gels may be produced with furcellaran.

Cooked Flans or Custards.—Historically, these products were made with sugar, flavor, eggs, and milk. Baking results in coagulation of the egg proteins and provides a "set." Carrageenan-prepared custards eliminate the need for eggs since the set is provided by the carrageenan. In Europe, South America and Japan, custards or "flans" as they are referred to in certain countries, are much more popular than in the United States which favors starch puddings.

Preparation of these custards simply involves adding a dry mix containing sugar, carrageenan, flavor and color to whole milk, bringing to a boil with stirring, pouring into molds and allowing to set. An alternate composition involves the use of milk solids in the dry mix and reconstitution with water as per the aforementioned procedure. At times, small quantities of starch are also incorporated for thickening and texture modifying effects.

Kappa-carrageenans may be used for the preparation of custards, usually at levels of from 0.5–1.0 gm per pint of milk. Since the milk gel developed with *kappa* as the sole gelling agent tends to be somewhat brittle and develop excessive syneresis, it has been found that a virtually syneresis-free product with excellent smoothness may be prepared by using combinations of *kappa* and *iota* with added tetrasodium pyrophosphate (TSPP). The *iota* and TSPP function synergistically to impart a creamy mouthfeel and reduce syneresis. This combination is also very well suited for preparing ready-to-eat pasteurized custards which are sold refrigerated in the dairy case.

In Europe and South America, furcellaran is most widely used for the preparation of custards or flans, and this application represents the major use of this sulfated polysaccharide. As was previously mentioned, furcellaran resembles certain *kappas*, a major difference being the normal 12–15% ester sulfate as opposed to 20% or more for all carrageenans. Furcellaran produces very smooth-textured, low syneresis milk gels at concentrations of about 1.5 gm per pint of milk. The texture of the furcellaran milk gel may be approximated with a high 3,6-AG *kappa* used in combination with certain calcium chelating or precipitating agents. Removal of calcium ions from the milk apparently reduces the magnitude of the carrageenan/casein interaction, resulting in a smoother, creamier gel. This system does have the disadvantage of poorer flavor release than the furcellaran gel and some individuals find the taste of the auxiliary salts objectionable.

It is of interest to note that the noncaloric nature of carrageenan and furcellaran lends itself to the preparation of low calorie custards without starch or sugar, wherein an artificial sweetener such as saccharin is used.

Cold Prepared Flans or Custards.—This type of product has not received wide acceptance in the United States although it is very popular in European countries. It is generally based on reconstituting a dry mix with cold milk, or, alternatively, reconstituting with water a dry mix containing an appropriate level of milk solids. The possibilities here are *kappa-iota-*, or *lambda*-carrageenan products, usually in combination with TSPP.

Puddings and Pie Fillings (Starch Base).—The most popular type of pudding consumed in the United States today is based on starch. There are a number of available types, i.e. dry mix to be cooked with milk, pasteurized ready-to-eat, sterilized ready-to-eat, and cold-prepared from dry mix.

Dry Mix Cooked with Milk.—These puddings consist mainly of sugar, starch and flavors (cocoa for chocolate types). Historically, these have been prepared by adding milk to the dry mix, stirring while bringing to a boil, pouring into molds and cooling. Many of the pudding mixes now sold contain carrageenan as an adjunct to the starch. Since the consistency of the finished pudding is much dependent on the gelatinization of the starch, any overcooking or undercooking will result in too wide a variation in the finished pudding consistency. Carrageenan serves to maintain a more uniform set despite less than ideal cooking conditions. Moreover, carrageenan may be used to impart a light set characteristic to those products which may also be employed as pie fillings. High 3,6-AG *kappa*-carrageenans with good water gelling properties are most frequently used at levels of from 0.2–0.5 gm per pint of milk.

Starch based pasteurized milk puddings containing carrageenan are also marketed as refrigerated items sold in the dairy case.

Ready-to-eat.—The single serving canned pudding has achieved a high degree of popularity in recent years. This type of product may conveniently be prepared by in-can sterilization techniques or by aseptic methods of packing as previously described. High 3,6-AG *iota*-carrageenan at levels of from 0.10–0.20% has been used most widely to minimize syneresis development as well as to reduce the level of starch required. By replacing a portion of the starch, the viscosity of the mix at elevated temperatures is reduced, thus improving heat transfer. With regard to the finished product, improved flavor release as well as controlled degree of set may also be achieved with carrageenan.

Cold Prepared.—Puddings of this type are invariably prepared by reconstituting dry mixes. Starches are used for bodying in combination with phosphate salts such as TSPP which have the unique ability to gel cold milk. *Lambda* type products may be used to modify texture and contribute to syneresis control.

Whipped Products

Whipped Cream.—Natural cream to which carrageenan is added will show an improved stabilization when whipped. If the carrageenan is added to the cream cold, *lambda* at 0.05–0.15% is preferable. If added to the cream during pasteurization, *kappa* at 0.02–0.05% is recommended.

Aerosols.—Both natural and artificial creams benefit from the incorporation of carrageenan which functions to stabilize the fat emulsion in the container without developing excessive viscosity. The carrageenan should also contribute to foam stability. When natural cream is used, moderate 3,6-AG, high ester sulfate *kappa*-carrageenan is used at levels of about 0.03%. When sodium caseinate is employed in place of milk solids and vegetable fats replace the butterfat of natural creams, the concentration of carrageenan required is somewhat higher, e.g. about 0.05–0.10%. Very frequently, locust bean gum is used in combination with the carrageenan at levels of about 0.10%.

Cold Prepared Drinks and Shakes

In recent years, the concept of instant breakfast has achieved a degree of popularity. Dry mixes containing a mixture of nutritionally balanced ingredients are available for reconstitution with cold milk. Without carrageenan the finished milk is very watery and, in the case of the popular chocolate types, the cocoa quickly settles. It is a requirement that the carrageenan hydrate very quickly in cold milk, impart a rich mouthfeel and suspend insoluble ingredients without creating a gelled effect. This function is best accomplished by very fine mesh (through 250) *lambda* at levels of about 0.10% by weight of the finished product.

Another class of products is the shake. This may be marketed in the form of a dry mix which is simply shaken with cold milk. In addition to performing the functions previously mentioned, fine mesh *lambda* at levels of about 0.20% is also very effective in stabilizing the overrun.

Acidified Milk Systems

Until now, all of the milk applications discussed have involved an essentially neutral pH. The category of acidified milk systems is a relatively new one. As was previously pointed out, carrageenan is not normally compatible with proteins at or below the isoelectric point of the protein. The ester sulfate groups of the carrageenan react with the basic groups of the protein and precipitation usually results. This produces a grainy structure, and wheying off frequently takes place. Under certain conditions, this effect may be avoided. The preparation of certain types of yogurt may be used as an illustration.

Yogurt is most frequently prepared by pasteurizing milk, cooling to incubation temperature, adding a culture, and allowing the acidity to develop. This causes coagulation of the milk casein and results in a "set."

Yogurt may be cultured in bulk or in the finished container itself. One method of producing yogurt with fruit involves adding the fruit to the container first, followed by the milk to which culture has been added. The whole is then incubated. This type of product suffers from the disadvantage that the fruit is at the bottom of the container and the yogurt has to be broken in order to reach the fruit. Another method of producing yogurt with fruit is to first culture the yogurt in bulk. The curd is then mechanically broken and the fruit added. In the absence of an appropriate stabilizer, the broken curd would be very thin and the fruit would settle out. The stabilizer is normally dissolved in the milk at the beginning of the process.

While gelatin is most popular in this application, carrageenan may also function effectively. This type of yogurt is commonly referred to as Swiss style. It has been found that a combination of carrageenan at 0.10% and locust bean gum at 0.30% by weight of the finished product will form a smooth-textured yogurt at pH of 3.8–4.4. It is postulated that locust bean gum acts as a protective colloid in preventing precipitation of the carrageenan/casein complex. A high 3,6-AG *kappa*-carrageenan with about 20% ester sulfate is most effective.

WATER APPLICATIONS

Dessert Gels

This class of products with carrageenan may conveniently be divided into two classes, i.e., dry powders intended for reconstitution with hot water and finished ready-to-eat types. For the former, gelatin is far and away the most popular gelling agent and carrageenan is employed only in applications which preclude the use of gelatin. One example is that of institutional dessert gel preparations where carrageenan is preferred to gelatin because carrageenan sols do not require refrigeration for gelation and the gels do not harden appreciably with age. Another example involves preparations where gelatin is prohibited because of religious considerations.

A ready-to-eat dessert gel which is marketed in cans or plastic cups and does not require refrigeration during the distribution process cannot satisfactorily be prepared with gelatin since gels made with the latter exhibit sufficient softening during storage so that considerable hydrolysis takes place. There are currently two basic carrageenan systems which may effectively be used in this application.

One system involves the use of a high 3,6-AG *kappa*- and *iota*-carrageenan combination. It was previously pointed out that *kappa* produces a brittle gel which is subject to syneresis and that *iota* produces an elastic gel which is not subject to syneresis. By varying the ratio of these two carrageenan products, syneresis control may be achieved along with the desired degree of rigidity. In these water gel systems, *kappa*-carrageenans of ester sulfate as low as 20% may be used just as effectively as those with higher ester sulfate.

Another composition involves combinations of *kappa*, clarified locust bean and *iota*. The *kappa* and locust bean gum work synergistically to develop the gel structure while the *iota* serves to control syneresis development. Until the last few years it was not feasible to market clear carrageenan gels contining native locust bean gum since the latter contains insoluble impurities that impart an undesirable opacity and grainy structure to the gel. Refined locust bean gums have been developed by the manufacturers of carrageenan who responded to the need for such a product. As a result, clear, nongrainy gels may easily be produced.

Since the pH of these dessert gels is generally between 3.5 and 4.0, it is mandatory that hydrolysis be minimized during the sterilization process in order to retain as much gel strength as possible. The use of HTST processing systems most effectively meets this requirement. Rapid heating, holding at sterilization temperature for a short period and rapid cooling followed by filling into presterilized containers conveniently holds hydrolysis to a minimum. Another practical method involves heating a solution of all of the ingredients with the exception of the acid and then metering this nonacidified solution into a small surge tank along with a solution of the acid, followed by filling. Definitely not recommended is adding the acid to the large tank, even if it is added at the end of the cooking cycle. The time it takes to empty out a large tank during filling of containers results in nonuniformity, with the product filled at the tail-end being much weaker than the product filled at the beginning.

Low Calorie Jellies

Since carrageenans are nonassimilable, they are particularly well suited to the preparation of low calorie products. Jellies of the low calorie type may contain about 1/2 the concentration of sugar as their high sugar counterparts or may contain no sugar at all if an artificial sweetener is used.

Conventional high sugar jellies contain approximately 65% of soluble solids and are set by high methoxyl pectin which requires both acid and high concentrations of sugar for gelation. High methoxyl pectin, however, cannot be employed for low calorie jellies because it will not set with reduced sugar levels. In the case of low sugar jellies, setting is accomplished with either low methoxyl pectin or carrageenan. The latter usually has the advantage of being less critical to use in production and of producing jellies having a lower degree of syneresis. As in the case of dessert gels *kappa*- and *iota*-carrageenans are frequently used in combination, as well as *kappa*-carrageenan and refined locust bean gum.

Pet Foods

The use of carrageenan products in pet food applications has grown considerably in the past few years. Carrageenan is used primarily in retorted canned pet food products. High 3,6-AG *kappa*-carrageenan with ester sulfate of about 20% is frequently employed at about

0.20-0.50% in combination with similar concentrations of locust bean gum to prevent fat separation during processing and to impart a richness to the gravy so that it is not thin and runny. Some manufacturers particularly like to obtain particles of gel throughout the product.

Fish Gels

In these applications, primarily *kappa*- and *iota*-carrageenans are used in combination to gel the broth surrounding the fish packed in cans or jars. Flavor is preserved, condiments are maintained uniformly suspended and the gelled broth palatability is increased. Total carrageenan level is usually 0.5-1.0% by weight of the broth.

Syrups

Most of the syrups which employ carrageenan are chocolate syrups intended for dilution with milk in order to make chocolate milk or for variegating ice cream. The major ingredients of these syrups which normally contain total solids of 65% or more are water, sugar, cocoa, and carrageenan. Moderate to high 3,6-AG *kappa*-carrageenans having ester sulfate of about 25% are most commonly employed. The need for high ester sulfate, particularly in syrups intended for dilution with milk, is caused by the fact that suspension of the cocoa is required upon dilution with milk. It was previously pointed out that low ester sulfate *kappa*-carrageenans are not effective as cocoa suspending agents in milk.

Syrups are generally prepared by heating a mixture of the aforementioned ingredients to obtain solution. When these syrups are sold to dairies for chocolate milk production, 1 part of syrup is generally added to 10-12 parts of milk. Most commonly, the syrups are heated with milk which, after pasteurization, is cooled to refrigeration temperatures with agitation and filled into bottles or containers. The function of the carrageenan is to suspend the cocoa in the finished milk as explained previously. Considering that approximately 0.03% is required in the finished milk for suspension of cocoa, this means that prior to dilution with milk the syrup has somewhat more than 0.30% of carrageenan and the concentration in the aqueous phase approaches 1.0%. If, during preparation, these syrups were allowed to cool quiescently after heating, undesirable gelation would take place. This would be attributable not only to the high concentration of carrageenan in the aqueous phase but also the fact that most of the cocoa used has been processed with alkaline potassium salts (Dutch process). As previously mentioned, the presence of potassium ions serves to enhance carrageenan gelation. For these syrups, it is customary to agitate the syrup while cooling to below the gelling temperature of the carrageenan. This "interrupted gel" technique allows a syrup to be prepared which can easily be pumped. Moreover, the finished syrup is easily miscible with the milk with which it is diluted.

Another class of syrups intended for the preparation of chocolate milk is based on the dilution of a hot prepared syrup with *cold* milk. Syrups of this type require the use of *lambda*-carrageenan since only the latter has the ability to suspend cocoa when the syrup is added to *cold* milk. The insensitivity of *lambda* to potassium ions also permits hot packing of the syrup without fear of gelation. The level required for stabilization in the finished milk is generally about 0.05%. While *lambda* could be used in those syrups intended for reconstitution with *hot* milk, it is generally not as economical to employ as *kappa* nor does it provide quite as much flexibility.

In the case of syrups intended for variegating ice cream, these may be either chocolate or fruit flavored. Carrageenans, usually *kappa* types, function to prevent the syrup from "feathering" into the ice cream.

Lambda-carrageenan has been used successfully in low calorie maple syrups to impart mouthfeel and bodying at levels of from 0.2–0.5%.

Iota-carrageenan is not employed in syrups containing high sugar concentrations because of its limited solubility as previously pointed out.

Fruit Drink Powders and Frozen Concentrates

Either a dry mix is reconstituted with cold water or a frozen concentrate is diluted with water. *Lambda* and the sodium salt of *kappa* are both soluble in cold water and may be used to provide bodying and mouthfeel characteristics at concentrations of from 0.2–0.5%. A *kappa* of controlled potassium and calcium content can produce a swelling characteristic which imparts a pulpy mouthfeel. Frequently both types of carrageenan are used in combination (Glicksman 1968).

In the case of the dry mix, the carrageenan is quite stable if the packaging is done properly and the reconstituted drink is not stored long enough for hydrolysis to be a problem. In the case of the frozen concentrates, the ingredients are dissolved in cold water and the whole is then frozen. Under these conditions, the carrageenan is quite stable in the presence of the acid. Again, the reconstituted finished product is usually not held for more than one or two days and the fact that it is refrigerated minimizes hydrolysis fairly well.

Because of continuing hydrolysis, carrageenan is not generally recommended for use in acidified fluid systems requiring prolonged shelf-life.

Relishes, Pizza and Barbecue Sauces

Even though these are low pH applications, moderate 3,6-AG *kappa-* or *iota*-carrageenans may satisfactorily be used at concentrations up to 0.5% if some gelation is produced. In addition, those products containing tomato seem to impart a higher degree of stability to the carrageenan for some as yet unexplained reason.

Imitation Milk

This is a nondairy application in that sodium caseinate and/or soy protein are used in place of milk solids and vegetable fat replaces butterfat. Due to the reduced carrageenan/protein interaction it is necessary to utilize somewhat higher levels of carrageenan than would be required in a milk-based system. *Iota* and *lambda* are generally used at about 0.05% to provide bodying and stabilization of the fat emulsion.

Imitation Coffee Creams

These products are extremely popular, having replaced natural cream to a great extent. They utilize vegetable fats, emulsifiers and proteins such as soy and sodium caseinate. They have developed to the point where they are virtually indistinguishable from natural cream. Moreover, they have an excellent shelf-life compared to natural cream.

Spray-dried mixes intended for direct addition to coffee do not require carrageenan as a stabilizing agent. On the other hand, liquid preparations frequently contain *lambda* at levels of from 0.1–0.2% by weight of the finished product which may be purchased either refrigerated or frozen. Carrageenan functions to stabilize the emulsion as well as to impart excellent freeze/thaw stability in the frozen types.

Whipped Toppings (Artificial)

These nondairy products customarily utilize sugar, vegetable fats, sodium caseinate and soy protein, emulsifiers and stabilizers. Again, these have been categorized as nondairy even though they are substitutes for natural cream. With regard to the functionality of the carrageenan, the system is considered as aqueous-based rather than as one containing milk. High 3,6-AG *kappa-* or *iota*-carrageenans at levels of from 0.02–0.05% have been used in this application for emulsion stabilization and maintenance of overrun. Carrageenan also imparts excellent freeze/thaw stability characteristics to frozen products of this type which at present constitute a large portion of the market.

Puddings (Nondairy)

The most common of these is the frozen pudding based on vegetable fats, emulsifiers, sodium caseinate and soy protein. No milk is present in the products which are thawed prior to consumption. They will withstand repeated freeze/thaw cycles without developing excessive syneresis. A *kappa*-carrageenan of high 3,6-AG with ester sulfate of about 25% functions to impart freeze/thaw protection at levels of from 0.10–0.30%. This may appear to be unusual since aqueous gels of *kappa* are highly susceptible to syneresis after freezing and thawing. While the mechanism is not fully understood it is undoubtedly related to the fat globule size of 1 μ or less obtained by emulsification and homogenization. The dispersed fat serves to interrupt the gelling tendency of the carrageenan and to prevent a continuous gel from forming.

BIBLIOGRAPHY

ANDERSON, N. S., DOLAN, T. C. S., and REES, D. A. 1968A. Carrageenans. Part III. Oxidative hydrolysis of methylated κ-carrageenan and evidence of a masked repeating structure. J. Chem. Soc. (C) 1968, 596–601.

ANDERSON, N. S. *et al.* 1968B. Carrageenan. Part V. The masked repeating structure of λ and μ-carrageenans. Carbohydrate Res. 7, 468–473.

ANDERSON, N. S. *et al.* 1968C. Carrageenans. Part IV. Variations in structure and gel properties of κ-carrageenan and the characterisation of sulphate esters by infrared spectroscopy. J. Chem. Soc. (C) 1968, 602–606.

DOLAN, T. C. S., and REES, D. A. 1965. The Carrageenans. Part II. The Positions of the Glycosidic Linkages and Sulphate Esters in λ-Carrageenan. J. Chem. Soc. *1965*, 3534–3539.

GLICKSMAN, M. 1968. U.S. Pat. 3,395,021, July 20.

LAWSON, C. J., and REES, D. A. 1970. An enzyme for the metabolic control of polysaccharide conformation and function. Nature *227*, 392–393.

LIN, C. F., and HANSEN, P. M. T. 1968. Stabilization of calcium caseinates by carrageenan. J. Dairy Sci. *51*, 945.

LIN, C. F., and HANSEN, P. M. T. 1970. Stabilization of casein micelles by carrageenan. Macromolecules *3*, 269–274.

McCANDLESS, E. L., CRAIGIE, J. S., and WALTER, J. A. 1973. Carrageenans in the gametophytic and sporophytic stages of *Chondrus crispus*. Planta (Berlin) *112*, 201–212.

MUELLER, G. P., and REES, D. A. 1967. *In* Drugs from the Sea. Univ. of Rhode Island, Kingston.

NATL. ACAD. SCI. 1974. First Supplement to the Food Chemicals Codex, 2nd Edition. National Academy of Sciences, Washington, D.C.

O'NEILL, A. N. 1955. 3,6-Anhydro-D-galactose as a constituent of κ-carrageenan. J. Am. Chem. Soc. 77, 2837–2839.

O'NEILL, A. N. 1955. Derivatives of 4-*O*-β-D-galactopyranosyl-3,6-anhydro-D-galactose from κ-carrageenan. J. Am. Chem. Soc. 77, 6324–6326.

PERNAS, A. J., SMIDSRØD, O., LARSEN, B., and HAUG, A. 1967. Chemical heterogeneity of carrageenans as shown by fractional precipitation with potassium chloride. Acta. Chem. Scand. *21*, 98-110.

REES, D. A. 1969. Structure, conformation, and mechanism in the formation of polysaccharide gels and networks. *In* Advances in Carbohydrate Chemistry, Vol. 24, M. L. Wolfrom, R. S. Tipson, and D. Horton (Editors). Academic Press, New York.

SMITH, D. B., and COOK, W. H. 1953. Fractionation of carrageenan. Arch. Biochem. Biophys. *45*, 232-233.

STANCIOFF, D. J., and STANLEY, N. F. 1969. Infrared and chemical studies on algal polysaccharides. Proc. Intern. Seaweed Symp., *6*, 595-609.

STANLEY, N. F. 1963. (to Marine Colloids, Inc.) U.S. Pat. 3,094,517.

STOLOFF, L. 1959. Carrageenan. *In* Industrial Gums, R. L. Whistler (Editor). Academic Press, New York.

CHAPTER 9

A. Jerome Ganz

Cellulose Hydrocolloids

INTRODUCTION

The primary objective of this chapter is to describe some of the properties of cellulose derivatives which are important to their use in foods. Four water soluble, chemically modified products and one water insoluble, unmodified product prepared from cellulose and used in the food industry will be discussed.

WATER SOLUBLE, CHEMICALLY MODIFIED PRODUCTS

Sodium Carboxymethyl Cellulose

Sodium carboxymethyl cellulose, often referred to as CMC, is an anionic polyelectrolyte and is the deterivative of cellulose most widely used in the food industry. The purified, food grade is often referred to as cellulose gum. While the sodium salt is the most commonly used product, other salts, e.g., K, Ca, NH_4, have been made for specific nonfood uses.

Sodium carboxymethyl cellulose was first produced commercially in Germany shortly after World War I. The product was impure and was used primarily as a starch, natural gum, or gelatin substitute. Interest in it gradually developed with the finding that it facilitated whiteness retention when incorporated into laundry detergent formulations. Purified sodium carboxymethyl cellulose was first produced in the United States in 1943 and the rate of production and usage has grown rapidly. Sodium carboxymethyl cellulose is of particular interest because of its ability to thicken and modify the texture, i.e., structure, of water and because of its reactions with proteins. Currently, the sole manufacturer of a food grade product in the United States is Hercules Incorporated.

Hydroxypropyl Cellulose

Hydroxypropyl cellulose is a nonionic ether derived from cellulose and sold under the trademark, Klucel.[1] It is a high purity prod-

[1] Registered trademark of Hercules Incorporated.

uct which was first introduced into the trade by Hercules Incorporated in 1963. It is of particular interest because of its surface activity, solubility in polar organic solvents, thermoplasticity, and the insensitivity of viscosity to changes in pH.

Methyl Cellulose and Hydroxypropyl Methyl Cellulose

These high purity nonionic ethers derived from cellulose were introduced into the food industry in 1939, under the trademark, Methocel.[2] Derivatives of cellulose containing only methyl groups as well as methyl plus varying amounts of hydroxypropyl groups are available. These are of particular interest because of their insensitivity to pH, surface activity, gel forming capacity on heating, solubility in some organic solvents, thermoplasticity in the presence of plasticizers, and film-forming characteristics.

Methyl Ethyl Cellulose

Methyl ethyl cellulose is a high purity, nonionic, mixed ether derived from cellulose which was introduced into the food industry in 1961 under the trademark Edifas A.[3] It is of particular interest because of its surface activity.

WATER-INSOLUBLE, CHEMICALLY UNMODIFIED PRODUCT

Microcrystalline Cellulose

Microcrystalline cellulose is a specially hydrolyzed, unmodified form of α-cellulose which was introduced to the food industry in 1961 under the trademark, Avicel.[4] Subsequently, a codried product, Avicel RC, consisting of Avicel and sodium carboxymethyl cellulose, was introduced to overcome some of the difficulties encountered in dispersing pure Avicel. The level of sodium carboxymethyl cellulose may vary from 8–11%. It is of interest because of its bulking and water-absorbing properties, ability to orient at oil-water interfaces, and thermal stability.

The use of these cellulose derivatives in the food industry has become widespread. In 1968, approximately 9,000,000 lb were used in the foods. Of this amount, about 8,000,000 lb were sodium carboxymethyl cellulose (Glicksman 1969). The products sold to the food industry are physiologically inert, i.e., they are not metabolized. This

[2] Registered trademark of The Dow Chemical Company.
[3] Registered trademark of Imperial Chemicals Industries, Ltd.
[4] Registered trademark of Food Machinery Corporation, Inc.

provides the basis for acceptance for use in foods by the FDA, subject of course, to any limitations imposed by the Standards of Identity which exist for any specific food product.

The preparation of these products, their chemical composition, their properties in water and in the presence of other solutes, their stability and special features (e.g., surface tension, thermoplasticity, film properties, influence on sensory effects, etc.) will be discussed in this chapter. Comparable data for each derivative cannot be given in detail since such data are not available in the literature. Applications are not listed in detail; however, a few of the important uses are mentioned.

In most instances, detailed references are not given since these are available from other sources (Whistler and BeMiller 1973; Klose and Glicksman 1968; Glicksman 1969; Mark *et al.* 1965; Jirgensons 1958; Savage 1971). Detailed information is also available in trade literature available from manufacturers of these products (Hercules, Inc. 1971A, 1971B; Dow Chemical Co. 1962; Imperial Chemical Industries Ltd. 1961; FMC Corp. 1971). More complete information on applications will be found in the references above.

PREPARATION

Cellulose is the most abundant of all natural organic substances. It occurs mixed or combined with varying amounts of waxes or fats, inorganic substances, and polymeric organic substances such as pentosans, lignin, and proteins. These amounts vary with the source. Thus, wood tissue from pine contains 27% lignin, while cotton fiber contains only traces.

Cellulose is a highly polar, hydrophilic, but water insoluble, linear, high molecular weight polymer consisting of ordered, crystalline areas connected by disordered, amorphous areas (generally about 35%). The density of a completely dry cellulose fiber (1.57) is very close to that calculated from X-ray data (1.60 ± 0.05) indicating that even the amorphous areas are tightly packed.

Cellulose bears many structural similarities to the amylose fraction of starch as shown in Fig. 9.1 (Karlson 1963). Cellulose is made up of anhydroglucose units linked through a β,1-4-glycosidic linkage (cellobiose). The forward edges of the anhydroglucose units are represented by the heavier lines. The $-CH_2OH$, $-OH$, $-H$ groups are alternately above and below the plane of the anhydroglucose ring. Positions above the plane are also represented by heavier lines. The individual molecules in a cellulose fiber are, for the most part, very

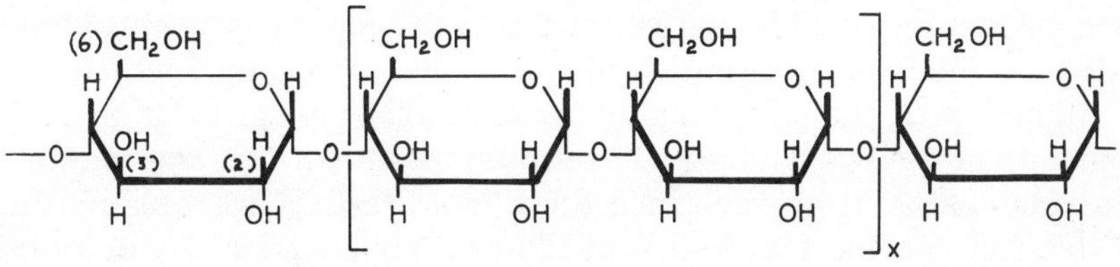

AMYLOSE

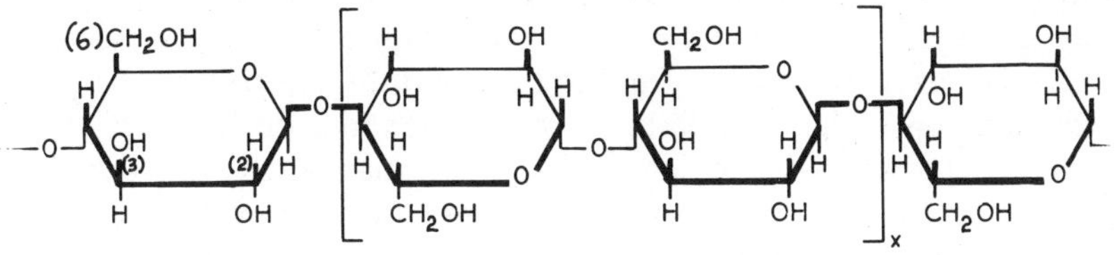

CELLULOSE

Courtesy of Hercules Incorporated

FIG. 9.1. STRUCTURAL FORMULAS FOR AMYLOSE AND CELLULOSE

tightly packed through hydrogen bonding and are, therefore, insoluble in water at any temperature. Amylose is made up of anhydroglucose units linked through an α,1-4-glycosidic linkage (maltose). The $-CH_2OH$ groups all lie in the same plane and have the same relationship to the backbone of the polymer. They serve to keep the polymer molecules spatially separated or loosely packed so that even the crystalline regions are soluble in hot water.

The anhydroglucose units of the cellulose molecule contains three hydroxyl groups, a primary hydroxyl at the 6-position and two secondaries at the 2 and 3 positions. The 3- and 6-hydroxyls are sterically oriented to the same side of the ring; the 2-hydroxyl is oriented to the opposite side. By reaction with these hydroxyl groups, water soluble derivatives are formed. "DS," or degree of substitution, is the average number of substituted hydroxyls per anhydroglucose unit. The maximum possible DS is 3.0. In practice, the DS levels of commercially available materials are considerably lower.

The added substituent may contain a hydroxyl group such as hydroxypropyl, which reacts with additional reagent more readily than does a hydroxyl of the anhydroglucose unit. If so, "chaining-out" occurs, and extended side chains result. For a proper description of such products not only the DS but also the MS must be considered. "MS" is the average number of molecules of reactant which have combined with the anhydroglucose unit; MS is equal to or generally higher than the DS. The MS:DS ratio represents the average length of the side chains. However, since there is no readily available method of determining the DS in such products, they are generally described in terms of MS.

In general, water soluble derivatives are prepared from purified cellulose by swelling the cellulose with alkali to break down the crystalline areas, followed by introduction of the substituent. The swelling process is rarely achieved completely and the final product generally contains more or less crystalline areas depending on how efficiently the swelling process is carried out. This influences the uniformity of substitution. Thus, some anhydroglucose units may contain more than one substituent molecule, and some might contain none depending on the conditions of preparation. The uniformity of substitution influences the rheological properties of solution of these derivatives. The derivatives are susceptible to coiling-up depending on environmental conditions, e.g. presence of salts.

As substituents are introduced into alkali cellulose it becomes more nearly water soluble. A point is reached at which it can be considered truly water soluble. As the degree of substitution is still further increased, nonionic derivatives gradually become soluble in polar organic solvents.

The molecular weight or degree of polymerization (DP) of the water soluble derivatives is controlled through choice of raw materials, e.g., cotton linters, wood pulp, from which the cellulose is derived before swelling in alkali or by careful treatment of the derivatives with oxidizing agents to reduce the DP to the desired level. Whereas in native cellulose, the DP may be 5000 or more, the derivatives discussed herein have DP's of about 100-4000.

Sodium Carboxymethyl Cellulose

The cellulose-derived hydrocolloid most widely used in the food industry is the water soluble sodium carboxymethyl cellulose (CMC) or cellulose gum. In preparing sodium carboxymethyl cellulose the cellulose is treated with sodium hydroxide and is then reacted with sodium monochloroacetate. The product is purified by washing to remove the salts. The reaction is:

$$R_{cell}(OH)_3 + NaOH + ClCH_2COONa \longrightarrow R_{cell}(OH)_2(OCH_2COONa) + NaCl + H_2O$$

In Fig. 9.2 is shown an idealized cellulose gum molecule of DS 1. By virtue of the introduction of the carboxymethyl group, the product is ionic. It is the salt of a weak acid having a dissociation constant of 5.0×10^{-5}, comparable to acetic acid. The sodium salt is not soluble in any organic solvent except glycerol and even then, heat must be applied to achieve solution.

Sodium carboxymethyl cellulose becomes water soluble at about DS 0.4. Products having DS of 0.4–1.2 are commercially available. In this chapter, sodium carboxymethyl celluloses will be appended with a number which is 10 times the actual DS, e.g., CMC of DS = 0.4 will be referred to as CMC-4. Products having varying viscosities (e.g. high, medium, low) are available, the viscosity being determined by the DP.

Sodium carboxymethyl cellulose is classified under "substances that are generally recognized as safe" (GRAS) by Title 21, Code of Federal Regulations (CFR) 121.101. The FDA defines it as the sodium salt of carboxymethyl cellulose, not less than 99.5% on a dry-weight basis, with a maximum substitution of 0.95 carboxymethyl groups per anhydroglucose unit and with a minimum viscosity of 25 cps in a 2% (by weight) aqueous solution at 25°C.

More uniformly substituted types as well as less uniformly substi-

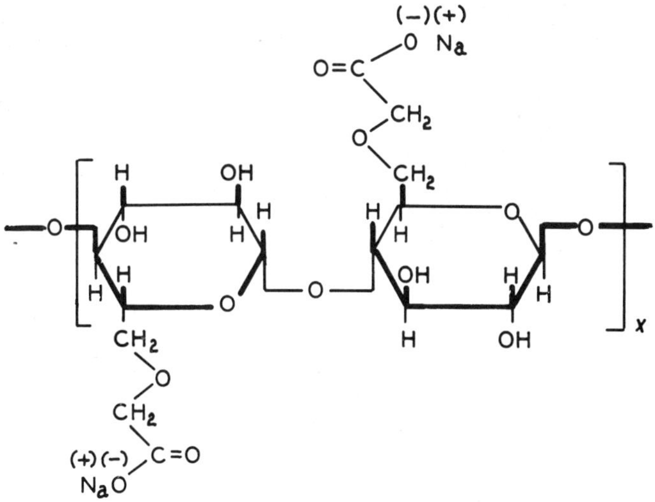

Courtesy of Hercules Incorporated

FIG. 9.2. STRUCTURAL FORMULA FOR SODIUM CARBOXYMETHYL CELLULOSE

tuted types are also available. Physical characteristics, e.g., particle size, density may also be varied.

Methyl and Hydroxypropyl Methyl Cellulose

Methyl cellulose, a methyl ether derivative of cellulose, is prepared by treating alkali cellulose with methyl chloride. The reaction involved is as follows.

$$R_{cell}(OH)_3 + CH_3Cl + NaOH \longrightarrow R_{cell}(OH)_2(OCH_3) + NaCl + H_2O$$

At low levels of substitution, the products are soluble only in alkali; at high levels, they are soluble in polar organic solvents. An idealized formula of DS 1.5 is shown in Fig. 9.3. The mixed ethers of cellulose containing both hydroxylpropyl and methyl substituents are also widely used. These are prepared by reacting cellulose with methyl chloride and propylene oxide. A wide variety of types are available (Table 9.1).

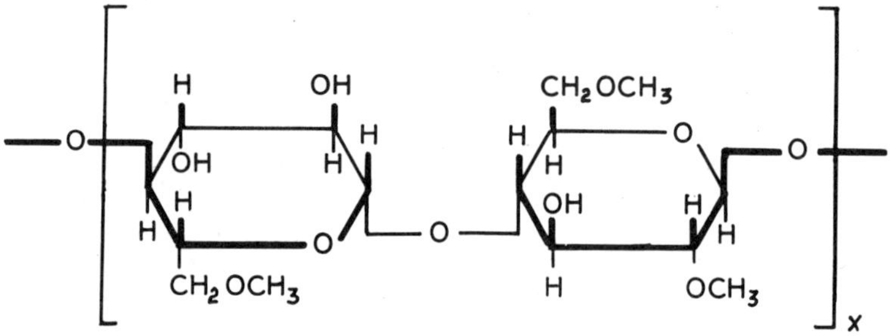

Courtesy of Hercules Incorporated

FIG. 9.3. STRUCTURAL FORMULA FOR METHYL CELLULOSE

TABLE 9.1

CHEMICAL COMPOSITIONS OF METHYL CELLULOSE AND HYDROXYPROPYL METHYL CELLULOSES

Methoxy (%)	Hydroxypropoxyl (%)	Soluble In	Nominal Gel Temperature (°C)
27.5-32.0	0	Water	54-56
26-30	7-12	Water and organic solvents[1]	60
27-29	4.0-7.5	Water	65
19-24	4-12	Water	70-90

Source: Printed with permission of Dow Chemical Company.
[1] Thermoplastic.

Methyl cellulose for use in foods is classified under the GRAS listing by Title 21, CFR 121.101. It is defined as U.S.P. methyl cellulose, except that the methoxy content shall not be less than 27.5% and not more than 31.5% on a dry weight basis. The use of hydroxypropyl methyl cellulose is covered in Title 21, CFR 121.1021.

Hydroxypropyl Cellulose

Hydroxypropyl cellulose is a nonionic, water soluble derivative prepared by reacting alkali cellulose with propylene oxide.

$$R_{cell}(OH)_3 + \overset{\quad O}{CH_2\!-\!CH}\!-\!CH_3 + OH^-$$

$$R_{cell}(OH)_2[OCH_2CH(OH)(CH_3)]$$

An idealized formula is shown in Fig. 9.4. The product shown has a DS of 2 and an MS of 4. The commercial products have an MS of about 4 and are available in a variety of viscosity types.

Hydroxypropyl cellulose for use in foods is defined in Title 21 CFR 121.1160, as a cellulose ether containing propylene glycol groups attached by an ether linkage and containing, on an anhydrous basis, not more than 4.6 hydroxypropyl groups per anhydroglucose unit. The additive has a minimum viscosity of 145 cps for a 10% by weight aqueous solution at 25°C.

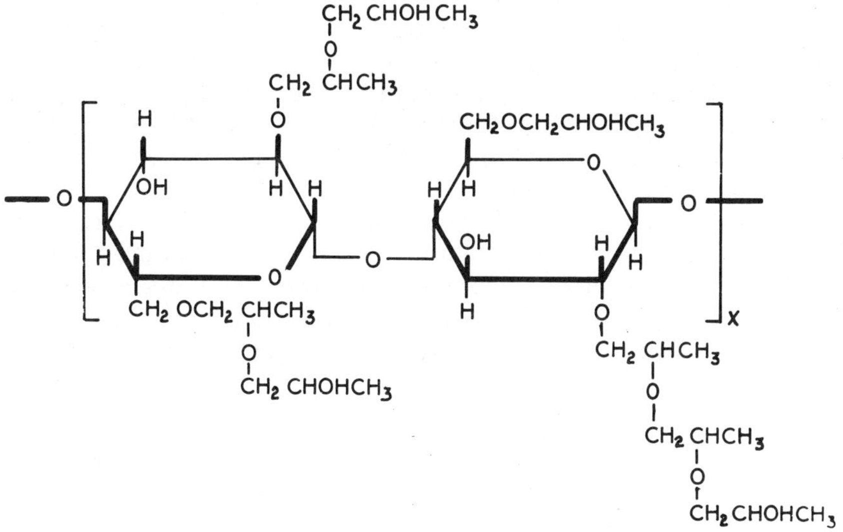

Courtesy of Hercules Incorporated

FIG. 9.4. STRUCTURAL FORMULA FOR HYDROXYPROPYL CELLULOSE

Methyl Ethyl Cellulose

Methyl ethyl cellulose is a nonionic, water soluble derivative of cellulose prepared by reacting alkali cellulose with ethyl and methyl chlorides.

$$R_{cell}(OH)_3 + CH_3Cl + CH_3CH_2Cl \longrightarrow R_{cell}(OH)(OCH_3)(OCH_2CH_3)$$

The DS is about 0.4 with respect to the methyl and about 0.9 with respect to the ethyl substituents. Methyl ethyl cellulose is defined in Title 21, CFR 121.1112. For use in foods the methoxy content shall be not less than 3.5% and not more than 6.5%, calculated as OCH_3, and the ethoxy content shall be not less than 14.5% and not more than 19%, calculated as OC_2H_5, both measured on the dry sample. The viscosity of a 2.5% solution in water at 20°C is 20–60 cps.

Microcrystalline Cellulose

This product is a specially hydrolyzed α-cellulose, containing no substituent groups. α-cellulose is high grade cellulose insoluble in 17.5% NaOH at 20°C. While α-cellulose, as found in nature, is fibrous and does not adsorb significant amounts of water, microcrystalline cellulose is nonfibrous and has significant water adsorptive capacity. In preparing this product, α-cellulose is hydrolyzed into two fractions, one acid soluble and the other not. The acid insoluble fraction is recovered as microcrystalline cellulose. By this process, the amorphous regions are hydrolyzed completely, leaving microcrystallites which are referred to as level-off DP cellulose, i.e., further hydrolysis would not lower the DP. The product is insoluble in water, dilute acids, oils, and common organic solvents.

Dispersion and hydration of this product is difficult; consequently special products consisting of codried microcrystalline cellulose and sodium carboxymethyl cellulose have been introduced under the trademark Avicel RC. These products are readily dispersible and hydratable. They also have a higher bulk density, finer particle size and higher ash content than pure microcrystalline cellulose.

PREPARATION OF SOLUTIONS

If precautions are not taken to ensure proper dispersion and solution of the cellulosic hydrocolloids, erroneous conclusions can be drawn about their properties and their effects in food products. There are two main factors to consider: (1) ease of dispersion and (2) rate of solution. It is convenient to use a large particle size gum since the large particles are readily dispersed merely by spooning them into water; however, these dissolve slowly. For faster solution, a finer grind is required; however, if these fine grind materials are merely spooned into water, lumps having an outer layer of hydrated, jelly-like gum and a dry core result. Therefore, fine grind material should be added to water slowly so that the individual particles remain separate or the gum should be premixed with a diluent as sugar, or suspended in carriers in which the gum is insoluble.

There are alternative procedures which can be used to aid dispersion. The first involves the use of a mixing device (Fig. 9.5) widely used in industry. Essentially, this consists of a funnel to hold the dry powder, an air inlet and an eductor which is a water jet pump used as mixing device. As water enters at the side and the powder, partially dispersed by air, flows down, intimate contact with water is achieved in the zone of high turbulence developed in the throat.

Another procedure involves prehydration. As an example sodium carboxymethyl cellulose can be suspended in a carrier in which it is

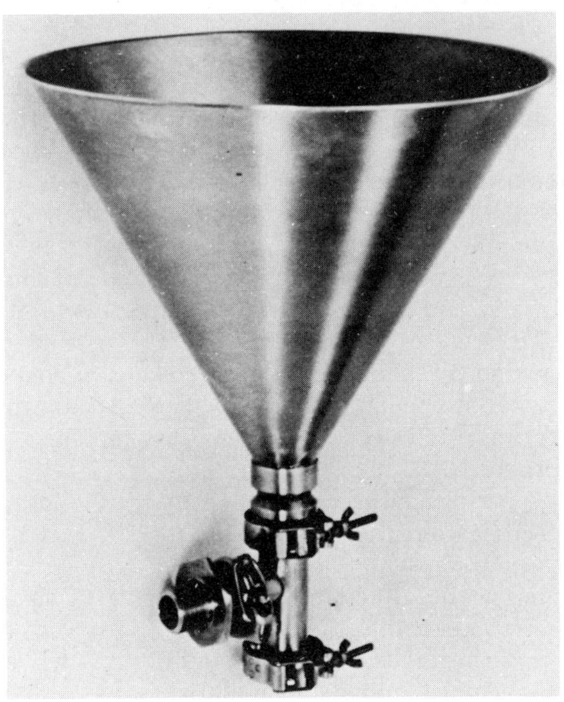

Courtesy of Hercules Incorporated

FIG. 9.5. MIXING DEVICE

insoluble, such as propylene glycol. A small amount of water is added. The gum becomes hydrated but does not dissolve. When this suspension is added to additional water, almost instantaneous solution occurs. Salt may be added to repress thickening of the dispersion which occurs with time; surfactant may be added to inhibit settling and tight packing of the hydrated gum. Settling in this slurry can also be inhibited by thickening the propylene glycol with a gum such as hydroxypropyl cellulose (Ganz 1969).

Other methods for achieving rapid dispersion and consequently solution have been described in the literature. These include coating each particle with a surfactant such as a mono/diglycerides, coating with sugar or other materials.

Faster dispersion and solution are favored by high DS, low molecular weight, and absence of water binding solutes such as sugar.

PROPERTIES IN WATER AND IN THE PRESENCE OF NONPOLYMERIC SOLUTES

There are several ways in which cellulosic hydrocolloids perform in foods. First, they provide water with a structure of varying degrees of rigidity. Structure can be achieved by water molecules being tightly bound by hydrogen bonds to the polar groups of the hydrocolloid; water molecules, probably in layers, being associated by hydrogen bonding to the first layer; slightly associated water being immobilized by the geometry of the hydrocolloid molecule; water becoming immobilized through gel formation or by cross-linkages and electrostatic forces between chains even though it is not attached. To summarize: the texture or structure of the water phase is influenced by restricting the mobility of the water molecules. The structure of water and the influence of solutes upon the structure of water is well discussed elsewhere (Ward 1963). The net effect is generally observed by rheological measurements. Such measurements should also take into account the properties of the water, by itself, e.g., viscosity, and the effect of temperature, presence of salts, etc., upon these properties.

Second, the performance of cellulosic hydrocolloids can be influenced by specific interaction of the hydrocolloid with other components of the system, such as proteins, salts, etc.

Third, there can be effects, such as lowering of surface tension, adsorption effects, heat gelation, etc. Often several of these effects may be operating simultaneously. In addition, such attributes as mouthfeel, appearance, cutting characteristics, gloss, and luster may be considered. Detailed discussions of these factors can be found in the bibliography references.

All water soluble gums and microcrystalline cellulose thicken water to varying degrees. Their effectiveness in thickening water is determined primarily by their concentration and, except for microcrystalline cellulose, by the degree of polymerization (DP) which may vary from 100 to several thousand anhydroglucose units. Figure 9.6 illustrates this for sodium carboxymethyl cellulose and Fig. 9.7 for methyl cellulose. The increase in viscosity (the term viscosity used throughout this chapter indicates apparent viscosity unless otherwise stated) is not directly proportional to changes in concentration. At the lower concentrations, the molecule can be completely expanded and there is no competition between polymer chains for available water. Also, the water attached to the molecule may be serving as a lubricant, permitting the polymer chains to slide past one another. As the concentration of polymer increases, there is not enough water to permit complete expansion of the chains or sliding; the molecules can become entangled or aggregated, water can be trapped, and larger kinetic units of flow result. Consequently, the viscosity rises rapidly. From this, it follows that it is more accurate to think of gum concentrations in a formulation in terms of the water present, than of total formula weight.

Because of its insolubility in water, microcrystalline cellulose

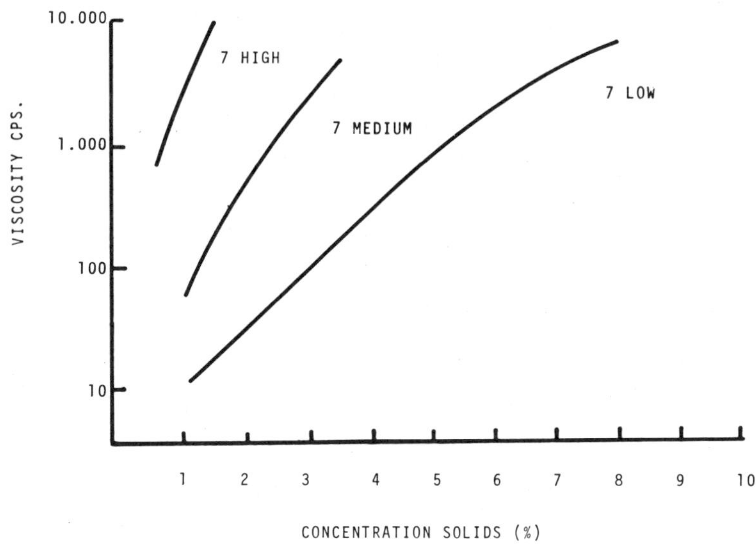

Courtesy of Hercules Incorporated

FIG. 9.6. SODIUM CARBOXYMETHYL CELLULOSE: VISCOSITY VERSUS CONCENTRATION

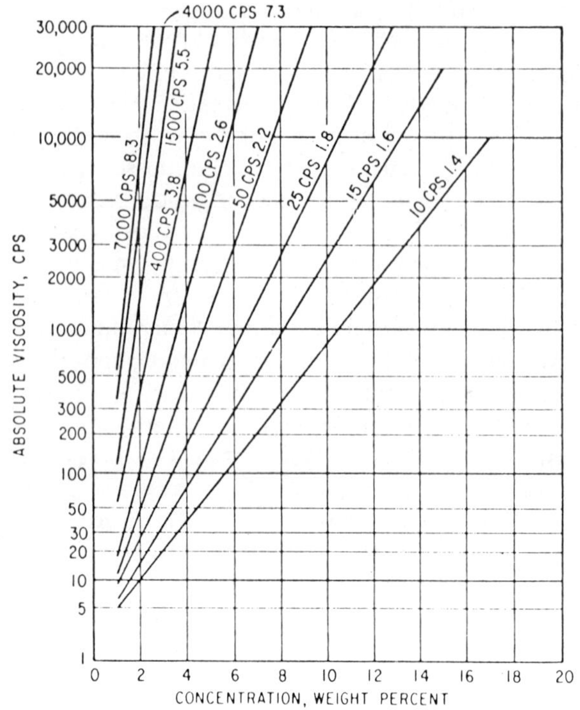

Courtesy of the Dow Chemical Company

FIG. 9.7. VISCOSITY-CONCENTRATION CHART FOR METHYL CELLULOSES

Notations on lines are the viscosity of a 2% solution in water at 20°C and the intrinsic viscosity at 20°C.

shows a somewhat different behavior. At concentrations of under 1%, colloidal dispersions are formed; above 1%, thixotropic gels are formed (Fig. 9.8). The term thixotropy is applied to a system which becomes thinner under fixed flow conditions. Thixotropy arises when a reversible sol-gel system is possible and is a consequence of a weak, three dimensional structure. This structure breaks down temporarily upon stirring but reforms on standing.

Temperature has a marked effect upon the viscosity of water solutions of these gums. This is illustrated in Fig. 9.9 which shows the decrease in viscosity which occurs upon heating solutions of sodium carboxymethyl cellulose. Presumably, as the thermal energy of the water molecule increases, hydrogen bonds are broken, and less water is attached to the polymer molecule at the elevated temperature. Also, the viscosity of the water phase decreases with increasing tem-

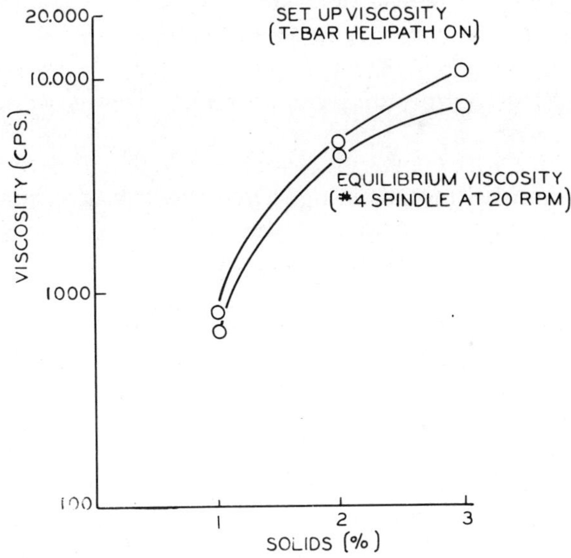

Courtesy of Food Machinery Corp., Avicel Division

FIG. 9.8. TYPICAL VISCOSITIES OF MICROCRYSTALLINE CELLULOSE (AVICEL RC-581) DISPERSIONS

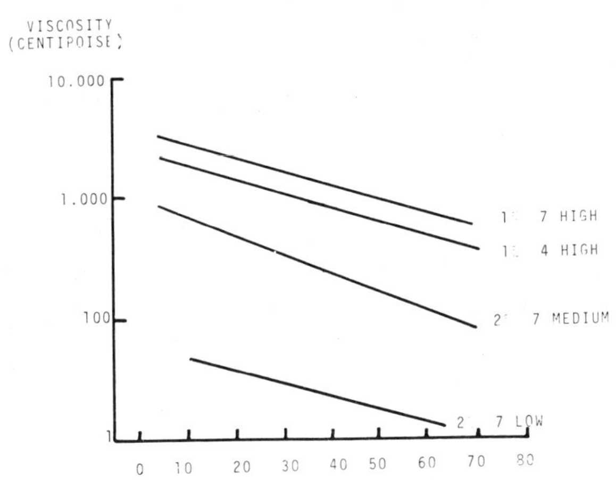

Courtesy of Hercules Incorporated

FIG. 9.9. EFFECT OF TEMPERATURE ON VISCOSITY OF SODIUM CARBOXYMETHYL CELLULOSE SOLUTIONS

perature. Consequently, the viscosity is lowered. However, there is still sufficient water attached to keep the gum dissolved. This viscosity drop is reversible, provided heating is not so severe or prolonged as to cause hydrolytic degradation, i.e. cleavage of the polymer molecule.

Solutions of hydroxypropyl cellulose or methyl ethyl cellulose perform differently upon heating. First, there is the usual viscosity decrease. Upon continued heating, the hydroxypropyl cellulose precipitates at about 40°C and the methyl ethyl cellulose at about 60°C. Apparently, so much water has been lost from the polymer that there is no longer enough attached water to keep the polymer in solution. These "cloud points" can be lowered by adding solutes which compete for the water and raised by the addition of water miscible organic solvents in which the gums are soluble.

Solutions of methyl cellulose and hydroxypropyl methyl cellulose perform differently from solutions of hydroxypropyl cellulose and methyl ethyl cellulose. As the temperature is increased, there is the usual decrease in viscosity. Upon continued heating more of the attached water molecules are removed from the sphere of influence of the polymer. Because of their spatial configuration, the polymer chains interlock, possibly through hydrophobic bonding of the methyl groups or hydrogen bonding, and a thermoreversible gel forms. It has been suggested that gelation results from degradation of micelles and the association of chain molecules through "quasi-crosslinkages" (Iso and Yamamoto 1970). The character of the gel can be changed by continued heating and by additives, molecular weight, and concentration; in the case of hydroxypropyl methyl cellulose it can also be changed by altering the ratio of substituents. Gelling can occur from about 50–90°C. The gelling temperature is also dependent upon the molecular weight, DS, and proportions of methoxyl and hydroxypropyl substituents (Table 9.1). The gelling temperature is lowered by most electrolytes, by sugar, and, in general, by additives which decrease the hydration of the gum or which bind water. This property of thermogelling is of advantage in such diverse uses as egg yolk substitutes, binders, and grease holdout, and in the preparation of baked goods made with flours other than the usual wheat flour. Very uniformly substituted methyl celluloses which can be heated almost to the boiling point without gelation can be prepared, but they are laboratory specialties and are not available commercially.

Microcrystalline cellulose loses some viscosity upon heating but not nearly to the same extent as the other gums (Fig. 9.10). Also, there is less hydrolytic degradation than occurs with other cellulosic hydrocolloids which are water soluble.

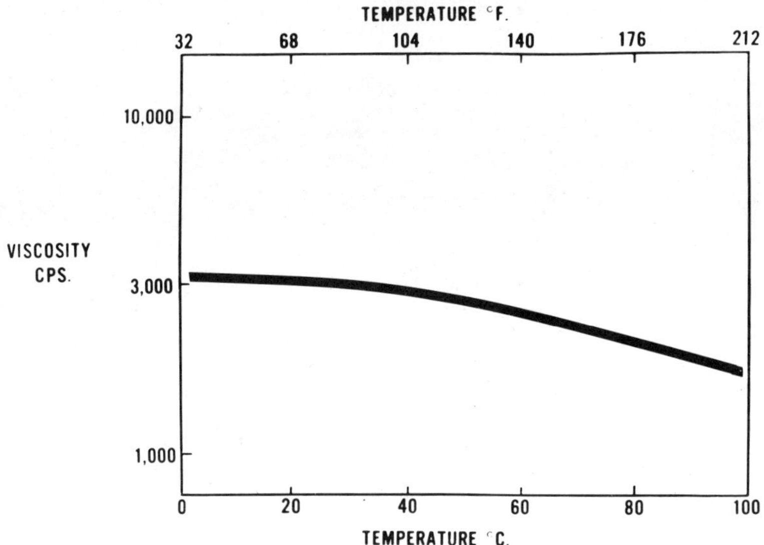

Courtesy of Food Machinery Corp., Avicel Division

FIG. 9.10. EFFECT OF TEMPERATURE ON VISCOSITY OF A 3% MICROCRYSTALLINE CELLULOSE (AVICEL RC) DISPERSION (BROOKFIELD VISCOMETER, RVT, #5 SPINDLE, 20 RPM)

Solutions of cellulose hydrocolloids are non-Newtonian, i.e., the viscosity varies with the rate of shear. They are pseudoplastic or shear thinning, i.e., the apparent viscosity decreases with increasing rate of shear. Pseudoplasticity is commonly associated with the orientation and elongation of swollen, random coils of linear polymer molecules. A useful way of picturing the flow behavior of water soluble cellulosic polymers is shown in Fig. 9.11, where the apparent viscosity of sodium carboxymethyl cellulose is plotted against shear rate. A Newtonian system would give a straight line of zero slope; thus, the lower molecular weight material is less pseudoplastic than the higher. This figure also shows how the shear rate varies with conditions of use. Normally, the viscosity is measured at an intermediate shear-rate range (1–10 sec^{-1}) with, for example, a Brookfield rotational viscometer. However, the differentiation obtained by "Brookfield measurements" in this shear rate range may not hold under other conditions. Thus, if high viscosity at high shear rate is wanted, use of a higher concentration of the type of gum usually designated as "low or medium viscosity," when measured in the intermediate shear range, might be desirable.

In addition to being pseudoplastic, some solutions of water soluble cellulose hydrocolloids are smooth flowing and some are thixotropic.

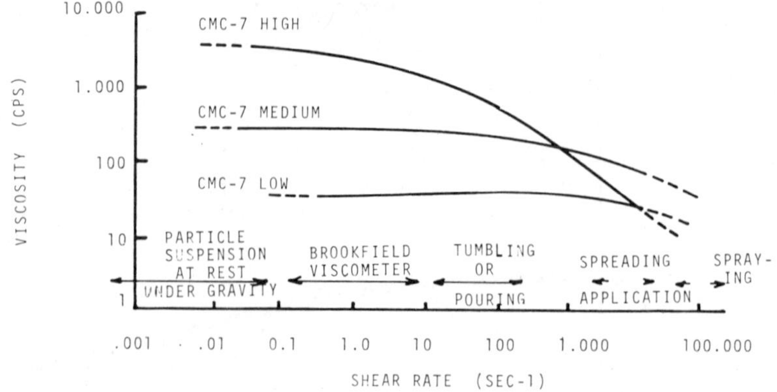

Courtesy of Hercules Incorporated

FIG. 9.11. TYPICAL FLOW BEHAVIOR OF SODIUM CARBOXYMETHYL CELLULOSE SOLUTIONS

As stated earlier, the substituent groups can be substituted uniformly or unevenly along the polymer chain. Smooth flowing solutions are believed to occur when the substituent groups are substituted evenly along the chain; structured or thixotropic solutions are believed to result when the substitution is uneven. Smoothly flowing solutions are promoted by the use of water soluble gum of sufficiently high DS in which the substituent groups are evenly distributed along the chain. Thixotropic solutions are promoted by the use of water soluble gum of low DS in which the substituent groups are unevenly distributed, leading to retention of some of the original crystalline areas along the polymer chains. Low DS sodium carboxymethyl cellulose (4.0) gels when the solution is subjected to high shear and allowed to stand.

DS influences the steady shear properties of water soluble cellulose hydrocolloids as was shown in a recent study employing the Weissenberg rheogoniometer, an instrument which permits the measurement of parameters not readily obtained with other instruments. Two types of rheological measurements were made using a cone and plate configuration: steady shear and dynamic. In steady shear measurements the solution is subjected to a large strain, the magnitude of which may be changed by changes in the applied shear rate and time of shearing. Any weak, three dimensional structure present will be broken down. Also, orientation and distortion of suspended particles or components can occur. Upon cessation of shear, the structure of the system may or may not reform, slowly or rapidly. Essentially,

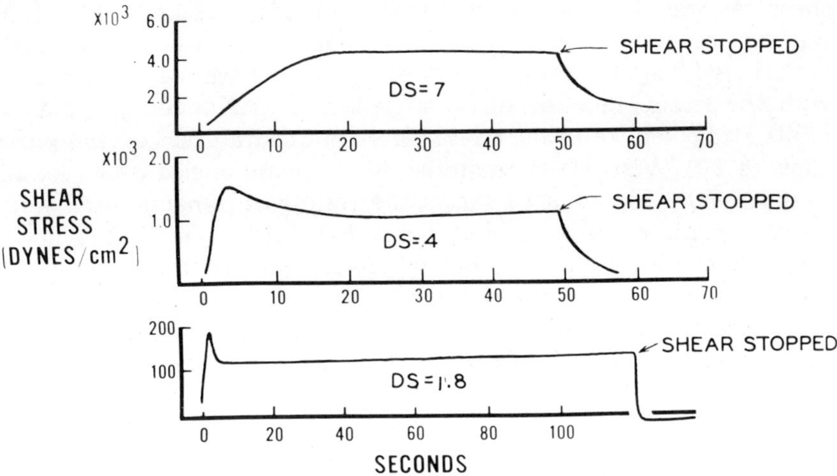

Courtesy of Hercules Incorporated

FIG. 9.12. EFFECT OF D.S. ON SHEAR STRESS OF 5% SODIUM CARBOXYMETHYL CELLULOSE SYSTEM (STEADY SHEAR 0.211 SEC^{-1})

these measurements show the response of the system to flow. In dynamic measurements the stress response to an imposed sinusoidal strain of low amplitude is observed. The total strain is small and structure breakdown is minimal. Thus, the dynamic method measures the response of the system under conditions which approach rest.

Figure 9.12 illustrates the effect of DS on the steady shear properties of 5% sodium carboxymethyl cellulose solutions prepared at high shear (Waring blender). As stated above, the rise in shear stress is measured over a period of time; then the shear is stopped, and the decrease in shear stress is measured. The curve for DS 7 material is typical of a viscoelastic system. Any peaking of the stress-time curve is vanishingly small, and the rise in the stress is relatively gradual, as is the stress decay. The curve for a sample having a DS of 4 shows a sharper stress rise and a definite peak indicative of moderate elasticity and gel structure. The sample having a DS of 1.8 gives a curve characteristic of a gel. It has a sharp peak and an extremely rapid stress decay. These gels of low DS sodium carboxymethyl cellulose have been judged organoleptically to be unctuous, whereas the other was not. Steady shear curves similar to those obtained with low DS sodium carboxymethyl cellulose have been obtained with butter.

Dynamic oscillatory measurements have also been made on the solution of low DS sodium carboxymethyl cellulose prepared at high shear. As stated above, in these measurements, a sinusoidal strain is imposed on the sample. Under these conditions, a perfectly elastic material will show a resulting sinusoidal stress wave which is in phase with the strain, i.e., the phase angle is 0°C. A Newtonian fluid exhibits strain and resulting stress waves in quadrature, i.e., the phase angle is 90°. Viscoelastic materials show phase angles which lie between 0–90°. Figure 9.13 shows the results of dynamic oscillatory measurements on a gel of low DS sodium carboxymethyl cellulose prepared at high shear. It is out of phase with the strain, and, in addition, has the square wave form which has been shown to be one of the characteristics of unctuous materials (Elliott and Ganz 1971; Elliott and Green 1972). Comparable data are not available for other water soluble cellulose hydrocolloids.

Cellulose derivatives in solution may exist in various states of aggregation. When a water soluble cellulose derivative is added to a liquid and brought to equilibrium, the gum may do one of several things. It may remain as a suspended powder, swell until all the liquid is imbibed into the particles; or dissolve completely into a true molecular solution.

The capacity of the liquid, which can be a single substance or a mixture, to act as a solvent is one of the important factors in determining the state of the gum, that is, its position on an aggregation-disaggregation curve (Fig. 9.14). If the gum does not swell or dissolve, it is in State I; in State II, there is maximum swelling or maximum solvent imbibition, thus maximum viscosity. In going from State I to

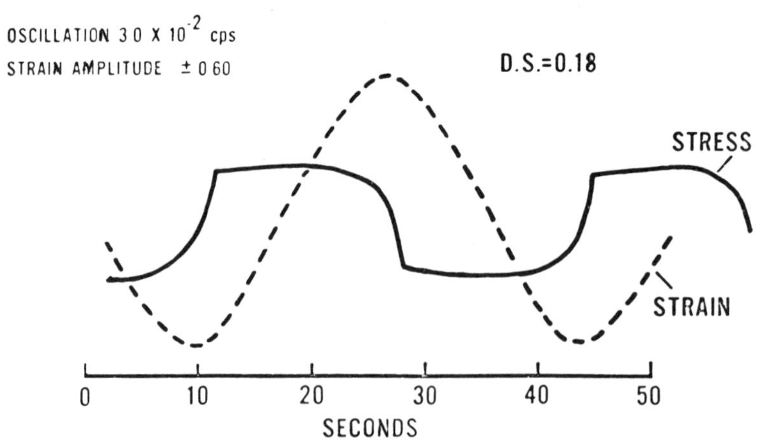

Courtesy of Hercules Incorporated

FIG. 9.13. DYNAMIC MEASUREMENTS OF A 5% SODIUM CARBOXYMETHYL CELLULOSE (D.S. 1.8) SYSTEM

LIBRARY
ATHROFA GOGLEDD DDWYRAIN CYMRU
THE NORTH EAST WALES INSTITUTE
of higher education CONNAH'S QUAY

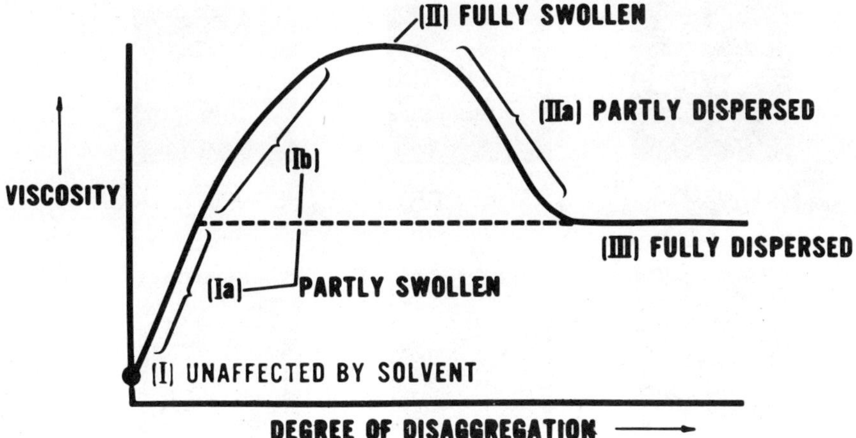

Courtesy of Hercules Incorporated

FIG. 9.14. EFFECT OF DEGREES OF DISAGGREGATION UPON THE VISCOSITY OF SOLUTIONS OF WATER-SOLUBLE CELLULOSE ETHERS

State II, some, but not all, of the internal bonds are broken. In going from State II to State III, more of the internal bonds are broken, the particle becomes more readily deformed and begins to approach molecular dispersion. In State III the molecule is completely dispersed. Aggregation-disaggregation is still further illustrated in Fig. 9.15. Thus, the flow properties of solutions of cellulosics will vary with the degree of aggregation. Aggregation, in turn, is affected by such variables as the nature and composition of the solution and its shear history. Solution rate and molecular dispersion of water soluble cellulose hydrocolloids are enhanced by a higher degree of substitution and lower molecular weight.

Ionic and nonionic cellulosic hydrocolloids respond to environment changes such as the addition of salts. In some instances, the responses are similar, in others the responses are different. In general, salts decrease the viscosity of gum solutions by decreasing the hydration of the gum. In addition, they have specific effects on anionic derivatives. Figure 9.16 shows the effect of the addition of HCl and the chlorides of mono-, di-, and trivalent cations on the viscosities of solutions of anionic sodium carboxymethyl cellulose. Acid reduces the viscosity, presumably through protonation of the carboxyl group. This protonation reduces the water-binding capacity of the gum and also leads to a more coiled up configuration of the polymer. The viscosity of high viscosity, sodium carboxymethyl cellulose (DS 7) at pH 4 is only 60% of its viscosity at pH 8. In addition, the viscosity at a given pH may vary somewhat depending on the acid

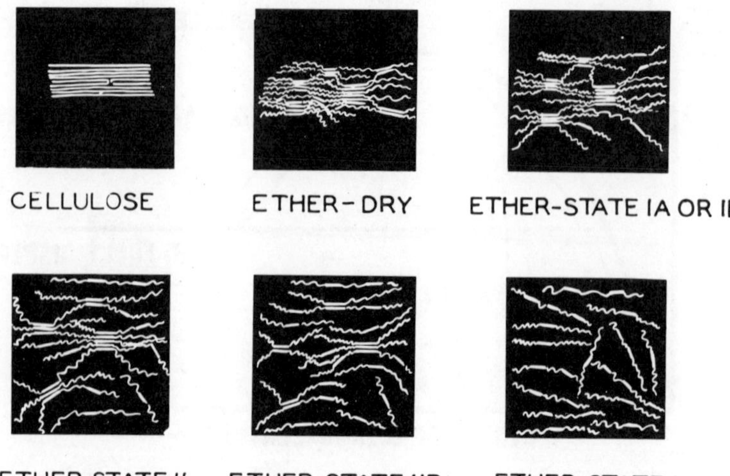

Courtesy of Hercules Incorporated

FIG. 9.15. STATES OF AGGREGATION OF WATER-SOLUBLE CELLULOSE ETHERS

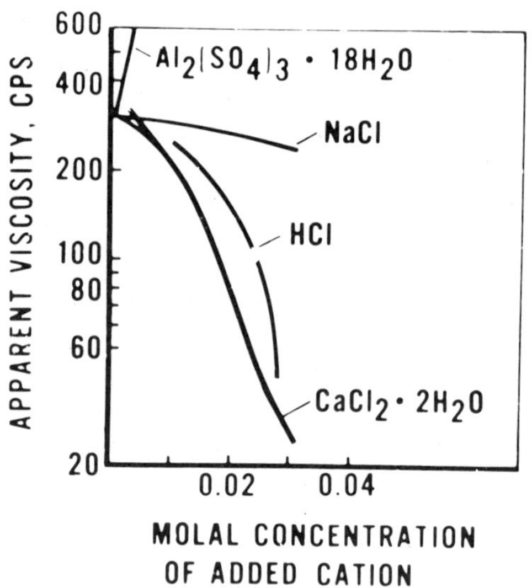

Courtesy of Hercules Incorporated

FIG. 9.16. EFFECT OF CONCENTRATIONS OF CATIONS ADDED TO 1% SODIUM CARBOXYMETHYL CELLULOSE SOLUTIONS (7 HIGH VISCOSITY TYPE)

employed. Monovalent cations provide a screen of counterions around the carboxyl group, thereby leading to a reduction of the repulsive forces between the carboxyl groups and causing the polymer to adopt a more coiled up configuration. Divalent cations reduce the viscosity through the formation of salts of borderline solubility. The addition of trivalent ions can lead to the formation of precipitates, highly thixotropic solutions, or gels. To achieve gel formation, high local concentrations of trivalent ion must be avoided. The formation and properties of the aluminum-sodium carboxymethyl cellulose gels is strongly influenced by the presence of chelants (Podlas 1973). Acids have little effect upon the viscosity of nonionic gums in the pH range of 3-10.

The order of addition of salts can influence the viscosity. This is illustrated in Fig. 9.17 which shows how the order of addition influences the viscosity of sodium carboxymethyl cellulose. When the salt is added after the gum is dissolved, the viscosity is much higher than when the gum is dissolved in salt solution. This behavior is related to the presence of residual crystalline areas in the polymer and is noticed less with evenly substituted types than with unevenly substituted types. Consequently, for full viscosity development, the salts are best added after dissolving the gum.

Salts flocculate dispersions of microcrystalline cellulose. This flocculation can be inhibited by the presence of sodium carboxymethyl cellulose or hydroxypropyl methyl cellulose which serve as protective colloids.

Larger than expected viscosities can also result from the introduction into the system of solutes which tie up water. This is illustrated

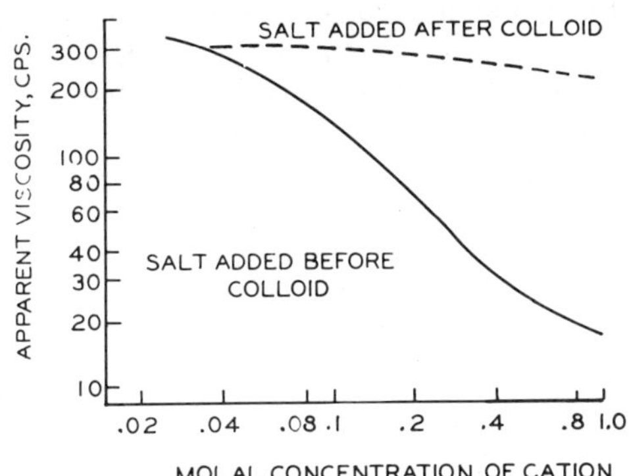

Courtesy of Hercules Incorporated

FIG. 9.17. EFFECT OF POINT OF ADDITION OF NaCl UPON VISCOSITY 1% SODIUM CARBOXYMETHYL CELLULOSE (7 HIGH VISCOSITY TYPE) SOLUTIONS

in Table 9.2, the effect of sugars on the viscosities of sodium carboxymethyl cellulose is shown. The ratio of gum to water is the same for all systems. However, the viscosities and therefore the textures are different due to differences in the amount of free water in the system. By free water, is meant water not bound by solute. The gum is functioning as if it were more concentrated than 1%. However, the possibility of some H bonding between the gum and sugars cannot be excluded. Also, the increased viscosity may be caused in part by an increase in viscosity of solvent. The implication is that as any water-binding ingredient in a food system is varied, changes in flow properties and texture are to be expected. Microcrystalline cellulose also exhibits this phenomenon.

The water-binding capacity of water soluble gums varies with the viscosity type. The high viscosity types appear to hold water more effectively than lower viscosity types. While this behavior is an advantage in some instances, e.g., in the use of gums in baked goods where gums promote the introduction of more water and sugar into the system, retention of freshness, and improved mouthfeel, it can sometimes be a disadvantage. Thus, when systems containing gums are spray dried, the gum appears to increase the intermolecular structuring of water thereby reducing its rate of diffusion and evaporation during terminal stages of drying. Consequently, adjustments in drying conditions may be necessary (Berlin *et al.* 1973).

Gums also influence crystallization. The viscosity of the gum is the primary consideration. The low viscosity type of sodium carboxymethyl cellulose is more effective in controlling the crystallization of sucrose than the high viscosity type at the same level, indicating that the rate-controlling mechanism involves surface adsorption rather than diffusion. This finding is of value in confections.

Gums inhibit ice crystal growth. In this function, the high viscosity types are more effective than the low viscosity types at the same gum concentration. This property has encouraged their widespread use in frozen products, particularly the widespread use of sodium carboxy-

TABLE 9.2

VISCOSITIES OF SODIUM CARBOXYMETHYL CELLULOSE, HIGH VISCOSITY (DS 7) IN SUGAR SOLUTIONS

Composition of Solution (%)	% Conc. of CMC in Water	Viscosity 6 rpm
1.0 CMC-H_2O	1.0	2640
0.5 CMC-50% Sucrose	1.0	4700
0.5 CMC-50% Corn Syrup Solids	1.0	3700
0.5 CMC-40% Sucrose, 10% dextrose	1.0	4050

methyl cellulose in frozen desserts. In addition, gums can raise the ice recrystallization temperature, increase the amount of water not converted to ice, and can inhibit the precipitation of solutes (Mitchell 1969).

The stability of emulsions is enhanced by active gums such as sodium carboxymethyl cellulose which are not interfacially active under ordinary conditions. They function by increasing the viscosity of the system. Sodium carboxymethyl cellulose can lower the interfacial tension when the lipid phase is nonpolar and the aqueous phase is acidic, around pH 2 (Block and Lamy 1968). Nonionic water soluble cellulose derivatives stabilize emulsions through their ability to lower the interfacial tension. Advantage can be taken of this lowering of the interfacial tension in the preparation of flavor emulsions, proteinless whipped toppings, and similar products. Microcrystalline cellulose is effective in stabilizing emulsions even though it does not lower the interfacial tension. The particles are wetted by both oil and water and locate themselves at the oil-water interface producing a film of enhanced strength and thermal stability. In addition, the microcrystalline cellulose gives a yield stress to the water phase, inhibiting Brownian motion of the oil particles thereby decreasing their tendency to coalesce. This has encouraged its usage in aerated frozen products.

PROPERTIES IN PRESENCE OF WATER MISCIBLE ORGANIC SOLVENTS

Sodium carboxymethyl cellulose is soluble in water and insoluble in organic solvents. However, it is soluble in mixtures of water and water miscible organic solvents e.g. lower alcohols or glycerides, depending upon the concentration of solvent. The solubility of the gum is higher the lower the DP, i.e., with the low viscosity type of sodium carboxymethyl cellulose than with the higher viscosity types. The apparent viscosity of sodium carboxymethyl cellulose in mixtures of glycols and water varies with the composition of the solvent as shown in Fig. 9.18.

Hydroxypropyl cellulose is soluble in many polar organic solvents e.g., ethanol, methanol, per se as well as in water solutions of these solvents. The latter raise the "cloud point" in water. Thus, in the presence of 40% ethanol, the cloud point of hydroxypropyl cellulose (MS-4) is raised to about 70° C from about 42° C. The same type of effect is observed when ethanol is substituted by propylene glycol. There is no "cloud point" in pure organic solvents.

Methyl cellulose and hydroxypropyl methyl cellulose are also soluble in organic solvents. The solubility varies with the degree of substitution and the type of substituents. The higher the degree of

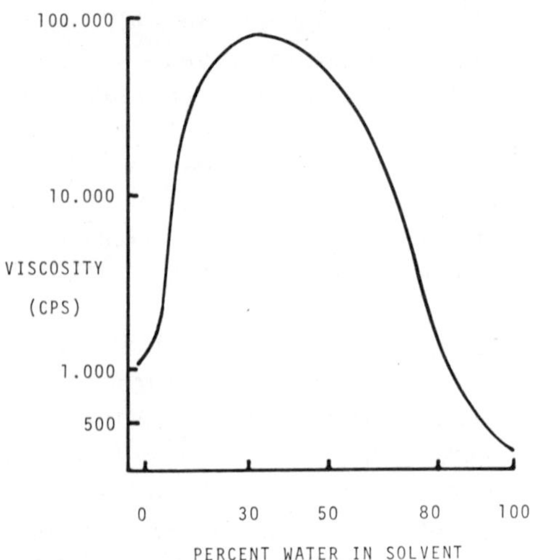

Courtesy of Hercules Incorporated

FIG. 9.18. VISCOSITY OF 1.75% SODIUM CARBOXYMETHYL CELLULOSE (7 MEDIUM VISCOSITY TYPE) IN WATER-GLYCERINE SOLUTIONS

substitution, the higher is the organic solubility. The gelation temperature of these cellulose ethers in water is raised by introducing water-miscible organic solvents, e.g., ethanol, glycols, into the system.

Microcrystalline cellulose does not form colloidal dispersions in alcohols or glycols. However, it does have an excellent tolerance for these water miscible solvents and, once a colloidal dispersion is formed, as much as 50-60% of the alcohol or glycol can be added without causing flocculation.

PROPERTIES IN THE PRESENCE OF POLYMERIC SOLUTES

By and large, hydrocolloids from cellulose are compatible with other natural gums. Some exceptions can be found. Thus, microcrystalline cellulose is not compatible with guar gum at a ratio of 9:1 (total solids = 1%). In some instances, synergistic viscosity effects are noted in solutions containing anionic and nonionic cellulose ethers (Table 9.3).

The presence of a gum during heating of starch suspensions influences the gelatinization profiles observed in a Brabender amylograph. It has been shown that in the presence of sodium carboxymethyl cellulose, the swelling of corn and wheat starch occurred in two steps (Crossland and Favor 1948). This behavior was not observed with tapioca and waxy starches.

TABLE 9.3

SYNERGISTIC VISCOSITY EFFECTS IN SOLUTIONS OF SODIUM CARBOXYMETHYL CELLULOSE AND HYDROXYPROPYL CELLULOSE

	Apparent Viscosity (CPS) of 1% Solution at 25°C
Sodium Carboxymethyl Cellulose	1500
Hydroxypropyl Cellulose	1640
Viscosity of Blend of equal parts	
Expected	1570
Actual	3280

These studies have been extended by amylograph studies on the gelatinization profiles of wheat starch in the presence of sodium carboxymethyl cellulose of varying DP and DS (Ganz 1969B). The peak viscosities achieved are much higher in the presence of sodium carboxymethyl cellulose (even when the viscosity of the gum is subtracted from the viscosity of the mixture), than in the absence of gum and vary with the type and level of gum present. The apparent gelatinization temperature of wheat starch is lowered to about 60°, and swelling shows up as a two-stage process. In addition, the gelatinization profiles vary with the presence of additives and the type and level of gum used. Higher peak viscosities are achieved with low DS rather than with high DS types, with thixotropic rather than with smooth types, and with higher rather than lower viscosity types. Cookout and setback are decreased in the presence of gum. There is no correlation between the Brookfield viscosities of the gum in dilute solutions and their effects on the GPs of starch. The simplest rationalization is that as the starch swells, it takes up water, thereby increasing the "effective" gum concentration. The properties of the system then begin to relate to the properties of concentrated, more or less aggregated sodium carboxymethyl cellulose solutions.

Under conditions of limited water content, methyl cellulose and sodium carboxymethyl cellulose inhibit the gelatinization process. The degree of inhibition varies with the temperature of the system (Watson 1964).

Often, when ionic water soluble cellulose derivatives are added to systems containing protein, unexpected textural effects are observed which do not correlate with those expected from gum viscosities measured in water. When the anionic polyelectrolyte, sodium carboxymethyl cellulose is added to a system containing protein, interaction has been observed through viscosity and other physical measurements. The reaction between gum and protein is undoubtedly primarily ionic in nature, even though hydrogen bonding or van der Waals forces may also be operative.

An ionic representation has been proposed as shown in Fig. 9.19 (Ganz 1969B). In or around the isoelectric region of the protein, a complex can form. Upon the addition of alkali, the positively charged amine group is neutralized and combination with the anionic polyelectrolyte is inhibited. As this occurs, the viscosity of the system approaches the viscosity to be expected from a blend. Upon dilution, or as salts are added and the ionic strength is increased, the ionic force between oppositely charged groups is reduced and again the viscosity of the system might be expected to approach the viscosity expected from a blend in the same environment. When acid is added, precipitates may form.

However, in addition to the effects of acid, alkali, and salts upon the complex, the effects of these additives upon both the protein and the sodium carboxymethyl cellulose must be taken into account in interpreting experimental results.

Figure 9.20 shows the effect of pH upon the viscosities achieved in systems containing gelatin and sodium carboxymethyl cellulose. Under these conditions, the viscosity of gelatin is under 10 cps and is not shown on the graph. The viscosities obtained above pH 4.2 are far higher than would be achieved if no interaction had occurred, indicating formation of a soluble complex. Higher viscosities are obtained with Type A gelatin prepared by acid pretreatment of pigskin or ossein and which has a higher isoionic point (pH 8.9) and a larger

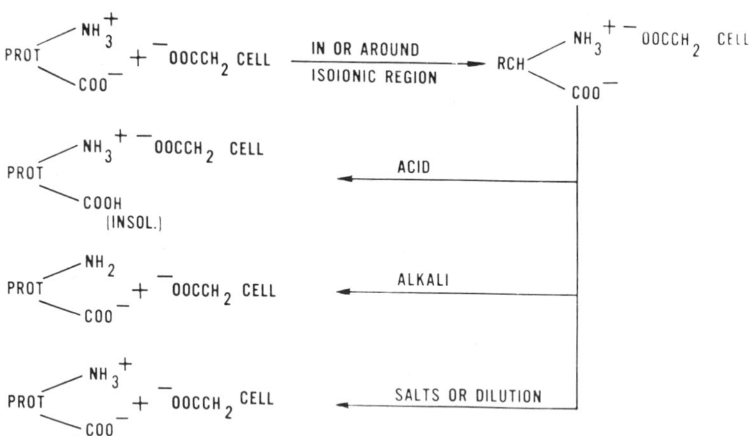

Courtesy of Hercules Incorporated

FIG. 9.19. IONIC REPRESENTATION OF REACTION OF PROTEINS AND SODIUM CARBOXYMETHYL CELLULOSE

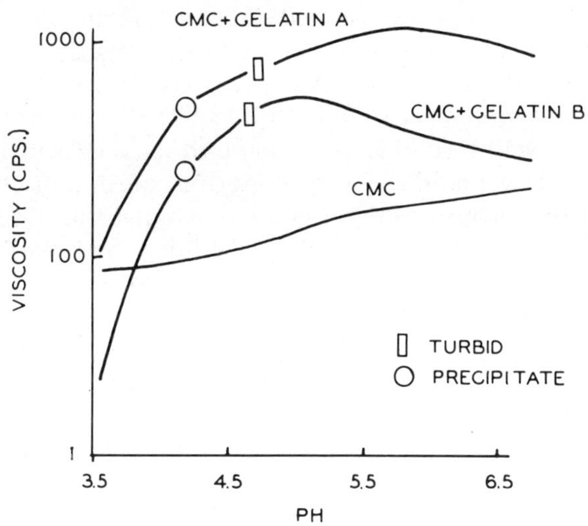

Courtesy of Hercules Incorporated

FIG. 9.20. VISCOSITIES OF 0.5% SODIUM CARBOXYMETHYL CELLULOSE (7 HIGH VISCOSITY TYPE) SOLUTIONS WITH AND WITHOUT 0.5% GELATIN, 200 BLOOM

number of positively charged amine groups than Type B gelatin prepared by treatment of any collagen stock with lime and having a lower isoionic point (pH 5). The pH optimum for viscosity development occurs below the isoelectric region for Type A gelatin, but around the isoelectric region for Type B gelatin. As noted, there are optimum pH regions for viscosity development. As the pH is lowered, the viscosity decreases. This decrease may reflect protonation of some of the carboxyl groups of the protein-gum complex. Protonation of some of the carboxyl groups of the uncomplexed cellulose gum is also likely, resulting in fewer sites for ion-pair formation. There may also be steric changes involved. As has been stated, protonation of the gum leads to a more coiled up configuration. This also results in a lower viscosity. When the pH is lowered sufficiently, insolubilization occurs. There is no indication of coacervate formation, i.e., neither liquid drops nor two liquid layers were formed.

Nonionic gums perform differently. Relatively little interaction occurs nor are precipitates observed in the pH range tested. The absence of interaction or precipitate formation precludes the possibility that the effect of sodium carboxymethyl cellulose is merely that of binding water, making it unavailable to the gelatin and forcing the gelatin into an aggregated state. Variables in this system include pH, ratios of protein and gum, molecular weights, concentration, etc. This reaction of gelatin and sodium carboxymethyl cellulose has been of interest in products as diversified as meringues, reduced fat spreads, and puddings.

Gelatin and sodium carboxymethyl cellulose can be codried to obtain a cold water soluble product. Ordinary gelatin is not cold water soluble. In general, the properties of a solution of codried gelatin and sodium carboxymethyl cellulose parallel the properties of a solution of the sodium carboxymethyl cellulose and gelatin prepared by first dissolving the gelatin in hot water.

The effect of pH on the reaction between soy protein and sodium carboxymethyl cellulose is shown in Fig. 9.21. Soy protein is insoluble in its isoelectric region. Sodium carboxymethyl cellulose inhibits the precipitation of soy protein in this pH range. Also, abnormally high viscosities are reached. Apparently the carboxyl groups of the gum are combining with positively charged groups of the protein, and the more polar complex formed is soluble. A network appears to be formed as indicated by the very high viscosities achieved. Similar results are obtained in skimmilk and with sodium caseinate. Thus, by use of these proteins can be extended with sodium carboxymethyl cellulose into a more acid pH region than currently used. The systems can be heated to 95°C and cooled without insolubilizing the protein.

In one series of studies, the interaction of sodium carboxymethyl cellulose with skimmilk proteins and isolated α-casein, β-casein, and

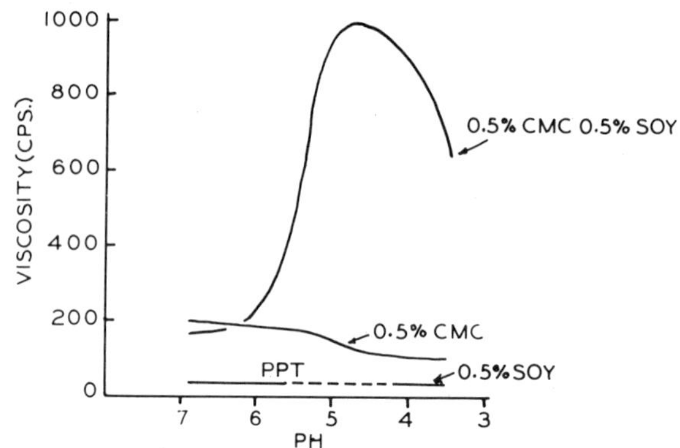

Courtesy of Hercules Incorporated

FIG. 9.21. VISCOSITIES OF SODIUM CARBOXYMETHYL CELLULOSE (7 HIGH VISCOSITY TYPE) SOLUTIONS WITH AND WITHOUT SOY PROTEIN

β-lactoglobulin was observed at various pH values and temperatures by electrophoretic analyses, optical density determinations and viscosity measurement. By adding the gum, the skimmilk proteins and solutions of α-casein were stabilized in the pH range of 4.0–5.0. In addition, the milk proteins but not the α-casein solutions were stabilized against the effect of heat (80° C, 30 min). Solutions of β-casein and β-lactoglobulin were stabilized in the pH range of 4.0–7.0 and were also stabilized against the effect of heat (Asano 1966). It has also been shown that unevenly substituted sodium carboxymethyl cellulose was more effective in stabilizing α-, α_s-, β-, and κ-casein than evenly substituted gum (Asano and Ishida 1971).

In other studies, conditions (pH, ionic strength, ratio of polymers) for the fractionation of whey proteins with sodium carboxymethyl cellulose were established (Hildago and Hansen 1971). β-lactoglobulin forms insoluble complexes with sodium carboxymethyl cellulose at pH 4 at an ionic strength of <0.1. The complex is dissociated by increasing the ionic strength. The binding ratio is 45.3 ± 4.6 moles of carboxyl groups per mole of β-lactoglobulin. Peptization of the insoluble complex results upon further addition of gum (Hildago and Hansen 1969).

The stabilizing effect of sodium carboxymethyl cellulose on the proteins of milk is used to advantage in the preparation of milk drinks containing fruit juices (Shenkenberg *et al.* 1972; Luck and Rudd 1972).

This effect of sodium carboxymethyl cellulose in solubilizing soy protein and casein in their isoelectric regions is to be differentiated from wheying off or serum separation, i.e., the precipitation of casein with sodium carboxymethyl cellulose at approximately neutral pH. As is well known in the dairy industry, the casein in milk can be precipitated with as little as 0.1–0.2% sodium carboxymethyl cellulose at pH 6.6. This reaction has been studied in detail, and it was concluded that the reaction involved bridging of the protein and gum through calcium (McClusky *et al.* 1969).

Another area of interest is in codried egg white solids-sodium carboxymethyl cellulose. In the absence of whipping aids such as sodium lauryl sulfate, the stability of egg white foams is enhanced by using egg whites codried with sodium carboxymethyl cellulose. Comparable results are not obtained with a physical blend. Also, the enhancement of stability is greater than would be expected if the sole effect of the gum were to compensate for the loss in viscosity which occurs on drying. In addition, the oil tolerance of the foam is improved (Ganz 1966).

SENSORY EFFECTS

The mouthfeel characteristics of gums have been correlated with their rheological behavior (Szczesniak and Farkas 1962). As shown in Fig. 9.22, gums which are very slimy in the mouth deviate only slightly from Newtonian viscosity. The degree of sliminess decreases with increasing deviations from the Newtonian character. Gums that exhibit a high degree of shear thinning are not slimy in the mouth.

Gums influence taste perception. The results vary markedly with the nature and level of gum and substrate. It has been shown that methyl cellulose raises the threshold values of caffeine, quinine, and saccharine (Mackey and Valassi 1956). Sodium carboxymethyl cellulose influences the threshold values of the basic taste sensations (Table 9.4). At this particular concentration (0.2%) and this type of gum (7-High), the threshold value of sugar was decreased, i.e., flavor perceptibility was increased, while the threshold values of the other taste sensations were increased (Ganz N.D.). In another study, it was shown that perceived sweetness in sugar (2-64%) and sodium saccharine (0.01-0.32%) solutions decreased with viscosity; also, that the sweetness varies as a power function of viscosity with an exponent between -0.20 and -0.25 (Arabie and Moskowitz 1971).

Low concentrations of sodium carboxymethyl cellulose (1% 7-High, smooth) also were reported to raise the threshold value of sucrose whereas at suprathreshold levels, the gum enhanced sweetness

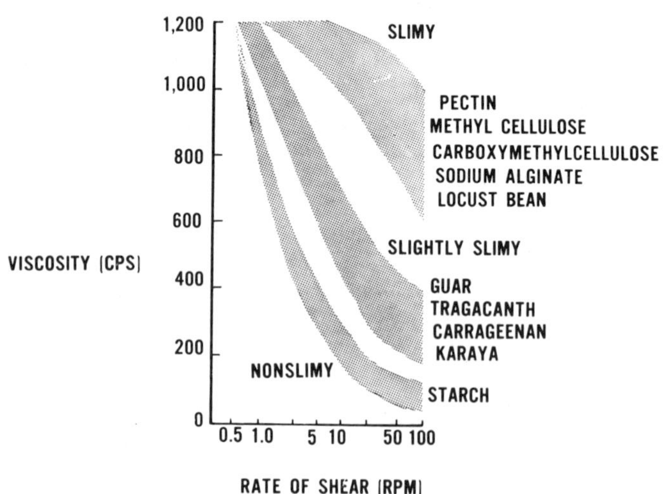

Courtesy of General Foods Corporation

FIG. 9.22. RELATIONSHIP OF RHEOLOGICAL BEHAVIOR WITH ORGANOLEPTIC PROPERTIES

TABLE 9.4

THRESHOLD VALUES (% BY WEIGHT)

	Sweet (sucrose)	Sour (lactic acid)	Bitter (caffeine)	Salty (salt)
Panel	0.60	0.006	0.01	0.007
With 0.2% CMC 7-high	0.51	0.02	0.04	0.084
Change	decrease	increase	increase	increase

(Stone and Oliver 1970). Other investigators reported that this gum diminished the taste intensity of glucose, NaCl, quinine sulfate, and citric acid (Moskowitz and Arabie 1970).

In another study, it was shown that, except for sucrose, the effects of hydrocolloids (sodium alginate, xanthan gum, sodium carboxymethyl cellulose, and hydroxypropyl cellulose) in the viscosity range of 4–72 cps on taste intensity was more dependent on the nature of the gum and of the taste substance than on the viscosity level. Viscosities above 16 cps depressed the sweetness intensity of sucrose. Of the gums tested, sodium carboxymethyl cellulose showed the greatest frequency and magnitude of significant taste-modifying effects (Pangborn *et al.* 1973). Also, it has been reported that hydrocolloid gels, which exhibit most shear thinning in the mouth also exhibit more sweetness than gels which exhibit less shear thinning (Vaisey *et al.* 1969). At very low concentrations, 0.02–0.05%, sodium carboxymethyl cellulose has a blending or smoothing out effect on flavors which is particularly noted in low calorie beverages.

In another study, the flavor and odor modifying effects of several hydrocolloids (sodium carboxymethyl cellulose, hydroxypropyl cellulose, sodium alginate and xanthan gum) were determined. The flavorants (acetaldehyde, acetophenone, butyric acid, dimethyl sulfide) represented polar and nonpolar, low and high boiling compounds. The hydrocolloids decreased the odor and flavor intensities, the effects being most noted with dimethyl sulfide (low boiling point, nonpolar) and least noted with acetophenone (high boiling point, nonpolar). The effects were specific for the gum/odorant combination and independent of viscosity (Pangborn and Szczesniak 1973).

FILM FORMING AND THERMOPLASTICITY

Sodium carboxymethyl cellulose, hydroxypropyl methyl cellulose and hydroxypropyl cellulose are used as film formers. Films can be prepared from solutions of these gums by evaporation. Food products can be coated with films by dipping and pan coating. Hydroxypropyl cellulose and hydroxypropyl methyl cellulose are both soluble in certain organic solvents and are thermoplastic, however, the hydroxypropyl methyl cellulose requires the presence of plasticizers. These gums can be extruded at elevated temperatures, injection molded, compression molded and can be hot cast. Sheets or products of various dimensions can be produced. The properties of these products can be modified by the inclusion of plasticizers which enhance flexibility and elongation, or by the addition of fillers e.g., starch, salts, protein. The films are strong, clear and are relatively unaffected by oils, greases, and most nonpolar organic solvents.

Heat sealable films of hydroxypropyl methyl cellulose can be prepared by extrusion and molding procedures. Heat sealability results from the proper choice of plasticizers.

A novel use for hydroxypropyl cellulose takes advantage of its thermoplasticity at low moisture contents. It is possible to blend hydroxypropyl cellulose and milk solids, extrude at elevated temperatures such as 125°C and upon cooling produce sheets which readily rehydrate in water or milk but which retain their integrity. This can also be done with proteins, crystallizable sugar, fruit, and meat. The products can contain as little as 2% hydroxypropyl cellulose (Ganz 1973).

These films have excellent barrier properties as is illustrated in Fig. 9.23 and 9.24 for hydroxypropyl cellulose.

STABILITY AND PRESERVATION

Under some conditions, the water soluble cellulosic hydrocolloids are susceptible to acid hydrolysis of the glycosidic linkages, resulting in scission of the long-chain molecules. Acid hydrolysis is accentuated by low pH and elevated temperature. An oxidative type of degradation occurs under alkaline conditions in the presence of oxygen; oxidative degradation is increased by heat or ultraviolet light. Exoenzymes produced by molds or bacteria can cause biodegradation although in the absence of such enzymes, solutions of these gums may be maintained almost indefinitely. Resistance to biodegradation varies with DS, the high DS types being more resistant than the lower. Also, uniformly substituted material is more resistant to biodegradation (Wirick 1968).

The common food preservatives (sorbic acid and its salts, sodium benzoate, sodium propionate, methyl and propylparabenzoate) are compatible with the cellulosic hydrocolloids with the exception of hydroxypropyl cellulose which is not compatible with the parabenzoates.

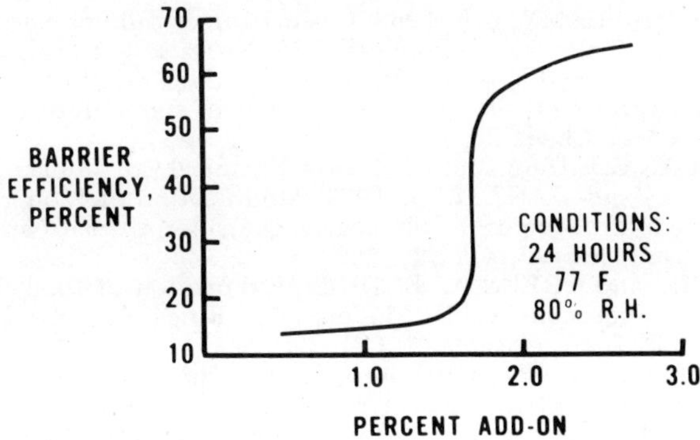

Courtesy of Hercules Incorporated

FIG. 9.23. EFFICIENCY OF HYDROXYPROPYL CELLULOSE (LOW VISCOSITY TYPE) IN INHIBITING MOISTURE PICKUP BY PEANUTS

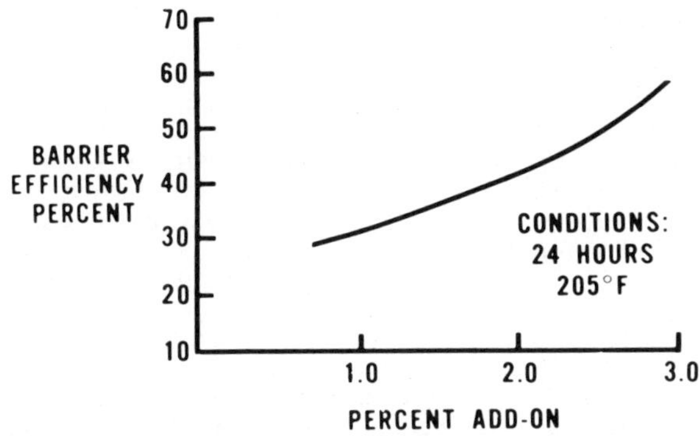

Courtesy of Hercules Incorporated

FIG. 9.24. EFFICIENCY OF HYDROXYPROPYL CELLULOSE (LOW VISCOSITY TYPE) IN INHIBITING OXIDATIVE RANCIDITY IN PEANUTS

BIBLIOGRAPHY

ARABIE, P., and MOSKOWITZ, H. R. 1971. The effects of viscosity upon perceived sweetness. Perception Psychophysics *9*, 410–412.

ASANO, Y. 1966. The interaction between milk proteins and carboxymethylcellulose in fruit flavored milk. Intern. Dairy Congr. Proc. 17th Munich *5*, 695–702.

ASANO, Y. 1970. Interaction between casein and carboxymethylcellulose in the acidic condition. Agr. Biol. Chem. *34*, 102–107.

ASANO, Y., and ISHIDA, Y. 1971. Chemical structure and protein stabilization acitivity of carboxymethylcellulose. Agr. Biol. Chem. *35*, 1018–1023.

BERLIN, E. *et al.* 1973. Water sorption of dried dairy products stabilized with carboxymethylcellulose. J. Dairy Sci. *56*, 685–689.

BLOCK, L. H., and LAMY, P. P. 1968. Orientation of cellulosic ethers at liquid/ liquid interfaces. Kolloid-Z. Z. Polymers *225*, No. 2, 164-166.
CROSSLAND, L. B., and FAVOR, H. H. 1948. Starch gelatinization studies. II. A method for showing the stages in the swelling of starch during heating in the amylograph. Cereal Chem. *25*, 213.
DOW CHEMICAL CO. 1962. Methocel. Dow Chemical Co., Midland, Mich.
ELLIOTT, J. H., and GANZ, A. J. 1971. Modification of food characteristics with cellulose hydrocolloids. I. Rheological characterization of an organoleptic property. J. Texture Studies *2*, 220-229.
ELLIOTT, J. H., and GREEN, C. E. 1972. Modification of food characteristics with cellulose hydrocolloids. II. The modified Bingham body—a useful rheological model. J. Texture Studies *3*, 194-205.
FMC CORP. 1971. Avicel RC. FMC Corp., Marcus Hook, Pa.
GANZ, A. J. 1966. Codried liquid egg white-carboxymethyl cellulose composition, process for the preparation thereof and food mix utilizing same. U.S. Pat. 3,287,139. Nov. 22.
GANZ, A. J. 1969A. Water-soluble cellulose ether or starch compositions. U.S. Pat. 3,485,651. Dec. 23.
GANZ, A. J. 1969B. CMC and hydroxypropylcellulose—versatile gums for food use. Food Prod. Develop. *3*, No. 6, 65-71.
GANZ, A. J. 1973. Method of makingg a thermoplastic food product. U.S. Pat. 3,769,029. Oct. 30.
GANZ, A. J. (N. D.) Unpublished results.
GLICKSMAN, M. 1969. Gum Technology in the Food Industry. Academic Press, New York.
HIDALGO, J., and HANSEN, P. M. T. 1969. Interactions between food stabilizers and β-lactoglobulin. Agr. Food Chem. *17*, 1089-1092.
HIDALGO, J., and HANSEN, P. M. T. 1971. Selective precipitation of whey proteins with carboxymethylcellulose. J. Dairy Sci. *54*, 1270-1274.
HERCULES INC. 1971A. Cellulose Gum: Chemical and Physical Properties. Hercules Inc., Willimgton, Del.
HERCULES INC. 1971B. Klucel-Hydroxypropylcellulose: Chemical and Physical Properties. Hercules Inc., Wilmington, Del.
IMPERIAL CHEMICAL INDUSTRIES LTD. 1961. Edifas. Imperial Chemical Industries Ltd. (Nobel Div.), Glasgow, Scotland.
ISO, N., and YAMAMOTO, D. 1970. Effects of sucrose and citric acid on the sol-gel transformation of methylcellulose. Agr. Biol. Chem. *34*, 1867-1869.
JIRGENSONS, B. 1958. Organic Colloids. American Elsevier Publishing Co., New York.
KARLSON, P. 1963. Introduction to Modern Biochemistry. Academic Press, New York. (German)
KARMAS, E. 1973. Water in biosystems. J. Food Sci. *38*, 736-739.
KLOSE, R. E., and GLICKSMAN, M. 1968. Gums. *In* Handbook of Food Additives, T. E. Furia (Editor). Chemical Rubber Co., Cleveland, Ohio.
LUCK, H., and RUDD, S. 1972. Milk flavored with natural fruit juice. S. African J. Dairy Technol. *4*, 153-158.
MACKEY, A. O., and VALASSI, K. 1956. The discernment of primary tastes in the presence of different food textures. Food Technol. *10*, 238-240.
MARK, H. F. *et al.* 1965. Encyclopedia of Polymer Science and Technology, Vol. 3. John Wiley & Sons, New York.
McCLUSKEY, F. J., THOMAS, E. L., and COULTER, S. T. 1969. Precipitation of milk proteins by sodium carboxymethyl cellulose. J. Dairy Sci. *52*, 1181-1185.
MITCHELL, W. A. 1969. Natural and synthetic gums in frozen products. Bakers Dig. *43*, 47-50.
MOSKOWITZ, H. R., and ARABIE, P. 1970. Taste intensity as a function of stimulus concentration and solvent viscosity. J. Texture Studies *1*, 502.

PANGBORN, R. M., and SZCZESNIAK, A. S. 1973. Effect of hydrocolloids and viscosity on flavor and odor intensities of aromatic flavor compounds. J. Texture Studies *4*, 467-482.
PANGBORN, R. M., TRAUBE, I. M., and SZCZESNIAK, A. S. 1973. Effect of hydrocolloids on oral viscosity and basic taste intensities. J. Texture Studies *4*, 224-241.
PODLAS, T. J. 1973. Process of preparing CMC gels. U.S. Pat. 3,719,503. Mar. 6.
SAVAGE, A. B. 1971. Derivatives of cellulose. *In* Cellulose and Cellulose Derivatives, Vol. 5, N. M. Bikales, and L. Segal (Editors). John Wiley & Sons, New York.
SHENKENBERG, D. R. *et al.* 1972. Milk-fruit juice beverage and process for preparing same. U.S. Pat. 3,692,532. Oct. 19.
STONE, H., and OLIVER, S. 1970. Effect of viscosity on the detection of relative sweetness intensity of sucrose solutions. J. Food Sci. *35*, 129-134.
SZCZESNIAK, A. S., and FARKAS, E. 1962. Objective characterization of the mouthfeel of gum solutions. J. Food Sci. *27*, 381-385.
VAISEY, M. *et al.* 1969. Some sensory effects of hydrocolloid sols on sweetness. J. Food Sci. *34*, 397-400.
WARD, A. G. 1963. The nature of the forces between water and the macromolecular constituents of foods. *In* Recent Advances in Food Science, Vol. 3, J. M. Leitch, and D. N. Rhodes (Editors). Butterworth's, London.
WATSON, C. A. 1964. Effect of two proteins and two cellulose derivatives on starch gelatinization. Ph.D. Thesis, Kansas State Univ., Manhattan.
WHISTLER, R. L., and BeMILLER, J. N. 1973. Industrial Gums. Academic Press, New York.
WIRICK, M. G. 1968. A study of the enzymic degradation of CMC and other cellulosic ethers. J. Polymer Sci. *6*, Part A1, 1965-1974.

CHAPTER 10

Denny B. Nelson
C. J. B. Smit
Raldon R. Wiles

Commercially Important Pectic Substances

INTRODUCTION

Pectic substances are a group of heterogeneous polysaccharides with high molecular weight and whose predominant structural subunit is D-galacturonic acid. These materials are a necessary component of photosynthesizing land plants. The pectic substances are an important food topic for two major reasons: (1) the influence of *in situ* pectic substances on the texture or consistency of many fruits and vegetables, and (2) the wide use in the food industry of commercially extracted pectins as gelling or thickening agents. Subjects related to commercial pectins form the bulk of this chapter.

Vanquelin has been credited with discovering pectin before the turn of the 18th century. Since then a large number of papers related to pectic substances have been published. The literature prior to 1950 was summarized by Kertesz (1951) in his comprehensive book on the pectic substances. The recent literature has been reviewed by such authors as Doesburg (1965, 1973), Worth (1967), McCready (1970), Pilnik (1970) and Glicksman (1969). These reviews provide additional information not covered in detail in the present chapter and contain extensive references to the original literature.

Occurrence

The pectic substances are found in the intercellular regions and the cell walls of all higher plants. In most plant tissue and certainly in immature fruit, the major portion of the pectic material is present in a water insoluble form commonly designated as protopectin (Joslyn 1962). During fruit ripening the insoluble protopectin is gradually transformed into a more water soluble form. This change in pectic solubility has an influence on the textural changes accompanying ripening (Pilnik and Voragen 1970; Doesburg 1965). As suggested by other authors (Joslyn 1962; Doesburg 1965; McCready 1970), the water insolubility of these native pectins (protopectins) of high molecular weight is probably due to a number of factors, such as primary and secondary bonding between pectic substances and other cell wall materials, formation of insoluble salts involving the carboxyl groups and multivalent cations, and mechanical enmeshing of the large polymeric molecules with other cell wall constituents.

Apparently the role of the pectic polysaccharides in plant cell walls will be revealed by experiments utilizing purified hydrolytic enzymes and improved analytical techniques. Such studies (Albersheim *et al.* 1973; Albersheim 1975) have led to a model describing the organizational relationship of the pectic, hemicellulosic and cellulosic components of the cell walls of suspension-cultured sycamore cells.

Large variations in pectic content exist among different plants, different plant varieties, and different parts of plants. In addition, the chemical makeup of the pectic substances will vary with the plant source. Both level and composition of the pectic materials vary with stages of plant growth. Table 1 gives representative pectin contents for a range of plant materials (Neukom 1967).

Nomenclature

The early literature on pectic substances is confusing because many different terms were used to designate the pectic materials extracted from plant sources. The nomenclature was standardized by a committee of the American Chemical Society (1944). As a result, *pectic substances* is the general term used by most authors to refer to the group of complex, colloidal carbohydrates occurring in, or prepared from plants, and containing a large proportion of anhydrogalacturonic acid units linked together in a chain-like fashion. More specifically, those pectic substances which contain very few methyl ester groups are designated *pectic acids* (salt forms = *pectates*). *Pectinic acids* are those polygalacturonic acids which contain more than a negligible level of methyl ester groups (salt forms = *pectinates*). Protopectin refers to the water insoluble parent pectic substances that occur in plants and yield pectin on hydrolysis. Pectin is the term most associ-

TABLE 10.1

APPROXIMATE PECTIN CONTENT OF VARIOUS PLANT MATERIALS

Plant Material	Total Pectic Substance as % of Dry Material
Potatoes	2.5
Carrots	10.0
Horseradish	15.0
Tomatoes	3.0
Apples	4–7
Apple pomace	15–20
Sunflower heads (without seeds)	25
Sugar beet pulp	15–20
Citrus peel (albedo)	30–35

Source: Neukom (1967).

ated with the product of commerce. *Pectin* refers to the water soluble pectinic acids, possessing various levels of methyl ester groups, which form gels under suitable conditions. High-methoxyl (*HM*) or *high-ester* pectins have *over* 50% of their carboxyl groups esterified, *i.e.*, the degree of esterification (DE) is above 50%. Such an ester level corresponds with a methoxyl content of around 6–7%. High ester pectins are capable of forming jellies in the presence of relatively high concentrations of sugar and acid. The degree of esterification strongly influences the time and temperature relationships required for gelation. Convenience in measuring a "setting time" has led to the classification of high-ester pectins as either *rapid-set* or *slow-set* pectins. Rapid-set pectins have a higher DE (over 68%) than slow-set pectins (near 60% DE). Commercial *low-methoxyl* (*LM*) or *low-ester pectins* usually have a DE between 30 and 50%. Low-ester pectins are capable of gel formation in the presence of suitable cations and do not require sugar and/or acid to form a gel.

Structure

The structural aspects of the mixture of acidic and neutral polysaccharides, which together constitute pectic substances, have been the subject of recent reviews (Aspinall 1970A, 1970B; Somers 1970; Rees 1970). The major structural feature of the pectic substances is the presence of acid polysaccharides composed largely of D-galacturonic acid units linked through α-(1 → 4) glycosidic bonds as shown in Fig. 10.1. Normally some of the carboxyl groups are esterified with methyl alcohol. The neutral sugar content is dominated by D-galactose, L-arabinose, and L-rhamnose with lesser amounts of D-xylose, L-fucose, 2-*O*-methyl-D-xylose and 2-*O*-methyl-L-fucose. Portions of the neutral sugars occur as neutral polysaccharides, arabans, galactans, and arabinogalactans. Some of these polysaccharides may be removed from pectic substances by redissolving the pectin in water and precipitating the pectin with alcohol. Part of the neutral polysaccharides tend to remain in solution, but removal of all the nonuronide residues by repeated precipitations is usually impossible. Considerable evidence suggests that some of the nonuronide materials are covalently linked to the galacturonan chain. Aspinall *et al.* (1968A, 1968B) has shown that 2-linked rhamnosyl units are distributed in the chains of α-(1 → 4) linked galaturonosyl residues.

Techniques currently used in the structure elucidation of pectic substances are quite different from the techniques reviewed by Kertesz (1951). Access to these methods can be gained from recent publications, e.g., Aspinall *et al.* (1968A, 1968B), Aspinall (1970A, 1970B) and Albersheim *et al.* (1973).

The techniques used to clarify the structural knowledge of the

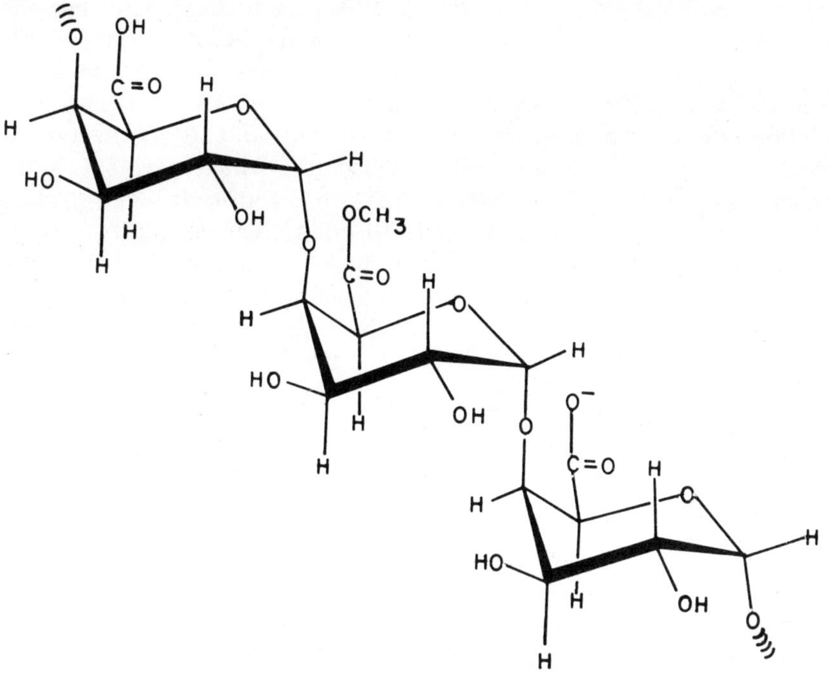

FIG. 10.1. REPRESENTATION OF THE GALACTURONAN SEGMENT OF A PECTIC SUBSTANCE IN THE C-1 CONFORMATION

pectic substances, since the review by Kertesz (1951), are based on advances in several areas. Chemical methods of carbohydrate chemistry (methylation analysis, hydrolysis methods, use of derivatives) have been refined. Physical methods for structure determination (infrared, nuclear magnetic resonance and mass spectroscopy, X-ray methods, methods for determining molecular size) have been developed that provide large amounts of structural information from small amounts of sample. Possibly the most important aid in developing a more firm knowledge of the structure of the heterogeneous pectic substances has been the advance in chromatographic techniques. The use of paper, thin-layer, ion-exchange, open-column and high-pressure-liquid chromatography together with vapor-phase chromatography and separations based on size or shape (as gel filtration with Sephadex) have allowed researchers to isolate polymers, oligomers and monomers in pure forms.

Recent work aimed at illuminating the biosynthesis of pectin has been summarized by Hassid (1970). Current evidence substantiates uridine 5′-(α-D-galactopyranosyluronic acid pyrophosphate), referred to as uridine diphosphate galacturonic acid or UDP-D-galacturonic acid, as the precursor of a nonesterified galacturonan chain (a pectic acid or polygalacturonic acid). UDP-D-galacturonic acid can be formed from UDP-D-glucose in a two-step sequence involving conversion to UDP-D-glucuronic acid (via dehydrogenase) followed by conversion to UDP-D-galacturonic acid (via epimerase). Alternately, a plant kinase can convert D-galacturonic acid and adenosine triphosphate to the uronic acid 1-phosphate which forms UDP-D-galacturonic acid by a pyrophosphorylase reaction utilizing uridine triphosphate. Apparently, the galacturonan chain is not formed if the nucleoside D-galacturonic acid diphosphate contains a hetrocylic base other than uracil. Neutral sugars, such as L-rhamnose, may be incorporated into the galacturonan chain through the UDP-nucleoside form, e.g., UDP-L-rhamnose (Barber 1962).

Formation of methyl ester groups seems to happen after polymerization of the galacturonan chain has begun. Enzymatic transfer of a methyl group from *S*-adenosyl-L-methionine to D-galacturonan has been demonstrated (Kauss *et al.* 1967).

COMMERCIAL PECTINS

Chemical Properties

The presence of free carboxyl groups causes pectin solutions to display an acidic pH. A 1% solution of a nonbuffered pectin (no neutralization of the carboxyl group) may give a pH ranging from 2.7–3.0. Before standardization most commercial pectins are buffered by partial neutralization so that the pH of a 1% solution will fall between pH 3–4.

When subjected to acidic conditions, the glycosidic bonds and the methyl ester linkages of pectin are prone to hydrolysis leading eventually to galacturonic acid. Hydrolysis of acid sensitive glycosidic links, e.g., arabans, usually lowers the nonuronide content of acid-treated pectins. Severe acid treatment leads to the decarboxylation of the carboxyl group and extensive degradation. During acid treatment at reduced temperature, the rate of glycosidic hydrolysis is much slower than the rate of deesterification; therefore, preparations of low ester pectins with little chain cleavage (Smit and Bryant 1968) is possible.

Treatment with alkali at elevated temperatures results in very rapid saponification and degradation; even at room temperature, dilute alkali rapidly removes methyl ester groups and causes degradation. Cleavage of the glycosidic linkage of pectin by alkali has been suggested to occur by a *trans*-β-elimination reaction (Neukom and Deuel 1958; Albersheim *et al.* 1960) as shown in Fig. 10.2. BeMiller and Kumari (1972) propose that the elimination reaction proceeds via an ElcB mechanism.

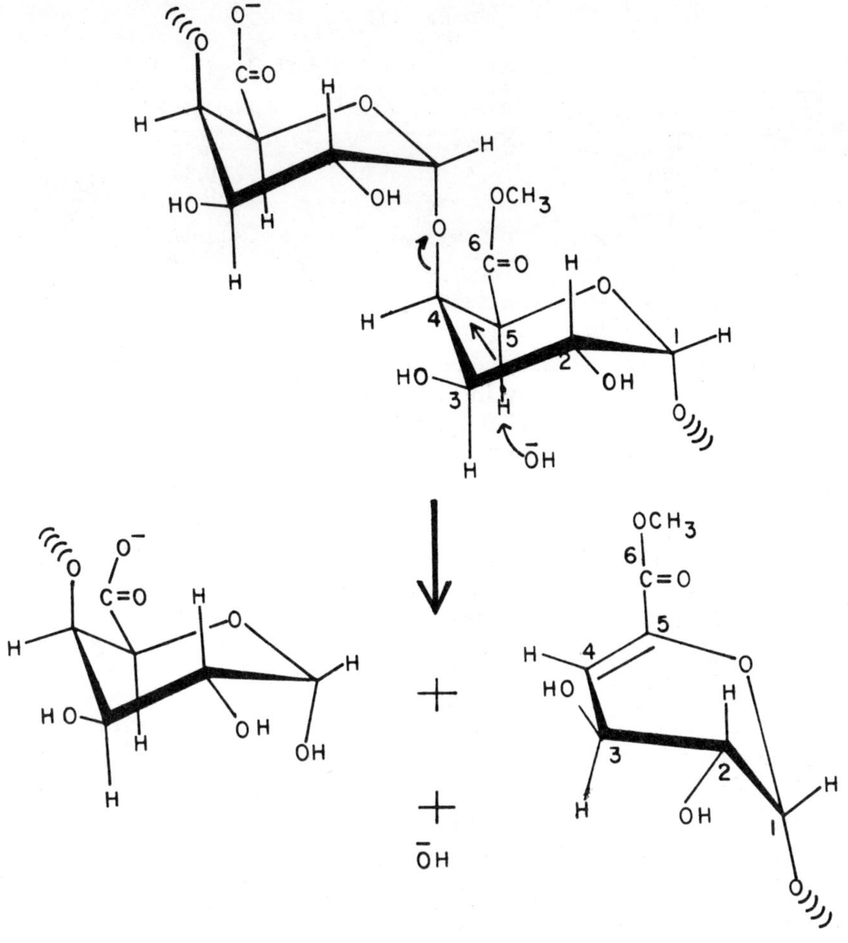

FIG. 10.2. CLEAVAGE OF THE GLYCOSIDIC LINKAGE OF PECTIN VIA β-ELIMINATION IN THE PRESENCE OF ALKALI

An ionized carboxyl group at carbon-6 is less able to stabilize anionic character at carbon-5 than is the methyl ester, hence, low-ester pectins are more resistant to degradation under alkaline conditions than are high-ester pectins. Pectin degradation via β-elimination can occur at a pH as low as 6.8 (Albersheim 1959) making pectin less desirable as a thickening agent in heat-processed foods of little acidity (pH approaching 7).

Pectic substances are degraded in the presence of oxidizing agents such as peroxide, periodate, permanganate, dichromate, free halogens and ascorbic acid.

Some of the secondary hydroxyl groups of the galacturonic acid units of pectin are esterified with acetic acid. McComb and Mc-

TABLE 10.2

ACETYL CONTENT OF SEVERAL PECTINS

Source of Pectin	Acetyl (%)
Raspberry	0.25
Citrus	0.23
Cherry	0.18
Apricot	1.36
Strawberry	1.47
Sugar beet	2.50

Source: McComb and McCready (1957).

Cready (1957) reported the acetyl levels given in Table 10.2. The presence of acetyl groups in pectin has a marked effect on gel-forming ability. Pippen *et al.* (1950) showed that jelly made with pectin containing about 5% acetyl did not set in several weeks. Samples containing 3.5–4.0% acetyl gave weak gels while jellying power was restored at levels around 2.4% acetyl. The acetyl level of commercial apple and citrus pectins is well under 1% (Schultz 1965) and has little influence on gel behavior. Unless processed to lower acetyl levels, sugar beet pectins may suffer decreased jellying ability.

Physical Properties

The solubility, viscosity and gel-forming ability of a pectin depend on the chemical characteristics of the pectin, e.g., degree of esterification, molecular weight, plus the presence and level of chemical entities that may be part of the pectin molecule. The properties of pectin in solution are also influenced by conditions inherent to the solvent, e.g., pH, and soluble materials such as cations. The physical properties of two pectins may be quite different even though the molecular weight, DE, and other factors are the same. Neither average molecular weight nor degree of esterification indicates the homogeneity of a pectin sample, i.e., those measurements give no indication of the distribution of functional groups or the distribution of polymer sizes. Fractionation studies have demonstrated that heterogeneity markedly influences the jellying ability of high-ester pectins (Smit and Bryant 1967). Enzyme deesterified low-ester pectins seem to have different gelling characteristics than acid- or alkali-demethylated pectins of similar low ester level. The textural distinctions may depend on differences in the distribution of ester and carboxyl groups. Deesterification via acid or alkali is a more random process than enzymatic deesterification.

Water is the best solvent for pectin; aqueous solutions containing up to 4% pectin (w/w) are commonly prepared. Ultimately, viscosity acts to limit the level of pectin that can be conveniently dissolved. Solubility in water decreases as carboxyl acid level increases (decreased DE). Decreased solubility can be overcome by converting the carboxyl groups to the soluble salt forms (salts of monovalent cations).

Upon contact with water, the outer portion of the pectin particles softens; the partially hydrated particles tend to stick to each other. Hence, special care must be used when powdered pectins are added to water or troublesome clumping will result. To avoid clumping; pectins are usually mixed with additional sugar, which acts as a spacer, and then dissolved during vigorous mechanical stirring. A soluble coating of the pectin particles may help to avoid problems in solubilization by preventing pectin contacts during dispersion. Similarly, wetting with alcohol or acetone before addition of water helps to avoid clumping by slowing the initial rate of pectin solubilization.

Pectin is generally insoluble in organic solvents; it can be precipitated from aqueous solutions by addition of water miscible solvents such as methanol, ethanol, isopropanol and acetone. Precipitation of pectic substances from solution by addition of polyvalent cations has been used for many years. Quaternary ammonium compounds (Scott 1965), water soluble basic polymers (Stutz and Deuel 1955) and proteins (Doesburg 1965) also can be used to precipitate pectins. Pectin is soluble to some degree in formamide, dimethylsulfoxide, dimethylformamide and hot glycerol, but this behavior has no present industrial application.

The viscosity of a pectin solution is dependent on the degree of polymerization of the pectin. The viscosity is also markedly influenced by the pH of the solution, the presence of electrolytes, the temperature, the pectin concentration and the shear rate (Neukom 1967). Viscosity measurements (Christensen 1954) are usually employed to determine the molecular weight of pectins, and molecular weights by this method may range from about 50,000 to around 300,000.

The interrelationships of pectin physical properties and gel formation are presented in a later section.

Manufacture

It is probably safe to estimate that the annual world pectin production is in the neighborhood of 10 million pounds of nonstandardized pectin. In the United States the major sources of commercial pectin are citrus peel and apple pomace, fresh citrus peel being the more important of the two. Dried peel or apple pomace is also used extensively in Europe or in areas where fresh citrus peel from processing

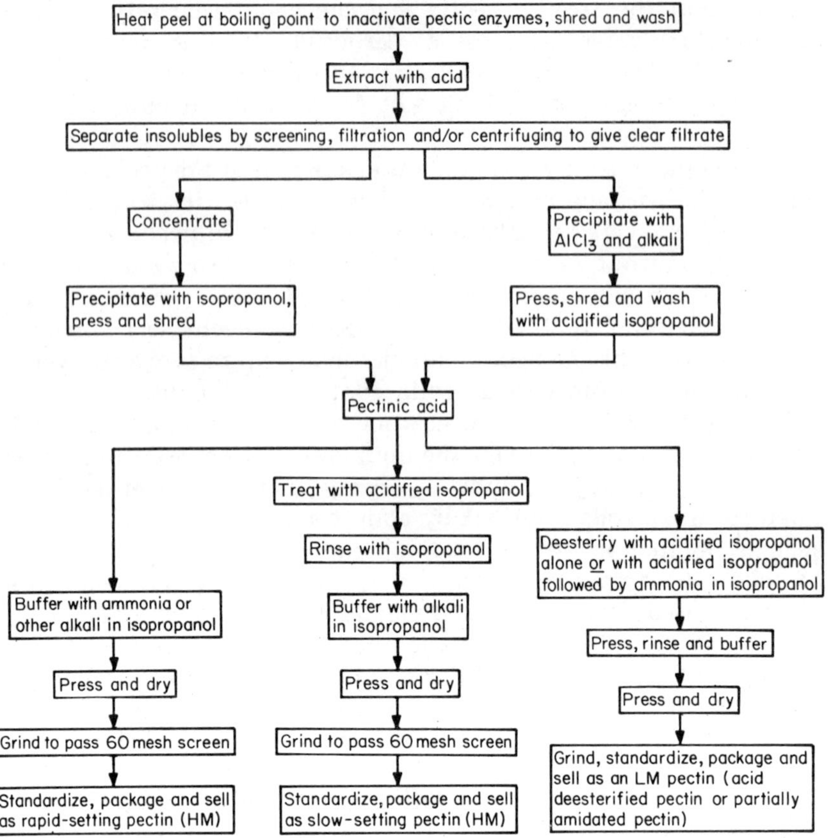

FIG. 10.3. MANUFACTURE OF PECTINS

factories is not readily available. Some manufacturers have a preference for certain types of citrus peel, but lemon, orange, lime and grapefruit peel have all been used for pectin production.

The flow sheet in Fig. 10.3 illustrates the main steps during the manufacture of HM and LM pectin from citrus peel. The process principally involves the extraction, filtration, precipitation, washing, drying and standardization of the pectin. The manufacturing conditions are chosen to give the maximum amount of standardized pectin at the lowest cost. When the starting material is readily available, no attempt is made to extract all the pectin or to prepare pectin with the maximum gel-forming capacity as this would add to production costs. Manufacturing conditions may differ among manufacturers and when pectins are being made for specific uses.

A heat treatment to inactivate enzymes, plus comminution and washing, usually precedes the extraction step. Pectin extraction is carried out with acid at elevated temperatures; a variety of dilute acids may be used such as sulfuric, nitric, sulfurous or hydrochloric acid. Mild agitation is used during extraction, and care is taken to avoid materials which may be subject to corrosion. Extraction is followed by pressing and/or filtration. In the case of pectin preparation from apple pomace, starch is removed before final filtration by treating the extract with a suitable amylase preparation. The clear filtrate may be concentrated, buffered, preserved and sold as a liquid pectin preparation. Normally, the pectin extract is precipitated and eventually sold as a powder. The pectin extract may be added to an alcohol (conventionally isopropanol) to give a fibrous precipitate, or the extract may be coprecipitated with $Al(OH)_x$ by adding an aluminum salt and alkali to give a curdy coagulum. After pressing, the aluminum is removed by washing with acidified isopropanol.

The pressed, acid-washed pectin is rinsed with an alcohol and sometimes buffered with a suitable alkali to neutralize residual mineral acid and help establish the solution pH of the final pectin product. The pressed and shredded pectin is finally dried, ground, evaluated for jelly grade, blended with sugar to give a standard jelly grade and packaged. Avoiding excessive exposure to high temperatures and acid conditions in the manufacturing process results in a rapid-set pectin. To prepare a slow-set pectin, the precipitated pectin is suspended in an acidified alcohol solution. The deesterification is controlled by temperature and time to give a slow-set pectin having a DE near 60%. In some manufacturing sequences, the slow-set pectin is generated *in situ* by "pickling" the pectin source material in acid prior to extraction.

Preparation of low-ester or LM pectin involves more extensive deesterification. Ester cleavage can be achieved chemically by acid or base catalyzed reaction and enzymatically using a pectin methylesterase. Low-ester pectins prepared using enzyme or base tend to form gels having undesirable textural qualities. Enzymatic deesterification gives LM pectins that produce gels of poor texture unless the calcium levels used in gel preparation are closely controlled. Alkaline deesterification is a rapid reaction, but depolymerization via β-elimination will give a pectin of low molecular weight (LM) that forms gels prone to syneresis. Careful control of added alkali and use of cold temperature can result in the preparation of useful low-ester pectins. Less need for control and less need for cooling together with good retention of molecular weight has made acid deesterification a preferred industrial method, even though the reaction is slow in terms of manufacturing time. Higher temperatures speed up the reaction, but increased temperature results in increased depolymerization of the pectin chains with an adverse effect on gel quality.

A considerable amount of LM pectin is presently manufactured by the procedure of Joseph *et al.* (1949) in which pectinic acid is treated with ammonia—isopropanol. This process gives a pectin in which 15-25% of the methyl ester groups are converted to primary amide groups. A recent patent (Wiles and Smit 1971) gives details for the preparation of such an amidated LM pectin which is particularly suitable for the preparation of superior low-solids gels.

Gel Formation with High-ester Pectin

Kertesz (1951) stated that the formation of gels from pectic materials is a complex phenomenon and that the exact mechanism of gel formation is not fully understood. If a gel can be described as an organized state of matter that is intermediate between liquid and solid, then a pectin jelly is an organized matrix of pectin and solution lying between a pectin solution and precipitation of the pectin. Beyond the level and gel forming ability of the dissolved pectin, two requirements must be met to form a high solids jelly. First, the soluble solids (sugar or polyol) level must be over 50%; best results demand levels over 60%. Bender (1949) considered sugar to function as a dehydrating agent. Other authors (Speiser and Eddy 1946; Hinton 1939) suggest sugar, pectin and water interact through hydrogen bonding. Rees (1970) does not accept hydrogen-bonded bridging involving sucrose as part of the gel mechanism; instead, he suggests the main function of sucrose to be control of water activity. A gel model has been proposed by Rees (1972), *vida infra.* The second requirement, acid level, must be adjusted to approach pH 3 in order to provide a significant level of un-ionized carboxyl groups. When both conditions are met and the dissolved level of commercial pectin is near 0.5%, gelation will occur. In most high ester pectin uses, gel formation is initiated by lowering the pH of the hot solution which contains the other ingredients.

Pectins with a degree of esterification of 70% or higher will form satisfactory gels with sugar at pH levels of 3.0-3.4 and relatively high temperatures (near boiling). At ester levels of 50-70%, pectins will form good gels with sugar at a pH of 2.8-3.2 and somewhat lower temperatures. Pectins with a DE over 85% can form gels with sugar and without addition of acid.

Smit and Bryant (1968) found that setting time increased as the percentage of esterification decreased until an ester content near 50% was reached. Below 50% DE setting time decreased again very sharply. Pectins with an ester content of 65.9-73.7% showed little difference in gel forming ability at pH values under 3, but lost jellying ability rapidly above a pH of 3.0. As the ester content of samples decreased further, the gel forming ability at low pH values increased while the pH at which maximum gel strength was obtained decreased. The best gel forming ability was obtained at pH 2.2 with a pectin having an ester content of 43.7%.

Solms (1960) suggested that pectin macromolecules, because of their structural regularity, are able to form zones of attachment along the molecules. In a cross-linking reaction, the size of the zones of attachment is not controlled by the constitution of the molecules but by the reaction conditions. If association becomes complete, one no longer has a stable gel, but a complete coagulum.

Gel Formation with Low-ester Pectins

Low-ester pectins have the ability to form gels upon the addition of multivalent cations, usually calcium. Various viewpoints concerning the mechanism of gel formation with LM pectin and the resulting gel structure have been expressed. Harvey (1960) stated that the formation of gels from solutions of long chain polymers such as pectin arises from the cross-linking of neighboring macromolecules forming a continuous network of high mechanical stability. He further stated that the type of cross-link depends on the chemical characteristics of groups suitably spaced along the chains as well as on their environment. The cross-links could involve hydrogen bonding, electrovalent attraction between ionized carboxyls and a bridging cation, and possibly, covalent bonding.

Speiser *et al.* (1946) described the gelation of a polymer solution as being the transformation of a liquid solution incapable of withstanding shear stress into a solid "solution" which is rigid and elastic. This transformation occurs when the polymer molecules possess the ability to form an extended three dimensional network which is rigid enough to resist shear yet is capable of incorporating mobile solvent molecules into the structure. These workers attributed the brittle nature of calcium gels to the strength of a relatively short bond between calcium ions and carboxyls as compared to longer hydrogen bonds in high-ester gels. However, brittleness is more characteristic of gels formed with pectins of very low-ester content. If cross-linking involves more than one pectin chain, the pectin molecules must approach each other at these bonding points.

According to Kertesz (1951), the strength of an LM pectin gel depends *both* on the extent of esterification and the distribution of the ester groups on the pectinic acid chain. For desirable gel properties, low-ester pectins should have ester levels between 30–50% DE (Speiser and Eddy 1946; Kertesz 1951).

Rees (1970) discounted calcium bridging between carboxyls or chelation of calcium as being principally responsible for low-methoxyl gel formation by stating that these mechanisms could not contribute sufficiently to gel stability. Rees suggests (1972) that high-sugar gels (HM pectin) and low-sugar gels (LM pectin) are formed by a mechanism that involves accumulation of three-dimensional stacked zones. For gels made with HM pectin, the segments of the pectin chain that associate to give a chain-stacked zone involve largely uronic ester residues plus occasional uronic acid residues. Gels made with LM pectin involve chain-stacking of zones dominated by uronic acid anions and Ca^{2+} with lesser levels of un-ionized uronic acid and uronic amide.

Differences observed in the gel strengths of acid and enzyme demethylated pectins are thought to be due to differences in the distribution of the methyl ester groups in the two types of pectins. A more even distribution of these groups in the low-ester pectins is thought to produce stronger gels (Speiser and Eddy 1946; Hills *et al.* 1949). Heri (1962) demonstrated that patterns of ester distribution of hydrochloric acid and sodium hydroxide deesterified pectins were similar, but different from those of enzyme demethylated pectins. Separation of pectins prepared with acid and alkali produced two major pectin fractions with degrees of esterification slightly higher and slightly lower than that of the original material applied to the column. Pectins deesterified by enzyme were separated into several fractions having degrees of esterification between 92–94% with only 30% of the separated pectin possessing a degree of esterification equivalent to that of the original pectin.

A third characteristic which plays a role in determining strength of a calcium pectinate gel is molecular weight (Hills *et al.* 1949). While some authors have felt that the strength of such gels are less influenced by molecular weight than are high-solids jellies (Speiser and Eddy 1946), recent work has pointed out the importance of molecular weight in assuring good gel texture in low-solids gels (Wiles and Smit 1971).

LM pectins are quite suitable for use in gelled products within a pH range of 2.5–6.5. They may be used in fruit gels at low pH values or in gelled-milk products at higher pH values.

Calcium is the cation usually employed in low-methoxyl pectin gels. As the ester level of a given pectin is lowered, the amount of calcium required to produce a gel of given firmness decreases (Owens *et al.* 1949). The calcium requirement for a given gel varies with the method of deesterification used to produce the gel-forming pectin. Doesburg (1965) reported optimum calcium concentrations, expressed as milligrams of calcium per gram of pectin, of 4–10 for enzymatically prepared pectins, 15–30 for pectins produced by the ammonia process and 30–60 for acid deesterified pectins. The way in which the calcium is added to the gel is important. Joseph *et al.*

(1949) stated that the ideal way to obtain a gel would be to dissolve the pectin in the absence of calcium to enhance solubility and then to add a solution containing calcium. If the pectin and calcium are added simultaneously, they recommended the use of slowly soluble calcium salts such as calcium phosphate, monobasic.

Gel Characterization of Commercial Pectins

The pectins of commerce, especially those used in jam and jelly manufacture, are nearly always standardized with respect to gel-forming ability (jelly grade) by the addition of dextrose or other sugars and sometimes buffer salts. The whole question of industry-wide standardization has been difficult. The difficulties hinged on the effect of end use conditions (pH, soluble solids level, presence of cations, time and temperature during and after gel preparation) on pectin performance. Pectin suppliers and manufacturers of jams and jellies disagreed on the conditions under which the test jelly would be prepared, on the physical characteristics that would define a standard test jelly and on the method of measuring these characteristics. The most significant approach to the problem was reported by a special committee of the Institute of Food Technologists (1959). While there is still some disagreement on the basis for standardizing high-ester pectins, the IFT method is probably the most widely used basis for defining the "jelly grade" of pectin. The method involves preparation of a pectin-sugar-acid-water jelly of 65% soluble solids at a pH of 2.2–2.4 which, after a storage period of 18–24 hr, is examined with a Ridgelimeter. The jellies, having been formed in glasses of fixed dimensions, are turned out on glass plates and the amoung of "sag" deformation is measured and related through factor tables to jelly grade. Detailed instructions for jelly preparation and evaluation are given in the IFT report (1959).

Low-methoxyl pectin gels made with calcium salts can be evaluated by similar sag measurements but very little agreement on the standardization is found among the different suppliers and users of the low-methoxyl pectins. A method for sag evaluation of LM pectin is found in a recent Food Chemicals Codex (Natl. Acad. Sci. 1972). In this procedure concentrations of citric acid, sodium citrate, calcium chloride and tartaric acid are incorrectly given. In all instances, grams of reagent per 1000 ml of water should rather read grams of reagent per 1000 ml of solution.

Analysis of Commercial Pectins

The chemical and physical properties of pectin most commonly measured are those which correlate with pectin behavior under end-use conditions. Current monographs (Natl. Acad. Sci. 1972; American Pharmaceutical Assoc. 1970) list identity tests and provide a description of some chemical methods such as determination of the levels of anhydrogalacturonic acid and ester. Detailed descriptions of additional chemical and physical determinations can be found in recent reviews, e.g., McCready (1970).

Chemical characterization of a pectin sample usually requires removal of standardizing sugars and buffer salts. Removal of standardizing substances may be accomplished by careful washing of the sample in a mixture of 60% (v/v) isopropanol, 35% water and 5% concentrated hydrochloric acid. After a series of rinsings to remove excess HCl (test filtrate via $HNO_3/AgNO_3$), the sample is dried prior to analysis. Knowledge of sample weight before and after washing together with data on percentage of volatiles allows estimation of the level of substances added for purposes of standardization.

Determination of the level of carboxyl groups can be done by simple titration. Conditions that have proven to be convenient and workable involve use of 0.5 gm of washed and dried sample, solubilization in 100 ml of carbon dioxide-free water, and titration using 0.1 *N* NaOH. Phenol red, phenolphthalein and Hinton's reagent —0.4% bromothymol blue (1 part), 0.4% phenol red (3 parts), 0.4% cresol red (1 part) and distilled water (1 part)—have been used as indicators of equivalence. After equivalence has been reached, an excess of alkali (20 ml of 0.5 *N* NaOH) is added; the solution is allowed to sit at ambient temperature for 20-30 min for saponification to proceed to completion; an equivalent amount of acid (20 ml of 0.5 *N* HCl) is added and titration of the carboxylic acid groups formed on ester cleavage is continued with 0.1 *N* NaOH and additional indicator.

Amide level is determined by cleaving the primary amide to yield ammonia, sweeping the ammonia into an HCl trap and back titrating the remaining HCl. The ammonia can be swept into the HCl solution by reflux alone or by reflux in combination with a nitrogen sweep. Conditions used may involve 0.5-1.0 gm of washed and dried sample, an excess of >2 *N* NaOH, reflux in a round bottom flask fitted with either a nitrogen inlet and a Kjeldahl trap linked to a gas dispersion tube or fitted with a Kjeldahl trap linked to a condensor and a gas dispersion tube. The gas dispersion tube is immersed in a cylinder containing 20-30 ml of 0.1 *N* HCl. After collection of $\approx$100 ml of distillate using the reflux method or $\approx$30 ml of distillate using a N_2 sweep, the HCl is titrated with 0.1 *N* NaOH using methyl red as indicator.

The analytical data is in the form of milliequivalents (meq) of anhydrogalacturonic acid, ester or amide per unit of weight used in each analysis. When the milliequivalents are converted to a common weight basis, such as per gram, the meq level for each group may be divided by the meq total to give the % distribution of each functional group at carbon-6 of the anhydrogalacturonan residues. The weight portion of the pectin sample, which is accounted for by titration, can be found by totaling the milligrams of anhydrogalacturonic acid (1 meq = 176 mg), anhydrogalacturonic acid methyl ester (1 meq = 190 mg) and anhydrogalacturonic acid amide (1 meq = 175 mg). Alternately, the meq total has been conventionally used to determine the monomeric galacturonic acid that could be formed by hydrolysis of the pectin sample. For this determination, each meq of the meq total is considered to equal 194 mg. A second convention, long associated with pectin analysis, involves evaluation of the ester level in terms of percentage of methoxyl. To calculate percentage of methoxyl, each meq of ester is considered in terms of mg of methoxyl (OCH_3 = 31 mg/meq); the percentage is then obtained by dividing the milligrams of OCH_3 by the sample weight. For comparing similar pectin samples, percentage of methoxyl has been a useful tool.

The determination of acetyl groups is most associated with sugar beet pectins since the level of acetyl groups present in apple or citrus pectin is low and has little influence on gelation. To avoid degradation of the pectin, which leads to formation of volatile acids, a mild alkaline saponification (pH 12, 25°C, 1 hr) of the acetate esters linked to carbon-2 or carbon-3 of the anhydrogalacturonic acid residues is preferred (Schultz 1965; McCready 1970). After saponification the solution is acidified, the acetic acid is steam distilled and the distillate is titrated (phenol red) with dilute NaOH.

Molecular weight determinations are most often based on measurement (Ostwald-Cannon-Fenske pipette) of the relative viscosity using a dilute (approximately 0.1%) solution of the purified pectin in a 1% sodium hexametaphosphate solution. Christensen (1954) and McCready (1970) list slightly different methodologies. Relative viscosities are then converted to intrinsic viscosity which is related to molecular weight in the following way (Bettelheim, 1970).

$$\text{intrinsic viscosity} = [\eta] = KM_v^{\,a}$$

In this equation K is a constant characteristic of the polymer, and M_v is viscosity-average molecular weight. The factor a is related to the shape of the polymer. Christensen (1954) uses $K = 4.7 \times 10^{-5}$ and a as one (assumes polymer is an elongated rod).

TABLE 10.3

FOOD APPLICATIONS OF PECTINS

Type	Typical uses	Approximate amount (%) in final product	Important Factors
Regular HM pectin (150 grade)	Fruit jellies, jams, preserves, also baker's jellies	0.1–0.8	Soluble solids for jellies 65% jams, 60%; pH 2.8–3.2
	Confectionery jelly pieces	0.85–1.25	Soluble solids 80–82%; buffer salt needed; pH (50–50 mixture jelly, water) 3.4–3.7; equal weights glucose and sucrose
	Home jelly making mixtures	Usually 3 oz for 6–8 glasses	Mixtures of dextrose, pectin, fruit acid, about 10 grade
	Thickeners for low-calorie fruit syrups and beverages	Under 0.5	Care needed in preparing solution when sugar is not included as spacing agent
	Flavor emulsions, salad dressings	2–3 of water phase	Best for oil contents 15–20% and higher
	Cream whipping aids, baker's glazes, malted milk thickeners	—	These and other specialty uses require specific instructions
Low-methoxyl pectin (LM)	Salad and dressing gels (a) Immitation flavor and color (for home consumption)	0.8–1.5	$Ca(H_2PO_4)_2 \cdot H_2O$, 8–16% weight of pectin; sometimes sodium citrate and fruit acid are added
	(b) Fruit and vegetable juices (canned gels)	1.0–1.8	$CaCl_2 \cdot 2H_2O$, 6–14% weight of pectin; pH 3.6–3.8; pectin in solution before adding calcium salt
	Milk gels and puddings	0.8–1.5	No calcium salt needed
	Low-calorie, jam-like fruit gels for dietetic use	0.8–1.5	Need proper balance of pectin and calcium; small amounts sorbitol and glycerol desirable
	Frozen strawberries	0.1–0.15 of berry weight	Most effective when in sugar syrup added to packed berries
	Fruit and berry gels for use in ice cream	0.8–1.5	40–50% sugar; fruit acid, sometimes calcium salt.

Source: Joseph (1953).

Uses of Pectin

High-ester pectins are principally used in the manufacture of jams, jellies, marmalades and jellied candies of high quality. Neukom (1967) estimated that some 80-90% of the pectin produced commercially is used for these purposes. A small amount of pectin can form a gel under the conditions found in jams and jellies—high sugar and low pH—much better than other related polysaccharides and gums. Therefore, pectin is not likely to be replaced in these products by another gelling agent in the near future. Joseph (1953) summarized the uses of high- and low-ester pectins in foods (Table 10.3). Additional uses for pectins have been suggested from time to time, but widespread application of pectin has been restricted by competition with less expensive, water soluble gums.

In the nonfood area, pharmaceutical applications are important. Pectin is used to treat diarrhea; according to Neukom (1967), it has been used for the treatment of wounds, as a hemostatic agent, as a blood-plasma substitute, as a detoxicant of poisonous heavy metals and for the formation of complexes with insulin, penicillin, etc. to prolong their action.

BIBLIOGRAPHY

ALBERSHEIM, P. 1959. Instability of pectin in neutral solutions. Biochem. Biophys. Res. Commun. *1*, 253-256.

ALBERSHEIM, P. 1975. The walls of growing plants. Sci. Am. *232*, 81-95.

ALBERSHEIM, P., NEUKOM, H., and DEUEL, H. 1960. Splitting of pectin chain molecules in neutral solutions. Arch. Biochem. *90*, 46-51.

ALBERSHEIM, P., TALMADGE, K. W., KEEGSTRA, K., and BAUER, W. D. 1973. The structure of plant cell walls. I. The macromolecular components of the walls of suspension-cultured sycamore cells with a detailed analysis of the pectic polysaccharides. Plant Physiol. *51*, 158-173.

AMERICAN CHEMICAL SOCIETY. 1944. Report of the committee for the revision of the nomenclature of pectic substances. Chem. Eng. News *22*, 105-106.

AM. PHARMACEUTICAL ASSOC. 1970. The National Formulary, 13th Edition. American Pharmaceutical Assoc., Washington, D.C.

ASPINALL, G. O. 1970A. Polysaccharides, Pergamon Press, New York.

ASPINALL, G. O. 1970B. Pectins, plant gums, and other plant polysaccharides. *In* The Carbohydrates, Chemistry and Biochemistry, Vol. IIB, W. Pigman, and D. Horton (Editors). Academic Press, New York.

ASPINALL, G. O., CRAIG, J. W. T., and WHYTE, J. L. 1968. Lemon-peel pectin. Part I. Fractionation and partial hydrolysis of water-soluble pectin. Carbohyd. Res. *7*, 442-452.

ASPINALL, G. O., GESTETNER, B., MOLLOY, J. A., and UDDIN, M. 1968. Pectic substances from lucerne, *Medicago Sativa*. Part II. Acidic oligosaccharides from partial hydrolysis of leaf and stem pectic acids. J. Chem. Soc. (C), 2554-2559.

BARBER, G. A. 1962. The enzymic synthesis of uridine diphosphate-L-rhamnose. Biochem. Biophys. Res. Commun. *8*, 204.

BeMILLER, J. N., and KUMARI, G. V. 1972. *beta*-Elimination in uronic acids: evidence for an EIcB mechanism. Carbohyd. Res. *25*, 419-428.

BENDER, W. A. 1949. Grading pectin in sugar jellies. Anal. Chem. *21*, 408-411.

BETTELHEIM, A. 1970. Introduction to polysaccharide chemistry. *In* The Carbohydrates Chemistry and Biochemistry, Vol. IIA, W. Pigman, and D. Horton (Editors). Academic Press, New York.

CHRISTENSEN, P. E. 1954. Methods of grading pectin in relation to the molecular weight (intrinsic viscosity) of pectin. Food Res. *19*, 163-172.

DOESBURG, J. J. 1965. Pectic substances in fresh and preserved fruits and vegetables. Inst. for Research on Storage and Processing of Horticultural Produce, I.B.V.T. Commun. *25*, Wageningen, The Netherlands.

DOESBURG, J. J. 1973. The pectin substances. *In* Phytochemistry, Vol. 1, L. P. Miller (Editor). Van Nostrand Reinhold Co., New York.

GLICKSMAN, M. Gum Technology in the Food Industry, Academic Press, New York.

HARVEY, H. G. 1960. Gels—with special reference to pectin gels. Soc. Chem, Ind. (London) Monograph 7, 29-63.

HASSID, W. Z. 1970. Biosynthesis of sugars and polysaccharides. *In* The Carbohydrates Chemistry and Biochemistry, Vol. IIA, W. Pigman, and D. Horton (Editors). Academic Press, New York.

HERI, W. 1962. Chromatographic fractionation of pectin by diethylaminoethylcellulose. Dissertation *3172*, Eidgenössischen Technischen Hochschule, Zurich.

HILLS, C. H., MOTTERN, H. H., NUTTING, G. C., and SPEISER, R. 1949. Enzyme demethylated pectinates and their gelation. Food Technol. *3*, 90-94.

HINTON, C. L. 1939. Fruit pectins: their chemical behavior and gelling properties. Dept. Sci. Ind. Res. Brit. Food Invest. Spec. Rept. *48*.

INSTITUTE OF FOOD TECHNOLOGISTS. 1959. Pectin standardization: Final report of the IFT committee. Food Technol. *13*, 496-500.

JOSEPH, G. H. 1953. Better pectins. Food Eng. *25*, No. 6, 71-73, 114.

JOSEPH, G. H., KIESER, A. H., and BRYANT, E. F. 1949. High-polymer, ammonia-demethylated pectinates and their gelation. Food Technol. *3*, 85-90.

JOSLYN, M. A. 1962. The chemistry of protopectin, a critical review of historical data and recent developments. *In* Advances in Food Research, Vol. 11, C. O. Chichester, E. M. Mrak, and G. F. Stewart (Editors). Academic Press, New York.

KAUSS, H., SWANSON, A. L. and HASSID, W. Z. 1967. Biosynthesis of the methyl ester groups of pectin by transmethylation from *S*-adenosyl-L-methionine. Biochem. Biophys. Res. Commun. *26*, 234-240.

KERTESZ, Z. I. 1951. The Pectic Substances. Interscience Publishers, New York.

McCOMB, E. A., and McCREADY, R. M. 1957. Determination of acetyl in pectin and in acetylated carbohydrate polymers. Anal. Chem. *29*, 819-821.

McCREADY, R. M. 1970. Pectin. *In* Methods in Food Analysis, M. A. Joslyn (Editor). Academic Press, New York.

NATL. ACAD. SCI. 1972. Food Chemicals Codex, 2nd Edition. National Academy of Sciences, Washington, D.C.

NEUKOM, H. 1967. Pectic substances. *In* Encyclopedia of Chemical Technology, 2nd Edition, Vol. 14. John Wiley & Sons, New York.

NEUKOM, H., and DEUEL, H. 1958. Alkaline degradation of pectin. Chem. Ind. (London) *1958*, 683.

OWENS, H. S., McCREADY, R. M., and MacLAY, W. D. 1949. Gelation characteristics of acid-precipitated pectinates. Food Technol. *3*, 77-82.

PILNIK, W., and VORAGEN, A. G. J. 1970. Pectic substances and other uronides. *In* The Biochemistry of Fruits and Their Products, A. C. Hulme (Editor). Academic Press, New York.

PILNIK, W., and ZWIKER, P. 1970. Pektine. Gordian *70*, 202-204, 252-257, 302-305, 343-346.

PIPPEN, E. L., McCREADY, R. M., and OWENS, H. S. 1950. Gelation properties of partially acetylated pectins. J. Am. Chem. Soc. *72*, 813-816.

REES, D. A. 1970. Structure, conformation and mechanism in the formation of polysaccharide gels and networks. *In* Advances in Carbohydrate Chemistry and Biochemistry, Vol. 24, M. L. Wolfrom, R. S. Tipson, and D. Horton (Editors). Academic Press, New York.

REES, D. A. 1972. Polysaccharide gels: A molecular view. Chem. Ind. (London) *19*, 630-636.

SCHULTZ, T. H. 1965. Determination of acetyl in pectin. *In* Methods in Carbohydrate Chemistry, Vol. V, R. L. Whistler, J. N. BeMiller, and M. L. Wolfram (Editors). Academic Press, New York.

SCOTT, J. E. 1965. Fractionation by precipitation with quaternary ammonium salts. *In* Methods in Carbohydrate Chemistry, Vol. V, R. L. Whistler, J. N. BeMiller and M. L. Wolfram (Editors). Academic Press, New York.

SMIT, C. J. B., and BRYANT, E. F. 1967. Properties of pectin fractions separated on diethylaminoethyl-cellulose columns. J. Food Sci. *32*, 197-199.

SMIT, C. J. B., and BRYANT, E. F. 1968. Ester content and jelly pH influences on the grade of pectins. J. Food Sci. *33*, 262-264.

SOLMS, J. 1960. Some structured aspects of gel formation of pectins and related polysaccharides. *In* Physical Functions of Hydrocolloids. Advances in Chemistry Series, No. 25, American Chemical Society, Washington, D.C.

SOMERS, P. J. 1970. Plant polysaccharides. *In* Carbohydrate Chemistry, Vol. III, R. D. Guthrie, R. J. Ferrier, T. D. Inch, and P. J. Somers (Editors). Burlington House, London.

SPEISER, R., COPELY, M. J., and NUTTING, G. C. 1946. Effect of molecular association and charge distribution on the gelation of pectin. J. Phys. Colloid Chem. *51*, 117-133.

SPEISER, R., and EDDY, C. R. 1946. Effect of molecular weight and method of deesterification on the gelling behavior of pectin. J. Am. Chem. Soc. *68*, 287-293.

STUTZ, E., and DEUEL, H. 1955. Polyampholytes of various charge distribution. Helv. Chim. Acta. *38*, 1757-1763.

WILES, R. R., and SMIT, C. J. B. (Assignee Sunkist Growers, Inc.). 1971. Method for producing pectins having high resistance to breakage and high capability for gelling in the presence of calcium. U.S. Pat. 3,622,559.

WORTH, H. G. J. 1967. The chemistry and biochemistry of pectic substances. Chem. Rev. *67*, 465-473.

CHAPTER 11

Ian W. Cottrell
Peter Kovacs

Algin

INTRODUCTION

Algin was first discovered by the British chemist E. C. C. Stanford as a component of seaweed about 1880 (Stanford 1881, 1883, 1884). Commercial utilization of algin in the food industry did not occur on an industrial scale until 1934 when the Kelco Company in California introduced milk soluble algin as an ice cream stabilizer (Lucas 1937A, 1937B; Lucas and Green 1937; Goodman 1935). Since that time the uses of algin within the food industry have increased rapidly, and with the development of propylene glycol alginate in 1944, algin and its derivative have become two of the most important hydrocolloids used in the food industry (Steiner 1947; Steiner and McNeely 1949, 1950A, 1950B, 1951; Kelco Co. 1952).

Algin occurs in all brown seaweeds, Phaeophyceae, as a structural component of the cell walls analogous to cellulose and pectin in the cell walls of terrestrial plants. In the algal cell wall algin occurs as the insoluble mixed calcium, magnesium, sodium, and potassium salt of alginic acid. Alginic acid is a high molecular-weight linear glycuronan consisting solely of D-mannuronic acid and L-guluronic acid units.

Considerable research has been carried out in recent years to determine the fine structure of alginic acid isolated from different seaweeds using both chemical and physical techniques. This detailed structural information can be used to explain the differing behavior of algin from various seaweeds in diverse food applications. The many uses of algin in the food industry result from the unique colloidal behavior exemplified by the thickening, stabilizing, emulsifying, suspending, film-forming, and gel-producing properties. Several comprehensive reviews describing the chemistry and properties of algin have been published in recent years (Steiner and McNeely 1954; Smith and Montgomery 1959; Maass 1959; Haug 1964; Okazaki 1971; Percival and McDowell 1967; McNeely and Pettitt 1973).

PRODUCTION OF ALGIN

Sources

Algin occurs in all of the species of the brown seaweeds, Phaeophyceae, that have been examined, but recently similar types of polysaccharides have been isolated from bacteria (Linker and Jones 1966; Gorin and Spencer 1966; Imrie 1973). However, most of the alginate produced in the United States is isolated from the giant kelp, *Macrocystis pyrifera*, which grows in the off-shore waters of the West Coast of the North American continent ranging from Baja California, Mexico, to Monterey, California (McNeely and Pettitt 1973). The kelp beds range in size from 50 ft to 1 mile in width, several miles in length, and 25-80 ft in depth. The kelp is anchored to the rocky ocean floor by means of a holdfast, a root-like structure, which can be several feet in diameter. From the holdfast grow the stipes, which are analogous to the stems of land plants and which bear the fronds or leaves. Each frond carries a small float bulb which allows the kelp to float at the surface of the ocean, and it is this feature which makes *Macrocystis pyrifera* amenable to mechanical harvesting, Fig. 11.1. The kelp is cut several feet below the ocean surface by mechanical cutters and conveyed into the hold of the boat for transport to the

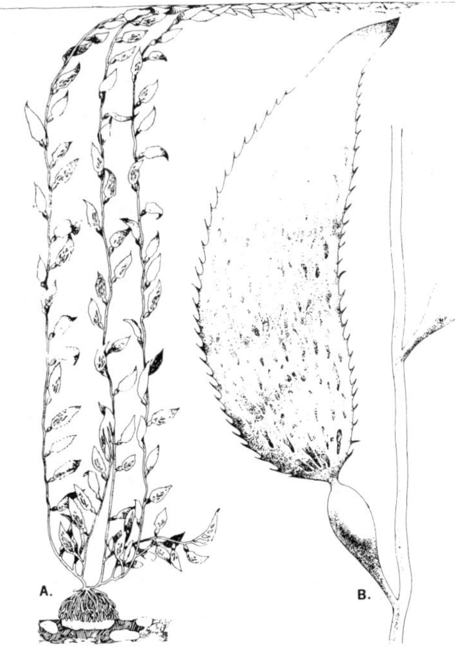

FIG. 11.1. MACROCYSTIS PYRIFERA

A—1/64 natural size. B—¼ natural size. The giant kelp is shown in the left part of the plate in a natural pose with the long leafy stipes rising to the sea surface from the massive holdfast. On the right is one of the leaf-like fronds showing the gas-filled float-bladder at its base and the distinctive teeth along the margin.

production facility. This cutting is beneficial to the growth of the kelp in that it permits greater penetration of sunlight to the younger plants below and prevents the canopy from rotting (North and Hubbs 1968). *Macrocystis pyrifera* can grow at a rate of 12–24 in. per day, thereby promoting sufficient growth for harvesting to be carried out several times per year.

Other commercially important brown algae are *Laminaria digitata*, *Laminaria hyperborea*, *Ascophyllum nodosum*, *Ecklonia maxima*, *Ecklonia cava*, and *Eisenia bicyclis.* The alginic acid content of some of these commercially important brown algae is shown in Table 11.1.

The major algin-producing countries are the United States, Great Britain, Norway, France, Japan, and Canada. It has been estimated that the annual production of algin is in excess of 10,000 tons but exact figures have not been published by the industry (Chapman 1970; McDowell 1969).

Processing

The basic principle of algin processing is the conversion of the naturally occurring insoluble mixed calcium, magnesium, sodium, and potassium salt of alginic acid into a soluble salt such as sodium alginate. This can be accomplished by either of two basic procedures. One procedure is to convert the insoluble salt into alginic acid using a dilute acid treatment followed by treatment with an alkali such as soda ash to produce a soluble alginate such as sodium alginate. The other procedure is to convert the insoluble salt directly into a soluble salt of alginic acid, again using an alkali such as soda ash. Recovery of the sodium alginate from the dilute solution is carried out by several procedures.

TABLE 11.1

ALGINIC ACID CONTENT OF THE COMMERCIALLY IMPORTANT BROWN ALGAE

	Alginic Acid Content (%) on a Dry Weight Basis	
Algae Species	A	B
Macrocystis pyrifera	—	13–24
Ascophyllum nodosum	—	20–30
Laminaria digitata	—	15–40
fronds	14.5–26.5	—
stipes	27–33	—
Laminaria hyperborea	—	14–24
fronds	8.5–19	—
stipes	18.5–23.5	—
Ecklonia maxima	—	29.6–38

Source: A—Data of Black (1950). B—Data of Hoppe and Schmid (1962).

Many improvements have been made on Stanford's original process and many patents and publications have issued, but the basic principle of an aqueous alkaline extraction has not changed (Bashford *et al.* 1950; Black and Woodward 1954; Heydt 1959). Although the commercial methods for producing algin have not been disclosed, it is quite likely that the processes are related to the patents of Green (1936) and of LeGloahec and Herter (1938).

Green's process consists of leaching the kelp (e.g., *Macrocystis pyrifera*) in dilute acid solution to reduce the salt content. The leached kelp is digested with soda ash at pH 10 for times ranging from 1-24 hr. Both leaching and digestion are carried out in the cold at 10° C. The digested kelp is diluted with several volumes of water, and the insoluble residue consisting mainly of cellulose is removed by standard solid-liquid separation techniques such as filtration. The clarified liquor is treated with calcium chloride to convert the soluble sodium alginate into insoluble calcium alginate. Bleaching of the calcium alginate with sodium hypochlorite can be carried out at this stage in the process. The bleached calcium alginate is converted to alginic acid by an acid treatment. The alginic acid is partially de-watered and converted to the required salt by treatment with the appropriate carbonate or hydroxide. The resulting paste is dried, milled, screened, and packed.

LeGloahec and Herter's process consists of leaching the kelp (e.g., *Laminaria digitata*) with aqueous calcium chloride to remove laminaran and mannitol followed by several water washes to remove calcium salts. Digestion of the finely milled kelp is carried out with soda ash at 40° C until a homogeneous paste is produced. The resultant paste is diluted and vigorously aerated, causing the fine, insoluble particles of cellular debris to float to the surface. After several hours the floating mat is sufficiently firm for the clear alginate liquor to be removed. The liquor is bleached with hypochlorite before the alginic acid is isolated by precipitation with mineral acid. The precipitation is carried out by intimately mixing the liquor stream with the mineral acid stream. Because of the vigorous mixing conditions, air is entrained in the alginic acid, which floats to the surface, where it is skimmed off and de-watered. The alginic acid is neutralized with the appropriate base, dried, milled, screened, and packed.

The only organic derivative of algin to have found widespread use in the food industry is propylene glycol alginate, which is prepared by the reaction of partially neutralized fibrous alginic acid with propylene oxide under slight pressure (Steiner 1947; Steiner and McNeely 1949, 1950A, 1950B, and 1951; Kelco Co. 1952). Many other derivatives including the acetate (Schweiger 1962), sulfate (Schweiger

1967), amide (Kohler and Dierichs 1959), and carboxymethyl ether (McNeely and O'Connell 1959) of alginic acid have been prepared, but none has been commercialized.

Alginic acid can be neutralized with cationic compounds such as amines (McNeely 1954; Boyle 1960) and quaternary ammonium compounds (McNeely 1961; National Lead Co. 1961) to produce amine salts of alginic acid and quaternary ammonium alginates. However, none of these alginates has found use in the food industry.

STRUCTURE OF ALGIN

Historical

As mentioned earlier, algin was first isolated by Stanford in 1881 from brown algae. He established that this material, which he termed "algic acid," was a weak organic acid which would form salts with bases. Stanford also believed that the "algic acid" contained nitrogen, but it was shown by Krefting (1896) that purified alginic acid does not contain nitrogen.

It was shown by Atsuki and Tomoda (1926) and independently by Schmidt and Vocke (1926) that uronic acid is a component of alginic acid. The identity of the uronic acid was determined by Nelson and Cretcher (1929), Miwa (1930), and Bird and Haas (1931) and shown to be D-mannuronic acid. However, only a 45% yield of mannuronic acid could be obtained from the acid hydrolyzate of alginic acid although decarboxylation with 19% hydrochloric acid indicated that alginic acid was a polymer consisting entirely of uronic acid residues.

Lunde *et al.* (1938) suggested that anhydromannuronic acid was the monomeric unit and, based on X-ray diffraction data, that the structure of alginic acid was similar to the structures of cellulose and pectic acid (Kringstad and Lunde 1938).

Determination of the linkage between the anhydromannuronic acid units was carried out by Hirst *et al.* (1939) by methylation of a degraded alginic acid. They isolated the methyl ester of methyl 2,3-di-*O*-methyl-D-mannuronoside by methanolysis of the methylated alginic acid and concluded that alginic acid consists of D-mannuronic acid units linked through carbon atoms 1 and 4. The configuration at the glycosidic linkage was assumed to be β based on the high negative specific rotation, $[\alpha]_D$ -120° to -150°. These conclusions were confirmed by Chanda *et al.* (1952) using a less degraded alginic acid sample.

Therefore, by 1952 the alginic acid structure which had emerged from the extensive studies outlined above had been shown to consist of a chain of β-D-mannuronic acid units linked through C_1 and C_4.

However, in 1955 it was shown by Fischer and Dörfel (1955) using paper chromatographic techniques that L-guluronic acid as well as D-mannuronic acid was present in alginic acid. The L-guluronic acid content ranged from 30–70% dependent upon the sample of alginic acid and the brown algae from which it was isolated. Whistler and Kirby (1959) showed that 20–40% of the alginic acid isolated from *Macrocystis pyrifera* was composed of L-guluronic acid. Other investigators had also confirmed the presence of L-guluronic acid in alginic acid (Hirst 1958; Drummond *et al.* 1958). The possibility that L-guluronic acid was an artifact produced by epimerization of D-mannuronic acid in the alkaline medium used to extract the alginic acid was shown to be unlikely by isolating L-guluronic acid from the acid hydrolyzate of whole algae.

The question now arises as to whether alginic acid is a mixture of polymannuronic acid and polyguluronic acid or whether both mannuronic acid and guluronic acid occur in the same molecule. Vincent (1960) obtained several oligouronides containing both mannuronic acid and guluronic acid by partial acid hydrolysis and concluded that at least some of the alginic acid molecules contain both mannuronic acid and guluronic acid. This was confirmed by Hirst *et al.* (1963, 1964) by reducing alginic acid to the neutral glycan followed by partial acid hydrolysis and isolation of crystalline mannosylgulose from the hydrolyzate. Therefore, by 1964 alginic acid had been shown to be a polysaccharide containing D-mannuronic acid and L-guluronic acid with both uronic acids occurring to some extent in the same molecule and with the uronic acids being linked through C_1 and C_4.

Improvements in the techniques for the hydrolysis, separation, and analysis of alginic acid allowed accurate determinations of the composition of alginic acid from different sources. Table 11.2 shows the composition of alginic acid obtained from commercially important brown algae and from the bacteria *Azotobacter vinelandii*. All the alginic acids with the exception of the alginic acid from *Laminaria hyperborea* stipes have similar composition with D-mannuronic acid predominating. The alginic acid from *Laminaria hyperborea* has a high L-guluronic acid content. The table also shows the range in composition which can be obtained for the alginic acid isolated from a single species of algae. This variation in M:G ratio depends upon the location, season, and part of the plant from which the alginic acid was extracted.

The presence of three kinds of polymer segments in alginic acid from various sources has been shown by mild acid hydrolysis (Haug *et al.* 1966, 1967A, 1967C). One segment consists essentially of D-mannuronic acid units, a second segment consists essentially of L-

TABLE 11.2

MANNURONIC ACID (M) AND GULURONIC ACID (G) COMPOSITION OF ALGINIC ACID OBTAINED FROM COMMERCIAL ALGAE

Species	Mannuronic Acid (%)	Guluronic Acid (%)	M:G Ratio	M:G Ratio Range
Macrocystis pyrifera	61	39	1.56[1]	—
Ascophyllum nodosum	65	35	1.85[1]	1.40–1.95[1]
Laminaria digitata	59	41	1.45[1]	1.40–1.60[1]
Laminaria hyperborea (stipes)	31	69	0.45[1]	0.40–1.00[1]
Ecklonia cava and *Eisenia bicyclis*	62	38	1.60[1]	—
Azotobacter vinelandii[2]	60	40	1.50[3]	—

[1] Data of Haug (1964) and Haug and Larsen. (1962).
[2] An acetylated extracellular polysaccharide consisting of mannuronic acid and guluronic acid is produced by certain bacteria.
[3] Penman and Sanderson (1972).

guluronic acid units, and the third segment consists of alternating D-mannuronic acid and L-guluronic acid residues. The monomer structure and polymer structures are shown in Fig. 11.2.

The proportions of the three polymer segments in alginic acid samples from different sources have been determined by Haug and co-workers (1966, 1967A) using partial acid hydrolysis to separate the alginic acid into homopolymeric and alternating fractions. Recently Penman and Sanderson (1972) have determined the proportion of polymannuronan and polyguluronan segments by proton magnetic resonance spectroscopy. Table 11.3 shows the proportions of polymannuronan, polyguluronan, and alternating segments for several alginic acid samples.

The proportions of the three segments shown in the table were determined for single samples. As with the M:G ratio, the proportions of the three segments from the alginic acid isolated from a single species of algae vary dependent upon the location, season, and part of the plant from which the alginic acid was isolated.

The alginic acids from *Macrocystis pyrifera* and *Ascophyllum nodosum* are very similar in structure, containing approximately equal proportions of polymannuronan segments and alternating segments and small proportions of polyguluronan segments. The alginic acid from *Laminaria hyperborea* is quite different from the alginic acid isolated from *Macrocystis pyrifera* and *Ascophyllum nodosum* in that it contains a high proportion of polyguluronan segments, an intermediate proportion of alternating segments, and a small proportion of polymannuronan segments. The alginic acid from *Azotobacter vinelandii* is different from either of the above

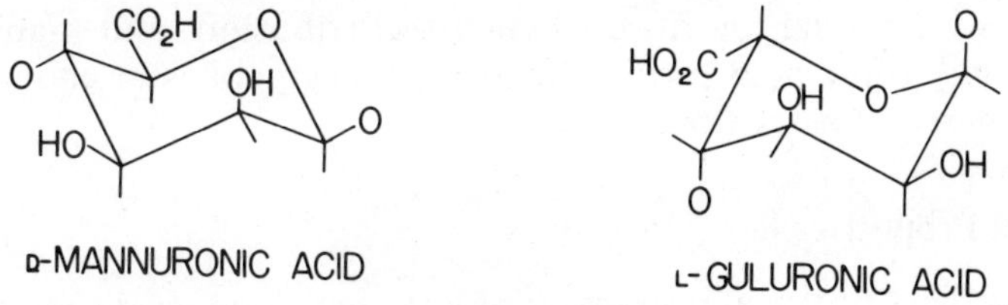

D-MANNURONIC ACID L-GULURONIC ACID

...-M-M-M-M-M-M-M-M-...
POLYMANNURONIC ACID SEGMENT

...-G-G-G-G-G-G-G-G-...
POLYGULURONIC ACID SEGMENT

...-M-G-M-G-M-G-M-G-...
ALTERNATING SEGMENT

FIG. 11.2. STRUCTURE OF MANNURONIC ACID, GULURONIC ACID, AND POLYMER SEGMENTS

TABLE 11.3

PROPORTIONS OF POLYMANNURONAN, POLYGULURONAN, AND ALTERNATING SEGMENTS IN ALGINIC ACID FROM DIFFERENT SOURCES

Source	Poly-mannuronan	Poly-guluronan	Alternating
Macrocystis pyrifera	40.6	17.7	41.7
Ascophyllum nodosum	38.4	20.7	41.0
Laminaria hyperborea	12.7	60.5	26.8
Azotobacter vinelandii	17.8	0.5	81.7

Source: Data of Penman and Sanderson (1972).

types of alginic acid being almost entirely composed of alternating sequences. These differences in structure account for the difference in properties and functionality of alginates isolated from different brown algae.

PROPERTIES OF ALGINATES

The different types of alginates available for use in the food industry are the sodium, potassium, ammonium, mixed ammonium-calcium, and mixed sodium-calcium salts of alginic acid as well as propylene glycol alginate. These water soluble alginates are produced in many forms varying in molecular weight, calcium content, particle form (i.e., granular or fibrous), mesh distribution, and mannuronic acid to guluronic acid ratio. The propylene glycol ester can also vary in the degree of esterification.

Physical Properties

The typical physical properties of a refined food-grade sodium alginate are as follows (Anon. 1972B).

Moisture Content	13%
Ash	23%
Powder color	Ivory
Specific gravity	1.59
Bulk density (lb/ft^3)	54.62
Browning temperature (°C)	150.0
Charring temperature (°C)	340, 460
Ashing temperature (°C)	480.0
Heat of combustion (Cal/gm)	2.5

Alginates have excellent dry-storage stability at moderate temperature, 24°C or lower, but at 32°C dry-storage stability decreases for alginate salts and propylene glycol alginate tends to become insoluble. In view of this, alginates should be stored in a cool place. Alginates, being hydrophilic colloids, absorb moisture from the atmosphere. The equilibrium moisture content is dependent upon the relative humidity, and this relationship is shown in Fig. 11.3 for two commercial alginates.

The physical properties of a 1% distilled water solution of a typical refined food-grade sodium alginate are as follows.

Heat of solution (Cal/gm)	0.080
Refractive index (20°C)	1.3343
pH	7.5
Surface tension (dynes/cm)	62.0
Freezing point depression (°C)	0.035

Solution Properties

The 1% water viscosity of commercially available alginates at ambient temperature varies from 10–2000 cps (Fig. 11.4) depending upon the molecular weight and calcium content of the alginate. As with most other polymer solutions, the viscosity of alginate solutions decreases with increasing temperature. This decrease is reversible provided that high solution temperatures are not maintained for long periods. Excessive temperatures cause partial depolymerization of the molecule with resultant loss of viscosity. The viscosity of an algi-

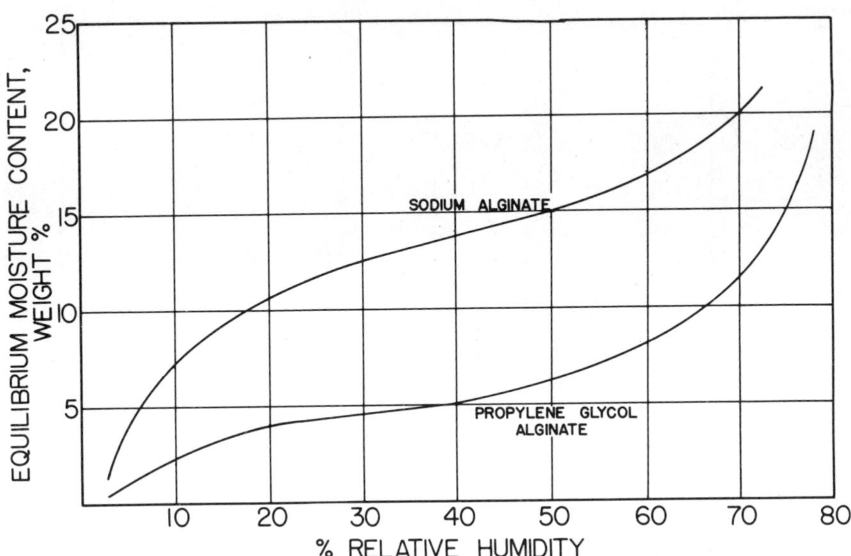

FIG. 11.3. EQUILIBRIUM MOISTURE CURVES FOR TWO COMMERCIAL ALGINATES

nate solution increases exponentially as the concentration of alginate is increased (Fig. 11.4).

The viscosity of alginate solutions is essentially constant in the pH range 5-10 (Fig. 11.5). Below a pH of 4.5 a significant viscosity increase occurs, and below pH 3 insoluble alginic acid is precipitated. At a pH above 11.5 sodium alginate forms a gel. In low-pH systems, pH 2-3, propylene glycol alginate is more soluble than sodium alginate, but at a pH above 6.5 propylene glycol alginate degrades rapidly. Alginate salts are stable at pH 5-10 at room temperature for extended periods, but outside of this pH range degradation is considerable. Propylene glycol alginate has good stability at pH 3-4 at room temperature for extended periods, but below pH 2 propylene glycol alginate solutions lose viscosity over a short period due to acid degradation of the polymer backbone. In alkaline solutions, propylene glycol alginate rapidly loses viscosity due to depolymerization of the polymer (Haug *et al.* 1967B). However, if the pH is reduced before substantial depolymerization takes place, a substantial viscosity increase occurs (McDowell *et al.* 1968; McDowell 1970).

Alginates are also depolymerized by free radicals formed by autooxidation of reducing compounds such as phenolics, ascorbic acid, and some metal ions (Haug 1964; Smidsrød *et al.* 1963; Haug *et al.* 1969; Smidsrød *et al.* 1967).

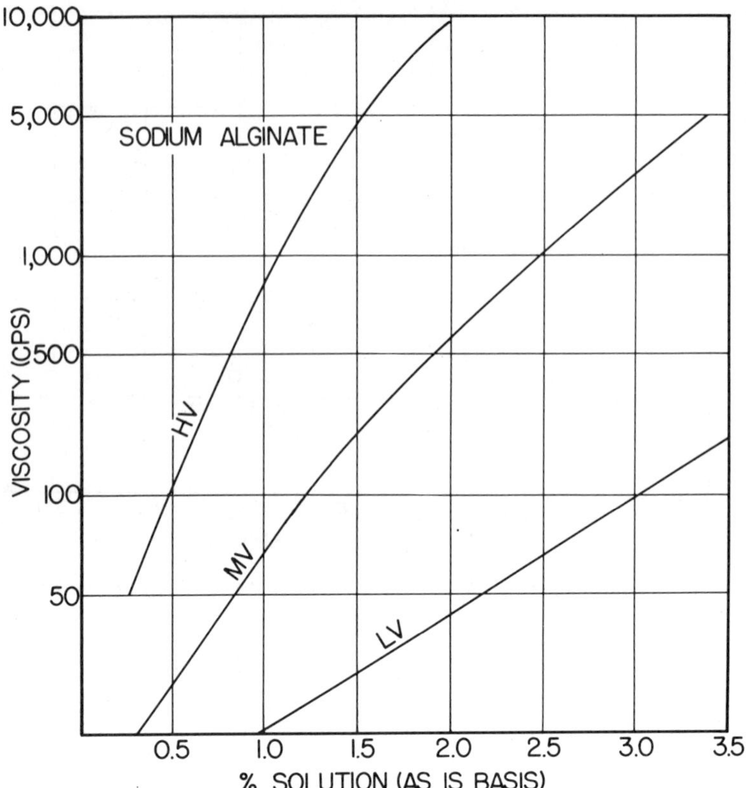

FIG. 11.4. EFFECT OF CONCENTRATION ON THE VISCOSITY OF HIGH, MEDIUM AND LOW VISCOSITY ALGINATES AT 25°C

Alginate solutions can be preserved from microbial degradation by addition of appropriate food-grade preservatives.

The viscosity of sodium alginate solutions is slightly depressed by the addition of monovalent salts. As is frequently the case with polyelectrolytes, the polymer in solution contracts as the ionic strength of the solution is increased. The maximum viscosity effect is obtained at about 0.1 N salt concentration (Fig. 11.6).

The flow properties of sodium alginate solutions are concentration dependent. A 2.5% medium-viscosity sodium alginate solution is pseudoplastic especially at high shear rates, 10–10,000 sec^{-1}, whereas a 0.5% solution of the same sodium alginate is Newtonian at low shear rates and pseudoplastic only at high shear rates, 1,000–10,000 sec^{-1}.

Sodium alginate in solution is compatible with a large number of compounds used in the food industry. Some of these compounds are shown in Table 11.4.

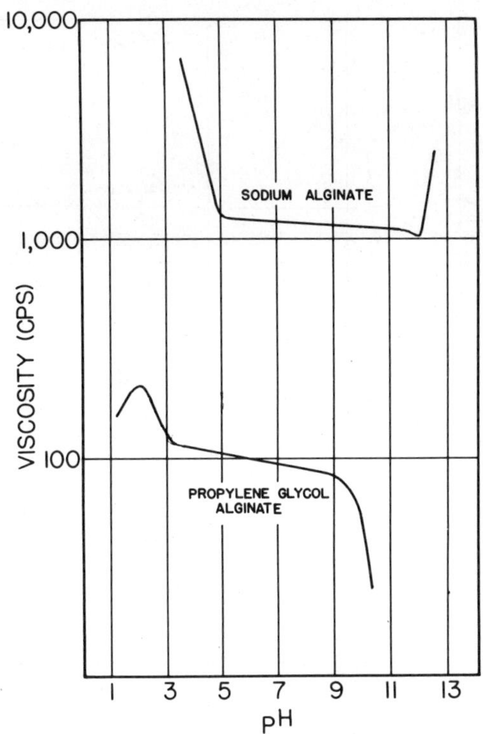

FIG. 11.5. EFFECT OF pH ON THE VISCOSITY OF ALGIN SOLUTIONS

Gelation

One of the most important and useful properties of alginates in the food industry is the ability to form edible gels by reaction with calcium salts. These gels, which resemble a solid in retaining their shape and resisting stress, consist of almost 100% water (actually, 99.0–99.5% water and 0.5–1.0% alginate).

A gel, in classical colloid terminology, is defined as a system which owes its characteristic properties to a cross-linked network of polymer chains which form at the gel point (Hermans 1949). A considerable amount of research has been carried out in recent years to elucidate the nature of the cross-links and determine the structure of alginate gels.

Prior to the recent research it had been suggested that the cross-links are formed either by simple ionic bridging of two carboxyl groups on adjacent polymer chains with calcium ions or by chelation of single calcium ions by hydroxyl and carboxyl groups on each of a pair of polymer chains. Although these bonds may play a role in the

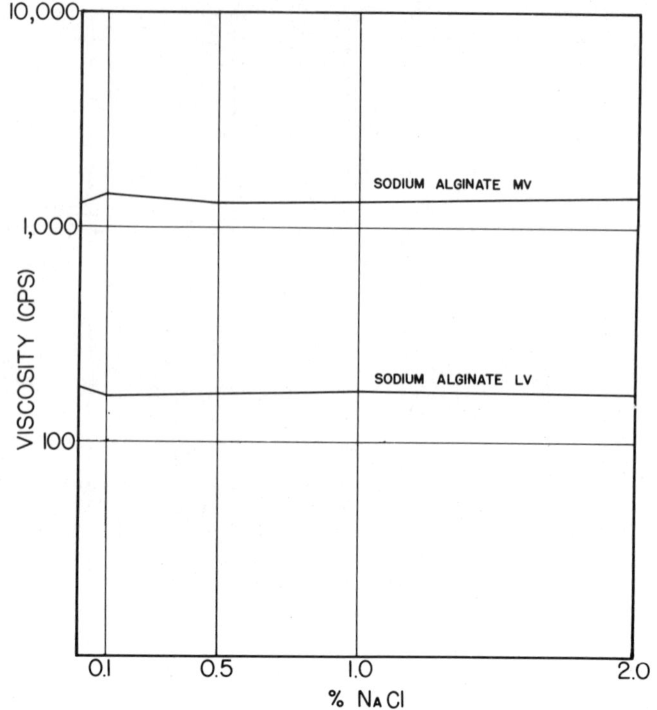

FIG. 11.6. EFFECT OF SODIUM CHLORIDE ON THE VISCOSITY OF SODIUM ALGINATE

TABLE 11.4

COMPATIBILITIES OF SODIUM ALGINATE WITH SEVERAL COMPOUNDS USED IN THE FOOD INDUSTRY

Compatible	
Carbohydrates	Starch, dextrin, sucrose, dextrose, invert sugar, carboxymethyl cellulose, and water soluble cellulose ethers
Water soluble gums	Acacia, carrageenan, guar, karaya, locust bean, pectin, tragacanth, xanthan gum
Polyhydric alcohols	Glycerol, sorbitol, mannitol
Proteins	Casein (acid-precipitated), gelatin, soy bean, egg albumin
Partially Compatible	
Water miscible solvents	Ethanol (25% by volume)
Salts	Sodium sulfate (10–12% of solution weight), sodium chloride (about 4%)
Incompatible	
Acids	Those sufficiently strong to reduce the pH below 3.5

Source: McNeely and Pettitt (1973).

gelation mechanism, they are not sufficiently energetically favorable to account for the gelation of alginate. A detailed discussion of this subject was presented by Rees (1969).

It has been shown, on the basis of fiber diffraction data and model-building calculations, that the shape of both the polymannuronic acid segments and the polyguluronic acid segments of alginic acid is ribbon-like and extended and that these extended ribbons can stack together in sheets (Atkins *et al.* 1970, 1971). On the basis of these data and the properties of gels, Rees (1969) has suggested that cooperative association of either polymannuronic acid segments or polyguluronic acid segments is involved in the formation of the cross-linked network of polymer chains. The proposed structure of an alginate gel in which calcium ions are bound between the associated segments of the polymer chains is shown in Fig. 11.7.

Circular dichroism studies have shown that the calcium ions react preferentially with the polyguluronic acid segments before reacting with the polymannuronic acid segments (Morris *et al.* 1973). It is quite likely that the alternating segments play no direct role in the gelation with calcium except to join the associated segments and hence provide a three-dimensional network of chains within the gel.

The nature of the interaction between the polyguluronate segments and calcium ions has been further refined using both the known coordination geometries of model compounds and the requirements for cooperative association. In this interaction the polyguluronate segments associate into aggregates with interstices into which the calcium ions fit, the "egg-box model," Fig. 11.8 (Grant *et al.* 1973).

These data can be used to predict the observed gelling characteristics of alginates from different sources. For example, the alginate

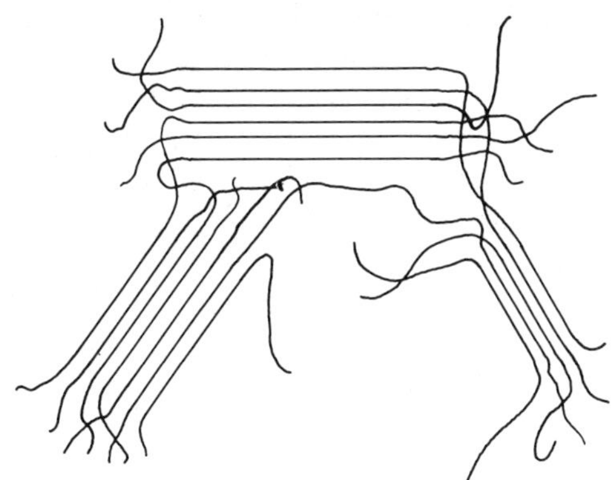

FIG. 11.7. STRUCTURE OF CALCIUM ALGINATE GEL (PROPOSED)

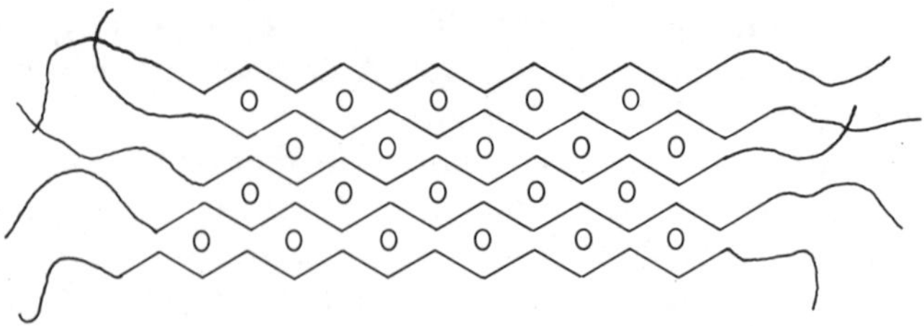

FIG. 11.8. THE EGG-BOX MODEL

from *Laminaria hyperborea* with a large percentage of polyguluronate segments forms rigid, brittle gels which tend to undergo syneresis, whereas the alginates from *Macrocystis pyrifera* or *Ascophyllum nodosum* form elastic gels which can be deformed (Penman and Sanderson 1972; Penman *et al.* N. D.)

The preparation of gels from soluble alginates such as sodium alginate can be accomplished by the gradual release of multivalent cations or hydrogen ions or combinations of both. Gel textures ranging from soft to hard may be produced by adjusting the ratio of salts to acid. Some acid gel-formers are D-glucose-1,5-lactone, adipic acid, fumaric acid, and citric acid. However, as mentioned previously, the most commonly encountered algin gel in the food industry is composed of calcium alginate. The usual calcium sources for the formation of calcium alginate consist of calcium carbonate, calcium sulfate, calcium chloride, calcium phosphate, or calcium tartrate (Miller and Rocks 1969; Freedman 1967; Gibsen 1959). The speed of gel formation as well as the quality and texture of the resultant gels can be controlled by the solubility and availability of the calcium source. While calcium chloride, being readily water soluble, causes the instantaneous formation and precipitation of calcium alginate, dicalcium phosphate dihydrate, for instance, fails to release the calcium until the temperature is raised to 93–107°C. Consequently, delayed thickening or gelling can be accomplished in this manner. The former system is widely used in several fabricated gel concepts. The latter has been applied to several canning systems (McDermott 1966).

The algin gel system can be further controlled by the use of sequestrants which are capable of reacting preferentially with the polyvalent metal ions. Sequestering agents are also used to remove calcium inherently present in all alginates or as protective devices to remove polyvalent metal ions from makeup waters which may upset the calcium balance of the algin system. The most commonly used sequestering agents consist of polyphosphates such as sodium hexametaphosphate or sodium tripolyphosphate. As shown in Fig. 11.9, the effect of the sequestrant on the viscosity of the algin solution depends on the inherent calcium content of the alginate. When an alginate with a low calcium content is used, the effect of the sequestrant is minimized, while a significant lowering of the viscosity occurs when the solution contains an alginate with a high calcium content. Stoichiometrically, calcium is required at 7.2% of the weight of sodium alginate for complete substitution. Gels are formed at about 32% of this amount and thickened, flowable solutions at about 15% or less.

Alginates also possess excellent film-forming properties. Films can be formed by thin-layer drying of a solution or by treating a wet thin-layer solution with acid or a complexing agent such as a calcium salt (McDowell 1960).

In summary, the basic properties of alginates which have led to their widespread use in the food industry are based on the colloidal nature of alginates, especially their polyelectrolyte character. The production of viscous solutions at low concentrations, the formation of gels with calcium ions, the formation of films, and the stabilizing and suspending properties are all based on the colloidal nature of alginates.

FOOD USES

The uses of alginates in the food industry have been reviewed (McDowell 1960; Glicksman 1969; McDermott 1962). Algin has been consumed as part of seaweed for centuries and for 40 yr as a constituent of processed foods.

Frozen Desserts

Algin-based stabilizers have been used for many years in the production of ice cream. The main function of algin in ice cream is to regulate the formation of ice crystals and thereby ensure smooth texture as well as desirable melt-down and overrun (Sperry 1955). Most algin-based ice cream stabilizers consist of a polyphosphate-treated algin, which enhances solubility in the high-calcium environment, blended with a dispersing aid such as dextrose or sucrose and a small amount of a protective colloid containing varying amounts of carrageenan, guar, and locust bean gum. Other high-quality ice cream stabilizers contain propylene glycol alginate (Sperry 1953), which is soluble in milk without the use of polyphosphates. The normal algin concentration in ice cream ranges from 0.1–0.5%.

Alginate-based stabilizers are also used in ice milk, water ices, sher-

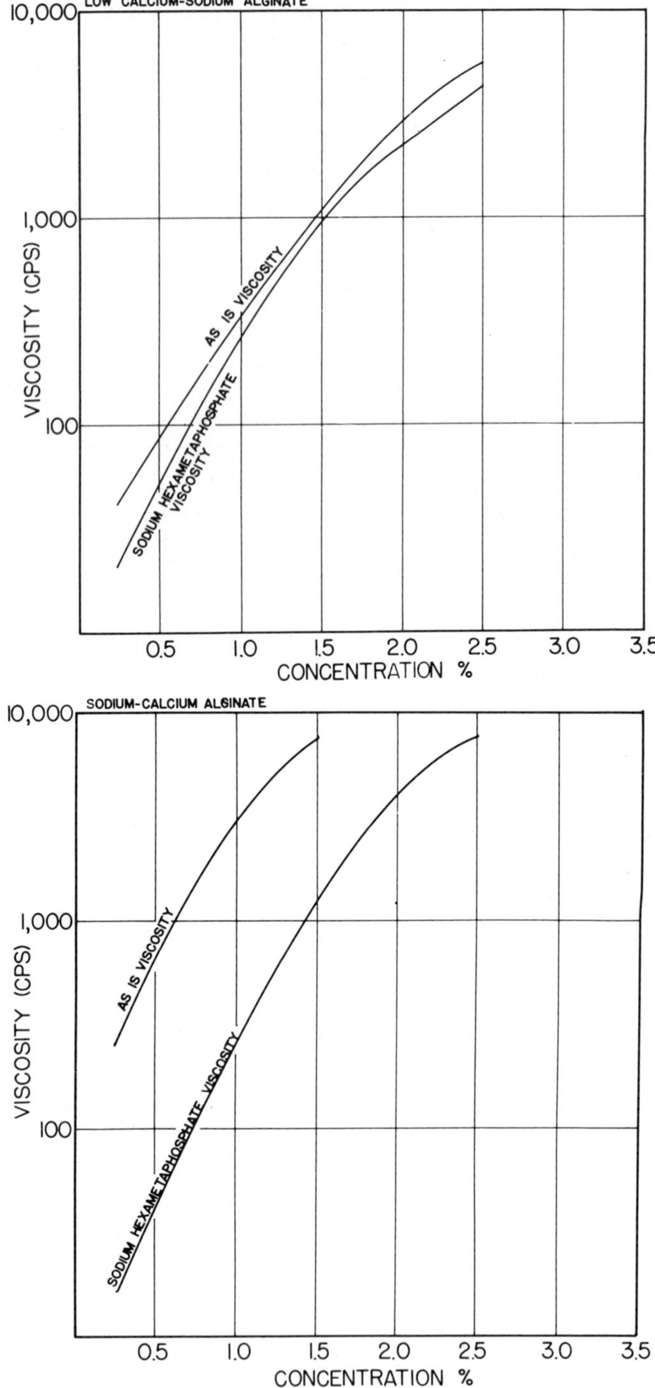

FIG. 11.9. EFFECT OF SEQUESTRANTS ON THE VISCOSITY OF ALGIN SOLUTIONS

bets and milk shakes. Ice milks which are stabilized with algin have good dryness and stiffness and are slow melting. In addition, delayed churning as well as processing viscosity control can be accomplished. The concentration of stabilizer may range from 0.2–0.5%. Milk soluble algin-based stabilizers provide good secondary overrun and excellent icy eating properties for hard-frozen ice milks and milk shakes. Sherbets stabilized with propylene-glycol-alginate-based stabilizer blends possess clean flavor release, as well as a smooth and refreshing texture. Sugar syrup separation is also prevented by the use of these stabilizers.

Miscellaneous Dairy Products

The use of alginate-based stabilizers is also well established in process cheese, cream cheese, sour cream, other cultured dairy products, whipped cream, and chocolate milk.

In process and cream cheese, the algin stabilizer, which consists mainly of propylene glycol alginate, is used at a level ranging from 0.3–0.8%. In whipped cream about 0.15% sodium alginate is sufficient for stabilization and thickening. To stabilize chocolate milk suspensions, algin-carrageenan compositions are used at less than 0.25% concentration. Milk soluble algin stabilizers are also used as thickeners and bodying agents for cultured dairy products.

Bakery Products

The stabilization of bakery toppings and icings with algin is well established in the bakery industry. Algin is used to stabilize icings for bakery products such as layer cakes (Steiner and Rothe 1949), to prevent sticking to wrapping paper (Edlin 1967; Ellis 1967), and to prevent the icing from cracking and drying out during storage (Gibsen and Rothe 1955). These uses are based on the film-forming and water-holding properties of alginates.

Syneresis is a serious problem with meringues, which consist of egg white, sugar, and water and contain as much as 300% overrun, but 0.2% propylene glycol alginate has been found to provide effective stabilization. The recommended algin concentration for fruit pie fillings is 0.3–0.5%, while 0.7–1.5% algin in combination with a whipping agent is recommended for chiffon-type neutral pH fillings (McDermott 1963). A novel chiffon pie filling based on an instant mix in cold water has also been described (Hunter 1961). Commercial fruit pie fillings are also effectively stabilized with alginates at 0.2–0.8% levels.

Beverages

Noncarbonated fruit beverages are considered to be of the highest quality when the fruit pulp remains in suspension. Due to its pH stability as well as effective thickening properties, propylene glycol alginate has been found to be one of the most effective stabilizers for this purpose. The slight viscosity increase contributed by propylene glycol alginate enhances the mouthfeel and flavor release of the product. In dietetic beverages, propylene glycol alginate also furnishes the body characteristics of drinks with a high sugar content. Concentrations usually range from 0.10–0.25%.

The use of propylene glycol alginate at the trace level of 40–50 ppm provides foam stability for beer. The foam stabilized with propylene glycol alginate resists breakdown even in the presence of soaps and detergents. The foam is creamy with an attractive, nonsticky, natural-appearing lace (Steiner 1953; Nordman and Mohr 1963).

Gels

Sodium alginate can be used to prepare dessert gels which are not temperature reversible and which are also clear and firm. The gel will set rapidly under refrigeration but will also set at room temperature. The set time as well as the gel consistency may be controlled by the sodium alginate concentration, pH, release of calcium, and the use of sequestering agents (Steiner 1948A). This type of dessert gel may also be prepared in a freeze-thaw-stable form (Glicksman 1962). A number of improvements and variations to this gel system relate to the preparation of more readily dispersible products (Poarch and Tweig 1957), utilization of slightly soluble dibasic acids (Gibsen 1959), and to the utilization of glucono-delta-lactone as an acid-releasing agent (Merton and McDowell 1960).

Quick-setting dessert puddings and custards containing milk or milk solids can be prepared with milk soluble algins (Hunter and Rocks 1960) at a level of 0.5%. These algin-gelled dessert puddings are usually supplied as prepackaged dry mixes which are prepared in their final form by dissolving the alginate in cold or hot milk and allowing the added calcium, as well as the calcium in the milk, to react with the algin. The resultant gels, which are smooth and creamy in consistency, give a natural body and full flavor release without pasty or starchy eating properties. The surface of the pudding remains soft, and no hardened skin formation occurs.

Salad Dressings and Sauces

The low-pH solubility and stability as well as the emulsifying and stabilizing properties of propylene glycol alginate make this algin derivative an excellent emulsion stabilizer for bottled pourable and spoonable salad dressings. Propylene glycol alginate is most effective at a 0.1-0.2% level as a corn starch modifier of spoonable salad dressings. The resultant body resists cracking and oil separation even after prolonged storage. The unique combination of emulsifying and thickening properties of propylene glycol alginate has resulted in the widespread utilization of this product in pourable dressings (Steiner 1948B).

Pourable dressings stabilized with this algin derivative have an extended shelf-life and resist emulsion breakdown at both refrigerated and elevated temperatures.

The stabilization of gravies, meat sauces, and barbecue sauces with propylene glycol alginate is also widely practiced in the food industry. Concentration levels may range from 0.15-0.50%.

Fabricated Foods

The calcium reactivity of sodium alginate is widely utilized for the preparation of fabricated foods. This concept has been successfully adopted for the manufacture of fabricated onion rings using fresh diced onions. Miscellaneous fabricated meat snacks have also been prepared using this process (Katz 1973; Smadar 1972). The preparation of these products involves the mixing of the small onion pieces or meat pieces with a matrix mixture containing the sodium alginate. The resultant composition is extruded into a calcium chloride bath, which causes the instantaneous gelation of the fabricated food particles. These particles are quick frozen and eventually deep fried prior to consumption.

Several years ago a process for the preparation of fabricated fruit pieces was developed (Peschardt 1946). This process consists of carefully extruding a syrup containing sodium alginate and a high concentration of sugar into a calcium chloride or acid bath to form a hard surface film which prevents the disintegration of the molded particle. Eventually the calcium or acid penetrates to the center of the particle and produces a uniform texture. These fabricated fruit particles may be used as produced or dried to about 70-80% solids to enhance chewability. The main advantage of this concept is that these fabricated fruit pieces are heat resistant and can be used effectively in or on baked goods such as muffins.

The film-forming properties of sodium alginate have been utilized in the manufacture of fabricated sausage casings. The sausage casing is prepared by extruding a sodium alginate solution into a calcium chloride bath and plasticizing the film with glycol and calcium lactate (Weingand 1957). A number of attempts have been made to correct several short-comings of this concept (Langmaack 1961; Hartwig and Reissman 1960), but alginate sausage casings have not yet been adopted on a wide scale in the United States.

The preparation of simulated fruit and vegetable pieces (Szczesniak 1968) has been accomplished by dialyzing the weak acid salts of calcium or magnesium into a sodium alginate solution. This procedure allows the calcium or magnesium ions to react with the alginate at the interface and produce a thin film which serves as a dialysis membrane to further control the supply of reactive calcium or magnesium ions. This uniform and gradual gel formation produces an edible, crisp, nonuniform cellular network simulating the textural qualities of fruits and vegetables.

MISCELLANEOUS APPLICATIONS

The shelf-life of quick-frozen fish and meat has been lengthened by coating them with an airtight film of calcium alginate (Earle 1968). The film is prepared by dipping the meat or fish in a sodium alginate solution containing dextrose followed by immersion in a calcium chloride bath prior to freezing. The coating, which is strong and freeze-thaw stable, prevents the introduction of air and development of oxidative rancidity.

A number of fountain syrups and toppings are stabilized and thickened with alginates. The alginate protects the syrups from drainage, separation, and loss of texture and provides a smooth, creamy texture and uniform appearance.

For canned foods containing sauces and gravy, the reduction in heat processing times has been made possible by replacing most of the starch normally used by 0.3–0.8% sodium alginate and a slightly soluble calcium salt. The delayed release of the calcium permits rapid heat penetration. As the temperature drops after the heating cycle is completed, the calcium reacts with the algin and the desired final consistency is obtained (McDermott 1966; Messina and Pape 1966).

SAFETY AND REGULATORY STATUS

The toxicology of alginates has been the subject of numerous reports and literature reviews (Anon. 1972A; McNeely and Kovacs 1974). Toxicological data obtained from long-term feeding studies of mice, rats, guinea pigs, and dogs, a three-generation reproduction study in the rat, as well as digestibility and absorption studies, have indicated that alginates are poorly absorbed and that their consumption causes no harmful effects. Sodium alginate, ammonium alginate, calcium alginate, and potassium alginate are included in a list of stabilizers that are generally recognized as safe under 21 CFR 121.101. Propylene glycol alginate is approved as a food additive under CFR 121.1015 for use as an emulsifier, thickener, and stabilizer. All these products may be used without specific quantity limitations in accordance with good manufacturing practice except where standards of identification regulations prohibit such use.

BIBLIOGRAPHY

ANDREW, T. R., and MacLEOD, W. C. 1970. Application and control of the algin-calcium reaction. Food Prod. Develop. *4*, No. 5, 99, 102, 104.

ANON. 1972A. GRAS (Generally Recognized as Safe) Food Ingredients—Alginates. PB-221 226. N.T.I.S.

ANON. 1972B. Kelco Algin, Hydrophilic Derivatives of Alginic Acid for Scientific Water Control. Kelco Company, San Diego, Calif.

ATKINS, E. D. T., MACKIE, W., and SMOLKO, E. E. 1970. Crystalline structure of alginic acids. Nature (London) *225*, 626-628.

ATKINS, E. D. T., MACKIE, W., PARKER, K. D., and SMOLKO, E. E. 1971. Crystalline structures of poly-D-mannuronic and poly-L-guluronic acids. J. Polymer Sci., Part B9, 311-316.

ATSUKI, K., and TOMODA, Y. 1926. Studies on seaweeds of Japan. I. On the chemical constituents of *Laminaria.* J. Soc. Chem. Ind. Japan *29*, 509-517.

BASHFORD, L. A., THOMAS, R. S., and WOODWARD, F. N. 1950. Manufacture of algal chemicals. I. Production of alginates from brown algae. J. Soc. Chem. Ind. (London) *69*, 337-342.

BIRD, G. M., and HAAS, P. 1931. Nature of the cell wall constituents of *Laminaria* spp. Mannuronic acid. Biochem. J. *25*, 403-411.

BLACK, W. A. P. 1950. Seasonal variation in weight and chemical composition of the common British *Laminariaceae.* J. Marine Biol. Assoc. U. K. *29*, 45-72.

BLACK, W. A. P., and WOODWARD, F. N. 1954. Alginates from common British brown marine algae. *In* Advances in Chemistry Series No. 11, American Chemical Society, Washington, D.C.

BOYLE, J. L. 1960. Improvements in or relating to compounds of alginic acid. Brit. Pat. 835,009.

CHANDA, K., HIRST, E. L., PERCIVAL, E. G. V., and ROSS, A. G. 1952. Structure of alginic acid. Part II. J. Chem. Soc. 1833-1837.

CHAPMAN, V. J. 1970. Seaweeds and Their Uses, 2nd Edition. Methuen and Co. Ltd., London.

DRUMMOND, D. W., HIRST, E. L., and PERCIVAL, E. 1958. The presence of L-guluronic acid residues in alginic acid. Chem. Ind. (London) 1088.

EARLE, R. D. 1968. Algin coating for extending shelf-life of seafood and meat. U.S. Pat. 3,395,024.

EDLIN, R. L. 1967. Algin products in bakery foods. Baker's Dig. *41*, No. 6, 49-51.

ELLIS, P. E. 1967. Process for manufacture of chocolate water-icings. U.S. Pat. 3,332, 784.

FISCHER, F. G., and DÖRFEL, H. 1955. The polyuronic acids of brown algae. Part I. Z. Physiol. Chem. *302*, 186-203.

FREEDMAN, J. C. 1967. Gel-forming alginate products and method. U.S. Pat. 3,349,079.

GIBSEN, K. F. 1959. Algin gel composition and method. U.S. Pat. 2,918,375.

GIBSEN, K. F., and ROTHE, L. B. 1955. Algin: Versatile food improver. Food Eng. *27*, No. 10, 87-89.

GLICKSMAN, M. 1962. Freezable gels. U.S. Pat. 3,060,032.

GLICKSMAN, M. 1969. Gum Technology in the Food Industry. Academic Press, New York.

GOODMAN, C. 1935. Technical control of ice cream with sodium alginate. Ice Cream Rev. *18*, 42-44, 48.

GORIN, P. A. J., and SPENCER, J. F. T. 1966. Exocellular alginic acid from *Azotobacter vinelandii.* Can. J. Chem. *44*, 993-998.

GRANT, G. T. *et al.* 1973. Biological interactions between polysaccharides and divalent cations: The egg-box model. FEBS Letters *32*, 195-198.

GREEN, H. C. 1936. Process for making alginic acid and product. U.S. Pat. 2,036,934.

HARTWIG, M., and REISSMAN, H. 1960. Method of reducing the swelling capacity of synthetic alginate skins. U.S. Pat. 2,965,498.

HAUG, A. 1964. Composition and Properties of Alginates. Rept. *30*, Norwegian Institute Seaweed Research, Trondheim, Norway.

HAUG, A., and LARSEN, B. 1962. Quantitative determination of the uronic acid composition of alginates. Acta. Chem. Scand. *16*, 1908-1918.

HAUG, A., LARSEN, B., and SMIDSRØD, O. 1966. A study of the constitution of alginic acid by partial acid hydrolysis. Acta. Chem. Scand. *20*, 183-190.

HAUG, A., LARSEN, B., and SMIDSRØD, O. 1967A. Studies on the sequence of uronic acid residues in alginic acid. Acta. Chem. Scand. *21*, 691-704.

HAUG, A., LARSEN, B., and SMIDSRØD, O. 1967B. Alkaline degradation of alginate. Acta. Chem. Scand. *21*, 2859-2870.

HAUG, A., MYKLESTAD, S., LARSEN, B., and SMIDSRØD, O. 1967C. Correlation between chemical structure and physical properties of alginates. Acta. Chem. Scand. *21*, 768-778.

HAUG, A., LARSEN, B., SMIDSRØD, O., and PAINTER, T. 1969. Development of compositional heterogeneity in alginate degraded in homogeneous solution. Acta. Chem. Scand. *23*, 2955-2962.

HERMANS, P. H. 1949. Gels. *In* Colloid Science, Vol. 2, H. R. Kruyt (Editor). Elsevier, Amsterdam.

HEYDT, G. 1959. Process technology of alginic acid and sodium alginate. Stärke *11*, 38-42. (German)

HIRST, E. L. 1958. Polysaccharides of the marine algae. Proc. Chem. Soc. 177-187.

HIRST, E. L., JONES, J. K. N., and JONES, W. O. 1939. Structure of alginic acid. Part I. J. Chem. Soc. 1880-1885.

HIRST, E. L., PERCIVAL, E., and WOLD, J. K. 1963. Structural studies of alginic acid. Chem. Ind. (London) 257.

HIRST, E. L., PERCIVAL, E., and WOLD, J. K. 1964. The structure of alginic acid. 4. Partial hydrolysis of the reduced polysaccharide. J. Chem. Soc. 1493-1499.

HOPPE, H. A., and SCHMID, O. J. 1962. Marine algae as modern industrial products. Botan. Marina 3 Suppl. 16-66. (German)

HUNTER, A. R. 1961. Neutral-type instant chiffon pie filling and method. U.S. Pat. 2,987,400.

HUNTER, A. R., and ROCKS, J. K. 1960. Cold milk puddings and method. U.S. Pat. 2,949,366.

IMRIE, F. K. E. 1973. Polysaccharide production—By culture of *Azotobacter vinelandii* microorganism having properties and uses similar to alginic acid. Brit. Pat. 1,331,771.

KATZ, A. 1973. Extrusion process shapes bigger sales for frozen onion rings. Quick Frozen Foods *35*, No. 7, 32-34.

KELCO CO. 1952. Alkylene glycol esters of alginic acid. Brit. Pat. 676,618.

KOHLER, R., and DIERICHS, W. 1959. Amides of alginic acid. U.S. Pat. 2,881,161.

KREFTING, A. 1896. An improved method of treating seaweed to obtain valuable products therefrom. Brit. Pat. 11,538.

KRINGSTAD, H., and LUNDE, G. 1938. Alginic acid. II. X-ray investigation of spun threads of alginic acid. Kolloid Z. *83*, 202-203.

LANGMAACK, L. 1961. Methods of producing synthetic sausage casings. U.S. Pat. 2,973,274.

LeGLOAHEC, V. C. E., and HERTER, J. R. 1938. Method of treating seaweed. U.S. Pat. 2,128,551.

LINKER, A., and JONES, R. S. 1966. A new polysaccharide resembling alginic acid isolated from pseudomonads. J. Biol. Chem. *241*, 3845-3851.

LUCAS, H. J. 1937A. Milk soluble alginate compound and process. U.S. Pat. 2,097,228.

LUCAS, H. J. 1937B. Ice cream mix and process. U.S. Pat. 2,097,231.

LUCAS, H. J., and GREEN, H. C. 1937. Ice cream or ice milk mixture and process. U.S. Pat. 2,097,229.

LUNDE, G., HEEN, E., and ØY, E. 1938. Alginic acid. I. Constitution of alginic acid. Kolloid Z. *83*, 196-202.

MAASS, H. 1959. Natural Plant Hydrocolloids. Strassenbau Chemie & Technik Verlagsgesellschaft, Heidelberg. (German)

McDERMOTT, F. X. 1962. Algin: Multi-use colloid. Food Eng. *34*, No. 5, 66-73.

McDERMOTT, F. X. 1963. Alginates: Improvers of flavor quality in pie fillings. Baker's Dig. *37*, No. 6, 66-68.

McDERMOTT, F. X. 1966. Process for preparing a canned food product. U.S. Pat. 3,257,214.

McDOWELL, R. H. 1960. Applications of alginates. Rev. Pure Appl. Chem. *10*, 1-19.

McDOWELL, R. H. 1969. The manufacture and properties of alginates. CIBA Rev. *1*, 3-12.

McDOWELL, R. H. 1970. New reactions of propylene glycol alginate. J. Soc. Cosmetic Chemists *21*, 441-457.

McDOWELL, R. H., BOYLE, J. L., COPES, W., and BRYDEN, W. 1968. Method of modifying alkylene glycol alginates. Brit. Pat. 1,135,856.

McNEELY, W. H. 1954. Compositions of alcohols with hydrophilic gum colloids. U.S. Pat. 2,688,598.

McNEELY, W. H. 1961. Derivatives of water-soluble gums. U.S. Pat. 2,979,499.

McNEELY, W. H., and KOVACS, P. 1975. The Physiological Effects of Alginates and Xanthan Gum. *In* Physiological Effects of Food Carbohydrates, A. G. Jeanes, and J. Hodge (Editors). American Chemical Society Symposium Series 15, American Chemical Society, Washington, D.C.

McNEELY, W. H., and O'CONNELL, J. J. 1959. Carboxymethyl alginate product and method. U.S. Pat. 2,902,479.

McNEELY, W. H., and PETTITT, D. J. 1973. Algin. *In* Industrial Gums, 2nd Edition, R. L. Whistler, and J. N. BeMiller (Editors). Academic Press, New York.

MERTON, R. R., and McDOWELL, R. H. 1960. Powdered alginate jelly composition and method. U.S. Pat. 2,935,409.

MESSINA, B. T., and PAPE, D. 1966. Ingredient cuts heat-process time. Food Eng. *38*, No. 4, 48-51.

MILLER, A., and ROCKS, J. K. 1969. Algin gel compositions and method. U.S. Pat. 3,455,701.

MIWA, T. 1930. Alginic acid. J. Chem. Soc. Japan *51*, 738-745.

MORRIS, E. R., REES, D. A., and THOM, D. 1973. Characterization of polysaccharide structure and interactions by circular dichroism. Order-disorder transition in the calcium alginate system. J. Chem. Soc., Chem. Commun., 245-246.

NATIONAL LEAD CO. 1961. Organic salts, capable of gelling and dispersing organic liquids. German Pat. 1,103, 336.

NELSON, W. L., and CRETCHER, L. H. 1929. The alginic acid from *Macrocystis pyrifera*. J. Am. Chem. Soc. *51*, 1914-1922.

NORDMAN, H. E., and MOHR, W. H. 1963. Algin: Beer foam's best friend. Am. Brewer *96*, 22-25.

NORTH, W. J., and HUBBS, C. L. 1968. Utilization of kelp-bed resources in Southern California. Fish Bull. *139*, Dept. Fish and Game, Calif.

OKAZAKI, A. 1971. Seaweeds and Their Uses in Japan. Tokai Univ. Press, Japan.

PENMAN, A., and SANDERSON, G. R. 1972. A method for the determination of uronic acid sequence in alginates. Carbohydrate Res. *25*, 273-282.

PERCIVAL, E., and McDOWELL, R. H. 1967. Chemistry and Enzymology of Marine Algal Polysaccharides. Academic Press, New York.

PESCHARDT, W. J. S. 1946. Manufacture of artificial edible cherries, soft sheets and the like. U.S. Pat. 2,403,547.

POARCH, A. E., and TWIEG, G. W. 1957. Gel-forming composition and method. U.S. Pat. 2,809,893.

REES, D. A. 1969. Structure, conformation, and mechanism in the formation of polysaccharide gels and networks. *In* Advances in Carhohydrate Chemistry and Biochemistry, Vol. 24, M. L. Wolfrom, and R. S. Tipson (Editors). Academic Press, New York.

SCHMIDT, E., and VOCKE, F. 1926. Polyglycuronic Acids (I.) Chem. Ber. *59*, 1585-1588. (German)

SCHWEIGER, R. G. 1962. Acetylation of alginic acid. I. Preparation and viscosities of algin acetates. J. Org. Chem. *27*, 1786-1789.

SCHWEIGER, R. G. 1967. Process of preparing sulfate esters of algin. U.S. Pat. 3,349,078.

SMADAR, Y. 1972. Extruded food products and method. U.S. Pat. 3,650,766.

SMIDSRØD, O., HAUG, A., and LARSEN, B. 1963. Degradation of alginate in the presence of reducing compounds. Acta. Chem. Scand. *17*, 2628-2637.

SMIDSRØD, O., HAUG, A., and LARSEN, B. 1967. Oxidative-reductive depolymerization: a note on the comparison of degradation rates of different polymers by viscosity measurements. Carbohydrate Res. *5*, 482-485.

SMITH, F., and MONTGOMERY, R. 1959. The Chemistry of Plant Gums and Mucilages. Van Nostrand Rheinhold, New York.

SPERRY, G. D. 1953. Algin stabilizers in chocolate mixes. Ice Cream Rev. *37*, No. 3, 74, 146-148, 150-153.

SPERRY, G. D. 1955. Stabilizers and HTST. Ice Cream Field *65*, No. 6, 80, 82, 84-87.

STANFORD, E. C. C. 1881. Manufacture of useful products from seaweed. Brit. Pat. 142.

STANFORD, E. C. C. 1883. On algin: A new substance obtained from some of the commoner species of marine algae. Chem. News *96*, 254-257.

STANFORD, E. C. C. 1884. On algin. J. Soc. Chem. Ind. (London) *3*, 297-303.

STEINER, A. B. 1947. Manufacture of glycol alginates. U.S. Pat. 2,426,125.

STEINER, A. B. 1948A. Algin gel-forming compositions. U.S. Pat. 2,441,729.

STEINER, A. B. 1948B. Production of oil-in-water emulsions. U.S. Pat. 2,455,-820.

STEINER, A. B. 1953. Method of stabilizing foam forming on malt beverages. U.S. Pat. 2,659,675.

STEINER, A. B., and McNEELY, W. H. 1949. Substituted alkylene glycol esters of alginic acid. U.S. Pat. 2,463,824.

STEINER, A. B., and McNEELY, W. H. 1950A. High-stability glycol alginates and their manufacture. U.S. Pat. 2,494,911.
STEINER, A. B., and McNEELY, W. H. 1950B. Higher alkylene glycol esters of alginic acid. U.S. Pat. 2,494,912.
STEINER, A. B., and McNEELY, W. H. 1951. Organic derivatives of alginic acid. Ind. Eng. Chem. *43*, 2073-2077.
STEINER, A. B., and McNEELY, W. H. 1954. Algin in review. *In* Advances in Chemistry Series No. 11, American Chemical Society, Washington, D.C.
STEINER, A. B., and ROTHE, L. B. 1949. Stabilizer for icing. U.S. Pat. 2,474,-019.
SZCZESNIAK, A. S. 1968. Artificial fruits and vegetables. U.S. Pat. 3,362,831.
VINCENT, D. L. 1960. Oligosaccharides from alginic acid. Chem. Ind. (London) 1109-1111.
WEINGAND, R. 1957. Improving the adhesion between synthetic sausage casings and contents. U.S. Pat. 2,785,074.
WHISTLER, R. L., and KIRBY, K. W. 1959. The composition of alginic acid of *Macrocystis pyrifera.* Z. Physiol. Chem. *314*, 46-48.

CHAPTER 12

Erwin Heckman

Starch and Its Modifications for the Food Industry

INTRODUCTION

The use of modified starches in the food industry has been almost limitless. As new foods, new starch modifications and new techniques for utilizing the property of starch are developed, new applications are being discovered continually. There is an extremely dynamic, ever-changing field of interaction between the food and starch technology.

Starch is the first visible product of photosynthesis in most green plants and is found in the chloroplasts and chromatopheres of green parts in the form of minute granules. This kind of starch is known as assimulation starch. Assimilation starch is dissolved during darkness within the chloroplasts by the action of a starch-splitting enzyme and passes into water-solution as glucose which is conveyed downward to those parts of the plant requiring food. In its descent, some is stored up in medullaryray cells, and in various parts of the xylem, phloem, pith and cortex in the form of small grains. Such starch has been termed translocation starch.

Considerable starch, however, is carried to the underground parts, such as rhizomes, tubers, corms, bulbs or roots, or into aerial storage tissues, such as the storage cells of seeds, where the leucoplasts store it in the form of larger-size grains called reserve starch. This type of starch is generally characteristic for the plant in which it is found and it constitutes stored up food for the plant during that period of the year when the vegetative processes are more or less dormant.

MICROSCOPIC IDENTITY

Starch is a carbohydrate polymer synthesized within the plant by the chemical interlinking of hundreds and thousands of individual glucose units to form long-chain molecules, in the form of starch granules. Starches are unique among the carbohydrates in occurring as discrete granules, whose characteristics vary from one plant source to another. The microscopist looks for the following to identify starch granules: shape of the grain, size in microns, position of the hilum, and appearance under polarized light, etc.

Starches can be obtained from most plants by filtering the pulped or ground plant on coarse cloth and settling the starch granules from the filtrate. They may be distinguished by a microscopic examination of the size and shape of the granules.

Rice Starch

The smallest granules are rice, about 3-8 μ in diameter and polygonal in shape which tend to aggregate in clusters (Fig. 12.1).

Corn

Corn granules are polygonal and rounded (Fig. 12.2). The average size is 15 μ (range from about 5–25 μ). The hilum in granules is located in the center. The waxy variety of corn has granules similar in shape to those of ordinary field corn. They can be distinguished by staining with iodine. Ordinary commercial corn starch stains deep blue, while the waxy varieties stain reddish.

Tapioca

The granules are rounded and truncated at one end to form kettle drum shapes (Fig. 12.3). The average size is about 20 μ (range 5–35 μ). Many of the granules show a hilum located at the center.

Wheat

Wheat starch appears to have flat, round or elliptical granules which tend to cluster in two size ranges (Fig. 12.4). The small ones range from about 2–10 μ and the large ones from 20–35 μ.

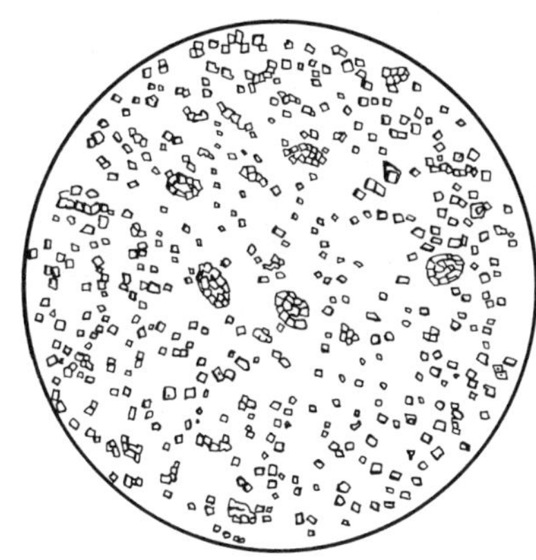

FIG. 12.1. RICE STARCH

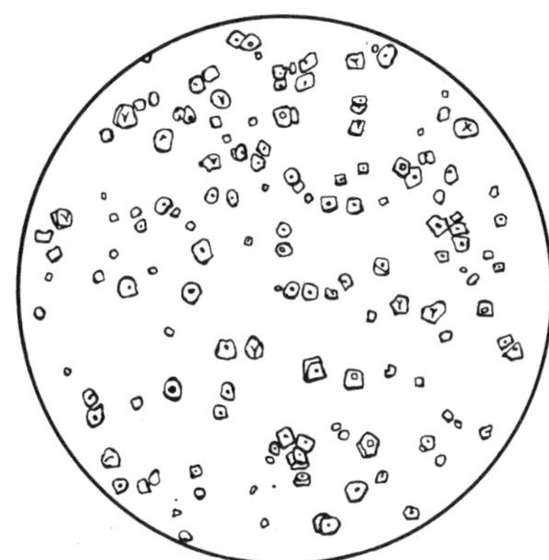

FIG. 12.2. CORN

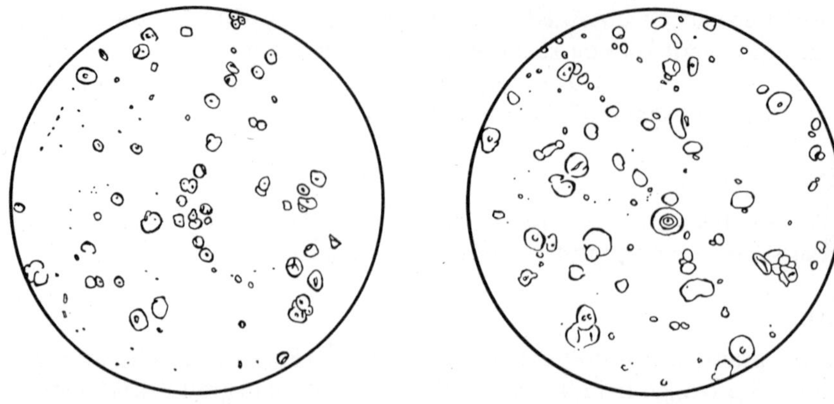

FIG. 12.3. TAPIOCA

FIG. 12.4. WHEAT

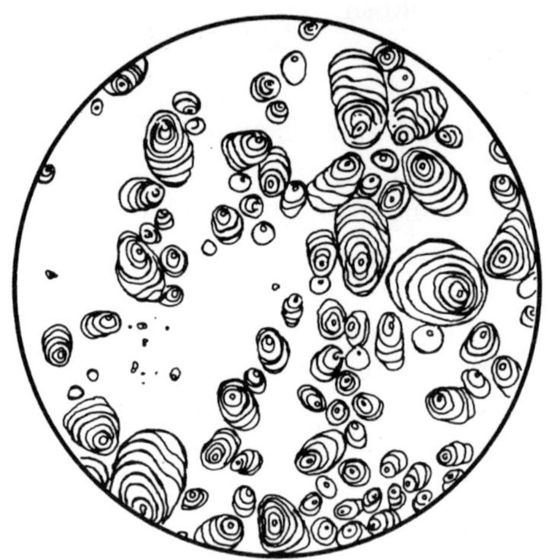

FIG. 12.5. POTATO

Potato

This starch has the largest granules of any of the common commercial starches. They range from about 15-100 microns and are oval and egg shaped (Fig. 12.5). This granule has the hilum located toward one end. Its granules also contain detectable striations.

COMPOSITION OF THE STARCH GRANULE

Starch grains are found scattered within the cytoplasm of plant cells (Fig. 12.6). They can easily be seen under an ordinary light

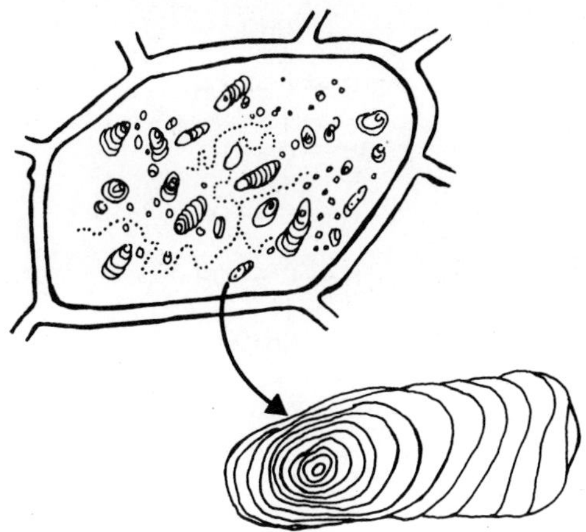

FIG. 12.6. CELL OF SOLANUM TUBEROSUM (POTATO) SHOWING RESERVE STARCH GRAINS

microscope. The inner portion of each grain is the soluble starch, amylose. The grain's outer layer is composed of the relatively insoluble starch, amylopectin. Both amylose and amylopectin can be split into the soluble glucose molecules of which they are composed.

Starch granules are largely composed of carbohydrate but contain minor constituents which may influence the property of the granules. Cereal starches contain approximately 0.5% fatty acids. These fatty acids are adsorbed on the starch. By treating with a suitable solvent, the acid may be completely extracted. Phosphorous in potato is bound as esterified phosphate at carbon atom 6 of the various D-glucose units.

Starch is a carbohydrate polymer having the chemical formula of $(C_6H_{10}O_5)n$ which is generally found as the visible product of photosynthesis in most green plants. The plant during photosynthesis uses two different mechanisms to synthesize its starch. It may form long linear units—amylose. A single glucose unit is illustrated in Fig. 12.7. When the amylose chain is built up to a length of approximately a dozen glucose units, the second mechanism may attach a glucose unit in a branching position. Glucose groups are then added to these branches until new branch points are started—amylopectin, a branched polymer of several thousand glucose units.

Amylose

Amylose is a linear polysaccharide, the straight chain component of naturally occurring starches. It is illustrated in Fig. 12.8 and 12.9. The repeat unit, derived from glucose $(C_6H_{10}O_5)$ is in the form of

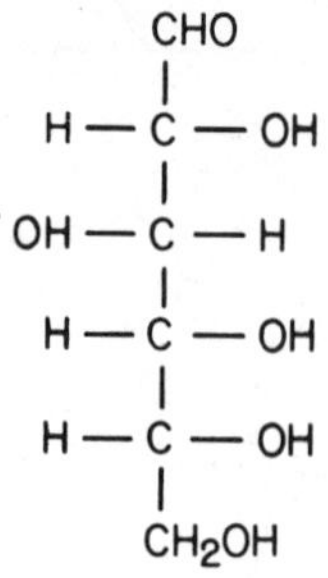

$$
\begin{array}{c}
\mathrm{CHO} \\
| \\
\mathrm{H-C-OH} \\
| \\
\mathrm{OH-C-H} \\
| \\
\mathrm{H-C-OH} \\
| \\
\mathrm{H-C-OH} \\
| \\
\mathrm{CH_2OH}
\end{array}
$$

FIG. 12.7. GLUCOSE

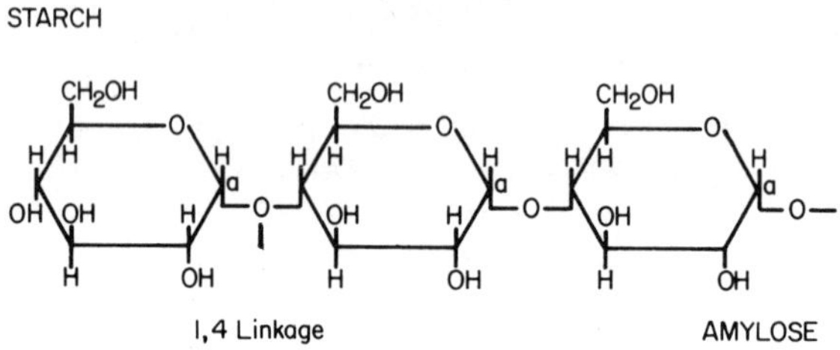

STARCH

1,4 Linkage

AMYLOSE

FIG. 12.8. AMYLOSE

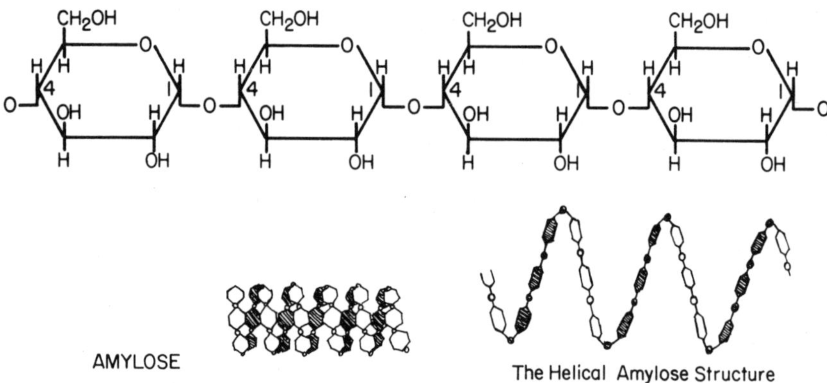

FIG. 12.9. AMYLOSE FRACTIONS SHOWING THE HELICAL STRUCTURE

glucopyranose. Successive glucose molecules are interlinked by oxygen bridge to form a polymer chain.

The oxygen bridge (C—O—C) is called a glycosidic bridge and in starch is a 1,4 type linkage. In this designation, alpha (α) indicates that the oxygen linkage is always located on the same side of the pyranose ring, and 1,4 refers to the positions of this oxygen linkage between the molecules.

Amylopectin

This fraction of starch is a highly branched, tumbleweed like configuration composed of linear chains similar to those of amylose. Branching occurs at a 1,6 linkages (Fig. 12.10). Branching is very effective in preventing gel formation by preventing the chains from becoming associated to produce a gel (Fig. 12.11).

THE HYDROGEN BOND AN ENERGY RELATIONSHIP BETWEEN ATOMS

All carbohydrates contain hydroxyl groups that attract water. Between these extremes there are interactions of intermediate energies which produce clusters of the molecular aggregates. The process

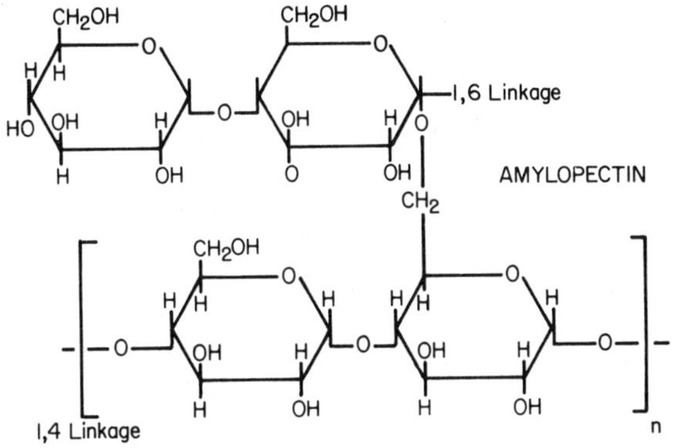

1,6 Linkage

1,4 Linkage

FIG. 12.10. AMYLOPECTIN

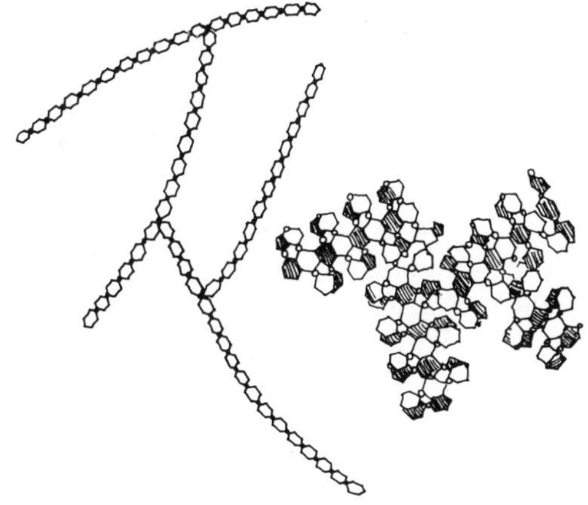

From Radley (1968)

FIG. 12.11. AMYLOPECTIN BRANCHING

TABLE 12.1

COMPARISON OF AMYLOSE AND AMYLOPECTIN

	Amylose	Amylopectin
Reaction with iodine	Intense Blue	Red Violet
Molecular weight	250,000	1,000,000
Chain length coverage		
Number of glucose residues per nonreducing end group	2,000 or more	20–30
X-ray analysis	High degree of crystalinity	Amorphorus
Solubility in water	Soluble	Insoluble
Stability in aqueous solutions	Retrogrades	Stable
Film (acetylated)	Tough	Brittle

Source: Radley (1968).

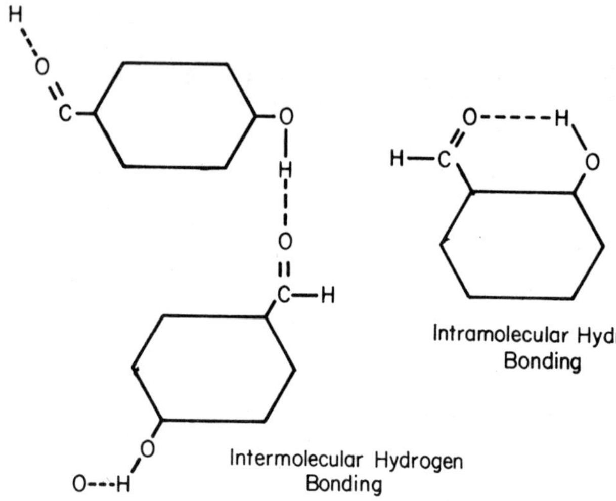

Intermolecular Hydrogen Bonding

Intramolecular Hydrogen Bonding

FIG. 12.12. MOLECULAR HYDROGEN BONDING

of forming these complexes is called association. The hydrogen bond is such an associative interaction. The energy breaking a hydrogen bond is of the order of a few kilocalories; clusters of molecules are held together in specific orientation and often in well defined numbers; the equilibria among the clusters and their component molecular parts are rapid and reversible.

The theory of the hydrogen bond remains one of the frontiers of our knowledge of chemical complexes. The importance of this chemical reaction is to be found in the hydrogen bonding properties of water.

The molecules of water that are in direct contact with the hydroxyls along the carbohydrate chains may be partially immobilized

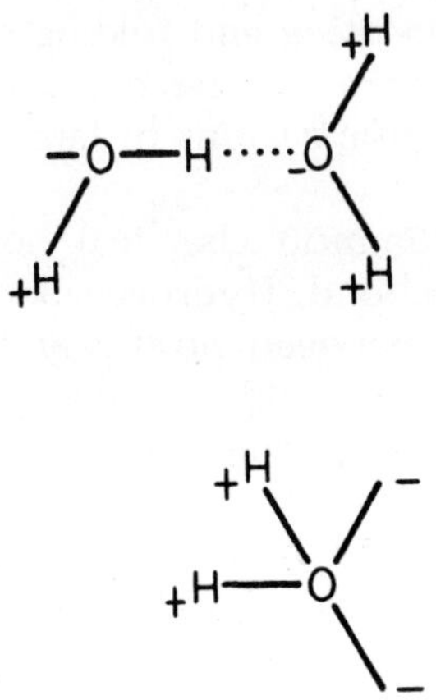

FIG. 12.13. WATER HYDROGEN BONDING

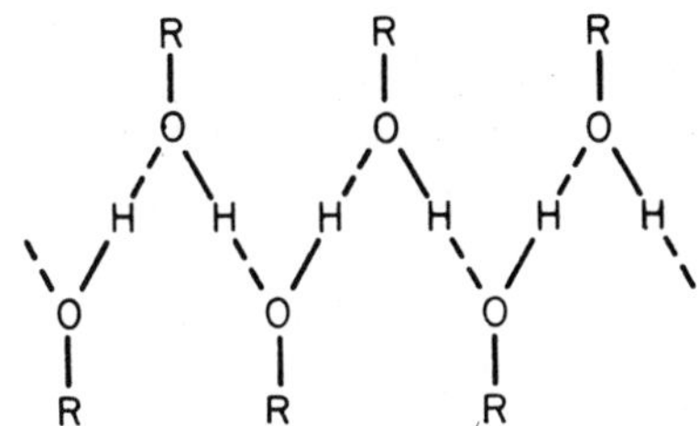

FIG. 12.14. ALCOHOL HYDROGEN BONDING

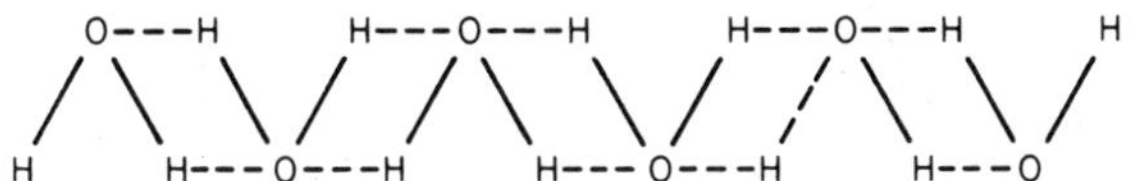

From Pimentel and McClellan (1960)

FIG. 12.15. POLAR REGIONS OF A MOLECULE

through hydrogen bridging. These intereactions become diffused throughout since they exist in aqueous systems. Many of the organic hydrated structures occur because of the hydrogen bond, so that the described structures are affected through hydrogen bonding by polar attraction.

Polarity is important in understanding both the geometry and the chemical characteristics of large molecules. The specific geometry exists in part because the polar groups on one part of the molecule attract polar groups on the other part of the same molecule. This produces the specific twisting and folding of the molecule which is all important to its chemical characteristics. Thus polarity brings small molecules, or specific regions of large molecules, into definite geometric relationship.

One of the most common chemical bonds produced by polar groups is the hydrogen bond. Hydrogen bonds are produced by the electrostatic attraction between positively charged hydrogen atoms on one part of the molecule and negatively charged atoms of oxygen on the same or another molecule. The oxygen atoms are negatively charged because their nuclei attract large numbers of electrons around them. Because the hydrogen bond occurs between polar regions of a molecule, it is like all polar attractions—weak.

BEHAVIOR OF STARCH POLYMERS IN SOLUTION

The hydroxyl groups impart hydrophilic properties to the polymer which lead to an affinity in water. The molecules of amylose are linear, having a tendency to become oriented parallel to one another and approach each other closely enough to permit association of one polymer with another through hydrogen bonding. This phenomenon of association between amylose molecules leads to retrogradation.

Amylopectin is a larger polymer than amylose. The size and branched nature of amylopectin interferes with the mobility of its molecules and their tendency to become oriented closely enough to

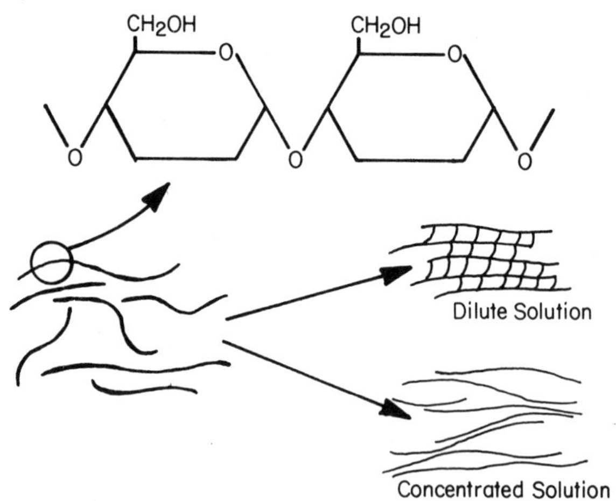

From Furia (1968)

FIG. 12.16. BEHAVIOR OF STARCH FRACTIONS IN SOLUTION: AMYLOSE

permit the extensive hydrogen bonding necessary for retrogradation to occur.

The majority of starches have nearly identical ratios of amylose to amylopectin (Table 12.2).

Throughout the granule, the linear branches are associated by hydrogen bonding. The relative arrangement in space of the hydroxyl group on C-2 and the oxygen connecting C-1 to the next glucose unit is the major structural difference between starch and cellulose, both of which are composed of D-glucose chains.

Cellulose
1,4 β linkage—Celloboise
Molecular weight—50,000–400,000

Starch
1,4 α linkage—Glucose
Amylose = 250,000 Mol. Wt.
Amylopectin = 1,000,000 Mol. Wt.

PROPERTIES OF STARCHES

Functional properties of starch during processing cooking are gelatinization temperature, rate of maximum viscosity measured by temperature and time, and general flow properties (short or long bodied starches). After processing the properties produced clarity, gel character, texture, flavor, water retention (syneresis), freeze-thaw stability, acid resistance and general storage stability. These properties can be best demonstrated by heating in water suspension as would be encountered in a model system.

TABLE 12.2

AMYLOSE CONTENT OF COMMON STARCHES

Starch	Amylose Content (%)
Potato	22
Tapioca	17
Arrowroot	20
Corn	20–28
Sorghum	23–28
Rice	16–17
Wheat	17–27
Oat	23–24
Waxy maize	0–7
Waxy sorghum	0–7
Waxy rice	0–7
Sago	27

Source: Glicksman (1969).

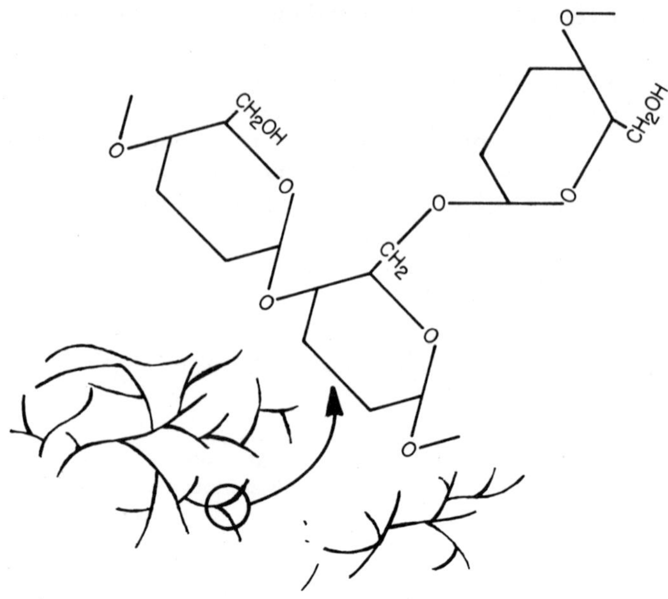

From Furia (1968)

FIG. 12.17. BEHAVIOR OF STARCH FRACTIONS IN SOLUTION: AMYLOPECTIN

As previously described, starch is insoluble in cold water because the molecular network is bound by H-bonds. When starch suspensions in water are heated, the following occurs:

(1) Time, temperature and shear exert sufficient caloric energy on the hydrogen bonds which become weakened so that water can be absorbed by the granules. Some of the granules begin to swell losing their characteristic polarization crosses (birefringence). If followed by microscopic inspection, it originates at the hilum and spreads rapidly to the periphery of the granule. To follow this granule change microscopically under controlled conditions is to observe the type of granule swelling. The following should be observed: a slide is made of unswollen starch and water and placed on the microscope slide. When the granules are in focus, slowly introduce 0.1 N $Ca(NO_3)_2$ solution under the cover slip. After the $Ca(NO_3)_2$ solution is diffused, a gas bubble appears and it rapidly expands until the granule is in total collapse, the granule now appears like broken glass fragments.

(2) After gelatinization begins with combined heat, clarity occurs with a marked increase in viscosity which reaches a peak. The

hot paste is a mixture of swollen starch granules and granule fragments with colloidally and molecularly dispersed starch molecules.

(3) When heat is discontinued, the starch solubles diffuse back and the system becomes a gel-like mass held together by associative bonds or retrograde to substantially increase the viscosity. Each starch species differs and produces different types of curves over the complete cycle of gelatinization, continued cooking and finally, cooling (see Fig. 12.18).

At temperatures above the gelatinization range, the hydrogen bonds continue to be disrupted and water molecules become attached to the liberated hydroxyl group. Parallel with the swelling of the granule, there is an increase in the starch solubility, paste clarity and viscosity. These properties are dependent on the pasting temperatures. Their quantitative relationship might be of interest. The starch granules require the excess of water in order to swell freely without mechanical disintegration. For given starch species, the swelling power and solubility curves are very similar, indicating a direct interrelation of these two functions.

On cooling, the clarity of the starch cook will sometimes decrease and the viscosity rises to varying degrees. In extreme cases, the starch sol will form an opaque, rigid gel. The exact way in which a starch

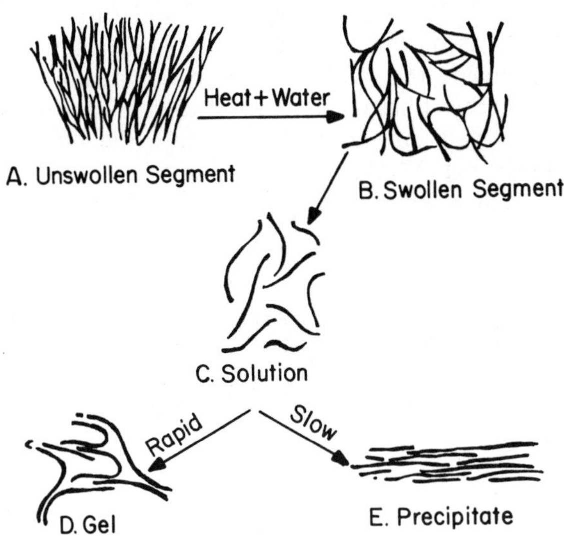

From Glicksman (1969)

FIG. 12.18. SWELLING AND SOLUBILITY BEHAVIOR OF STARCH

behaves during the process of gelatinization depends upon numerous factors; plant origin of the starch is one of the most important (see Table 12.3). The average increase in the diameter of granules of various starches going from anhydrous conditions to a water saturated atmosphere is indicated in Table 12.3, together with the observed increase in water sorption. The abundance of hydroxyl groups in the starch granules is a primary factor in their tendency to absorb moisture. Starches when equilibrated under normal atmospheric conditions usually contain the following amounts of hydroxyl groups:

Corn	10-12%
Tapioca	12-14%
Potato	16-18%

The concentration at which a starch is cooked influences its behavior particularly as it relates to the swelling power of granules (See Table 12.4). When concentrations are above this level the swollen granules will be capable of forming a continuous phase of swollen granules in which essentially all available water is entrapped. Below this concentration there will be free water and phase separation of the hydrated granules from the aqueous solution will occur (See Table 12.5).

With continued swelling of the granules, starch molecules that have become fully hydrated separate from the intricate nuclear network and diffuse with the surrounding medium. In a dilute starch paste, an equilibrium is reached where the concentration of soluble starch in the water inside the swollen granule is the same as that in the surrounding aqueous phase. In a concentrated starch paste, the individual granules gelatinize and swell freely while all the available water is imbibed (see Fig. 12.19).

Starches readily fall into three groups in decreasing order of association or increasing order of swelling: (1) cereal starches—corn, (2)

TABLE 12.3

EFFECT OF WATER-SATURATED ATMOSPHERE ON GRANULE DIAMETER AND WATER SORPTION OF VARIOUS STARCHES

Starch Source	Average Increase In Granule Diameter (%)	Water Sorption per 100 gm Dry Starch (gm)
Corn	9.1	39.9
Potato	12.7	50.9
Tapioca	28.4	42.9
Waxy Corn	22.7	51.4

Source: Furia (1968).

TABLE 12.4

CRITICAL CONCENTRATIONS OF COMMON STARCHES

Species	Critical Concentration Value[1]
Potato	0.1
Sago	1.0
Tapioca	1.4
Corn	4.4
Waxy Maize	1.6
Wheat	5.0
High Amylose Corn	20.0

Source: Glicksman (1969).
[1]These values represent the grams of dry basis starch required per 100 ml of water in order to produce a paste at 95°C in which the swollen granules occupy virtually the entire volume; thus, there is almost no "free" water between the swollen granules.

TABLE 12.5

STARCH GRANULE SIZES AND GELATINIZATION RANGES

Starch	Type	Granule Size (μ)	Gelatinization Range[1] (Kofler Hot-stage method) (°C)		
			Initiation	Midpoint	Completion
Corn	Cereal	5–25	62.0	67.0	72.0
High Amylose Corn (55%)	Cereal	—	67.0	80.0	?
Waxy Corn (Waxy Maize)	Cereal	10–25	63.0	68.0	72.0
Sorghum	Cereal	5–25	68.0	73.5	78.0
Waxy Sorghum	Cereal	6–30	67.5	70.5	74.0
Barley	Cereal	5–40	51.5	57.0	59.5
Rye	Cereal	5–50	57.0	61.0	70.0
Wheat	Cereal	2–45	58.0	61.0	64.0
Oat	Cereal	5–12			
Rice	Cereal	3–8	68.0	74.5	78.0
Potato	Tuber	15–100	59.0	63.0	68.0
Tapioca, Brazilian	Root	5–35	49.0	57.0	64.5
Tapioca, Dominican	Root	5–35	58.5	64.5	70.0
Sago	Pitch	15–70	?	66.2	70.0
Arrowroot	Root	15–50	66.2	66.2	70.0

[1]From Schoch and Maywald (1956, 1957).

root starches—tapioca, (3) tuber starches—potato. Potato starch swells rapidly and greatly while corn starch shows a relatively restricted swelling. Tapioca starch has immediate swelling power.

Starch origin is a governing factor in the behavior of the gel on cooling. The amylose content and the molecular size of the amylose fraction influence the tendency of starch sols to thicken or gel on cooling. Opacity is caused by the amylose molecules which associate through hydrogen bonding.

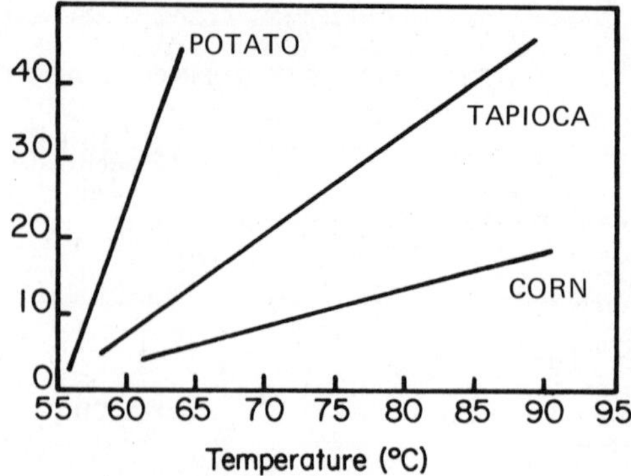

FIG. 12.19. SWELLING PATTERNS OF TYPICAL GRANULAR STARCHES

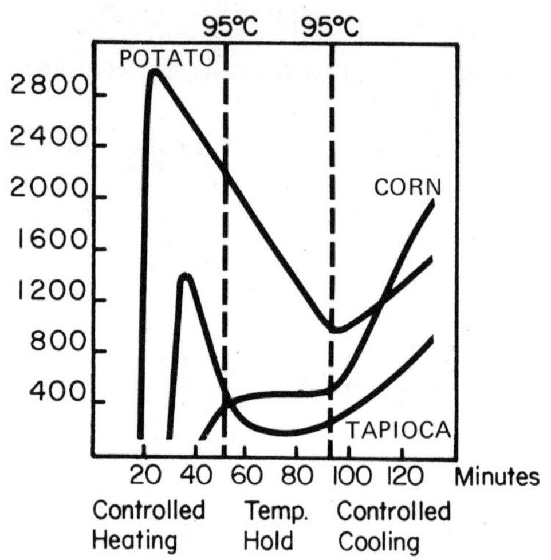

FIG. 12.20. BRABENDER SWELLING PROFILES AT 6% SOLIDS

TABLE 12.6

INFLUENCE OF STARCH ORIGIN ON BEHAVIOR OF GEL ON COOLING

Species	Amylose Content (%)	Type of Gel
Corn	28	Opaque, rigid
Waxy Corn	0	Nonrigid
Tapioca	18	Clear, nonrigid
Potato	23	Clear, nonrigid

Tapioca and potato starches have amyloses of higher molecular weights than corn starch amylose, which contains more branch points that prevent them from becoming oriented and approaching each other closely enough to prevent hydrogen bonding.

RETROGRADATION

Starch is a polymolecular compound exhibiting a colloid chemical property, retrogradation (denaturation by gradual loss of solution properties). The aqueous starch solutions during the process of aging become opalescent and turbid in time, and gradual precipitation begins. The viscosity decreases at the same time, and the resistance to enzymes increases. This is a slow, spontaneous aggregation due to Brownian movement. Aggregation occurs upon collision and the aggregates gradually grow bigger. The aggregates can be more or less oriented (crystalline) depending on the components; amylose aggregates more easily and forms particles with better orientation than amylopectin.

Retrogradation depends upon molecular weight (chain length); maximum effect is of medium chain length. Long macromolecules may be randomly branched and will not orient in bundle formation; and short chains will have intense Brownian movement which hinders intermolecular attraction. This mutual intermolecular attraction is caused by hydrogen bonding (see Fig. 12.21).

SYNERESIS

The structural stability of various gels is very different, depending on the structure, concentration and the affinity of the colloidal

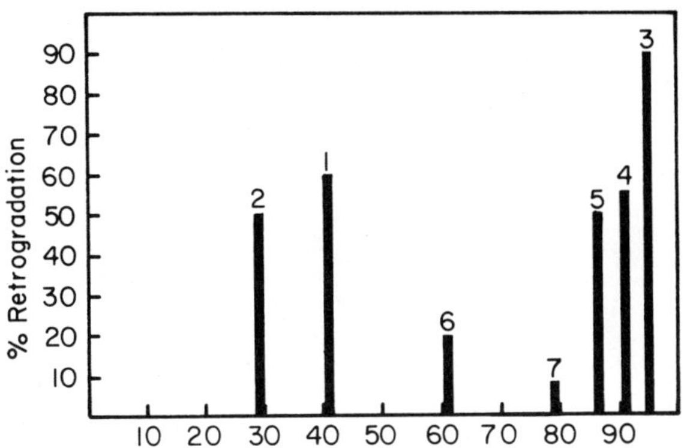

FIG. 12.21. RATE OF RETROGRADATION OF VARIOUS STARCHES

1. Corn. 2. Wheat. 3. Potato. 4. Sweet potato. 5. Arrowroot. 6. Tapioca. 7. Waxy maize at 2% starch solutions at just above freezing.

structural elements for the liquid. Syneresis is the reversal of swelling. Various ions promote syneresis in the same order as they promote gelation and in the same order as they retard swelling. Syneresis can be regarded as flocculation or as a result of flocculation of some gel components. It can be assumed that only part of the colloidal particles are linked into a coherent framework and that some particles are still relatively free. They eventually collide due to Brownian motion and cling together as a result of hydrogen bonding. The coarser the meshwork, the more easily the liquid will be liberated from the gel. Due to hydrogen bonding, the coarseness increases, the size of the intermolecular space also increases, and then there is a spontaneous liberation of the liquid from the gel termed syneresis (see Fig. 12.22).

STARCH WITH OTHER FOOD INGREDIENTS

Prepared foods are compounded mixtures of various food ingredients. The influence of some major food ingredients on starch during heating of liquid systems, and a single ingredient effect on the food system are important to the understanding of the prepared food using starch. These ingredients effect the inherent characteristics of the starch and can be best understood by an examination of the changes in a typical food processing system (see Fig. 12.23).

Sugars impede swelling of the starch granule, by competing for the water present. Concentration of sugars affects the gel strengths of the starch. Sugars extend the gelatinization range by indicating that

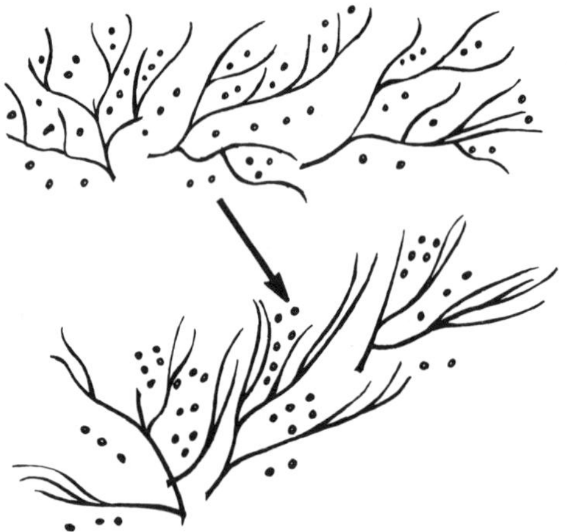

From Heckman (1959)

FIG. 12.22. SEPARATION OF WATER MOLECULES DURING FREEZING AND THAWING CYCLES

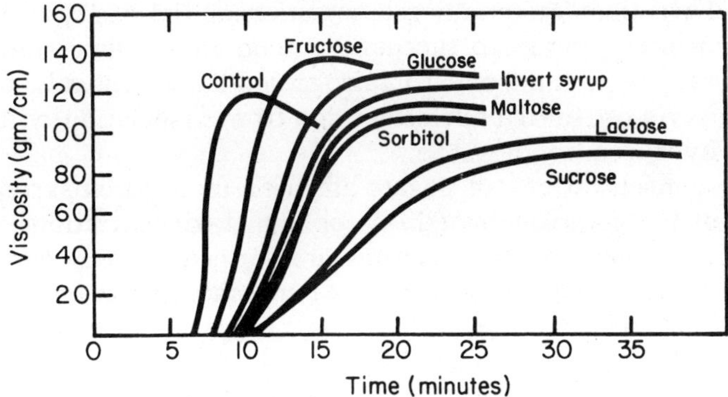

From Bean and Osman (1959)

FIG. 12.23. EFFECT OF DIFFERENT SUGARS ON GELATINIZATION

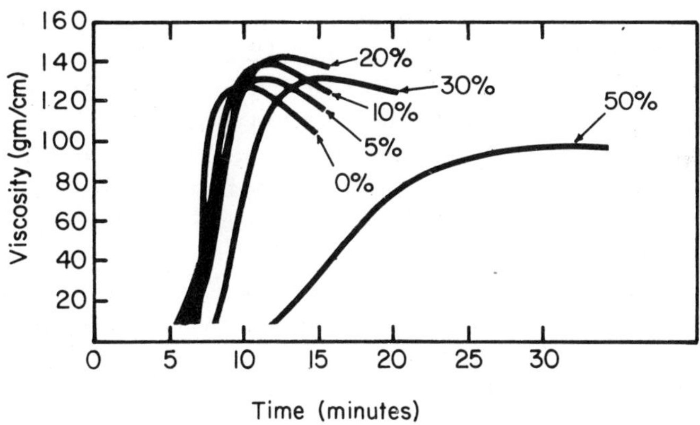

From Bean and Osman (1959)

FIG. 12.24. EFFECT OF DIFFERENT CONCENTRATIONS OF SUCROSE ON VISCOSITY

the birefrigerant granules are present at higher temperatures when viewed under the microscope. They also increase the apparent viscosity as recorded on the amylograph. These actions of sugars on starch prevent it from being cooked sufficiently to become properly hydrated. It is for this reason that it is best to cook the starch before adding the sugar to the food formulation (see Fig. 12.24).

The behavior of starch during cooking and cooling is influenced by the sucrose added. The higher the concentration of these water-soluble, hydroxyl-containing substances, the more will they exert increased inhibiting action on the swelling of the starch granule.

Sugars have their effect on gelatinization and viscosity decreases. These hydroxyl containing saccharides tend to stabilize starch gels and decrease their tendency to retrograde. They probably prevent the hydroxyls on the starch molecules from associating with each other by hydrogen bonding.

If high concentrations of sugars are used in a starch system, the strength of the gel is decreased on cooling. If concentrations of 10% or more are used, no gel structures are formed or only very weak ones. To overcome this effect, higher temperatures can be used with the aid of pressure cooking.

THE EFFECT OF pH ON STARCH

The normal range of pH for starches are 5.0-7.0 and cooking within this range, the variations are only minor in effect. Cooking at pH 5.0 or below and/or 7.0 or above tends to lower the gelatinization temperature and accelerate the whole cooking procedure. Under highly acidic pH's, hydrolysis of the glucosidic bonds may occur which will decrease the viscosity of the gel (see Fig. 12.25).

In sweetened products, the effect of acid is reduced to a considerable degree by the high sugar concentration. Addition of high sucrose concentrations reduce the swelling of the granules retarding the hydrolysis. The cause of a thin filling may be due to retardation of the starch because of insufficient cooking rather than overcooking with resulting acidic hydrolysis.

Acidic breakdown of the starch granule due to hydrolytic cleavage of the molecules usually results in a low viscosity during processing, followed by a rapid thinning. It also presents a shelf-life problem because of the poor resistance to hydrolysis which decreases viscosity

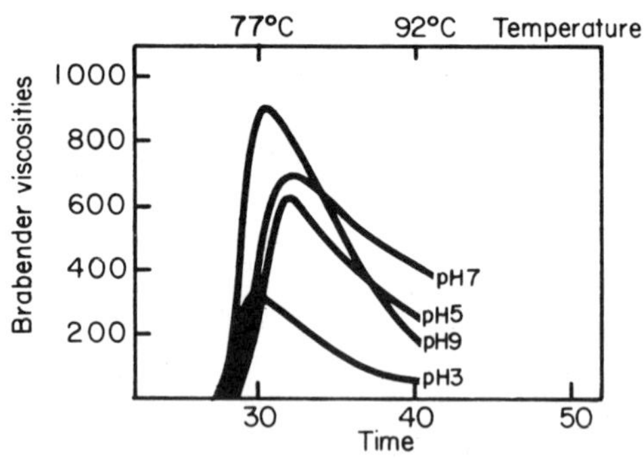

From Furia (1968)

FIG. 12.25. EFFECT OF pH ON STARCH

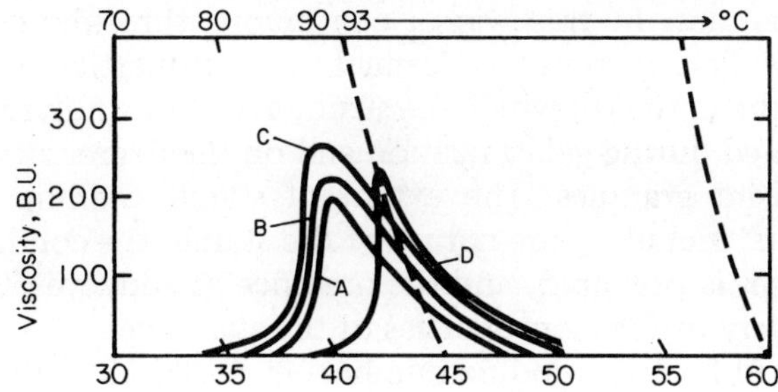

From Campbell and Briant (1957)

FIG. 12.26. EFFECT OF CITRIC ACID AND SUCROSE ON GELATINIZATION OF STARCH

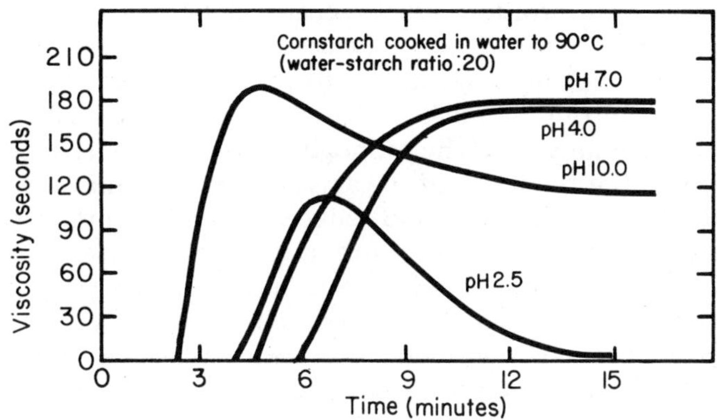

From Corn Industries Research Foundation (1964)

FIG. 12.27. pH EFFECT ON BREAKDOWN OF CORN STARCH

of the final gel and promotes hydrogen bonding to cause retrogradation so that syneresis also occurs (see Fig. 12.27).

THE EFFECT OF FATS AND OILS

Polymer molecules associate by means of secondary forces, such as hydrogen bonds or van der Waals' forces, and link up to form a continuous three-dimensional network. Gel formation by this process is usually thermally reversible, so that on warming the gel reverts back to a liquid. The formation of a starch gel occurs by a different mechanism. On heating above the gelatinization temperature, starch granules swell irreversibly. Some of the starch molecules become partly or completely detached from the granules, and on cooling, these molecules form hydrogen bonds with similar molecules from adjacent granules. In this way a continuous three-dimensional network of swollen granules is formed. The properties of such a gel depend on the extent to which the structures of the individual granules are maintained during gelatinization and on the degree of bonding between adjacent granules. The extent of starch swelling depends on many factors, including the nature of the starch, the conditions under which the gel is prepared, and the presence of additives. These all affect the widely differing properties of the starch gel.

Unsaturated fats with iodine numbers from 38–132 which are liquid to brittle fat (solid) have the same effect: none either increase or decrease the maximum viscosity when heated. All had the same effect in lowering the temperature at maximum viscosity. The strength of the final gels did not differ from control samples; however, some of the results were erratic because separated fat prevented formation of homogenous gel structure. Fats used were of 16–18 carbon atoms.

The addition of monoglycerides to processed foods which contain naturally occurring starches is based on the interaction of monoglyceride with amylose. The monoglyceride complexes with amylose and reduces the tendency of the amylose to leach out of the granule. This stabilizes the amylose and reduces its tendency to associate and retrograde. Thus it prevents gel formation, forms better textured gels, has less stickiness and retards the firming of bread (retrogradation) and other baked products for longer shelf-life stability.

However, there is considerable evidence that this complexing of amylose and surface-active agents does not explain the firming of bread upon standing. It has been shown that amylose becomes insoluble during baking and is not available to take part in the subsequent staling reaction. It is assumed that firming is caused by association of the amylopectin molecules occurring only between the ends of the branches, a few bonds being formed between adjacent molecules. With fewer bonds, the association cannot be as stable as between linear amylose molecules and can be broken more readily. Bread that becomes firm but not dry can be made to be soft by heating, since the bonds can be more readily broken between amylopectin molecules, but would not reverse the retrogradation of amylose. No mechanism that has yet been proposed appears adequate to explain all the observed facts. It is also possible that triglycerides with shorter fatty acids might differ in their effect on starch (see Fig. 12.28).

The inhibition of the swelling of starch granules may be attributed to the formation of a fatty layer around the granules which hinders the penetration of water into the granules. The mechanism of the effect produced by these ingredients is not clear and is still a matter of conjecture.

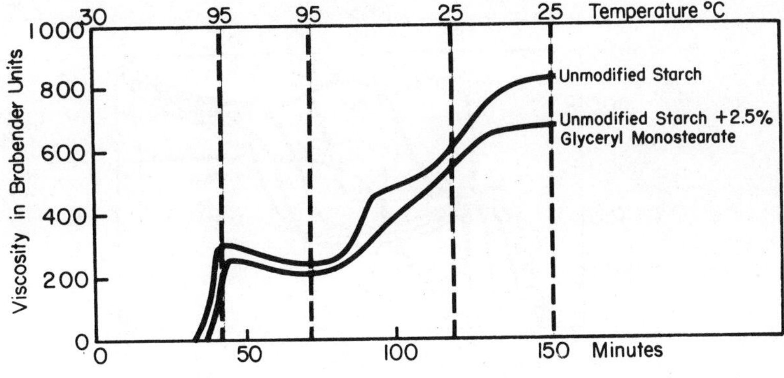

From Furia (1968)

FIG. 12.28. VISCOSITY EFFECT OF GLYCERYL MONOSTEARATE ON UNMODIFIED STARCH

EFFECT OF ELECTROLYTES ON STARCH

Electrolytes have a marked effect on swelling. The effect of different ions corresponds to their relative salting out effects on hydrophilic colloids in general. These ions can be classified in two series, as follows:

Anions
Sulfate, acetate, chloride, nitrate, chlorate bromide, iodide, and thiocyanate

Cations
Magnesium, lithium, sodium, potassium, and ammonium.

The two series are arranged in the order in which the ions increase swelling and decrease the gelatinization temperature. Sodium sulfate behavior with starch shown at different concentrations is given in Fig. 12.29.

In conclusion, there are many influencing factors that affect starch; those mentioned previously are some of the highlights. Those systems not reviewed in this chapter, i.e., protein, effect of mechanical injury to the granule, intense drying, etc., should be studied and may result in producing different effects in the processing of foods.

MODIFIED FOOD STARCHES

Starch is a principal ingredient of many foods, and it constitutes a major energy source, but it is also essential to the gross structure, texture and consistency of many food preparations. Food science has been limited in the past to the native starches, now many new types of

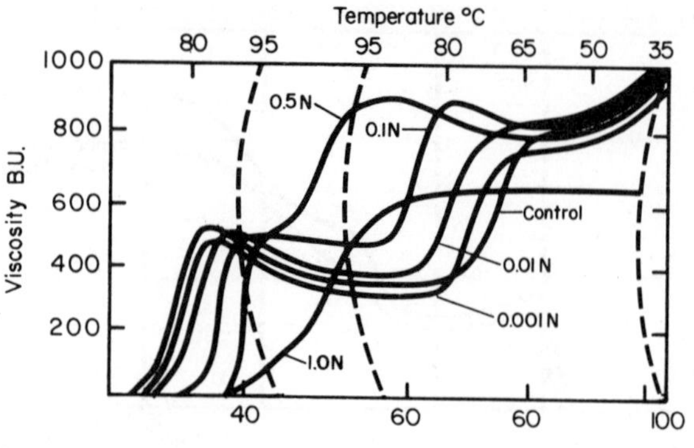

From Osman (1967)

FIG. 12.29. SODIUM SULFATE EFFECT ON VISCOSITY

modifications are becoming increasingly important due to the advancing food technology and requirements placed upon these new food preparations.

Pregelatinized Starches

Pregelatinized starches are prepared from starch slurries that are cooked and dried on heated drums. The dry film is then ground to a fine powder. This process renders the starch cold water swellable, and able to thicken and gelatinize; however, they show less thickening power and less tendency to set to a gel than the parent starch. This loss of gelling and thickening ability is thought to be due to retrogradation in the wet film when dried. When viewed under a microscope after reconstitution, there is evidence of destruction of the hydrated starch granules. They appear as collapsed balloons or broken glass instead of the characteristic granules.

Pregelatinized starches without being modified, exhibit extensive molecular reassociation similar to raw starch cooks. The gel gives a rough, grainy, opaque appearance which in many foods are unacceptable, especially in prepared instant puddings and sauces. There is evidence that these starch preparations show indications of reassociation with age, since they progressively give slower and lower cold-gel viscosity.

Use of Pregelatinized Starch in Food Applications

(1) Pregelatinized starch is used in instant-type puddings.

(2) Pregelatinized starch is used in cake mixes, as it increases water absorption and aids in air entrapment producing a moist cake with good volume. It also can add freshness and uniform texture.

(3) Pregelatinized starch aids in tolerance to varying additives of water and controls sugar crystal growth.
(4) When mixed with gelatin, pregelatinized starch can be used in whipped desserts.
(5) In prepared meat products, pregelatinized starch is used as a binder and moisture stabilizer.
(6) The starch protects fats by entrapping them, preventing oxidation in dry soap mixtures.
(7) This starch may be used as a glaze and for suspending cocoa.
(8) A pregelatinized starch properly treated and dried can yield gels that are smooth, glossy, nonstarchy or pasty, have good stability and excellent water-binding properties, preventing syneresis. The market for these pregelatinized starches is expanding very rapidly in easily prepared foods, frozen and canned complete meals, prepared dry mixes for baked items, instant foods and beverages.
(9) It is necessary to modify starches by chemical, physical, or mechanical means before drum drying (pregelatinizing) if smooth glossy, nongrainy, fast-hydrating starches with dry and wet stability for storage are desired.

Acid Modification

Acid thin boiling starches are obtained by treating starch with acid in water suspensions at sub-gelatinizing temperatures. The amount of acid and length of reaction time produces a controlled fluidity. After acid treatment, the starch is then washed free of acid and subsequently dried.

Differences When Compared to Its Parent Starch.—There is less hot paste viscosity, allowing the use of higher starch solids, as well as higher alkali number and lower cold to hot paste viscosity. In addition, acid-modified starch has lower iodine affinity, lower intrinsic viscosity, less granule swelling during gelatinization in hot water, increased stability in warm water below gelatinization temperature and higher critical absorption of sodium hydroxide. It resembles the parent starch in the following: it has the same physical insolubility in cold water and maintains its granule birefringence. It is used basically in the starch gum candies to give the product stronger-bodied short textured properties resulting in a softer tender gum by decreasing cooking time and the finished product, even at high moisture content, protects against syneresis.

Dextrins.—Dextrins are prepared by the action of heat and/or acid on starch by hydrolysis. The extent of the modification on degradation of the molecular structure depends on the temperature and time involved during processing. The general characteristics of heating the starch granule alone is illustrated by the different classes of dextrins. There are three important classes, and each of these has its own characteristic color, viscosity and solubility in cold water.

The structure of dextrins from starch: chemical evidence regarding the nature of the structural changes occurring during the thermal treatment of starch is limited. The inherent complexity of the native granule is difficult but some "model compounds" can be viewed to demonstrate the structural changes occurring in the solid.

TABLE 12.7

CHARACTERISTICS OF HEATING THE STARCH GRANULE AS ILLUSTRATED BY THE DIFFERENT CLASSES OF DEXTRINS

Type of Dextrin	Temperature (°C)	Catalyst	Time (Hr)	Color Development
British gum	54–104	none	10–20	Color before solubility (50%)
Canary	65–104	HCl[1]	6–18	Yellow before solubility (100%)
White	27–49	HCl[1]	3–8	No color (80%) solubility

Source: Wolfrom and Tipson (1967).
[1]Most often used commercially because it is a strong acid, disperses uniformly through the starch, and tends to volatilize during the last stages of dextrinization, so that neutralization of the dextrin may not be necessary.

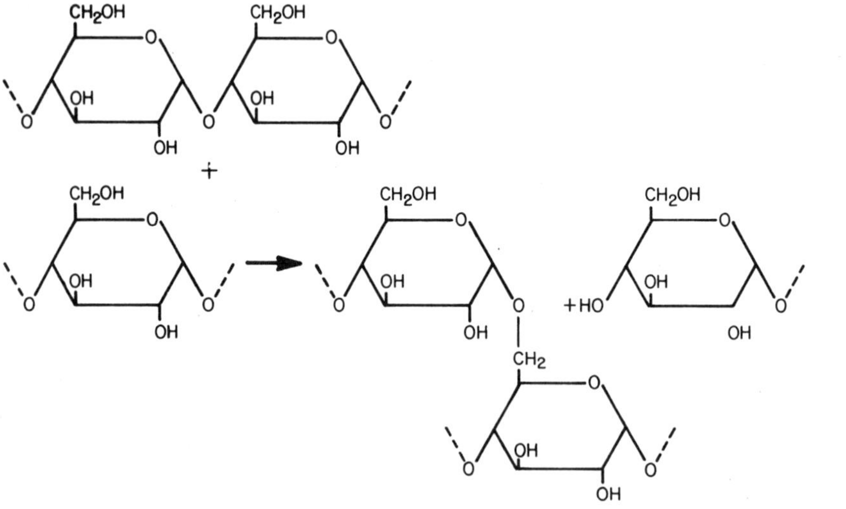

From Wolfrom and Tipson (1967)

FIG. 12.30. DEXTRINIZATION

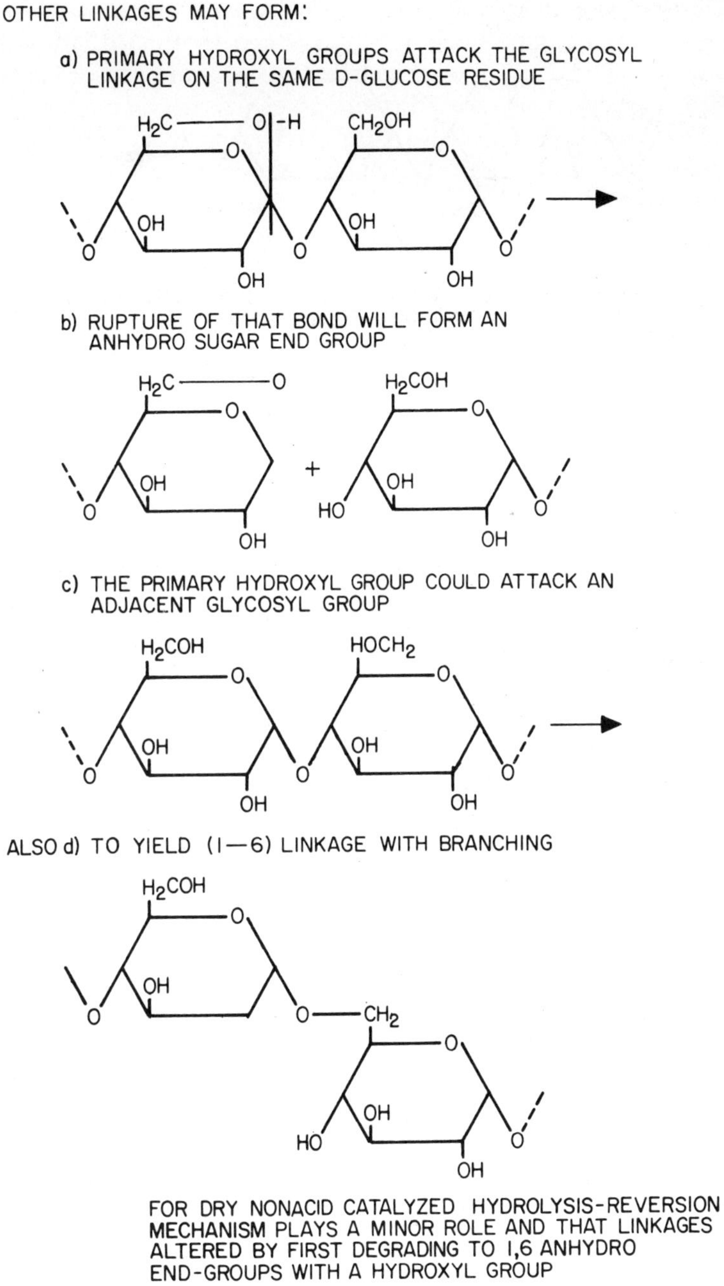

OTHER LINKAGES MAY FORM:

a) PRIMARY HYDROXYL GROUPS ATTACK THE GLYCOSYL LINKAGE ON THE SAME D-GLUCOSE RESIDUE

H_2C — O|-H CH_2OH
O   O
OH   OH
O   O   O
OH   OH

b) RUPTURE OF THAT BOND WILL FORM AN ANHYDRO SUGAR END GROUP

H_2C — O H_2COH
O   +   O
OH   OH
O   HO   O
OH   OH

c) THE PRIMARY HYDROXYL GROUP COULD ATTACK AN ADJACENT GLYCOSYL GROUP

H_2COH $HOCH_2$
O   O
OH   OH
O   O   O
OH   OH

ALSO d) TO YIELD (1—6) LINKAGE WITH BRANCHING

H_2COH
O
OH
O   O—CH_2
O
OH
HO   O
OH

FOR DRY NONACID CATALYZED HYDROLYSIS-REVERSION MECHANISM PLAYS A MINOR ROLE AND THAT LINKAGES ALTERED BY FIRST DEGRADING TO 1,6 ANHYDRO END-GROUPS WITH A HYDROXYL GROUP

From Wolfrom and Tipson (1967)

FIG. 12.31. DIFFERENT LINKAGE FORMATION DURING DEXTRINIZATION

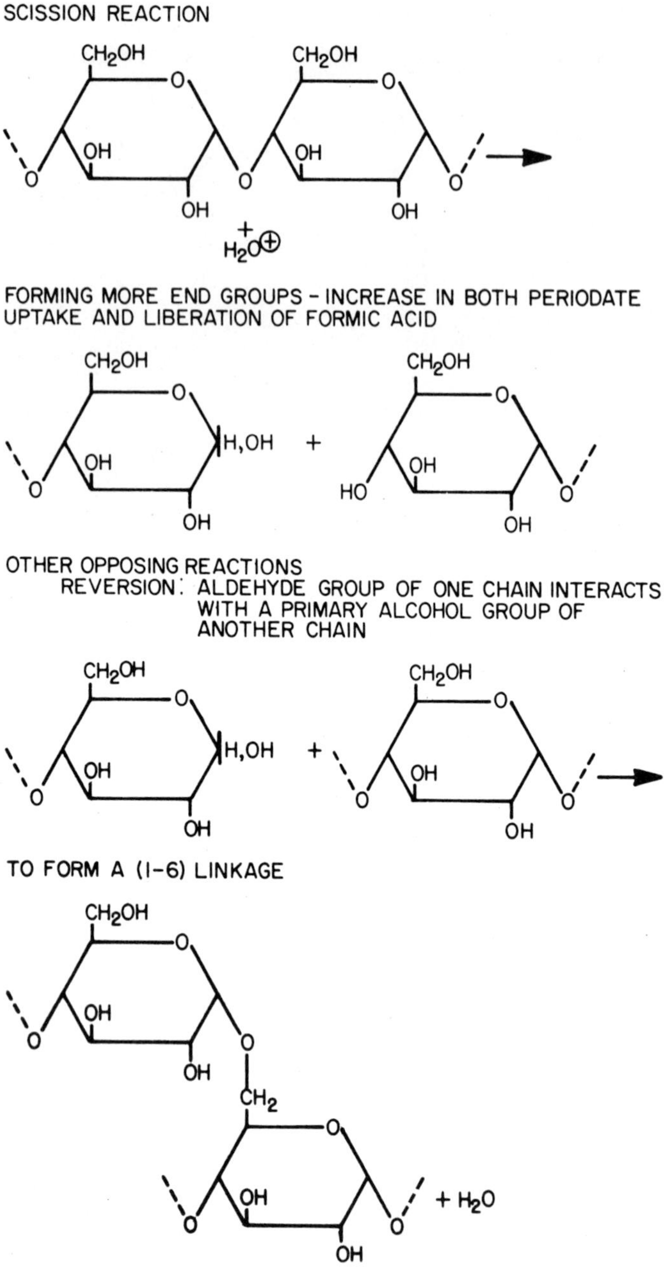

SCISSION REACTION

FORMING MORE END GROUPS - INCREASE IN BOTH PERIODATE UPTAKE AND LIBERATION OF FORMIC ACID

OTHER OPPOSING REACTIONS
REVERSION: ALDEHYDE GROUP OF ONE CHAIN INTERACTS WITH A PRIMARY ALCOHOL GROUP OF ANOTHER CHAIN

TO FORM A (1-6) LINKAGE

From Wolfrom and Tipson (1967)

FIG. 12.32. ACID-CATALYZED DEXTRIN REACTIONS

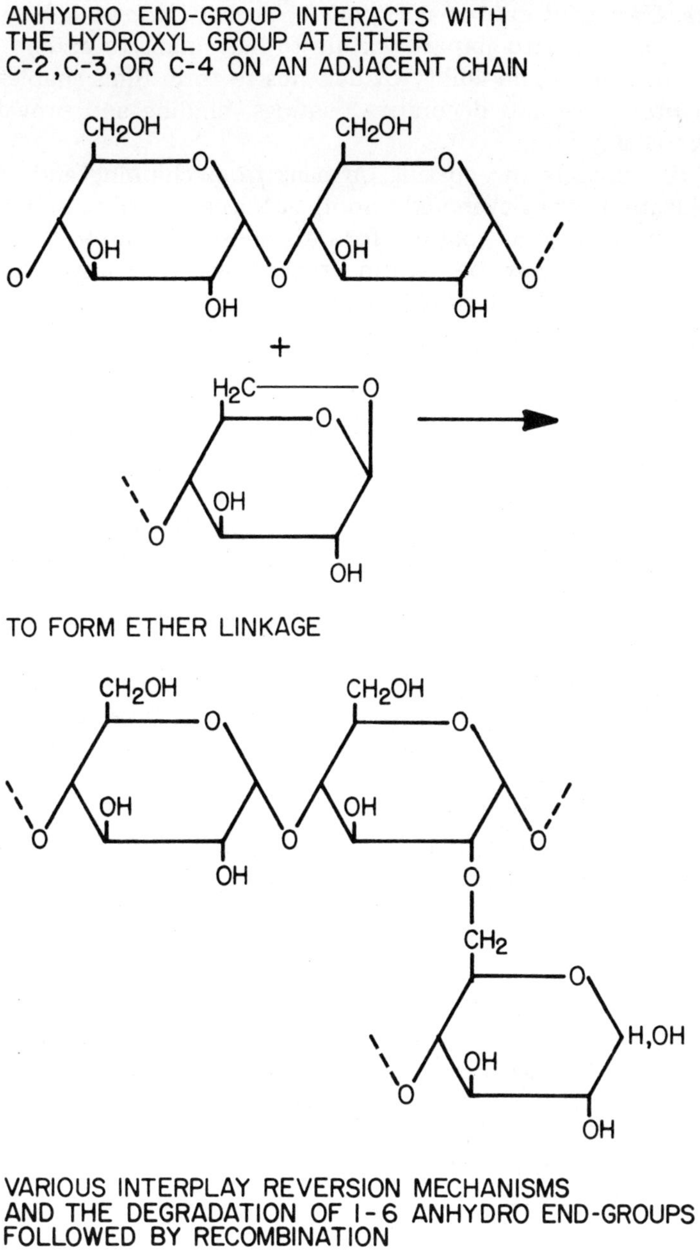
ANHYDRO END-GROUP INTERACTS WITH THE HYDROXYL GROUP AT EITHER C-2,C-3,OR C-4 ON AN ADJACENT CHAIN
CH2OH
CH2OH
OH
OH
OH
OH
O
O
O
O
O
+
H2C
O
O
OH
OH
O
TO FORM ETHER LINKAGE
CH2OH
CH2OH
O
O
OH
OH
O
O
O
OH
O
CH2
O
H,OH
OH
O
OH
VARIOUS INTERPLAY REVERSION MECHANISMS AND THE DEGRADATION OF 1-6 ANHYDRO END-GROUPS FOLLOWED BY RECOMBINATION

From Wolfrom and Tipson (1967)

FIG. 12.33. DIFFERENT RECOMBINATION FORMATIONS DURING DEXTRINIZATION

Some Uses of Dextrins in Foods.—Specially controlled dextrins (tapioca) are used to impart texture to hard gums to extend or replace gum arabic. The ability of dextrins to form films enables their use in protective and decorative coatings, binding and providing a matrix to carry foods.

Dextrin protects the coating on pans from chipping and inhibits crystallization giving even and smooth surfaces. Dextrin also provides physical protection as coatings for nuts and chocolate to prevent oil migration. It can also be used in French fries to improve crispiness by dipping them in the dextrin solution which coats the surface and reduces oil penetration during the frying step.

Dextrins are used to emulsify water-insoluble flavor oil and clouding agents. The emulsion of oil-water with the dextrin is homogenized to an emulsion particle of less than 5 μ. The emulsion is spray dried, yielding a dry product containing the flavor oil embedded in the dextrin matrix. The selection of the proper dextrin is important so that minimum loss of flavor by volatization or oxidative action will occur. When the dry flavor or fat is reconstituted in water it forms an oil in water emulsion producing an opaque solution.

The flavor mechanism for production of the oil is a molecular absorption process whereby molecules of the oil or fat are attracted to the dextrin molecules and held by extremely powerful forces. In this manner, the oil and fat particles are encapsulated by an impermeable physical barrier to prevent evaporation and oxidation.

Modifications by Cross-linking

Crosslinking starches was one of the great contributions of starch science to foods and its technology. If we examine the behavior patterns of native starches during their cooking process the following becomes evident.

(1) The unswollen starch behaves as a suspension in water.
(2) When the granules swell, the dispersion becomes a short, salve-like paste.
(3) With continued heating, it becomes elastic, rubbery in texture.
(4) The swollen granules rupture, forming dispersions of molecules and molecular aggregates.

Thus this stringiness and mucilagenous character is very undesirable because of poor texture and pasty mouthfeel. During processing the starch paste breaks down under stresses of shear, heat and acidity producing products of nonuniform consistencies that are long and cohesive instead of the preferred short and smooth consistency which has a better aesthetic appearance and has a platable taste.

Control of swelling can be achieved by the introduction of intermolecular bonds, which effectively ties the granules together. This crosslinking effect overcomes most of the undesirable characteristics by increasing the resistance of the paste to thin down with prolonged agitation, modifies the rubbery and cohesive gel character for improved texture that adds to the attractiveness of the food product.

It can be observed that as the starch is heated in water the granule swells but, even though the hydrogen bonds are replaced, the swollen granules are kept intact by the chemical bond. The high concentration level of crosslinking can reach where the swelling of the granules becomes so inhibited that they will no longer be able to hydrate and the viscosity will be decreased (see Fig. 12.35 and 12.36).

The chemically cross-bonded products undoubtedly represent the greatest single advance in food starches during the past 25 yr. Cross-bonding imparts stability not only with respect to mechanical agitation, but likewise resistance against supertemperature cooking and against thinning of the starch stabilizer with fruit acids.

The crosslinking agents used in food starches are either of the phosphate ester type, formed by the reaction of the starch with phosphorous oxychloride or water-soluble metaphosphate or of the ether type, formed by reaction with epichlorohydrin. Both types depend on the reaction of hydroxyl groups on two different starch molecules with the same molecule of reagent. These cross-bonding reactions apparently take place near the surface of the granule and, although the degree of crosslinking is very low, one crosslink, 200–1000 glucose units, the effect is quite pronounced (see Fig. 12.37 and 12.38).

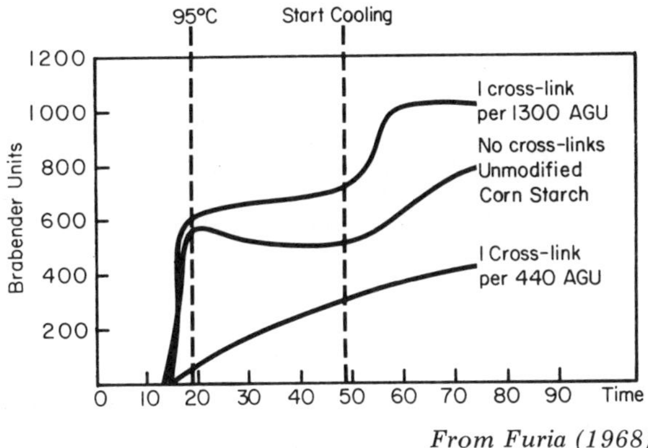

From Furia (1968)

FIG. 12.34. CROSS-LINKING EFFECT ON STARCH

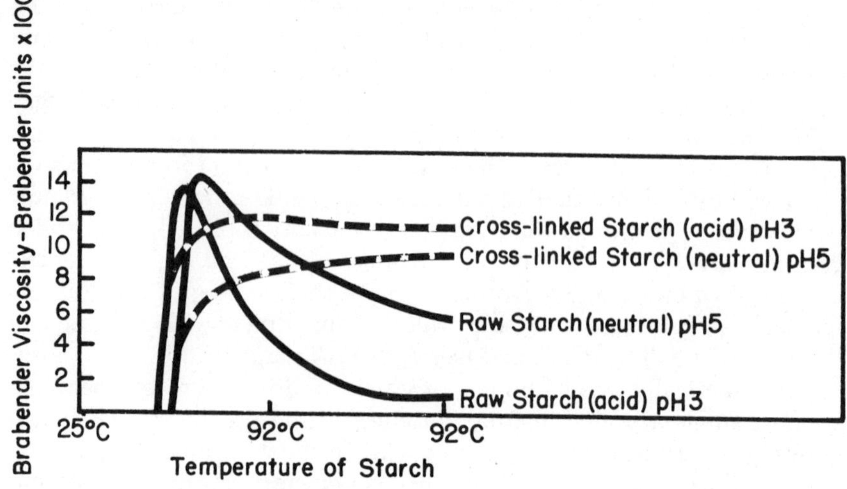

From Furia (1968)

FIG. 12.35. pH EFFECT ON CROSS-LINKED STARCHES

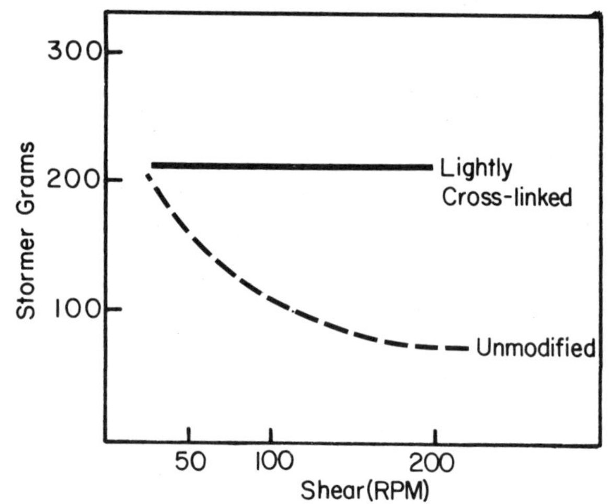

From Furia (1968)

FIG. 12.36. EFFECT OF SHEAR ON CROSS-LINKED STARCHES

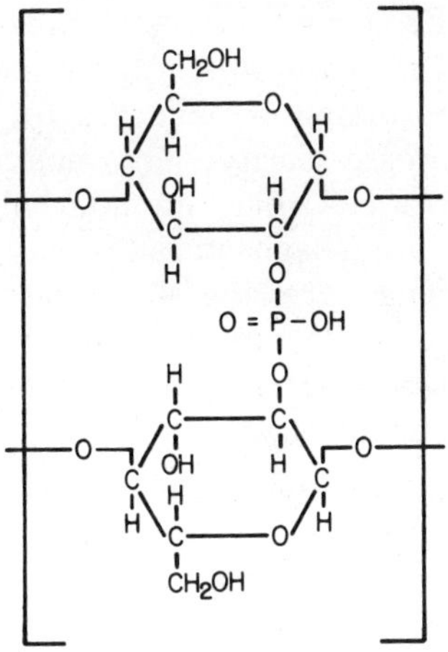

FIG. 12.37. DISTARCH PHOSPHATE

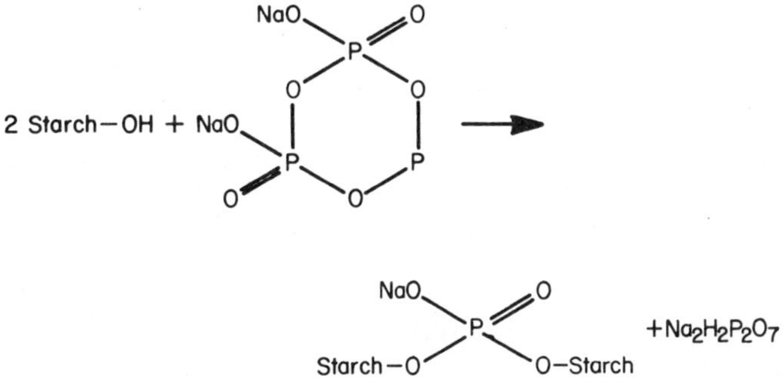

From deMan and Melnychyn (1971)

FIG. 12.38. REACTION OF STARCH WITH TRIMETAPHOSPHATE

This relationship between granule swelling and the character of the paste viscosity of native starches are cohesive or weakly elastic termed "viscoelasticity" or a more expressive adjective, "snotty or mucoid." The cross-bonded starches overcame this property to make them favorable for consumer taste in sauces, gravies, etc.

Uses in Foods.—(1) Processing systems such as heat exchanges and jet cookers where super-temperatures and high-shear stresses are needed in production for efficiency. Very few granules can survive this treatment unless they are reinforced by cross-linking.

(2) Fruit pie fillings (fresh, frozen and canned).
(3) Canned soups, gravies and sauces sterilized by retorting. Thickeners for Chinese foods to increase clarity, gloss and low gel tendencies to paste. Others include baby foods, cream-style corn, spaghetti sauces.
(4) Salad dressing thickeners. These require more crosslinking because of their pH accelerates the viscosity breakdown during preparation and storage.

These products are now widely used because of their functionality in food systems; however, one characteristic which should add to the potential use of starch phosphates, is their easy dephosphorylation by body phosphatases. Thus the starch phosphates should not produce foreign or harmful cleavage products during metabolism.

Starch Esters and Ethers

Starches are polyhydroxyl alcohols and offer the possibility of ester and ether formation with a variety of reagents. The only requirement of these reagents are that they are FDA approved for use in food systems.

The purpose of making a starch ester or ether is to modify some property of the starch granule or molecule so that it meets the requirements of specific applications. The granular starch derivatives are usually prepared in aqueous reactions where a 40–45% solids starch slurry is treated with reagents, usually with alkali catalysts. After they are reacted, unreacted chemicals and by-products are removed by washing, and the material is subsequently dried and ground to a powder.

Properties of the derivative depend upon the ester or ether used and the level of substitution. The major effect is to disrupt the orientation of the molecular segment in which substitution occurs, thus interfering with the tendency of the molecules to associate and so reduce or prevent retrogradation. The levels of substitution, by preventing association through hydrogen bonding, actually increase the ability of starch to hold water. If the substituent group is less hydrophilic than the hydroxyl groups, e.g., acetyl ester, the affinity of starch for water will be reduced and will eventually make the starch insoluble.

Another effect of derivatization on granular starch is to loosen the forces holding the granules intact. For example, the higher the de-

TABLE 12.8

ACETYLATED AND HYDROXYPROPYLATED STARCH

	Starch Ester	Starch Ether
Reagent	Acetic Anhydride	Propylene Oxide
Catalyst	Alkali Catalyst	Alkaline catalyst
Reaction pH	8.2–8.80	11.2–11.8
Solvent	Water	Water
Acetyl groups	0.2 AGU[1]	—
D.S. hydroxypropyl	—	0.027
Gelatinization temperature	Reduced	Reduced

Source: Radley (1968).
[1] AGU = Anhydroglucose unit.

gree of substitution of hydroxypropylation, the lower the gelatinization temperature; there is also an increase in the rate at which the starch cooks. If the substitution is high enough, the granules will begin to swell in cold water.

The structure of the starch plays an important part in derivatizing reactions; for example, in the granular state the reagents are only able to react readily with the molecules on the outside of the granule. The intra- and inter-molecular hydrogen bonding which occurs in the starch granule strongly resists the penetration of reagents of low polarity, and even the reactivity of the granule surface falls as a freshly prepared starch cake is dried. These starches are low-substituted derivatives and find broad applications in food products. These derivatives are used mainly when they are also crosslinked.

The use of these derivatives are mainly utilized when hot thick-boiling viscosity without breakdown during processing is required: They are used as binders in extruded pet foods and snack items, in salad dressings for stabilizing the oil emulsion, and as additives to modify textural properties to develop the proper texture and eating qualities. Their gels promote clarity, gloss and resistance to set back or gelling. Tapioca derivatives give a blander taste permitting better flavor development. Their gels do not exhibit syneresis or release water on storage or during freeze-thaw cycles.

In conclusion, the selection of starch for a given application will depend upon the texture and rheology which is desired as well as the conditions for use, aging properties needed and taste requirements.

ACKNOWLEDGEMENT

Figures 12.1–12.6 were drawn originally by Bonnie Heckman.

BIBLIOGRAPHY

ALBRECHT, J. J., NELSON, A. I., and STEINBERG, M. P. 1960. Characteristics of corn starch and its derivatives as affected by freezing, thawing and storage. Food Technol. *14*, 57-68.

BEAN, M. L., and OSMAN, E. M. 1959. Behavior of starch during food preparation. II. Effects of different sugars on the viscosity and gel strength of starch pastes. Food Res. *24*, 665-671.

CAMPBELL, A. M., and BRIANT, A. M. 1957. *Cited by* P. C. Paul, and H. H. Palmer. 1972. Food Theory and Applications. John Wiley & Sons, New York.

CARSON, B. G., MARRETT, L. F., and SELMAN, R. W. 1950. Polyoxyethylene monostearate as an anti-staling adjunct in bread. Cereal Chem. *27*, 438-451.

CORN INDUSTRIES RESEARCH FOUNDATION. 1964. Corn starch. Corn Industries Research Foundation, Washington, D.C.

deMAN, J. M., and MELNYCHYN, P. 1971. Symposium: Phosphates in Food Processing. Avi Publishing Co., Westport, Conn.

EDELMANN, E. C., CATHCART, W. H., and BERGQUIST, C. B. 1950. The effect of various ingredients on the rate of firming of bread crumb in the presence of polyoxyethylene (mono) stearate and glyceryl monostearate. Cereal Chem. *27*, 1-14.

FURIA, T. E. 1968. Handbook of Food Additives. The Chemical Rubber Co., Cleveland, Ohio.

GLICKSMAN, M. 1969. Gum Technology in the Food Industry, Academic Press, New York.

GREENWOOD, G. T. 1964. Structure, properties and degredation of starch. Food Technol. *18*, 138-144.

HECKMAN, E. M. 1959. Unpublished, given at a seminar at General Foods Corp., Tarrytown, New York.

LEACH, H. W., McCOWEN, L. D., and SCHOCH, T. J. 1959. Structure of the starch granule. Cereal Chem. *36*, 534-544.

MAZURS, E. G., SCHOCH, T. J., and KITE, E. E. 1957. Graphical analysis of the Brabender viscosity curves of various starches. Cereal Chem. *34*, 141-152.

NEWTON, J. M., and PECKHAM, JR., G. T. 1950. Oxidation of starch. *In* Chemistry and Industry of Starch, R. W. Kerr (Editor). Academic Press, New York.

OSMAN, E. M. 1967. Starch in the food industry. *In* Starch: Chemistry and Technology, Vol. II, R. L. Whistler, and E. F. Paschall (Editors). Academic Press, New York.

OSMAN, E. M., and DIX, M. R. 1960. Effects of fats and nonionic surface-active agents in starch pastes. Cereal Chem. *37*, 464-475.

PIGMAN, W. W., KERR, R. W., and SCHINK, N. F. 1952. The physical properties of starch gels. U.S. Pat. 2,609,326.

PIMENTEL, G. C., and McCLELLAN, A. L. 1960. Theory of the Hydrogen Bond. W. H. Freeman & Company, San Francisco.

RADLEY, J. A. 1953. Starch and Its Derivatives, Vol. 2. Chapman and Hall Ltd., London.

RADLEY, J. A. 1968. Starch and Its Derivatives, 4th Edition. Chapman and Hall Ltd., London.

SCHOCH, T. J. 1941. Physical aspects of starch behavior. Cereal Chem. *18*, 121-128.

SCHOCH, T. J., and GRAY, V. M. 1962. Effects of surfactants on the swelling and solubility of granular starches. Die Starke *14*, 239-245.

SCHOCH, T. J., and MAYWALD, E. C. 1956. Microscopic examination of modified starches. Anal. Chem. *28*, 382-387.

WHISTLER, R. L., and HILBERT, G. E. 1945. Separation of amylose and amylopectin by certain nitroparaffins. J. Am. Chem. Soc. *67*, 1161-1164.

WHISTLER, R. L., and JOHNSON, C. 1948. Effect of acid hydrolysis on the retrogradation of amylose. Cereal Chem. *25*, 418-424.

WHISTLER, R. L., and PASCHALL, E. F. 1967. Starch: Chemistry and Technology, Industrial Aspects, Vol. II. Academic Press, New York.

WILSON, E. J., SCHOCH, T. J., and HUDSON, C. S. 1943. The action of macerans amylose on the fractions from starch. J. Am. Chem. Soc. *65*, 1380-1383.

WOLFROM, M. L., and TIPSON, R. S. 1967. Advances in Carbohydrate Chemistry, Vol. 22. Academic Press, New York.

ATHROFA GOGLEDD DDWYRAIN CYMRU
THE NORTH EAST WALES INSTITUTE
of higher education CONNAH'S QUAY
DEESIDE, CLWYD CH5 4BR

CHAPTER 13

Peter Kovacs
Kenneth S. Kang

Xanthan Gum

INTRODUCTION

The use of hydrophilic colloids in the food industry is well established in hundreds of different applications and products (Glicksman 1969). While these water soluble gums are used effectively at low concentrations, the commercial demand for these products has greatly increased during the last few decades. As indicated by the GRAS List Review conducted by the Federal Food and Drug Administration, the use of locust (carob) bean gum, for instance, has more than doubled since 1960 (Anon. 1973A). Furthermore, new food processing techniques created a need for colloids with physical properties not associated with plant-derived polysaccharides. Consequently, this need for broader functionality range, coupled with the increasing commercial demand, has clearly indicated the necessity for the development of new colloidal stabilizers with well-established safety properties.

Since the functional properties of one gum over the other are based on very small, yet economically highly important differences, the development and commercialization of a new water soluble gum is a rather difficult task (McNeely and Kang 1973). This fact has been shown by the repeated and only partially successful attempts to commercialize dextran as a broadly acceptable water soluble gum (Jeanes 1952). Nevertheless, the research and development effort with dextran has shown that it is possible to develop water soluble gums with fermentation at reasonable costs.

While a number of biosynthetic gums have been evaluated for commercialization, to date only one of these products has met the stringent requirements for sustained commercial production at a substantial scale. This product is xanthan gum produced by fermentation of glucose by the microorganism *Xanthomonas campestris* (McNeely 1967; Kang and Kovacs 1974).

THE HISTORY AND ORIGIN OF XANTHAN GUM

Many species of microorganisms produce extracellular polysaccharides. These polysaccharides are continuously produced and, since they do not form covalent bonds with the cell walls, are secreted into liquid culture media (Wilkinson 1958). A vast array of polysaccharides of widely divergent physical and chemical properties has been obtained from these microorganisms. Among these microorganisms are a number of *Xanthomonas* species which have been reported in the literature (Stolp and Starr 1964). Four species of *Xanthomonas*, namely, *campestris*, *phaseoli*, *malvacearum*, and *carotae*, were reported to be the most efficient gum producers.

The Northern Utilization Research and Development Division of the United States Department of Agriculture conducted an extensive screening program for gum producers in their large microbial culture collection. Of the various biosynthetic polysaccharides produced in their laboratory, Polysaccharide B-1459 (xanthan gum) produced by *Xanthomonas campestris* NRRL-B-1459 was found to have characteristics that rendered it very promising as a commercial product (Jeanes *et al.* 1961).

Xanthomonas campestris is a naturally occurring bacterium originally isolated from cabbage plant and has long been known to produce viscid or gummy colonies (Breed *et al.* 1957). In 1958, it was reported that *Xanthomonas campestris* produced substantial quantities of gummy exudate or extracellular polysaccharides during the fermentation of a carbohydrate substrate (Lilly *et al.* 1958). This fermentation had a close similarity to the fermentation of *Xanthomonas campestris* in cabbage extracts, which contained the natural carbohydrates, proteins, and trace minerals of the cabbage as the sole source of fermentation substrate.

It has also been determined that the chemical components of xanthan gum produced by an industrial-type fermentation and those of xanthan gum produced on living cabbage tissues under natural conditions are the same (Sutton and Williams 1970). Furthermore, xanthan gum from both sources produced the same precipitin bands when reacted with an antiserum to the polysaccharide. The results of these highly specific serological tests, along with the chemical component analysis, provided very strong evidence that the gum obtained from a fermentation broth is identical to that obtained in the natural state. Additional tests demonstrated that the metal ion analysis profile and the rheological properties of both the commercially available food grade xanthan gum and the xanthan gum produced on a pure cabbage extract are similar (Kang 1972). This data indicates that the commercially produced xanthan gum is identical to the naturally occurring xanthan gum.

PRODUCTION OF XANTHAN GUM

On a commercial scale, xanthan gum is produced by a submerged aerobic fermentation by using a pure culture of *Xanthomonas cam-*

pestris in a suitable fermentation medium containing a carbohydrate substrate. After the fermentation, the xanthan gum is recovered and purified with isopropyl alcohol. These operations are followed by drying and milling. In the production of xanthan gum in the natural state on the cabbage plant by *Xanthomonas campestris*, the cabbage furnished the carbohydrate substrates, proteins, minerals, and buffering capacity for cell growth and gum production. In the commercial fermentation, carbohydrate, proteins, trace minerals, and pH buffering substance are supplied in a way that simulates the natural conditions.

The nutritional requirements for minimum growth of 30 different species of the genus *Xanthomonas*, including *Xanthomonas campestris*, were reported by Starr (1946). A mixture of glucose, ammonium chloride, a phosphate buffer, magnesium sulfate, and certain trace minerals was found to be sufficient for minimum growth. During the fermentation on cabbage, as well as during the commercial fermentation, the pH of the medium decreases due to the formation of organic acids and the xanthan gum, which also contains acidic functions. If these organic acids and the gum itself lower the pH below the critical point, such as 5.0, the gum production drastically decreases. With the slow fermentation on the cabbage, there is sufficient time for the buffer system of the plant to furnish cations to neutralize the carboxylic acid groups of the organic acids and the gum. If this plant buffer system is overwhelmed, the rate of gum production becomes negligibly small. In aerobic fermentation at around 28°C, 1-5% glucose concentrations were found to provide the best xanthan gum yields (Lilly *et al.* 1958). At higher glucose concentrations, an actual reduction in conversion efficiency was observed. Glucose, sucrose, and starch were found to be of equivalent efficiency in polysaccharide production.

In the commercial process (Fig. 13.1), with maintenance of optimum conditions for gum production, including a high degree of aeration and a constant temperature, the fermentation proceeds at a much faster rate than on the cabbage plant. As the xanthan gum production progresses in the commercial fermentors, the pH of the mixture gradually decreases. After a substantial fraction of the carbohydrate substrate has been consumed, the pH is maintained by adding a base to furnish metal ions for more xanthan gum production. Controlling the pH at near neutral levels allows the gum synthesis to continue until all the carbohydrate substrate is exhausted, at which point fermentation is complete. After pasteurizing the fermentation beer, the polysaccharide is recovered by precipitation with isopropyl alcohol followed by drying, milling, testing, and packaging.

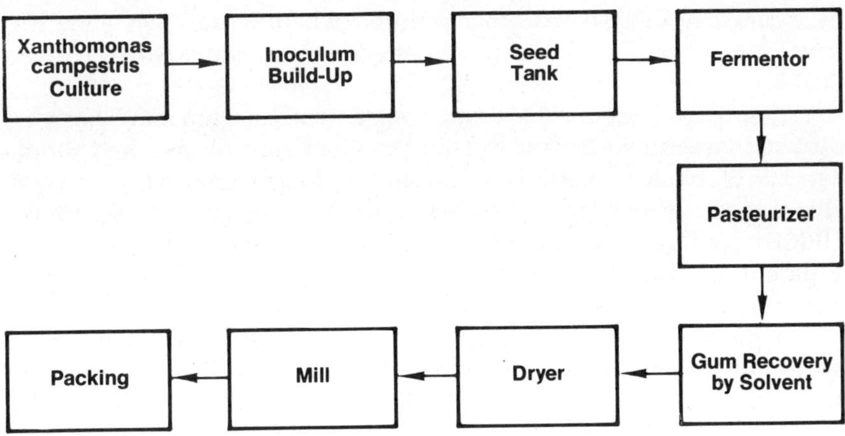

FIG. 13.1. FLOW SHEET DIAGRAM FOR THE XANTHAN GUM MANUFACTURING PROCESS

A number of alternative methods have been suggested for the recovery of xanthan gum from the fermentation liquor. Each process starts with the pasteurization of the liquor to kill the bacterial cells. Since xanthomonads are nonspore formers and the vegetative cells are sensitive to elevated temperatures, complete pasteurization can be readily accomplished.

Drum or spray drying of the fermentation beer yields a crude grade polysaccharide product (Rogovin *et al.* 1965). Long-chain quaternary ammonium salts have been proposed as precipitants (Rogovin and Allrecht 1964). It is doubtful that these methods would be acceptable for the manufacture of food grade products. Xanthan gum can be also recovered by precipitation with calcium ion. Washing the insoluble calcium complex with acid or salt would yield a pure gum (McNeely and O'Connell 1966).

A number of continuous fermentation methods which would offer significant economic advantages have been reported (Silman and Rogovin 1972; Patton and Lindblom 1962). However, the problem of maintaining sterility even in the batch fermentors indicates that the installation of a continuous fermentation system on a commercial scale would significantly increase the contamination risk under the fermentation conditions, including the type of carbohydrate substrate.

Because of the presence of nonviable bacterial cells, xanthan gum produces opaque aqueous solutions. In order to manufacture a product of clear solution properties, the fermentation liquor is diluted and clarified by filtration. A process whereby xanthan gum was treated with hypochlorite to produce a clear solution was patented by Colegrove (1970). Another method whereby xanthan gum was treated at high pH with high temperatures was patented by Patton (1973).

A number of chemical derivatives of xanthan gum have been reported. These include deacetylated xanthan gum (Jeanes and Sloneker 1961), carboxymethyl derivative (Schweiger 1966A), propylene glycol ester (Schweiger 1966C), cationic derivatives (Schweiger 1966B), and formaldehyde cross-linked products (Patton 1962). In a patent by Pettitt (1973), the graft copolymer of xanthan gum with acrylic monomers was described.

The biosynthesis of xanthan gum is genetically controlled as to both molecular structure and molecular weight. For this reason, xanthan gum with outstanding uniformity in its physical and chemical properties can be manufactured. The food grade xanthan gum produced according to the specifications set forth in the 2nd edition of the Food Chemicals Codex (National Research Council 1972) and the National Formulary, 14th Edition (Am. Pharm. Assoc. 1975) has been marketed under the trade name KELTROL since its approval as a general food additive by the Federal Food and Drug Administration (Anon. 1969).

THE STRUCTURE OF XANTHAN GUM

The exact chemical structure of xanthan gum has not yet been determined. It is a heteropolysaccharide with a molecular weight of several million (Leach *et al.* 1957; Dintzis *et al.* 1970). It contains D-glucose, D-mannose, and D-glucuronic acid in the molar ratio of 2.8:3:2.0 (Sloneker *et al.* 1964). The molecule contains about 4.7% acetyl and about 3% pyruvate (Sloneker and Orentas 1962). The pyruvate is attached to a single unit glucose side chain by a ketal linkage. The configuration of the pyruvic acid was also determined (Gorin *et al.* 1967).

The repeating-unit structure of xanthan gum, based on the most recent experimental evidence, is shown in Fig. 13.2 (Jansson *et al.* N.D.). As detailed in Fig. 13.2 each repeating block contains five sugar units consisting of two glucose units, two mannose units, and one glucuronic acid unit. The main chain of xanthan gum is built up of β-D-glucose units linked through the 1- and 4-positions; i.e., the chemical structure of the main chain of xanthan gum is identical to the chemical structure of cellulose. The side chain consists of the two mannose units and the glucuronic acid unit. The terminal β-D-mannose unit is linked glycosidically to the 4-position of β-D-glucuronic acid, which in turn is linked glycosidically to the 2-position of α-D-

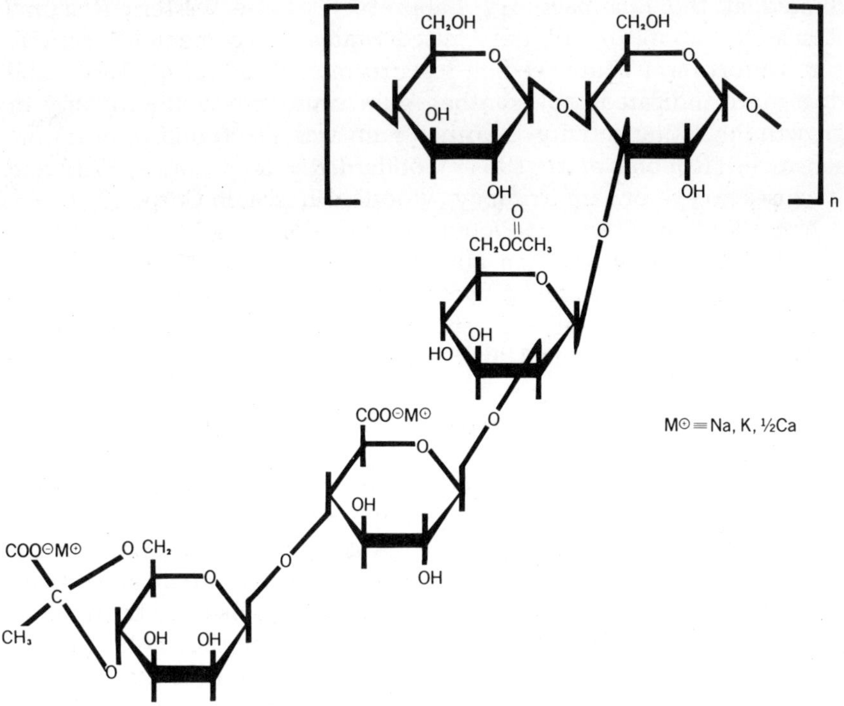

FIG. 13.2. STRUCTURE OF XANTHAN GUM

mannose. This three-sugar side chain is linked to the 3-position of every other glucose residue in the main chain. The distribution of the side chains is unknown. Also, about half of the terminal D-mannose residues carry a pyruvic acid residue ketalically linked to the 4- and 6-positions. The distribution of these pyruvate groups is unknown. The nonterminal D-mannose unit in the side chain contains an acetyl group at position 6.

This fact that the side chains shield the backbone of xanthan gum could be the major reason for the extraordinary enzymatic resistance of xanthan gum. Also unique are the unvarying chemical structure and the uniformity of chemical and physical properties.

THE SAFETY PROPERTIES AND REGULATORY STATUS OF XANTHAN GUM

With respect to safety properties, xanthan gum is one of the most extensively investigated polysaccharides (McNeely and Kovacs 1975). The initial short-term feeding studies on xanthan gum were conducted at the Pharmacology Laboratory of the Western Regional Research Laboratory of the United States Department of Agriculture. Short-term acute feeding tests to rats (Booth *et al.* 1963) and dogs have indicated that xanthan gum causes no acute toxicity or growth-inhibiting activity. Xanthan gum was also found to be a non-sensitizer (Robbins *et al.* 1964; Woodard Research Corp. 1973), and it causes no eye or skin irritancy (Woodard Research Corp. 1971).

According to a test conducted by the caloric availability method, the digestibility of xanthan gum was found to be zero. In an unpublished study conducted by the more reliable radioactive tracer method, the digestibility of xanthan gum was found to be approximately 15% (Gumbmann 1964). The theoretical caloric value of xanthan gum is approximately 3.78 Cal per gram. In view of the 15% digestibility factor, the approximate caloric value of xanthan gum is 0.5 Cal per gram.

To qualify xanthan gum as an approved food additive, additional short- as well as long-term safety studies were conducted on xanthan gum. The acute oral toxicity to rats was established as above 45 gm per kilogram (Woodard Research Corp. 1968A) and to dogs as 20 gm per kilogram (Woodard Research Corp. 1968B). Because the animals could not be administered higher doses within the 24-hr period, these LD-50 levels are only approximations. The long-term feeding studies on xanthan gum consist of two-year studies in rats and dogs and a three-generation reproduction study in the rat (Woodard *et al.* 1973). No significant effect on growth rate, survival, hematological values, or organ weights was found and no incidence of tumor was detected. Based on these feeding studies indicating that xanthan gum is safe for oral consumption, the Federal Food and Drug Administration issued a food additive order on March 19, 1969, permitting the use of xanthan gum in food products without any specific quantity limitations (Anon. 1969).

The subsequent successful amendment of the Federal identity standards for French dressing (Anon. 1971B) and certain process cheeses (Anon. 1973B) authorized use of xanthan gum in those products. Xanthan gum was proposed for approval, and later approved for use in various meat and poultry products by the United States Department of Agriculture, Animal and Plant Health Inspection Service, Meat and Poultry Products Inspection Division (Anon. 1973E).

To date, the countries of Canada (Anon. 1971A), Denmark (Anon. 1973C), and New Zealand (Anon. 1973D) have granted xanthan gum formal food additive approvals. Canada also included xanthan gum in her salad dressing and French dressing standards (Anon. 1972).

In 1974, the FAO/WHO Joint Expert Committee established an

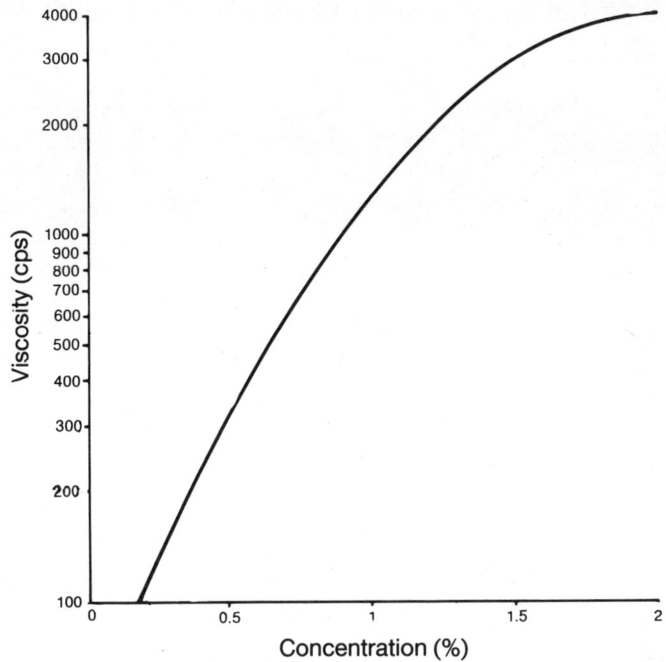

FIG. 13.3. EFFECT OF CONCENTRATION ON THE VISCOSITY OF XANTHAN GUM

ADI of 10 mg per kilogram for xanthan gum (Anon. 1975). As the Emulsifier/Stabilizer List of the European Economic Community (1974) has included xanthan gum among the Annex II substances, the product may now be considered for approval by member countries.

THE PHYSICAL PROPERTIES OF XANTHAN GUM

The physical properties of xanthan gum have been thoroughly investigated and reported (Rocks 1971; Kovacs 1973B). Xanthan gum dissolves readily in hot or cold water, producing relatively high-viscosity opaque solutions at low concentrations. The viscosity of xanthan gum as a function of concentration is shown in Fig. 13.3.

The solutions of xanthan gum exhibit pseudoplastic flow properties (Rao and Kenny 1975). As indicated in Fig. 13.4, the viscosity of the solutions decreases rapidly as increased shear rate is applied to the solutions. As the shear rate decreases, the original viscosity of the solution returns immediately. Due to its pseudoplastic rheology, xanthan gum is an excellent suspending agent at very low concentrations (Kelco 1975).

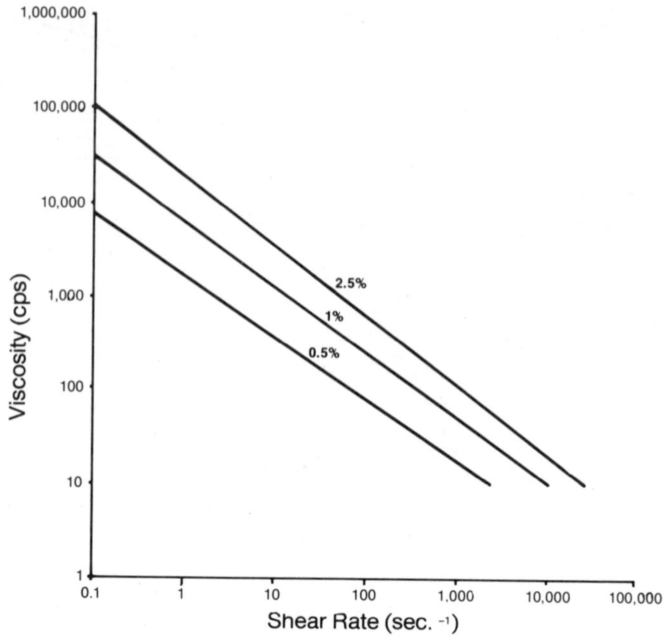

FIG. 13.4. EFFECT OF SHEAR RATE ON THE VISCOSITY OF XANTHAN GUM

The structural rigidity of the polymer, which is caused by the presence of β-(1→4) linkages and the specific nature of the branching, results in several of the unusual properties of xanthan gum. In contrast to the behavior expected from a typical anionic polysaccharide, addition of salts to a salt-free xanthan gum solution causes an increase in viscosity (above 0.15% gum concentration).

Most polysaccharide solutions show a decrease in viscosity when they are heated, whereas salt-free xanthan gum solutions in deionized water increase in viscosity after the initial viscosity decrease. This behavior of xanthan gum suggests that a conformation change is taking place. The technique of optical rotation can be used to probe conformation changes, and the measurement of the optical rotation of a salt-free xanthan gum solution has shown that the viscosity increase is exactly paralleled by a decrease in optical rotation (Rees 1972). This is consistent with the unwinding of an ordered conformation such as a helix into a random coil with a consequent increase in effective hydrodynamic volume and, therefore, viscosity.

The presence of 0.1 to 0.15% sodium chloride also improves the stability of xanthan gum solutions at high temperatures. Xanthan gum solutions with 1% distilled water lose about 25% of their vis-

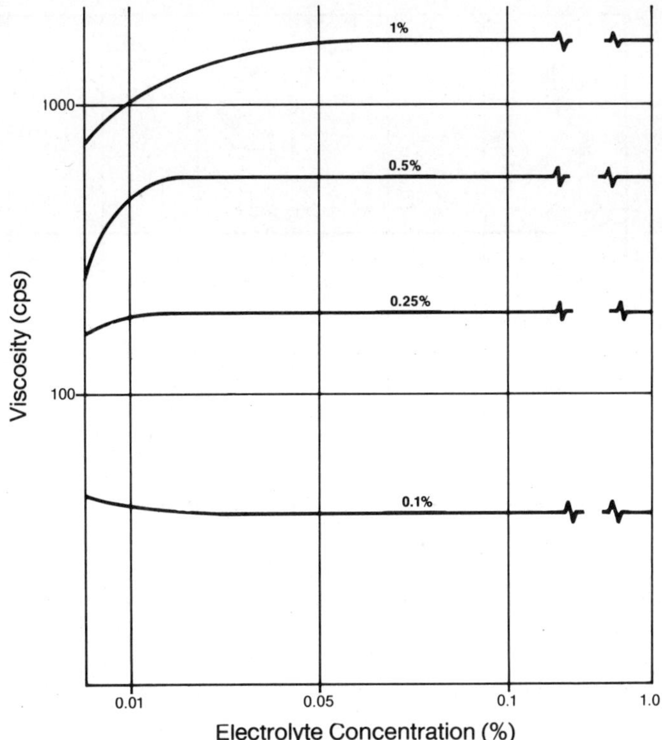

FIG. 13.5. EFFECT OF ELECTROLYTE CONCENTRATION ON THE VISCOSITY OF XANTHAN GUM

cosity when processed at 116°C. In the presence of 0.1% sodium chloride, however, this viscosity decrease is less than 10% (Kovacs 1970).

Unlike most other hydrocolloids (Pilnik and MacDonald 1968), the temperature and pH stabilities of xanthan gum are remarkable. As indicated in Fig. 13.6, the pH has almost no effect on the viscosity of xanthan gum solutions.

Figure 13.7 details the effect of temperature on xanthan gum solutions. In the case of 1% solutions, almost no viscosity decrease is noted as the temperature of the solutions is increased. At lower concentrations, the solution viscosity will decrease slightly as the temperature is increased. Precise measurements revealed that the relationship shown in Fig. 13.7 is not entirely constant when measured in electrolyte-free solutions (Rees 1972). This fact would indicate that the polysaccharide molecule is subjected to conformational shapes as the temperature of the solution is altered.

Xanthan gum has exceptional stability and compatibility with high

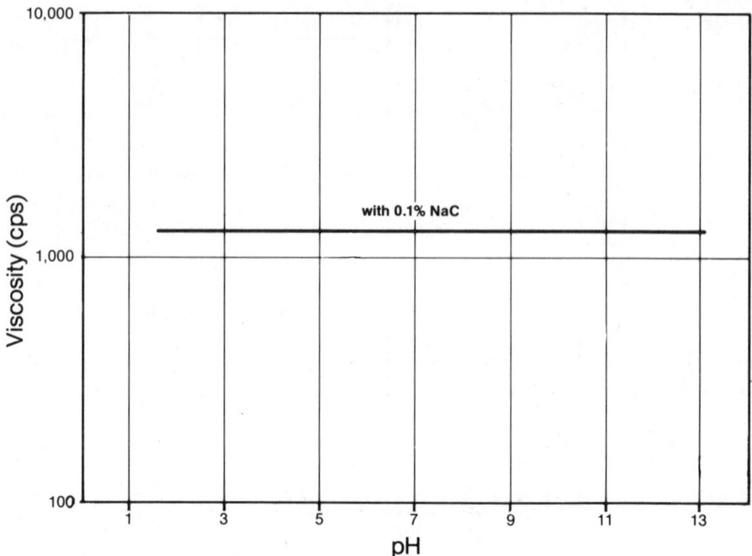

FIG. 13.6. EFFECT OF pH ON THE VISCOSITY OF 1% XANTHAN GUM SOLUTION

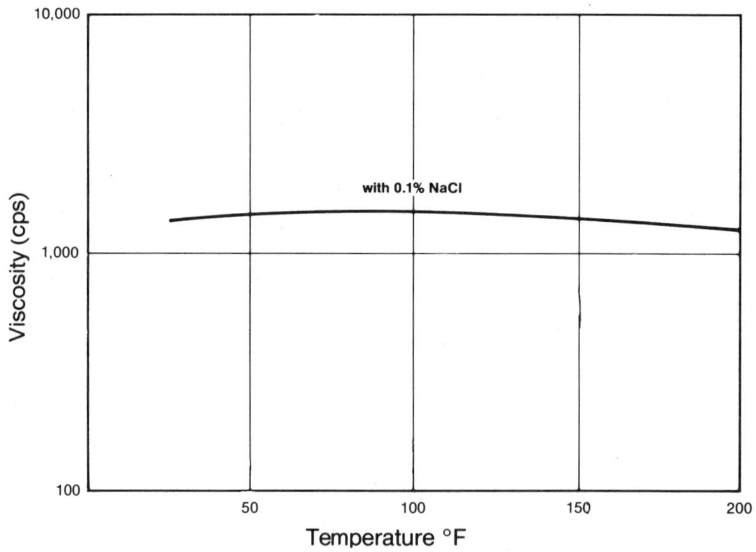

FIG. 13.7. EFFECT OF TEMPERATURE ON THE VISCOSITY OF 1% XANTHAN GUM SOLUTION

concentrations of many salts. Xanthan gum dissolves directly in 5% acetic acid, 10% hydrochloric acid, 25% phosphoric acid, 5% sodium hydroxide, and in up to 15% sodium and calcium solutions (Carnie 1964). The viscosity of these thickened acid and base solutions is reasonably stable for several months if the storage temperature is not excessively elevated.

While xanthan gum can be dissolved at 71°C in glycerin, it is insoluble in organic solvents and propylene glycol. Solutions of xanthan gum, however, will tolerate up to 50% water-miscible organic solvents such as ethyl alcohol.

Upon evaporation of a xanthan gum solution, it forms a water soluble film which strongly adheres to glass and many metal surfaces. Xanthan gum is also compatible with most hydrocolloids including starches.

Xanthan gum reacts with certain specific dextrins and exhibits incompatibility with highly concentrated gum arabic solutions below the pH of 5. Since xanthan gum is an acid polysaccharide, it reacts with milk proteins below their isoelectric point. Because of this reactivity, the isolation of β-lactoglobulin from acid whey can be accomplished with a number of anionic gums (Hidalgo and Hansen 1969), including xanthan gum.

While xanthan gum was found to be nonreactive with most other gums, it shows a definite reactivity with galactomannans such as locust bean gum (Schuppner 1971), guar gum (Jordan and Lester 1973), and tara gum (Glicksman and Farkas 1973, 1974). The reactivity with guar gum is manifested in a detectable viscosity increase, while aqueous combinations of xanthan gum with locust bean gum or tara gum produce strong cohesive and thermoreversible gel systems. The physical properties of a mixture of xanthan gum and locust bean gum were investigated and reported in detail (Kovacs 1973A). The reasons for the reactivity of xanthan gum with locust bean gum have not yet been conclusively determined. While many solution properties of hydrophilic gums are governed by hydrogen bonding, the specificity of this reactivity with galactomannans does not appear to be due to such a simple association.

Polymers frequently assume random coil formations in solutions. Certain polymers, especially polysaccharides, may have a more ordered structure. The gelling phenomenon is explained as a result of a complex network formation in which polymer molecules form highly ordered intermolecularly associated regions known as "junction zones" (Rees 1969). On the basis of this fact, it may be postulated that, upon cooling the solution of xanthan and locust bean gums, the gum molecules are rearranged from a more or less random coil formation to a more orderly conformation involving the junction zones. At the same time, specific stereochemical factors apparently allow the formation of a complex hydrogen-bonded network resulting in gelation. The strength of the xanthan gum/locust bean gum gel depends on the viscosity of the locust bean gum as well as the ratio of the two gums in the solution system.

In Fig. 13.8 1% xanthan gum/locust bean gum solutions were plotted against the colloid ratio. Maximum gel strengths were obtained with the xanthan gum/locust bean gum ratios of 6:4–4:6. This approximately 1:1 stoichiometric relationship would indicate that there are about an equal number of junction sites available on each polymer molecule.

If the colloid ratio is kept constant, the gel strength of the system is a function of total colloid concentration. Figure 13.9 demonstrates the effect of concentration on the gel strength at equal colloid

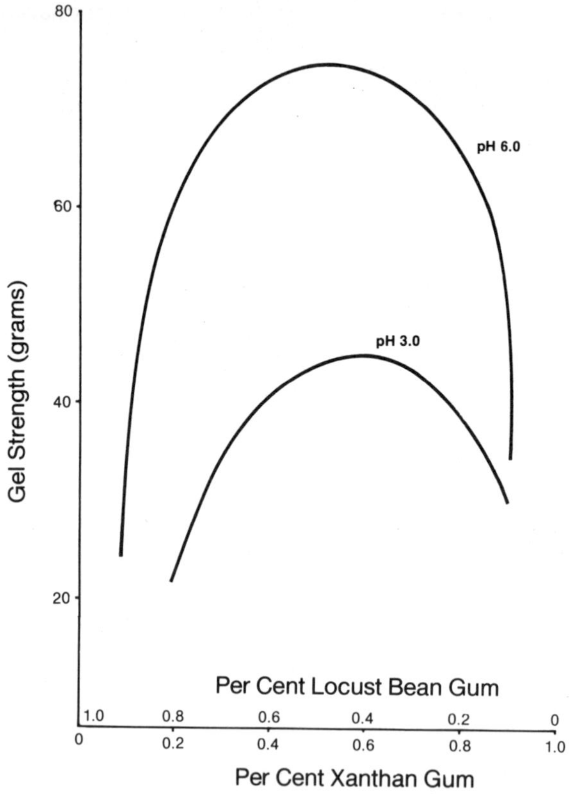

FIG. 13.8. EFFECT OF COLLOID RATIO ON THE STRENGTH OF THE XG/LBG GELS

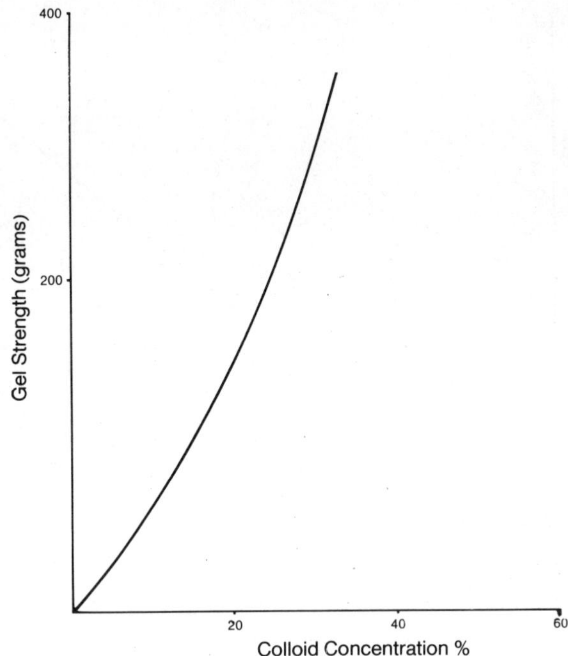

FIG. 13.9. EFFECT OF TOTAL COLLOID CONCENTRATION ON THE STRENGTH OF THE XG/LBG GELS

ratios. Since both xanthan gum and locust bean gums are relatively high-viscosity colloids, it is not possible to prepare more than 3% solutions containing these two colloids.

The viscosities of the xanthan gum/locust bean gum solutions at an equal colloid ratio and the solutions of other colloids widely used in the food industry are shown in Fig. 13.10. At low concentrations (0.005-0.1%) the viscosities of the xanthan gum/locust bean gum solutions upon heating and cooling are very high as compared to the other food colloids. At a 0.1% total colloid concentration, the viscosity of xanthan gum/locust bean gum solution as measured with a Brookfield Viscometer at 60 rpm is about 2000 cps, whereas the viscosity of a xanthan gum/locust bean gum solution as measured with a 9 cps, and guar gum about 4 cps, etc.

The xanthan gum/locust bean gum gel is thermally reversible with the melting or setting point between 49-54° C.

By plotting the stress and strain characteristics of the xanthan gum/locust bean gum gel, the highly elastic consistency of the gel system is shown. This elastic behavior can be at least partially elim-

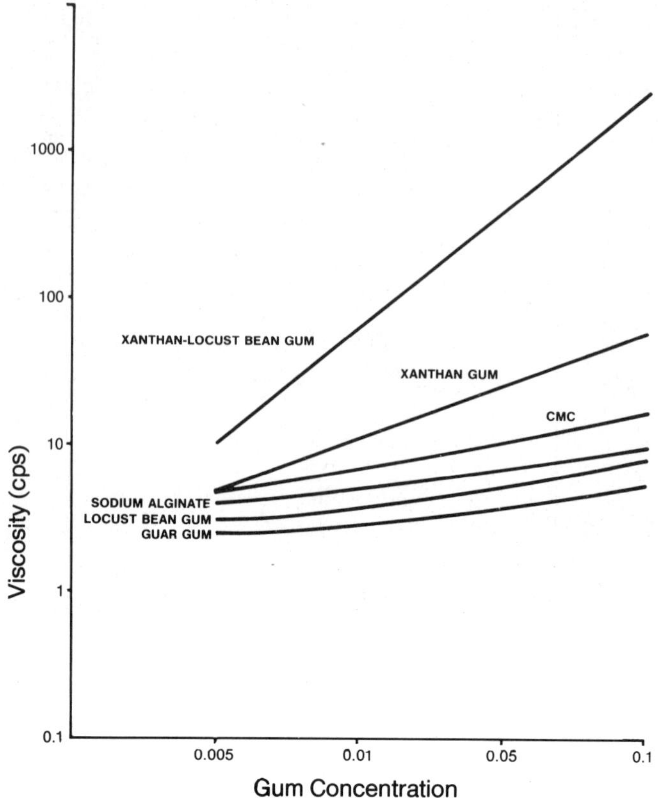

FIG. 13.10. EFFECT OF CONCENTRATION ON THE VISCOSITY OF LOW CONCENTRATION COLLOID SOLUTIONS

inated by the incorporation of starches and other colloids (Kovacs 1973A).

The reactivity of xanthan gum with locust bean gum can be used for the qualitative determination of xanthan gum or locust bean gum in food systems. The quantitative determination of xanthan gum in food systems was reported by Graham (1971).

APPLICATIONS OF XANTHAN GUM IN THE FOOD INDUSTRY

The applications of xanthan gum in the food industry are related to its previously described unique physical properties. Pseudoplastic rheology, heat, and pH stability coupled with high viscosities and good solubility properties are all responsible for the acceptance of xanthan gum by the food industry as an all-purpose stabilizer, thickener, and processing aid.

Emulsion Stabilization

The pseudoplastic rheology of xanthan gum results in the product being an efficient emulsion stabilizer for the food industry (O'Connell 1962). An emulsion may be described as a suspension system where the dispersed oil globules are suspended in a thickened aqueous medium. Since the yield point, namely the concentration where a solution at rest behaves like a solid, is very low for xanthan gum, unique stability properties for the resultant emulsions are obtained. Pharmaceutical mineral oil emulsions containing 0.25% xanthan gum were found to be stable with no sign of phase separation for more than six months (Araujo 1967). Likewise, French dressing compositions containing 0.4% xanthan gum were found to have storage stability for six months with no phase separation and less than 10% loss in the viscosity of the emulsion (Stanislav and Sheets 1971).

Cottage Cheese Creaming Emulsions

A blend of xanthan gum and locust bean gum has been used for the stabilization of cottage cheese creaming emulsions. A total gum combination of 0.10–0.12% provides excellent viscosity stability against drainage (Kovacs and Titlow N.D.).

Process and Cream Cheese Stabilization

An area where the higher than usual melting point of the xanthan gum/locust bean gum gel system is of advantage is in the stabilization of spreadable and sliceable process and cream cheese type products. A pasteurized process cheese product must maintain sufficient consistency for good body and sliceability at refrigerated and ambient temperatures along with proper melt-down characteristics when the sliced cheese is subjected to heat. The ideal situation requires a temperature-reversible system whereby the sliceability of the cheese spread is enhanced by the gel system at refrigerated or ambient temperatures, and melting is permitted at the desired temperature ranges.

Process cheese spreads have been prepared by using a stabilizer combination of xanthan gum, locust bean gum, and guar gum at 0.6–0.8% levels (Kovacs 1973B; Kovacs and Igoe N.D.).

Beverages

Xanthan gum gives citrus and fruit-flavored beverages enhanced mouth-feel with full-bodied taste and good flavor release (Schuppner 1968). Xanthan gum has also shown promise as a stabilizer for flavor emulsions employed in beverages. Concentrations of xanthan gum in the finished drink may range from 0.001–0.15%. Xanthan gum was also found useful in the stabilization of liquid chocolate confections (Arden 1974).

Canning

Due to its unusual heat stability, xanthan gum is very useful in heat-processed food systems containing sauces and gravies. Canned tuna, chicken, ham, potato, and macaroni salads have been formulated with retortable salad dressings containing xanthan gum. In the canning field, the gelling property of the xanthan gum/locust bean gum gel system has been adapted to the formulation of canned chip dip and imitation sour cream type products (Schuppner 1973) and to canned jellied pet foods. Xanthan gum alone or in combination with locust bean gum in canned pet foods was also found to be functional.

Frozen Foods

Frozen puddings or pie fillings stabilized with starches may show retrogradation and subsequent syneresis after two freeze-thaw cycles. The addition of 0.05–0.1% xanthan gum is able to maintain freeze-thaw stability for five or more cycles. The freeze-thaw stability of spoonable starch-based salad dressings can be also maintained by the addition of 0.1% xanthan gum (Wintersdorff 1972).

Relishes

In tomato, pickle, onion, or mustard relishes, xanthan gum is exceptionally effective in reducing liquor drainage. Neither the acid nor the salt content of these products is able to reduce the water-holding abilities of xanthan gum. Gloss and sheen in these products are excellent when the gum is used at levels of about 0.1%.

Instant Preparations

Xanthan gum is utilized for the stabilization of low calorie (Farkas 1974) and other aerated desserts (Katz 1971), instant milkshakes, breakfast drinks, soups, and sauces (Edlin 1972). In addition, the utilization of the xanthan gum/locust bean gum gel system for the preparation of instant puddings was also reported (Schuppner 1970).

Pastry Fillings

A bakery filling type product can be stabilized with only 0.2% xanthan gum, 0.2% locust bean gum, and about 2% starch. The spreadability of the product is excellent, and the eating quality is also improved due to the reduced stabilizer levels.

Frozen Desserts and Confections

The preparation in Japan of ice cream and sherbets stabilized with 0.2% xanthan gum with good heat shock and consistency was reported (Yaki and Okamura 1969). A combination of xanthan gum, locust bean gum, and guar gum at 0.08–0.10% levels has been found to be suitable for the preparation of experimental ice creams. The utilization of xanthan gum for coating of frozen confections, particularly ice cream, was also reported (Jenkinson and Williams 1973) along with the preparation of low fat imitation ice cream or ice milk (Gabby *et al.* 1974A) and other low calorie frozen desserts, toppings and spreads (Gabby *et al.* 1974B).

Organoleptic Effects

It was reported that in comparison with several other colloids, xanthan gum has a detectable depressing effect on the sweetness intensity of sucrose, while it has no effect on the taste of citric acid, saccharine, caffeine, and sodium chloride. This depressing effect on sucrose was attributed to a possible synergistic effect between the two compounds (Pangborn *et al.* 1973). In a study where the effect of flavorants on physical and sensory viscosity was determined, butyric acid was found to reduce both the physical and oral viscosity of xanthan gum. The flavor intensity of butyric acid was also reduced by xanthan gum, carboxymethyl cellulose, and sodium alginates.

When compared to the other hydrocolloids used in the food industry, xanthan gum is a relatively new product. An extensive research effort is continuously in progress to further explore the physical properties and the potential uses of xanthan gum which may eventually render xanthan gum the all-purpose colloidal stabilizer of the food industry.

BIBLIOGRAPHY

ANON. 1969. Xanthan gum. Federal Register *34*, No. 53, 5376–5377.
ANON. 1971A. Canada Gazette, Part II, 105.
ANON. 1971B. French dressing identity standard. Federal Register *36*, No. 168, 17333.
ANON. 1972. Canada Gazette, Part II, 106.
ANON. 1973A. Carob bean gum. Federal Register *38*, No. 143, 20041–20044.
ANON. 1973B. Certain cheese products; order listing xanthan gum as an optional ingredient. Federal Register *38*, No. 49, 6883–6886.
ANON. 1973C. The National Food Institute, Denmark.
ANON. 1973D. New Zealand Food and Drug Regulations.

ANON. 1973E. Xanthan gum. Federal Register *38*, No. 140, 19690-19692.

ANON. 1975. Toxicological Evaluation of Some Food Colours, Enzymes, Flavour Enhancers, Thickening Agents and Certain Other Food Additives. WHO Food Additives Series *6*. World Health Organization, Geneva.

AM. PHARM. ASSOC. 1975. National Formulary XIV. American Pharmaceutical Assoc., Washington, D.C.

ARAUJO, O. E. 1967. Emulsifying properties of a new polysaccharide gum. J. Pharm. Sci. *56*, 1141-1145.

ARDEN, S. 1974. Chocolate flavored liquid confections and methods of making the same. U.S. Pat. 3,784,715.

BOOTH, A. N., HENDRICKSON, A. P., and DE EDS, F. 1963. Physiologic effects of three microbial polysaccharides on rats. Toxicol. Appl. Pharmacol. *5*, 478-484.

BREED, R. S., *et al.* 1957. Bergey's Manual of Determinative Bacteriology, 7th Edition. Williams and Wilkins Co., Baltimore.

CARNIE, G. A. R. 1964. Evaluation of a new polysaccharide gum. Australasian J. Pharm. *45* (Suppl. 19), 580-583.

COLEGROVE, G. T. 1970. Treatment of *Xanthomonas* hydrophilic colloid. U.S. Pat. 3,516,983.

DINTZIS, F. R., BABCOCK, G. E., and TOBIN, R. 1970. Studies on dilute solutions and dispersions of the polysaccharide from *Xanthomonas campestris* NRRL B-1459. Carbohyd. Res. *13*, 257-267.

EDLIN, R. L. 1972. Method of producing a dehydrated food product. U.S. Pat. 3,694,236.

EUROPEAN ECONOMIC COMMUNITY. 1974. Official Publication of The European Economic Community, No. L 189/1, July 12.

FARKAS, E. 1974. Method for preparing low-calorie aerated structure. U.S. Pat. 3,821,428.

GABBY, J. L., CORBIN, D. D., and LOWE, J. B. 1974A. Ice milk or low fat imitation ice cream. U.S. Pat. 3,800,036.

GABBY, J. L., CORBIN, D. D., and LOWE, J. B. 1974B. Low calorie topping, spread, and frozen dessert. U.S. Pat. 3,809,764.

GLICKSMAN, M. 1969. Gum Technology in the Food Industry. Academic Press, New York.

GLICKSMAN, M., and FARKAS, H. E. 1973. Pudding compositions. U.S. Pat. 3,721,571.

GLICKSMAN, M., and FARKAS, H. E. 1974. Gum gelling system xanthan-tara dessert gel. U.S. Pat. 3,784,712.

GORIN, P. A. J., ISHIKAWA, T., SPENCER, J. F. T., and SLONEKER, J. H. 1967. Configuration of the pyruvic acid ketals, 4,6-*O*-linked to D-glucose units, in *Xanthomonas campestris* polysaccharide. Can. J. Chem. *45*, 2005-2008.

GRAHAM, H. D. 1971. Microdetermination of KELTROL (xanthan gum). J. Dairy Sci. *54*, 1622-1628.

GUMBMANN, M. R. 1964. Metabolism of carbon-14 polysaccharide, B-1459, by the rat. USDA Western Regional Lab., unpublished report.

HIDALGO, J., and HANSEN, P. M. T. 1969. Interactions between food stabilizers and β-lactoglobulin. J. Agr. Food Chem. *17*, 1089-1092.

JANSSON, P. E., KEENE, L., and LINDBERG, B. 1975. Structure of extracellular polysaccharides from *Xanthomonas campestris.* Carbohydrate Res. *45*, 275-282.

JEANES, A. 1952. Dextran, A. Selected Bibliography. U.S. Dept. Agr. (Peoria), Bur. Agr. and Ind. Chem. *AIC-288*.

JEANES, A. R., and SLONEKER, J. H. 1961. Method of producing an atypically salt-responsive alkali-deacetylated polysaccharide. U.S. Pat. 3,000,790.

JEANES, A. R., PITTSLEY, J. E., and SANTI, F. R. 1961. Polysaccharide B-1459: A new hydrocolloid polyelectrolyte produced from glucose by bacterial fermentation. J. Appl. Polymer Sci. *5*, 519-526.

JENKINSON, T. J., and WILLIAMS, T. P. 1973. Gel-coated frozen confection. U.S. Pat. 3,752,678.

JORDAN, W. A., and LESTER, W. H. 1973. Blends of *Xanthomonas* and guar gum. U.S. Pat. 3,765,918.

KANG, K. S. 1972. Unpublished data. Kelco Co., San Diego, Calif.

KANG, K. S., and KOVACS, P. 1974. New Microbial Polysaccharides as Potential Food Additives. Paper Presented at IVth International Congress of Food Science and Technology, Madrid, Spain.

KATZ, M. H. 1971. Edible mix composition for producing an aerated product. U.S. Pat. 3,582,357.

KELCO. 1975. Xanthan Gum/Keltrol/Kelzan/ A Natural Biopolysaccharide for Scientific Water Control, 2nd Edition. Kelco Co., San Diego, Calif.

KOVACS, P. 1970. Unpublished data. Kelco Co., San Diego, California.

KOVACS, P. 1973A. Useful incompatibility of xanthan gum with galactomannans. Food Techol. *27*, No. 3, 26-30.

KOVACS, P. 1973B. Xanthan gum, a new and unique colloidal stabilizer for the British Food Industry. Food Trade Rev. *43*, No. 11, 17-22.

KOVACS, P., and IGOE, R. S. (N.D.). The use of the xanthan gum—galactomannan system in the stabilization of cheese spreads. To be published.

KOVACS, P., and TITLOW, B. D. 1976. Xanthan gum, a unique stabilizer for cottage cheese creaming emulsions. Am. Dairy Rev. (In press).

LEACH, J. G., LILLY, V. G., WILSON, H. A., and PURVIS, M. R., JR. 1957. Bacterial polysaccharides: The nature and function of the exudate produced by *Xanthomonas* phaseoli. Phytopathology *47*, 113-120.

LILLY, V. G., WILSON, H. A., and LEACH, J. G. 1958. Bacterial polysaccharides. II. Laboratory-scale production of polysaccharides by species of Xanthamonas. Appl. Microbiol. *6*, 105-108.

McNEELY, W. H. 1967. Biosynthetic polysaccharides. *In* Microbial Technology, H. J. Peppler (Editor). Van Nostrand Reinhold Publishing Co., New York.

McNEELY, W. H., and KANG, K. S. 1973. Xanthan and some other biosynthetic gums. *In* Industrial Gums, 2nd Edition, R. L. Whistler, and J. N. BeMiller (Editors). Academic Press, New York.

McNEELY, W. H., and KOVACS, P. 1975. The physiological effects of alginates and xanthan gum. *In* Physiological Effects of Food Carbohydrates, ACS Symposium Series 15, A. Jeanes, and J. Hodge (Editors). American Chemical Society, Washington, D.C.

McNEELY, W. H., and O'CONNELL, J. J. 1966. Process for producing *Xanthomonas* hydrophilic colloid. U.S. Pat. 3,232,929.

NATIONAL RESEARCH COUNCIL. 1972. Food Chemicals Codex, 2nd Edition. National Academy of Sciences, Washington, D.C.

O'CONNELL, J. J. 1962. Edible compositions comprising oil-in-water emulsions. U.S. Pat. 3,067,038.

PANGBORN, R. M., TRABUE, I. M., and SZCZESNIAK, A. S. 1973. Effect of hydrocolloids on oral viscosity and basic taste intensity. J. Texture Studies *4*, 224-241.

PANGBORN, R. M., and SZCZESNIAK, A. S. 1974. Effect of hydrocolloids and viscosity on flavor and odor intensities of aromatic flavor compounds. J. Texture Studies *4*, 467-487.

PATTON, J. T. 1962. Thickening agent and process for producing same. U.S. Pat. 3,020,207.

PATTON, J. T. 1973. Process for synthesizing polysaccharides. U.S. Pat. 3,729,-460.

PATTON, J. T., and LINDBLOM, G. P. 1962. Process for synthesizing polysaccharides. U.S. Pat. 3,020,206.

PETTITT, D. J. 1973. Graft copolymers of *Xanthomonas* hydrophilic colloid and acrylic monomer. U.S. Pat. 3,708,446.

PILNIK, W., and MacDONALD, R. A. 1968. The stability of hydrocolloids. Gordian *68*, 531-533. (German)

RAO, M. A., and KENNY, J. F. 1975. Flour properties of selected food gums. Can. Inst. Food Sci. Technol. J. *8*, No. 3, 142-148.
REES, D. A. 1969. Structure, conformation, and mechanism in the formation of polysaccharide gels and networks. *In* Advances in Carbohydrate Chemistry and Biochemistry, Vol. 24, M. L. Wolfrom, R. S. Tipson, and D. Horton (Editors). Academic Press, New York.
REES, D. A. 1972. Shapely polysaccharides. Biochem. J. *126*, 257-273.
ROBBINS, D. J., MOULTON, J. E., and BOOTH, A. N. 1964. Subacute toxicity study of a microbial polysaccharide fed to dogs. Food. Cosmet. Toxicol. *2*, 545-550.
ROCKS, J. K. 1971. Xanthan gum. Food Technol. *25*, No. 5, 22-31.
ROGOVIN, S. P., and ALBRECHT, W. J. 1964. Recovering microbial polysaccharides from their fermentation broths. U.S. Pat. 3,119,812.
ROGOVIN, S. P., ALBRECHT, W. J., and SOHNS, V. 1965. Production of industrial-grade polysaccharide B-1459. Biotechnol. Bioeng. 7, 161-169.
SCHUPPNER, H. R., JR. 1968. Non-alcoholic beverage. U.S. Pat. 3,413,125.
SCHUPPNER, H. R., JR. 1970. Milk gel composition. U.S. Pat. 3,507,664.
SCHUPPNER, H. R., JR. 1971. Heat reversible gel. U.S. Pat. 3,557,016.
SCHUPPNER, H. R., JR. 1973. Acidified food products containing *Xanthomonas* colloid. U.S. Pat. 3,726,690.
SCHWEIGER, R. G. 1966A. *Xanthomonas* hydrophilic colloid ethers. U.S. Pat. 3,236,831.
SCHWEIGER, R. G. 1966B. Cationic ethers of *Xanthomonas* hydrophilic colloid. U.S. Pat. 3,244,695.
SCHWEIGER, R. G. 1966C. Method of improving viscosity characteristics of Xanthomonas hydrophilic colloids and esters produced thereby. U.S. Pat. 3,256,271.
SILMAN, R. W., and ROGOVIN, S. P. 1972. Continuous fermentation to produce xanthan biopolymer: effect of dilution rate. Biotechnol. Bioeng. *14*, 23-31.
SLONEKER, J. H., and ORENTAS, D. G. 1962. Pyruvic acid, a unique component of an exocellular bacterial polysaccharide. Nature *194*, 478-479.
SLONEKER, J. H., ORENTAS, D. G., and JEANES, A. 1964. Exocellular bacterial polysaccharide from *Xanthomonas campestris* NRRL B-1459. III Structure. Can. J. Chem. *42*, 1261-1269.
STANISLAV, L. R., and SHEETS, J. K. 1971. Improved French dressing formulation using xanthan gum stabilization. Food Prod. Develop. *5*, No. 6, 52-53, 56.
STARR, M. P. 1946. The nutrition of phytopathogenic bacteria. I. Minimal nutritive requirements of the genus *Xanthomonas.* J. Bacterial. *51*, 131-143.
STOLP, H., and STARR, M. P. 1964. Bacteriophage reactions and specification of phytopathogenic xanthomonads. Phytopathol. Z. *51*, 442-478.
SUTTON, J. C., and WILLIAMS, P. H. 1970. Comparison of extra-cellular polysaccharide of *Xanthomonas campestris* from culture and from infected cabbage leaves. Can. J. Botany *48*, 645-651.
WILKINSON, J. F. 1958. The extracellular polysaccharides of bacteria. Bacteriol. Rev. *22*, 46-73.
WINTERSDORFF, P. 1972. Method of preparing freeze-thaw-stable spoonable salad dressing. U.S. Pat. 3,676,157.
WOODARD, G. *et al.* 1973. Xanthan gum: Safety evaluation by two-year feeding studies in rats. Toxicol. Appl. Pharmacol. *24*, 30-36.
WOODARD RESEARCH CORP. 1968A. Xanthan Gum—Acute Oral Toxicity to Rats. Unpublished report. Woodard Research Corp., Herndon, Va.
WOODARD RESEARCH CORP. 1968B. Xanthan Gum—Acute Oral Toxicity to Dogs. Unpublished report. Woodard Research Corp., Herndon, Va.
WOODARD RESEARCH CORP. 1971. Safety Evaluation of Xanthan Gum by

I. Primary Skin Irritation and II. Test for Eye Irritation. Unpublished report. Woodard Research Corp., Herndon, Va.

WOODARD RESEARCH CORP. 1973. Keltrol—Intracutaneous Sensitization Potential in the Guinea Pig. Unpublished report. Woodard Research Corp., Herndon, Va.

YAKI, K., and OKAMURA, T. 1969. Manufacturing method for ice cream and sherbert. Japanese Pat. 44-10149.

CHAPTER 14

William A. Meer

Plant Hydrocolloids

The vegetable gums, or natural plant hydrocolloids, have served for centuries as foodstuffs, as well as thickeners and extenders of other foods. It is interesting to note the origin of the term "gum." The word is derived from an ancient Egyptian word meaning "a sticky substance." In hydrocolloid, the prefix "hydro" is the Greek word for water. The word colloid is derived from the French "col" meaning glue and "oid" meaning like. The hydrocolloids form molecular sols in most cases and are polymers of colloidal size and exhibit colloidal characteristics.

The natural plant hydrocolloids can be classified into three major categories, plant exudates, seaweed extracts and seed gums, as summarized in Table 14.1. The gums can also be chemically classified into two groups: anionic seaweed polysaccharides, and anionic exudate polysaccharides (Table 14.2).

Physically, these water soluble gums or colloids are best described as viscosity builders, forming colloidal sols or gels. Some, such as gum arabic, appear to form solutions and are soluble up to 50% concentration. The other gums form viscous sols at low concentration and gels at higher concentrations. Plant hydrocolloids, by definition, are water-loving materials which can influence the processing condi-

TABLE 14.1

CLASSIFICATION OF PLANT HYDROCOLLOIDS

Plant Exudates	
Arabic	*Acacia senegal*
Ghatti	*Anogeissus latifolia*
Karaya	*Sterculia* or *Cochlospermum*
Tragacanth	*Astragalus gummifer*
Seaweed Extracts	
Agar-Agar	*Gelidium gracilaria*
Irish Moss—Carrageenan	*Chondrus crispus*
	Gigartina mamillosa
Furcelleran	*Furcellaria fastigiata*
Alginates	*Phaeophycae*—Red Seaweeds
Seed Gums	
Guar	*Cyamonsis tetragonoloba*
Locust Bean Gum	*Ceratonia siliqua*
Tamarind	*Tamarindus indica*

TABLE 14.2

CHEMICAL GROUPINGS OF NATURAL PLANT HYDROCOLLOIDS

Anionic seaweed polysaccharides
- Agar—a linear polygalactose sulfuric acid ester
- Alginates—linear polymers of mannuronic and guluronic acids
- Carrageenan—a complex sulfate ester with high anhydrogalactose to galactose ratio, generally believed to be a combination of the *kappa, lambda* and *iota* fractions

Anionic exudate polysaccharides
- Arabic—complex of arabic acid, a highly branched polymer made up of galactose, rhamnose, arabinose and glucuronic acid
- Ghatti—complex of ghatti acid, composed of pentoses and hexuronic acids
- Karaya—a complex polymer of galactose, rhamnose and glucuronic acid, partially acetylated
- Tragacanth—a mixture of polysaccharides, tragacanthin and bassorin, polymers of fucose, xylose, arabinose and glucuronic acid

Nonionic seed polysaccharides
- Guar—a straight chain mannan grouping with relatively regular branching on every second mannose by a single galactose unit
- Locust bean gum—a straight chain mannan grouping with relatively regular branching on every fourth mannose group by single membered galactose units
- Tamarind—polysaccharide composed of galactose, xylose and glucose

tions and behavior of a food product in several ways, as it is evident that a change in the amount of water or in its physical state is largely responsible for alterations in the texture of processed food products.

The most important functions performed by gums in processed food formulations are (1) retention of water, (2) reduction in moisture evaporation rates, (3) alteration of freezing rates, (4) modification of ice crystal formation, (5) regulation of rheological properties or viscosity. Table 14.3 lists the typical characteristics of the plant hydrocolloids along with the major food applications for which they are currently being used.

Plant hydrocolloids are used in various food and flavor systems. Frozen, easy-to-prepare, baked goods have found a ready acceptance by the American housewife due to the ability of food processors to control the factors that lead to good flavor, color, texture, density, uniformity and shelf-life. The natural plant hydrocolloids are used to obtain smooth, uniform body and texture, to give complete flavor release, to improve whipping, thickening power, gel, stability and emulsification.

AGAR

The red alga (class Rhodophyceae), and more particularly the agarophytes, such as *Gelidium cartilagineum* (Linn.) Gaillon (family

TABLE 14.3

FUNCTIONS OF PLANT HYDROCOLLOIDS IN FOODS

Function	Example
Adhesive	Bakery glaze
Binding agent	Sausages
Calorie control agent	Dietetic foods
Crystallization inhibitor	Ice cream, sugar syrups
Clarifying agent (fining)	Beer, wine
Cloud agent	Fruit juice
Coating agent	Confectionary
Emulsifier	Salad dressing
Encapsulating agent	Powdered flavors
Film former	Sausage casings, protective coatings
Flocculating agent	Wine
Foam stabilizer	Whipped toppings, beer
Gelling agent	Puddings, desserts, aspics
Molding	Gum drops, jelly candies
Protective colloids	Flavor emulsions
Stabilizer	Beer, mayonnaise
Suspending agent	Chocolate milk
Swelling agent	Processed meats
Syneresis inhibitor	Cheese, frozen foods
Thickening agent	Jams, pie fillings, sauces
Whipping agent	Toppings, icings

Gelidiaceae), *Gracilaria confervoides* (Linn.) Greville (family Sphaerococcaceae), *Eucheuma isiforma* and their related species, are processed into agar. Production of agar reportedly dates back to the 17th century in Japan; but the material as we know it today has been commercially available only since the late 1800's. Agar served as a food in the Orient for ages, probably in the form of flavored or sweetened gels. Agar forms a translucent, porous mass which can be reconstituted into a gel by boiling in water and cooling (Tsen and Alexander 1946). This observation formed the basis for the commercial production of agar by Japanese industry (Selby and Selby 1959).

Agar is structurally a complex polysaccharide (Percival and Somerville 1937). Recent studies show agar to be composed of several polysaccharides with varying amounts of 3,6-anhydro-1-galactose and D-galactopyranose residues linked together (Smith and Montgomery 1959). It is presumably a mixture of at least two polysaccharides: agarose, the gelling agent which is composed of alternating residues of 1,3-β-D-galactopyranose and 1,4-3,6-anhydro-x-1-galactopyranose; and a very viscous, weak-gelling component (Araki and Hirase 1955; Araki and Araki 1957; Araki 1956). The complete structure has still not been established. Thus, the term agar must be used only in the generic sense, referring to polysaccharides which may differ structurally but which have certain properties in common.

Most commercial collection of agar seaweed occurs at low tide. The weeds are allowed to dry and bleach in the sun right on the beach after gathering. Agar is manufactured by hot water extraction followed by filtration and freezing. The freezing allows removal of soluble salts. Manufacture may involve the following steps: cleaning or washing of the raw material, bleaching or chemical treatment, extraction, filtration, gelation, freezing, thawing, filtration, drying, bleaching, washing and drying. The actual steps followed will depend on the country of origin, as well as on the degree of purity desired.

Agar is commercially available in bundles of thin, membranous, agglutinated strips or in cut, flaked, granulated, or powdered forms. The color varies from white to pale yellow to tan in color. It is either odorless or has a slight characteristic odor, and a mucilaginous taste. Agar is insoluble in cold water but soluble in boiling water. A 1.5% by weight solution congeals to form a firm, resilient gel which does not melt below 85°C. Many of its applications hinge upon this temperature difference between gelation and liquefaction.

Agar has a wide variety of uses. It is employed in the bakery, confectionery and dairy industries; in microbiological and culture media; in dentistry; in pharmaceuticals; in meat packing; and other miscellaneous applications. In food products agar is useful for its gelling and stabilizing properties. As a stabilizer it is utilized in pie fillings, piping gels, meringues, icings, cookies, cream shells, and similar products. In icings, a good use level will range from 0.2–0.5%. Agar is often used in combination with gum guar and locust bean gum in doughnut glaze stabilizers (Steiner and Rothe 1949). Agar is the preferred additive in jellied candies and in many specialty confectionery products, such as marshmallows and sugared fruit slices. It has been used as a filler in edible, rigid gels. Agar is used at levels of 0.05–0.85% in Neufchatel and cream cheeses to help reduce wheying off and improve body and slicing qualities. It is used as a thickening and gelling agent by poultry, fish and meat canners. Alone or in combination with gum guar, it also finds application in pet foods as well as meat pies.

TRAGACANTH

Gum tragacanth is one of the most widely used natural emulsifiers and thickeners available to the food, drug, and allied industries. The high viscosity imparted to water by the gum makes it useful for preparing aqueous suspensions of insoluble substances. Gum tragacanth was known and used in the days of Theophrastus, who described it in the 3rd century BC. It is listed in the *United States Pharmacopeia* (U.S. Pharmacopeal Convention 1975), and in the *Food Chemicals*

Codex (Natl. Acad. Sci.-Natl. Res. Council (1972)). Gum tragacanth is approved for food use and is on the GRAS list (generally recognized as safe for its intended use) under the Federal Food, Drug and Cosmetic Act.

Gum tragacanth is the dried gummy exudation of several species of *Astragalus* (family Leguminosae), a small, low, bushy perennial shrub characterized by a relatively large tap root, which, along with the branches, is tapped for the gum. The name "Tragacanth" is derived from the Greek tragos (goat) and akantha (horn) and probably refers to the curved shape of the ribbons, the best grade of commercial gum. It also occurs in flake form. The plants are common in certain sections of Asia Minor and in the semi-desert and mountainous regions of Iran, Syria, and Turkey.

When tragacanth is mixed with water, only the soluble fraction, called tragacanthin, dissolves to give a colloidal hydrosol whereas the insoluble fraction, consisting of 60–70% bassorin (Rowson 1937) swells to a gel. Chemically, tragacanthin is a complex mixture of acidic polysaccharides containing D-galacturonic acid. The other sugars produced on hydrolysis are D-galactose (Hilger and Dreyfus 1900), L-fucose(6-deoxy-L-galactose), D-xylose, and L-arabinose (Widtsoe and Tollens 1900; Jones and Smith 1949). Bassorin appears to be a methylated acidic polysaccharide. Demethoxylation of bassorin probably gives tragacanthin. Cellulose and starch are present in very small amounts. The *Food Chemicals Codex* has established standards for food-grade gum tragacanth.

Gum tragacanth has a high molecular weight (840,000), and the molecules have an elongated shape, 4500 Å by 19 Å, which accounts for its high viscosity (Graten and Karrholm 1950). Gum tragacanth solutions, at 0.5% and higher, exhibit structure viscosity, a phenomenon displayed by elongated molecules and recognized by the fact that the rate of flow of the solution in a capillary tube is not proportional to the pressure. Mechanical grinding of dry gums leads to molecular cleavage. Thus, the same concentration of gum solutions made from whole gum is more viscous than that from powdered gum. Gum tragacanth solutions reach maximum viscosity after being held for 8 hr at 40°C, and the viscosity is proportional to the methoxyl content of the gum.

An important property of gum tragacanth is its ability to produce solutions with high viscosity. A 1% solution of high grade gum has a viscosity of 3600 cps at 60 rpm using a Brookfield viscometer. At 25°C, the solution viscosity reaches a maximum at about 24 hr. This maximum may be obtained in about 2 hr at 50°C. A thick gel is produced at 2–4% concentration. The maximum initial viscosity of gum tragacanth solutions occurs at pH 8, but maximum stable viscosity is near pH 5 (Schwarz *et al.* 1958). Compared to other gums, tragacanth is quite stable over a wide pH range and is stable at about pH 2.

Gum tragacanth is widely used in many industries because of its stability to heat and acidity and because it is an effective emulsifying agent with an extremely long shelf-life. Gum tragacanth is widely used in the preparation of salad dressings; relishes; sauces; condiment bases; sweet pickle liquors; soft jellied products, such as gefilte fish; thick broths; beverage and bakery emulsions; ices and sherbets; bakery toppings and fillings; and confectionary cream centers (Glicksman 1969). Because of its acid resistance and its long shelf-life, gum tragacanth is useful in preparing stabilized French, Italian, Roquefort, and other creamy dressings. These are generally considered to be the pourable type and a few, such as the French and Italian types, are covered by standards of identity and must contain not less than 35% vegetable oil. The gum acts to thicken the water phase and prevents the oil phase from coalescing. Generally 0.4–0.75% gum, based on the total weight of dressing, is used. The preferred procedure is to wet the gum with a small amount of oil to inhibit lumping and then to disperse the mixture in water with rapid agitation. After all the ingredients have been added, the mixture is heated to approximately 72°C for 30 min. It is then homogenized in a colloid mill or another type of homogenizer.

Recently, there has been a shift to the preparation of these pourable dressings by a continuous process, using rigid process tolerances and techniques. Low calorie dressings have become an important part of the pourable dressings market. In these, the oil content is 1–5% and the gum content is 0.5–1.2% of the total dressing weight. The action of the gum is similar to that in a standard dressing, but a higher use level is required because of the larger aqueous phase.

The advantage of gum tragacanth in dressings is that it forms a creamier, more natural-looking dressing, has excellent shelf-life, and has good refrigerator stability in the home after the dressing has been opened by the homemaker.

Condiments and sauces are important product groups in which acid stability and long shelf-life are important. Vinegar is usually a primary ingredient in them, and gum tragacanth acts both as an emulsion stabilizer and thickener of the aqueous phase for such ingredients as spice flavorings and natural flavor extracts. Condiments are made (Burrell 1958) by heating a smooth mixture of ingredients to boiling. The gum may be added to the condiment at the end of the boiling period, and the mixture may be cooled in a heat exchanger. Generally between 0.4 and 0.8% of gum, based on total weight of sauce, barbecue sauce, or condiment, is used.

Gum tragacanth is used to stabilize bakery emulsions and fillings in which suspended fruit purées, natural flavor extracts and other flavors are used. The gum forms a creamy filling with good shine and transparency and gives a long shelf-life in conjunction with the fruit acids in the product. Recently, gum tragacanth has been used in a fruit topping for frozen cheese cakes in which the whole fruit is suspended in thick jelly giving clarity, brilliance, and improved texture. It is important to use a high grade gum for such an application to give the fruit a natural and rich appearance. Gum tragacanth is used in frozen pie-fillings and has been suggested as a cold-process stabilizer for meringues.

Gum tragacanth has been used as a stabilizer in ice cream mixes at concentrations of 0.2–0.35% (Potter and Williams 1950) giving smooth body and texture. A favorable characteristic in this application is the ability of tragacanth to maintain and increase its viscosity during heat processing. It is also used with frozen fruits that are to be suspended in the ice cream. It has been used as a stabilizer for water ices, ice pops, and sherbets, in concentrations of approximately 0.5%. In ice pops, gum tragacanth prevents the syrup from separating from the ice matrix.

Gum tragacanth is used in cream centers of candies that contain natural fruit acids. It has been found useful for the stabilization of vitamin C in aqueous solutions. A patent has been issued using gum tragacanth for preserving milk (Perin 1948).

KARAYA

Gum karaya was once introduced as a subsitute for gum tragacanth. However, it has many uses depending on its properties, as an adhesive, a binding agent, and its ability to absorb large amounts of water. It is the dried exudation of the tree *Sterculia urens* (Mantell 1947). Karaya is official in *The National Formulary* (Am. Pharm. Assoc. 1960) and the commercially available gum comes from India. The tree is large and is found on dry hills and plateaus in northern and central India. The best quality gum is collected in the spring and early summer before the monsoon rains. Karaya dries in the form of large, irregular tears. The dried tears are picked and brought to central collection areas. The tears are sorted on the basis of size, color and amount of adhering bark. The gum is available in several grades, #1, #2, #3 and siftings. The gum is further processed by air separation and micropulverization. The best grades contain the lowest percentage of bark and other foreign organic matter. The color of a fine powder ranges from off white to pink to tan. Gum karaya like tragacanth does not form a true solution but absorbs water, swelling greatly. Its solutions have a low pH.

Structurally gum karaya is a complex molecule. It contains approximately 38% uronic acid residues and 9% acetyl groups. Hydrolysis yields galactose, rhamnose and glucuronic and aldobiouronic acid residues (Aspinall and Nadir-ud-din 1965). In dilute solutions of gum, the viscosity increases linearly with concentration. Higher concentrations of about 4% form rubbery paste-like gels. Gum solutions show increased viscosity and ropiness above pH 7. Its excellent adhesive qualities become apparent when the gum is blended with an alkali, such as sodium borate or magnesium oxide. This forms the basis for many denture adhesives, both powders and gels.

In addition to its use as a denture adhesive, the pharmaceutical industry utilizes gum karaya as a bulk laxative. It is used in the form of a coarse crystal. The gum absorbs water and swells 70–100 times greater than its original volume. This mucilage acts as a laxative. The gum has the added advantage of not being metabilized or absorbed by the body.

The use of karaya in foods is limited. The gum prevents the bleeding of free water and retards ice crystal formation. Thus it is used in sherbets, slush ices and ice pops at concentrations of 0.2–0.5%. The acidic nature of karaya allows for its use in cheese spreads. In this application it improves spreadability and prevents water separation. It is a binder in sausage casings and ground meats. Gum karaya has been used in French dressings and meringues as a stabilizer.

GHATTI

Gum ghatti (Indian gum) is a complex water soluble polysaccharide. It is a plant exudate that has been long in use and whose name is derived from the word "ghats," which means passes, given to the gum because of its ancient mountain transportation routes. Gum ghatti is approved for food use and is in the GRAS (generally recognized as safe) list under the Federal Food, Drug and Cosmetic Act.

Gum ghatti is an amorphous exudate of the *Anogeissus latifolia* tree of the Combretaceae family. The tree is quite large and is found abundantly in the dry, deciduous forests of India and to a lesser extent, in Ceylon. The gum has a glassy fracture and occurs in rounded tears. The color of the exudate varies from very light to dark brown; the lighter the color, the better the quality and grade. Artificial incisions can be made in the tree bark to increase the yield; however, these incisions must be well planned so as not to destroy the tree. The gum is thought to act as a sealant when the bark is damaged.

Harvesting and grading of gum ghatti are done by methods similar to those used with gum karaya, near which it grows. The tonnage of gum karaya imported into the United States, however, far exceeds that of gum ghatti.

Ghatti tears are processed in the United States. This processing is mainly a grinding operation by which the gum is pulverized to a fine powder. However, various other mesh separations can also be made to satisfy the demands of the consumer. During the process of particle breakdown, impurities are removed from the gum by sifting, aspiration and density-table separation.

Ghatti has a bland taste and practically no odor. Only about 90% of the gum disperses in water and this portion forms a colloidal dispersion.

Gum ghatti occurs as a calcium-magnesium salt. Aspinall and coworkers (1955) have shown that it is composed of L-arabinose, D-mannose, D-xylose and D-glucuronic acid in a molar ratio of 10:6:2:1:2: and traces (below 1%) of 6-deoxyhexose. Partial hydrolysis gives two aldobiouronic acids, namely 6-*O*-(β-D-glucopyranosyluronic acid)-D-glactose and 2-*O*-β-D-glucopyranosyluronic acid)-D-mannose. A previous investigation reported the presence of 50% pentose and 12% galactose or galacturonic acid (Hanna and Shaw 1941). Further work (Aspinall *et al.* 1958A) has shown that the gum contains a backbone of (1→6) linked β-D-galactopyranosyl units and that partial acid hydrolysis (Aspinall *et al.* 1958B) affords two homologous series of oligosaccharides together with small amounts of 3-*O*-β-D-galactopyranosyl-D-galactose and 2-*O*-(β-D-glucopyranosyluronic acid) D-mannose. Acid-labile side chains are attached to the backbone through L-arabinofuranose resides (Aspinall and Christiansen 1961, 1965; Aspinall *et al.* 1965).

In solution, the molecules may have an overall rod shape (Elworthy and George 1963). A 1% dispersion of the gum has a pH of 2.63 and no buffering activity. The equivalent weight is 1340–1735 (Srivastava and Rai 1963). Gum ghatti forms viscous mixtures on dispersion in water to 5% concentration or greater. The dispersions are non-Newtonian in behavior, as is true with most of the water soluble gums, and their viscosity increases geometrically with concentration. The dispersions are less viscous than those of the gum karaya, although they are more viscous than those of gum arabic. Gum ghatti is a good emulsifying agent and can emulsify more difficult systems than can gum arabic.

Gum ghatti solutions may be slightly colored because of traces of pigment remaining in the gum. The normal pH of the dispersion is 4.8. There may be more incompletely dissolved material than observed in dispersions of either karaya or arabic gums. Insoluble matter can be removed from gum ghatti by filtration of its aqueous dispersion and spray-dried ghatti is currently available in which no insolubles are present. The viscosity of the spray-dried gum is somewhat lower than that of natural dry-milled gum.

The adhesiveness of gum ghatti dispersions is similar to that of gum arabic. Because of its higher viscosity, it is not possible to prepare dispersions as concentrated as with gum arabic. Gum ghatti does not form a true gel. Films prepared from gum ghatti dispersions are relatively soluble and brittle. Gum ghatti has a good emulsifying properties, which serve as the basis for most of its applications. Gum ghatti is used in applications also served by gum arabic. It is often used in pharmaceutical preparations as an emulsifying agent. In the United States, ghatti is employed in the preparation of stable, powdered, oil soluble vitamins (Dunn 1959). Gum ghatti has been used in combination with proteins as a means of stimulating formation of eosinophils. Gum ghatti is used in table syrup emulsion containing 2% butter to stabilize the emulsion. In such an application, about 0.4% ghatti is used in combination with 0.08% lecithin (Topalian and Elsesser 1966; Smith 1968). The refractive index of table syrup containing emulsified butter may be adjusted by additional quantities of ghatti to produce clarity.

Gum ghatti is used to emulsify petroleum and nonpetroleum waxes to form liquid and wax paste emulsions, which find wide uses in the paper industry as coatings and as barriers. Because of its high L-arabinose content, gum ghatti is hydrolyzed to prepare pure L-arabinose on a commercial scale. L-arabinose is used as a flavor adjunct in food products and in the preparation of nucleosides used as antitumor drugs.

GUAR GUM

Guar gum is derived from the seed of the guar plant, *Cyanaposis tetragonolobus*, family Leguminosae. The plant has been grown in India and Pakistan for centuries, where it is one of the most important crops, used as a food for both humans and animals. Guar was introduced to this country in the early 1900's when it was evaluated as a cover crop in Texas and Arizona.

The guar plant is a pod-bearing, nitrogen-fixing legume. The seeds of the plant are composed of the hull (15%), germ (45%) and endosperm (40%). Guar gum is produced by milling the endosperm after removal of the hull and germ. In practice, there is usually a residue of hull and germ, due to the difficulty encountered in a complete separation during processing. A typical analysis of gum guar is as follows: galactomannan 78–82%, moisture 10–15%, protein 4–5%, crude fiber 1.5–2.5% and ash 0.5–0.9%.

Guar gum is a galactomannan with linear chains of D-mannopyranosyl units with side branching units of D-galactopyranose attached by(1→6) linkages (Hirst and Jones 1948; Whistler and Durso 1951, 1952). Enzymic hydrolysis of guaran gives mannobiose, mannotriose and 6-*O*-α-D-galactopyranosyl-D-mannopyranose. The ratio of D-galactose to D-mannose in guar is 1:2. The great degree of branching in the gum is responsible for its ease of hydration as well as its greater hydrogen-bonding activity. Guar gum has a molecular weight of 220,000–300,000 (Hoyt 1966).

The chief property of guar gum is its ability to form viscous colloidal solutions when hydrated in cold water systems. Its solutions show the typical variance of viscosity versus shear rate of non-Newtonian fluids. Viscosities of 0.3% solutions change only slightly with increasing shear rates, while solutions of 1% or greater show marked changes. Viscosity is dependent on time, temperature, concentration, pH, ionic strength and the type of agitation as well.

The rate of hydration of gum guar varies. Rapid hydration requires a gum of fine particle size. Coarse particles, which are slow to hydrate, very often are easier to handle, as they show reduced tendency to lump up when dissolved. Full hydration is usually achieved within 2–4 hr.

Maximum viscosity of a guar gum solution is achieved at temperatures of 25–40°C. Higher temperatures speed up the rate at which maximum viscosities may be reached; however, excessive heating will degrade the gum. A 1% aqueous solution of gum guar has a viscosity of 2500–5000 cps. Doubling of the concentration from 1–2% results in a tenfold viscosity increase. In dilute solutions, the viscosity of the gum increases linearly with concentration up to about 0.5%. Gum guar is stable over a wide pH range. The optimum rate of hydration occurs between pH 7.5–9. Despite rate difference, maximum viscosities are the same between pH 1–10.5.

Guar gum is nonionic and is compatible with salts over a wide range of electrolyte concentration. Borate ion acts as a cross-linking agent with hydrated gum guar to form structural gels. Polysaccharides with numerous adjacent hydroxyl groups in the *cis* position can form three-dimensional borated gels (Deuel *et al.* 1948). Borated gels are reversible. Useful gels may be formed from other metal ions (Chrisp 1967).

The hydration and water-binding properties of guar gum are responsible for its use in food stabilization systems. Ice cream stabilizers, particularly high-temperature, short-time processes, use guar gum, at a concentration of 0.3%. It improves body, texture, chewiness and heat-shock resistance by binding free water present (Werbin 1950). It is on the list of ice cream standards (Federal Security Agency 1958). These same properties render it useful in the stabilization of ice pops and sherbets. Guar is allowed at levels up to 3.0% in cold-pack cheese foods. The gum helps to eliminate syneresis and weeping. In soft cheeses, guar increases the yield of curd solids and gives curds a better texture. Pasteurized process cheese spreads use a stabilizer consisting of guar, locust bean gum and emulsifier at 0.25–0.35% of total weight. Guar has been added to dough to retard dry out. It is useful in cake and donut mixes at levels under 1%. In pie fillings, guar thickens and prevents shrinking and cracking of the filling. Used in icings to absorb free water, guar gum is mixed at a level of 1 part gum to 250 parts sugar and 30 parts water. This prevents the icing from becoming sticky and adhering to the wrapper.

Gum guar has been used also as a thickener in salad dressings and pickle and relish sauces at 0.2–0.8%. Tragacanth, however, is superior in these applications as gum guar will break down more readily at low pH levels. Gum guar, however, is used at great advantage in meat sauces and gravies. The use of guar allows a significant reduction in the solids content of the product. Gum guar alone or in combination with agar at levels of 0.5% is useful in processing canned meat products. It prevents fat migration during storage and stops syneresis and water accumulation and reduces the tendency for voids to be present in the can.

Gum guar has good mouth feel and adds body to improve thin and watery products. Thus, it can be used in dietetic beverages, where sugar is absent. In addition, blends of guar gum and carrageenan are used in cocoa beverages as well as chocolate syrups. Guar gum is a versatile and effective suspending agent.

LOCUST BEAN GUM

The locust bean, or carob, is an ancient leguminous plant, botanically known as *Ceratonia siliqua*, family Leguminosae. It is indigeneous to the Mediterranean and Near East. The locust bean tree is a large evergreen growing 40–50 ft high. It can thrive in regions with little rainfall, and is an excellent shade tree. The tree begins to bear fruit 4–5 yr after budding.

Arabs used the carob seed as a unit of weight. The seed, called karat, became a standard (carats) against which gold and precious gems were weighed. The carob pod or fruit has been widely used as a foodstuff. It is very sweet and its flavor is further enhanced by roasting. It is commonly known as St. John's Bread, as the biblical food of St. John the Baptist. In German it is still called "Johannisbrot." The carob fruit is dark brown, 4–10 in. long and pod like. The pod contains the seed which is the source of locust bean gum. The seed is composed of the outer husk(30–35%), a central germ(25–30%), and the endosperm(35–45%). The semitransparent layer of endosperm yields the desired gum. In commerce the seed is dehusked, split and the germ removed. The separated endosperm is finely milled. The gum is an off-white powder consisting of 88% galactomannan, 3–4% pentosan, 5–6% protein, cellulose and ash (Griffiths 1952). It is available in several particle sizes and viscosity grades. The normal viscosity of a 1% solution is approximately 2800–3200 cps.

Locust bean gum has a molecular weight of 310,000 (Kubal and Gralen 1948). It is essentially a straight D-mannose polymer with branching on every fourth of fifth mannose group on C-6 by single D-galactose units. Ratios of D-galactose to D-mannose seem to vary with the origin of the gum or its growth conditions according to Whistler and Smart (1953). The structure is similar to that of gum guar.

Locust bean gum must be heated to achieve viscosity as it is only partly soluble in cold water. It is a neutral polysaccharide, and its viscosity is not effected by pH over the range of pH 3–11. Its solutions are cloudy, off-white and opaque. The chemical reactions of locust bean gum are the same as those of other neutral water soluble polysaccharides.

Locust bean gum can influence the gelling of other polysaccharides. Gels can be made in a wide range of solids content and pH with carrageenan and potassium chloride (Baker 1949, 1954; Baker *et al.* 1949). Gels of 85% agar and 15% carob gum are more elastic than those of agar alone (Deuel *et al.* 1950).

Locust bean gum is widely used today in many industries. It is an excellent ice cream stabilizer because of its ability to absorb water and high swelling power. The gum imparts a smooth melt-down and excellent heat-shock resistance to ice cream. It is not affected by lactic acid or calcium salts. It is a low cost stabilizer and does not mask the flavor of the product. Carob bean gum acts as a binder and stabilizer in processed meats, salami, bologna and pork sausages. It has a lubricating effect on the mix facilitating the extruding and stuffing operation. The gum yields a more homogenous product with better texture and also decreases weight loss during storage. In soft cheese manufacture, about 0.5% locust bean gum speeds up coagulation, increases the yield of curd solids by about 10%, and makes the curd easier to separate and remove. The resulting curd has good texture and the separated whey is limpid. The finished cheese has excellent body and structure. It is more homogenous, smoother, more resilient and exudation of water from fresh cheese is reduced. Locust bean gum is sometimes used to thicken soups at levels of 0.2–0.5%. In many similar food applications, the natural gums are preferred to starches. Locust bean gum is used as a thickener in pie fillings. The gum yields a clear, fruit like filling when used at a level of 1–2% of the weight of the fruit juice and water. In bakery products the use of high quality locust bean gum produces more uniform doughs. Flours vary and the gum produces doughs of constant properties, increased water-holding capacity, greater resiliency and higher yields. Addition of gum to cake and biscuit doughs also gives higher yields and savings of eggs. The cakes and biscuits are softer and maintain their freshness. They retain their shape better during processing and are easier to slice. Locust bean gum is utilized as a stabilizer and binder in many prepared foods such as soup bases, sauces, frozen batter, vegetable and fish dishes. It has been used to stabilize whipped cream, mayonnaise and tomato ketchup, as well as salad dressings.

GUM ARABIC

Gum arabic is the dried, gummy exudate obtained from *Acacia senegal* and various other Acacia species, family Leguminosae. Arabic was known over 4,000 yr ago and is the most common and universally used of all the natural gums. There are about 500 species of *Acacia*, but only a few are commercially important. The bulk of the available gum comes from the Sudan; however, Senegalese and Nigerian gum have recently become more plentiful. French West Africa and several other African countries account for the balance of the crop. The Sudanese gum or Kordofan sorts are considered best.

The tree is about 10–15 ft tall and has a life span of 25–30 yr. The exudation of gum arabic is a pathological condition attributed to unhealthy trees (Blunt 1926), as the gum is produced from breaks or wounds in the bark of the tree. It exudes to form balls or tears which when partially dried are collected and taken to central collection stations. There the gum is sun dried and sorted according to size and freedom from bark and dirt. The gum is commercially available in three grades of selected sorts, clean amber sorts and siftings. American importers and processors further sort and grade the gum, which is sold in various particle sizes ranging from coarse granules to fine powders, as well as in a spray-dried form. Spray-dried gum arabic is a purified product with low microbial counts, which has become very widely used by both the food and pharmaceutical industries.

Gum arabic is official in the *U.S. Pharmacopeia* (U.S. Pharmacopeal Convention 1975) as well as the *Food Chemicals Codex* (Natl. Acad. Sci.-Natl. Res. Council 1972). It is on the GRAS (generally recognized as safe) list under the Federal Food, Drug and Cosmetic Act.

Gum arabic is pale amber to off white in color. The powdered gum is white to yellowish white, practically tasteless and odorless. Typical specifications call for less than 4% total ash, less than 0.5% acid insoluble ash and less than 1% water insoluble residue. The whole tears contain 15-20% moisture, powder normally 12%, and spray-dried powder 8% moisture.

Gum arabic is a neutral or slightly acidic salt of a complex polysaccharide containing calcium, magnesium and potassium ions. The main molecular structure is a chain of β-galactopyranose units with some substitution at the C-6 position with various side chains. Complete hydrolysis yields four basic sugars: L-arabinose, D-galactose, L-rhamnose, and D-glucuronic acid (Hirst 1966).

The molecular weight of gum arabic is believed to vary from 250,000–750,000, and differs markedly depending on the type of measurement as well. Gum arabic is not normally used where swelling or thickening power is required. The gum is very soluble and the viscosity increases slowly up to 25% concentration. Above this concentration the viscosity increases much more rapidly and solutions of up to 50% can be prepared. Increasing temperature reduces viscosity considerably. The gum is soluble in up to about 50-60% ethanol. It is the ability of gum arabic to form highly concentrated solutions which is responsible for its emulsifying and stabilizing effects. A 10% solution has a pH of about 4.6; arabic acid is a strong monobasic acid. The viscosity of gum arabic solutions is maximized at pH 6–7 (Taft and Malm 1931).

Gum arabic is the most widely used of all the plant hydrocolloids. It is compatible with other gums, proteins, carbohydrates and starches. Electrolytes tend to reduce the consistency of gum arabic solutions. Some reagents give precipitates or heavy gels. This occurs upon the addition of borax, ferric chloride, basic lead acetate, gelatin, Millon's reagent, and Stoke's acid mercuric nitrate reagent. Gum arabic contains both oxidases and peroxidases that are inactivated by heating the gum in solution to 71°C or higher for about 1 hr. Enzyme inactivated gum is preferable for use in beverage emulsions.

The uses of gum arabic depend upon its action as a protective colloid or stabilizer, the adhesiveness of its water solutions, and as a thickener. The food industries consume about 50% of all the gum arabic imported into the United States. Gum arabic is the preferred emulsifier in flavor emulsions. It is commonly employed at a level of 2 lb per gallon of emulsion. It also serves to stabilize foams in the manufacture of soft drinks and beer. About 3 lb is enough to stabilize 50 barrels of beer. Gum arabic acts as an emulsifier in solution. The resultant emulsion can also be spray dried to encapsulate or fix the flavor. Thus essential oils and oleoresins may be entrapped at levels of 20–25% to yield excellent spray-dried flavors. These are used in dry mixed desserts, soup bases, beverage and cake mixes.

Gum arabic is also used in dairy products such as ices, sherbets and ice creams. Here it functions to retard both ice crystal formation and growth. A mixture of reacted arabic and carrageenan was patented as an ice cream stabilizer by Le Gloahec (1951).

Gum arabic is widely used in the candy industry. The gum is used to emulsify fats and retard sugar crystallization. It is used as a glaze in candy products and as a component of chewing gum, cough drops and candy lozenges. In jujubes and pastilles, where the sugar content is high and the moisture content comparatively low, it prevents sugar crystallization and keeps fats uniformly distributed throughout the product so as to prevent it from seeping to the surface and forming an easily oxidizable, greasy film. Here the gum is dissolved in water, the solution filtered, the sugar then added and boiled. The desired flavor should be added slowly without much stirring to minimize the formation of bubbles or opaque spots. Lozenges are also prepared from gum arabic by mixing finely ground or powdered sugar with a thick mucilage of gum arabic. Dietetic or sugarless hard candy drops and jujubes can be prepared from combinations of gum arabic, sorbitol and mannitol. Arabic at a level of 1–3% will give a hard candy, 5–35% a medium center, and a soft drop can be obtained utilizing approximately 70% gum. The greater the percentage of arabic, the softer and more chewable the candy. An excellent glaze for marzipan and buns, gum arabic acts also as a protective coating in panned confectionery goods, and may be used in coating nuts. It forms a thin, clear, transparent, adhesive film around the nut and seals in the oils and flavor. Gum arabic also improves the baking properties of rye and wheat flour at levels as little as 0.015% (Stemmer 1959). The gum extends the shelf-life of bread by rendering it softer. It truly has universal applications.

BIBLIOGRAPHY

AM. PHARM. ASSOC. 1960. The National Formulary, 11th Edition. American Pharmaceutical Association, Washington, D.C.

ARAKI, C. 1956. Structure of the agarose constituent of agar agar. Bull Chem. Soc. Japan *29*, 543; Chem. Abstr. *51*, 3462 (1957).

ARAKI, C., and ARAI, K. 1956. The chemical constitution of agar agar. Isolation of a new crystalline disacelaride by enzyme hydrolysis of agar agar. Bull. Chem. Soc. Japan *29*, 339; Chem. Abstr. *51*, 3465 (1957).

ARAKI, C., and HIRASE, S. 1954. Chemical constitution of agar agar. Exhaustive mercaptolysis of agar agar. Bull. Chem. Soc. Japan *27*, 105, 109; Chem. Abstr. *49*, 9517-9518 (1955).

ASPINALL, G. O., AURET, B. J., and HIRST, E. L. 1958A. Gum ghatti (Indian gum). Part II. The hydrolysis products obtained from the methylated degraded gum and the methylated gum. J. Chem. Soc. 221.

ASPINALL, G. O., AURET, B. J., and HIRST, E. L. 1958B. Gum ghatti (Indian gum). Part III. Neutral oligosaccharides formed on partial acid hydrolysis of the gum. J. Chem. Soc. 4408.

ASPINALL, G. O., BHAVANADAN, V. P., and CHRISTENSEN, J. B. 1965. Gum ghatti (Indian gum). Part V. Degradation of the periodute oxidized gum. J. Chem. Soc. 2677.

ASPINALL, G. O., and CHRISTENSEN, J. B. 1961. *Anogeissus schimperi* gum. J. Chem. Soc. 3461.

ASPINALL, G. O., and CHRISTIENSEN, J. B. 1965. Gum ghatti (Indian gum). Part IV. Acidic oligosaccharides from the gum. J. Chem. Soc. 2673.

ASPINALL, G. O., HIRST, E. L., and WICK-STROM, A. 1955. Gum ghatti (Indian gum). Part I. The composition of gum and the structure of two aldobioanonic acids derived from it. J. Chem. Soc. 1160.

ASPINALL, G. O., and NASIR-UD-DIN 1965. Plant gums of the genus *Sterculia*. Part I. The main structural features of *Sterculia urens* gum. J. Chem. Soc. 2710.

BAKER, G. L. 1949. Edible gelling composition containing Irish moss extract, locust bean gum and an edible salt. U.S. Pat. 2,466,146; Chem. Abstr. *43*, 5132 (1949).

BAKER, G. L. 1954. Gelling composition. U.S. Pat. 2,669,519; Chem. Abstr. *48*, 6051 (1954).

BAKER, G. L., CARROW, J. W., and WOODMANSEE, C. W. 1949. Food Ind. *21*, 617.

BLUNT, H. S. 1926. Gum Arabic with Special Reference to Its Production in the Sudan. Oxford Univ. Press, New York.

BURRELL, J. R. 1958. Food Manuf. *33*, 10.

CHRISP, J. D. 1967. Galactomannan gum gels for explosive composition and for textile or paper sizing. U.S. Pat. 3,301,723; Chem. Abstr. *67*, 92485 W.

DEUEL, H., HUBER, G., and SOLMS, J. 1950. Uber den einflub von und flohsomenschlein aug die mechanischen eigenschatten von agar-agar gelan. Experientia *6*, 138.

DEUEL, H., NEUKOM, H., and WEBER, F. 1948. Reaction of boric acid with polysaccharides. Nature *161*, 96.

DUNN, H. J. 1939. Stable oil-soluble vitamins. U.S. Pat. 2,897,119; Chem. Abstr. *53*, 20709 (1959).

ELWORTHY, P. H., and GEORGE, T. M. 1963. J. Pharm. Pharmacol. *15*, 781.

FEDERAL COMMUNICATIONS COMMISSION. 1972. Food Chemicals Codex, 2nd Edition, Publ. *1406*, Natl. Acad. Sci.-Natl. Res. Council, Washington, D.C.

FEDERAL SECURITY AGENCY. 1958. Service and Regulatory Announcements, Food and Drug Administration, Foods, Drugs, and Cosmetics. Title No. 2, Part 20 (March).

GLICKSMAN, M. 1969. Gum Technology in the Food Industry. Academic Press, New York.

GRATEN, N., and KARRHOLM, M. 1950. The physicochemical properties of solutions of gum tragacanth. J. Colloid Sci. *5*, 21.

GRIFFITHS, C. 1949. Mfg. Chem. *20*, 321.

GRIFFITHS, C. 1952. Food *21*, 58.

HANNA, D., and SHAW, JR., W. H. 1941. Proc. S. Dakota Acad. Sci. *21*, 78.

HILGER, A., and DREYFUS, W. E. 1900. Ueber arabinose, xylose and fucose aus tragacanth. Ber. *33*, 1178.

HIRST, E. 1966. The chemistry and rheology of water soluble gums. Soc. Chem. Monograph *24*, 3.
HIRST, E. L., and JONES, J. K. N. 1948. The galactomannan of carob seed gum (gum ghatto). J. Chem. Soc. 1278.
HOYT, J. W. 1966. Friction reduction as an estimation of molecular weight. J. Polymer Sci., Part B, *4*, 713.
JONES, J. K. M., and SMITH, F. 1949. Plant gums and mucilages. *In* Advances in Carbohydrate Chemistry, Vol. 4, M. L. Wolfrom, R. S. Tipson, and D. Horton (Editors). Academic Press, New York.
KUBAL, J. F., and GRALEN, N. 1948. Physicochemical properties of karaya gum and locust bean mucilage. J. Colloid Sci. *3*, 457.
LE GLOAHEC, V. C. E. 1951. Carrageenate-arabate coaceroate. U.S. Pat. 2,556,-282.
MANTELL, C. L. 1947. The Water Soluble Gums. Rheinhold Publishing Co., New York.
PERCIVAL, E. G. V., and SOMERVILLE, J. C. 1937. The acetylation and methylation of agar agar and the isolation of 2:4:6 trimethyl α-D-galactose by hydrolysis. J. Chem. Soc. *137*, 1615-1619.
PERIN, P. H. 1941. Preserving milk. French Pat. 860,210; Chem. Abstr. *42*, 6471 (1948).
POTTER, F. E., and WILLIAMS, D. M. 1950. Milk Plant Monthly *39*, No. *4*, 76.
ROWSON, J. M. 1937. J. Pharm. Pharmacol. *10*, 161.
SCHWARZ, T. W., LEVY, G., and KAWAGOE, H. H. 1958. J. Am. Pharmaceutical Assoc., Sci. Ed. *47*, 695.
SELBY, H. H., and SELBY, T. A. 1959. Agar. *In* Industrial Gums, R. L. Whistler (Editor). Academic Press, New York.
SMITH, F. 1968. Table sirup emulsion containing gum ghatti. U.S. Pat. 3,362,-833; Chem. Abstr. *68*, 67920 (1968).
SMITH, F., and MONTGOMERY, R. 1959. The Chemistry of Plant Gums and Mucilage. Reinhold Publishing Co., New York.
SRIVASTAVA, V. K., and RAI, R. S. 1963. Physicochemical studies on gum dhawa (*Anegeissus latifolic* wall). Kolloid-Z. *190*, 140.
STEINER, A. B., and ROTHE, L. 1949. Stabilizer for icings. U.S. Pat. 2,474,-019.
STEMMER, M. 1959. Improving the baking qualities of rye and wheat flour. German Pat. 1,051,755; Chem. Abstr. *55*, 8694 (1959).
TAFT, R., and MALM, L. L. 1931. Some physico-chemical properties of gum arabic—water systems and their interpretation. J. Phys. Chem. *35*, 874.
TOPALIAN, H., and ELSESSER, C. 1966. Table sirup emulsion U.S. Pat. 3,282,707; Chem. Abstr. *66*, 1748 (1967).
TSENG, C. K., and ALEXANDER, L. J. 1946. Colloid Chemistry, Vol. 6. Reinhold Publishing Co., New York.
U.S. PHARMACOPEAL CONVENTION. 1975. The United States Pharmacopeia, 19th Edition. Mack Publishing Co., Easton, Pa.
WERBIN, S. J. 1950. Ice cream stabilizer. U.S. Pat. 2,502,397; Chem. Abstr. *44*, 5496 (1950).
WHISTLER, R. L., and DURSO, D. F. 1951. The isolation and characterization of two crystalline disaccharides from partial acid hydrolysis of guaran. J. Am. Chem. Soc. *73*, 4189 (1951).
WHISTLER, R. L., and DURSO, D. F. 1952. A new crystalline trisaccharide from partial acid hydrolysis of guaran and the structure of guaran. J. Am. Chem. Soc. *74*, 5140.
WHISTLER, R. L., and SMART, C. L. 1953. Polysaccharide Chemistry. Academic Press, New York.
WIDTSOE, J. A., and TOLLENS, B. 1900. Mittheilungen, weber tragacanth ein beitrag zur kenntniss der planzenschleime. Ber. *33*, 132.

CHAPTER 15

Horace D. Graham

Analytical Methods for Major Plant Hydrocolloids

Plant hydrocolloids (phytocolloids) of many types are used widely in food products and formulations of diverse characteristics. In order to understand the role which they play in stabilization, emulsification, etc., methods for their isolation and quantitative determination have had to be developed. Another necessity for such microanalytical methods is for use in regulatory control.

In many respects, the quantitative detection of a plant hydrocolloid in a food product is like looking for a needle in a hay stack. This is so because, in most cases, the amounts added are very small (0.01–0.5%). Additionally, during processing and storage of the product (for example, ice cream, chocolate drink, evaporated milk, etc.) strong interactions with other food components may occur (for example proteins occur to form insoluble complexes, and if strong heating is applied, degradation to smaller units).

General tests for the classification and identification for most of the major food hydrocolloids have been developed by various workers. These have been amply reviewed and commented upon by Glicksman (1969). In most of these procedures severe overlapping of reactions occur between closely related hydrocolloids or many reactions are not definitive enough to allow for the unequivocal isolation and quantitative determination of a particular hydrocolloid. This chapter will deal with tests which have been developed for the isolation and determination of specific food gums. The ones to be given detailed attention are carrageenan, sodium carboxymethyl cellulose (CMC), alginates, and xanthan gum (Keltrol).

TESTS FOR CARRAGEENAN

Carrageenan, a strongly charged anionic sulfated polyelectrolyte, is used widely in many food products, especially dairy products, as a stabilizer. It exerts strong interaction in food mixtures due to the reactivity of its ionizable, anionic sulfate group. Methods for its isolation and quantitative determination are based up the reaction of the sulfate ion. These include the following.

(1) Prior dialysis of the food sample to remove inorganic sulfate followed by precipitation of the carrageenan ester sulfate by

barium and finally, gravimetric determination of the precipitated $BaSO_4$ (Hansen and Whitney 1960).

(2) Digestion of the food sample with a proteolytic enzyme, papain, to remove proteins followed by filtration, and determination of the liberated carrageenan in the filtrate.

Carrageenan in such a filtrate has been determined quantitatively by the following methods.

(1) Precipitation (concentration) of the carrageenan with a long chain quaternary ammonium compound (detergent) like cetypyridinium chloride. Then, this precipitate is dispersed by 30% H_2SO_4 and the carbohydrate content determined by the phenol-H_2SO_4 method (Graham 1968).

(2) The carrageenan in the filtrate was determined directly (without further treatment of the filtrate) by developing the color produced through its interaction with orthotolidine and sodium hypochlorite (Graham 1972A).

Several other reagents have been used to isolate and/or measure carrageenan content in several foods without prior enzymatic digestion of the proteins in the food. These include the following.

(a) Barium chloranilate (Graham 1966).
(b) Tetrazolium salts (Graham 1967).
(c) Polyvalent cobalt complexes (Graham and Williams 1966).

Electrophoretic methods

Recently, Chang and his co-workers (1974) have proposed the use of electrophoretic techniques to differentiate and possibly determine carrageenan in milk.

Of the tests mentioned previously, the gravimetric sulfate determination of Hansen and Whitney (1960), the method involving the use of a long chain quaternary ammonium detergent to precipitate the polymer from the papain digest of a food sample (Graham 1968) and the ortho-tolidine-hypochlorite method (Graham 1972A) seem to be most straightforward and useful procedures. Therefore, details of these procedures will be described.

Determination of Carrageenan Ester Sulfate in Milk Products

In this procedure, inorganic sulfate (free sulfate) in the milk sample was removed by dialysis. Then, the dialyzed sample was hydrolyzed with hydrochloric acid (1:20), decolorized with activated carbon and the sulfate content of the clarified filtrate determined gravimetrically as barium sulfate.

The method as outlined by Hansen and Whitney (1960) is as follows.

Reagents

Hydrochloric acid (conc.): (36.5-38.0% A.R.)

Hydrochloric acid (1:20): 1 vol. of concentrated HCl is diluted with 19 vol. of distilled water.

Activated carbon: Darco G-60, Atlas Powder Co., Wilmington, Delaware, or other activated carbon of low sulphate content.

Filter aid: Sil-O-Cel, Johns-Manville Sales Corp., Chicago, Illinois, or other filter aid of low sulphate content.

Barium chloride (5% w/v): 5 gm of $BaCl_2$ (A.R.) is dissolved in 100 ml of distilled water.

Procedure

Approximately 100 gm of milk, weighed to the nearest gram, is transferred to a 20-24 in. long Visking casing (No. 36/32) and the bag is closed so as to leave an air-filled headspace of not more than 25-30 ml, to prevent excessive dilution during dialysis. The sample is dialyzed in a cold room (5-6°C) for a minimum of 48 hr with 5-6 changes of precooled deionized water. After dialysis, the sample is transferred quantitatively to a 400-ml beaker (calibrated with a 200-ml mark). Ten milliliters of concentrated hydrochloric acid are added and the volume is made up to 200 ml with deionized water. The beaker is covered and placed in a covered water bath at boiling temperature. After some hydrolysis, 8 gm of activated carbon and 2 gm of filter aid are added, and the hydrolysis is continued for a total of 6 hr. The mixture is then stirred thoroughly and the insoluble material is allowed to settle. The supernatant is decanted through Whatman No. 42 filter paper in a 9-cm conical funnel. The precipitate in the beaker is suspended in 50 ml hydrochloric acid (1:20) and heated in the water bath, and the supernatant decanted through the filter. The washing of the precipitate is repeated and, when the filter has drained dry, the contents of the beaker are transferred to the filter and washed once with approximately 50 ml hydrochloric acid (1:20). The combined filtrates are collected in a 400 ml beaker and placed on top of the water bath at 65-70°C. When hot, an excess of hot 5% barium chloride is added rapidly with agitation, and the beaker is covered and allowed to stand for at least 6 hr at 65-70°C. The sample is then filtered through a tared Gooch crucible packed with acid-washed asbestos, and the beaker and crucible are washed with 100 ml of deionized water. A few drops of barium chloride are added to the filtrate, which is held for a few hours to insure complete removal of sulphate, and refiltered if necessary. In some instances, especially when the sulphate content is of the magnitude of only a few milligrams, postprecipitation of barium sulphate may occur after the passage through the Gooch crucible. This condition may be prevented by allowing a longer time for crystal growth if convenient. In this connection, it is advantageous to use the same crucibles for successive runs and not prepare fresh ones, since they would provide nuclei for retention of the last traces of barium sulphate in the filtrate. The crucible is ignited at 900°C and weighed after cooling to room temperature in a desiccator. The increase in weight is recorded as barium sulphate. Unless the sulphate content of the reagents is known, it is necessary to determine a reagent blank by the same procedure. The content of ester sulphate in the sample is calculated as follows.

$$\% \text{ ester sulphate} = \frac{(\text{wt. of BaSO}_4 \text{ in sample} - \text{wt. of BaSO}_4 \text{ in blank}) \times 41.15}{\text{wt. of sample}}$$

A modification of the above procedure is possible if a control without added carrageenan, but otherwise of the same identity as the sample, is available. In this case, the dialysis step may be omitted and, instead, the sample and the corresponding control may be hydrolyzed immediately with concentrated hydrochloric acid in the proportion of 1:20. The results in this case represent the total sulphate content of the carrageenan.

An exhaustive analysis of the effect of several variables on this test was made by the authors, as shown in Table 15.1, and good recoveries from several systems were realized (Table 15.2). This test, though somewhat lengthy, is reliable, involving no complicated equipment and requires only simple, readily available reagents.

Quantitative Determination of Carrageenan in Food Products Using Cetylpyridinium Chloride as a Precipitant

This procedure is based on the principle that carrageenan forms highly insoluble precipitates with cetylpyridinium chloride in the presence of relatively high concentrations of salts such as KCl (Scott 1956, 1060; Slack 1958). Conversely, other hydrocolloids such as sodium carboxymethyl cellulose, alginates, pectins and xanthan gum do not form precipitates at such high salt concentrations. Therefore, selective precipitation of the hydrocolloid from a clarified papain digest is possible.

Although the original method used cetylpyridinium chloride as the precipitant, it has been suggested (Johnson 1973) that other quarternary ammonium detergents, especially cetyltrimethylammonium bromide, will serve equally well.

Materials.—*Stock Solution of Carrageenan.*—Prepare a 0.5% (w/v) dispersion of carrageenan by dusting the weighed amount of the dried commercial product onto deionized, distilled water. While adding the powder, stir the mixture constantly. Heat the mixture and continue stirring until dispersion is complete. Cool the mixture, transfer it to a 500-ml volumetric flask and make the volume up to the mark with deionized, distilled water.

Buffer: 1.0 *M* sodium acetate-acetic acid, pH 6.5

Sulfuric acid: (Fisher Scientific Co.) 95-98%, sp. gr. 1.84

Celite 535: (Johns-Manville Co., New York)

Wash with distilled water, 1% sodium acetate, sodium chloride (1/4

TABLE 15.1

EFFECT OF VARIABLES

Variable	Investigated Levels	Source of Significant Variance[1] in terms of: mg % $SO_4^=$ in Sample	% $SO_4^=$ in Carrageenan	Selected Value	Control Necessary
Amount of activated carbon (gm)	4.0 8.0 12.0	Presence of carrageenan (1%) Carbon: Linear response[2] (1%) Quadratic response (5%)	None	8 gm	± 0.37 gm
Amount of filter aid (gm)	0.0 2.0 4.0 6.0	Presence of carrageenan (1%) Filter aid: Linear response[3] (1%)	None	2 gm	± 1.05 gm
Types of activated carbon	Darco G-60 Darco S-51 Nuchar C-190A Nuchar C-1000N	Presence of carrageenan (1%) Type (1%)	None	Darco G-60	Low sulphate content is recommended
Hydrolysis time (hr)	4.0 6.0 8.0 12.0	Presence of carrageenan (1%) Hydrolysis time: 4 vs. 6, 8, 10 (1%)	Hydrolysis time: 4 vs. 6, 8, 10	6 hr	At least 6 hr
Final volume of hydrolysis mixture (ml)	150.0 200.0	Presence of carrageenan (1%)		200 ml	Not important
Final acid strength of hydrolysis mixture (Volume of conc. HCl: final volume)	0.8:20 1.0:20 1.2:20	Final acid strength: 0.8:20 vs. higher concentrations (1%)	None	1.0:20	At least 1.0:20
Dialysis	None 40 hr		Dialysis (5%)	Dialysis	Exhaustive
pH during dialysis	3.0 4.0 5.0 6.0 7.0		Dialysis (1%)	Sample pH	Not important in the investigated range

Source: Hansen and Whitney (1960).

[1] All other sources of variance were found not to be significant.

[2] Linear regression coefficient: 0.71 mg $SO_4^=$ per gram of carbon.

TABLE 15.2

RECOVERY OF CARRAGEENAN ESTER SULPHATE FROM VARIOUS MILK PRODUCTS (DUPLICATES)

Sample	Concentration of Carrageenan (%)	Ester Sulphate in Sample (mg %)	Recovery[1] (%)
Water	0.0000	0.00	
	0.0295	6.85	97.3 ± 8.5
	0.0984	23.10	98.6 ± 2.6
Whole homogenized milk, commercial sample	0.0000	-0.25	
	0.0295	7.25	103.0 ± 8.5
	0.0984	24.10	102.9 ± 2.6
Chocolate milk			
2% fat	0.0000	-0.70	
0.85% cocoa	0.0295	6.95	98.7 ± 8.5
5.0% sucrose	0.0984	23.80	101.6 ± 2.6
8.0% m.s.n.f.			
Ice cream mix			
12.5% fat	0.0000	0.10	
10.0% m.s.n.f.	0.0295	6.75	95.9 ± 8.5
16.0% sucrose	0.0984	24.30	103.8 ± 2.6
0.5% egg yolk			
Ice cream mix			
Same as above, diluted	0.0000	0.30	
with an equal volume	0.0295	6.35	90.2 ± 8.5
of water	0.0984	24.00	102.5 ± 2.6
	Standard deviation ± 0.60 mg % $SO_4^=$		

Source: Hansen and Whitney (1960).

[1] Ester sulphate content of this carrageenan sample was established by eight independent determinations of water dispersions (0.10%).
Mean: 23.80% $SO_4^=$
Standard deviation of the mean: ± 0.25% $SO_4^=$

saturated) and finally with deionized water until the washings give a negative phenol-H_2SO_4 test.

Filtration Column.—The column, 18 mm in diameter and 40 cm high, is made from borosilicate glass and is fitted with a stopcock. Pack tightly with glass wool to a height of 10 cm. Filtration is done at the rate of 10–15 ml per min. Wash the glass wool consecutively with concentrated H_2SO_4, and deionized water until the washings give a negative phenol-H_2SO_4 test. After this, pour 25 ml of 0.1% CPC-0.05 *M* KCl over it, with the stopcock closed. Soak the column in this mixture for at least 30 min and keep it wet with this solution until ready for use.

Papain.—This is obtainable from the Mann research laboratories as a powdered concentrate assaying 72 milk clotting-units. Any other highly purified concentrate may be used.

Papain-EDTA-cysteine Hydrochloride Mixture.—Disperse papain (20.0 gm), cysteine hydrochloride (1.0 gm), and EDTA (1.0 gm) in 250 ml of warm (70°C) acetate buffer (pH 6.5).

Equipment.—Beckman DU-2 spectrophotometer or Bausch and Lomb Spectronic 20 spectrophotometer; pH meter, Beckman Zeromatic.

Wash all glassware thoroughly in water, soak in dichromate-H_2SO_4 cleaning solution, wash and finally rinse with distilled water.

Procedure.—*Digestion of the Protein with Papain.*—Place 5 gm of milk or milk product in a 250 ml centrifuge bottle. Add 1.25–10.0 mg of carrageenan, dispersed in distilled water.[1] Add enough solid sodium chloride to provide a 1.0 *M* solution and make the total volume up to 100 ml with distilled water. Shake the mixture and heat it for 10 min in a water bath (98 ± 1° C), cool to 70° C and add 5.0 ml of 1.0 *M* sodium acetate-acetic acid buffer, pH 6.5. Incubate the mixture at 70° C for 30 min and then add 25 ml of the papain-EDTA-cysteine hydrochloric mixture. Stopper the centrifuge bottle with a hard No. 6 rubber stopper, shake well and incubate the mixture at 70° C for 16 hr. At the end of this period add 2.0 ml of 1.0 *N* NaOH (or enough to adjust the pH of the mixture to 8.5), and 1.0 gm of Celite 535. Shake the mixture well and filter over glass wool. Wash the residue 3 times with 10 ml portions of distilled water, collect the washings and pool with the filtrate in a 250 ml centrifuge bottle. Discard the residue.

Precipitation of the Carrageenan.—To the filtrate containing the carrageenan add enough solid KCl to give a final concentration of 1.0 *M*, then add 10 ml of a 1% solution of CPC in water drop-wise, with constant shaking. Shake the mixture well and incubate for 30 min at 45° C. Add 1 gm of Celite 535. Shake the mixture and incubate for another 10 min. At the end of this period centrifuge the mixture for 10 min at 2,500 rpm and filter the supernatant over glass wool. Retain the residue containing the carrageenan-CP precipitate.

Removal of Sugars from the Carrageenan-CP Precipitate.—The residue described above contains the carrageenan-CP complex adsorbed onto the Celite. Admixed with this are sugars and probably other water soluble carbohydrates contained in the product. In order to free the precipitate of these interferences, wash the residue three times with 15-ml portions of 0.1% CPC-0.05% KCl and centrifuge at 2,500 rpm. Pour the supernatants over the column of glass wool. Test the filtrate from the column with the reagent and, if necessary, continue the washing until a negative test is obtained. Discard the washings.

Solution of the Carrageenan-CP Precipitate and Determination of the Carbohydrate Content.—To the sugar-free residue, add 20 ml of

[1] In carrying out a recovery test.

30% H_2SO_4. Stir the precipitate with a glass rod and hold mixture for 10 min in a water bath at 80°C. Centrifuge the mixture for 10 min at 2,500 rpm. Pour the supernatant over the column of glass wool and collect the filtrate in a 50 ml glass-stoppered volumetric flask. Wash the residue twice more with 12-ml portions of hot (80°C) 30% H_2SO_4, centrifuge, filter and collect the washings also. Make the filtrate and washings to 50 ml with 30% H_2SO_4, mix well, and use a 2-ml portion for determination of the carbohydrate content by the phenol H_2SO_4 method. The entire process is summarized in Fig. 15.1.

Phenol-H_2SO_4 Test.—The procedure used, outlined by Dubois *et al.* (1956), is as follows.

Place 2 ml of the carrageenan suspension, dispersed in 30% H_2SO_4 and containing 50–400 μg of the phytocolloid, in a series of 30 ml borosilicate test tubes. Add 1 ml of a solution of 5% phenol in water and shake the tubes to mix the contents well. Add 5 ml of concentrated H_2SO_4 rapidly from a burette, allowing the acid to fall directly into the center of the tube in order to obtain good mixing. Shake the

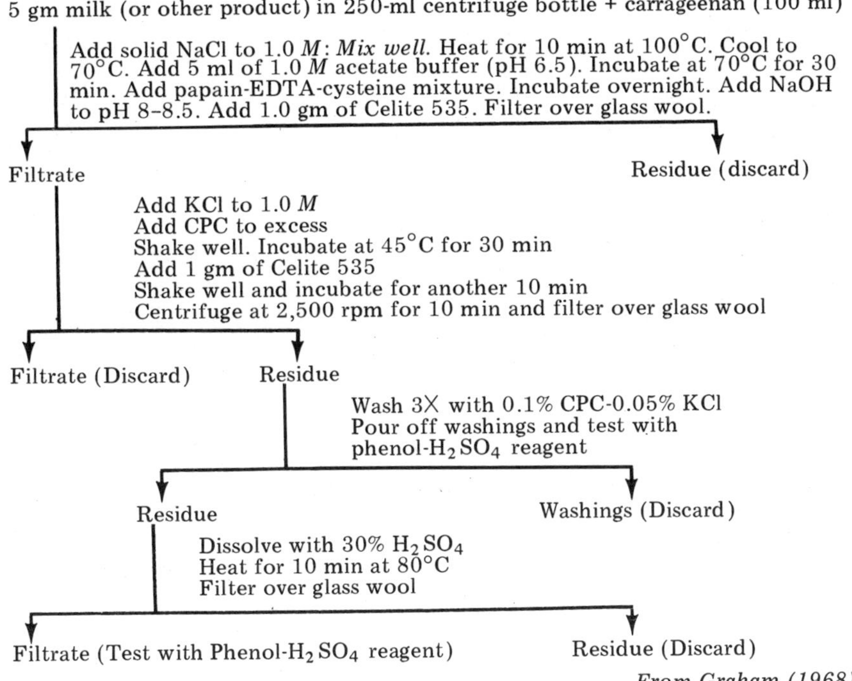

From Graham (1968)

FIG. 15.1. PROCEDURE FOR SEPARATION AND DETERMINATION OF CARRAGEENAN

tubes and leave them at room temperature (28–30° C) for 10 min. At the end of this period, place the tubes in a water bath for 10 min at 282 ± 2° C in order to cool the mixture. Measure the color developed at 490 nm against a reagent blank. This constitutes the standard curve.

For the determination of carrageenan recovered from milk or other products, 2.0 ml of the filtrate (appropriately diluted, if necessary) are used and the procedure for color development is the same. The blank is treated in the same way as the samples.

The effects of several variables are summarized in Table 15.3. Recovery data are shown in Table 15.4.

TABLE 15.3

INFLUENCE OF VARIABLES ON THE DETERMINATION OF CARRAGEENAN IN MILK

Variable Investigated	Range Investigated	Condition for Reproducibility		Value or Condition Selected
		Min.	Max.	
Final concentration of NaCl added (*M*)	0.0– 1.5	0.5	1.0	1.0
Time of pre-heating at 99 ± 1°C (min)	0.0–30	5.0	10.0	10.0
Final concentration of buffer during incubation (*M*)	0.0– 0.5	0.1	0.2	0.2
pH of buffered system during digestion	4.0– 7.0	6.0	6.5	6.5
Temperature of incubation (°C)	40–70	65	70	70.0
Time of incubation at 70°C (hr)	1.0–36	4.0	16	16.0
Amount of papain added (gm)	0.0– 5.0	1.5	2.0	2.0
Final adjusted pH of digest	6.5–10.0	8.0	8.5	8.5
Amount of Celite added before filtering off carrageenan (gm)	0.0– 2.0	0.5	1.0	1.0
Final concentration of KCl added (*M*)	0.0– 2.0	0.5	1.0	1.0
Amount of CPC added (mg)	1.0–100	25	100	100
Amount of Celite added to adsorb precipitate (gm)	0.0– 2.5	0.75	1.0	1.0
Temperature of incubation (°C)	28–60	37	45	37.0
Number of washings of precipitate	0.0– 6.0	Till negative to phenol-H_2SO_4 test		Till negative to phenol-H_2SO_4 test

Source: Graham (1968).

TABLE 15.4

RECOVERY OF CARRAGEENAN FROM VARIOUS PRODUCTS

Product	Amount of Product Added gm	Protein Content %	Carrageenan Recovered Range mg	Carrageenan Recovered Average[1] mg	Recovery (Average) %
Milk (commercial, homogenized, and pasteurized)	5.0	3.38	4.6–5.3	4.9	98.0
Chocolate milk (3% chocolate added to commercial milk)	5.0	3.44	4.2–4.6	4.5	90.0
Evaporated milk (commercial brand)	5.0	7.33	4.5–5.1	4.7	94.0
Ice cream (commercial, vanilla flavor)	5.0	3.88	4.5–5.2	4.8	96.0

Source: Graham (1968).
[1] Average of five different trials.

Use of Ortho-Tolidine and Sodium Hypochlorite for the Determination of Carrageenan

This procedure is quite suitable for rapid screening since it permits the determination of carrageenan in the papain digest without its prior isolation. It is very rapid and direct but is not as specific for carrageenan as are the previously described methods. Other ester sulfates such as dextran sulfate, heparin, starch sulfate mepesulfate (sodium salt of polygalacturonic acid methyl ester methyl glucoside), sodium lauryl sulfate and some sulfonated detergents, will react to varying degrees. However, these compounds are not permitted food additives and do not occur naturally in most food products. On the other hand, carrageenan is the most widely used permitted sulfated polysaccharide. Furcellaran, if present, will interfere.

Though not as sensitive as the CPC procedure, its sensitivity can be increased by reducing the volume of the filtrate collected to 100 ml instead of the normal 250 ml.

Color production in this test involves the formation of a relatively stable semiquinone and a less stable haloquinone, due to the reaction of sodium hypochlorite and ortho-tolidine in the buffered phosphate medium. When the sulfated hydrocolloid is added, it couples with the greenish semiquinone or haloquinone or both to yield a characteristic purplish or purplish green color having an absorption maximum at 570 nm. A probable mechanism for the reaction is shown in Fig. 15.2.

In performing the test, the buffer must be added first, secondly the hypochlorite, then the ortho-tolidine hypochlorite and finally the carrageenan or sample containing the carrageenan. This sequence of

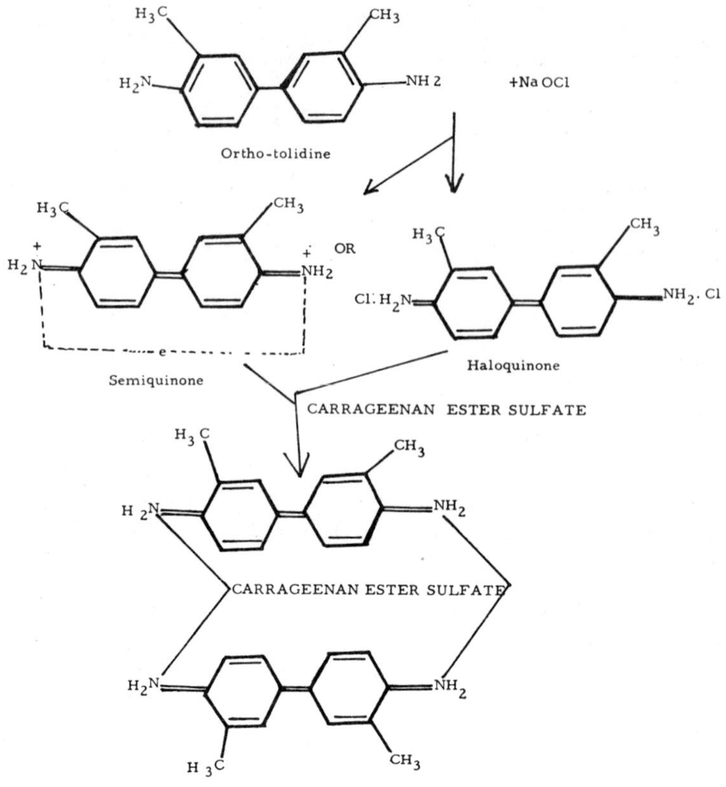

Ortho-tolidine

+Na OCl

Semiquinone

OR

Haloquinone

CARRAGEENAN ESTER SULFATE

PURPLISH OR PURPLISH-GREEN COMPLEX

From Graham (1972A)

FIG. 15.2. PROBABLE MECHANISM FOR THE REACTION OF CARRAGEENAN (OR OTHER ESTER SULFATES) WITH *O*-TOLIDINE AND NaOCl

addition of the ingredients is very important to achieve proper color development. The pH must be maintained at 2.5 ± 0.2 units. Metallic cations interfere with color formation, hence the use of a cation exchange resin to remove them and the use of diethanolamine to adjust the digest to an alkaline pH prior to the removal of the metallic cations.

Tables 15.5 and 15.6 respectively, show the influence of several variables on the test and recovery data.

Ortho-tolidine should be dispensed from a burette or from an automatic dispenser. Pipetting should not be done because of the suspected carcinogenic potential of this reagent.

Reagents and Materials.—Hydrocolloids. Carrageenan. Disperse and dialyze these as described previously.

TABLE 15.5

INFLUENCE OF VARIABLES ON COLOR INTENSITY (ORTHO-TOLIDINE-NaOCl TEST)

Variable	Range	Range Necessary for Reproducibility	Value or Level Selected
Phosphate buffer added			
pH of buffer (.5 M)	1.4-4.0	1.4-2.7	2.5
Molarity of buffer (pH 2.5)	.1-3.0	.5-1.0	.5
Ml of buffer (pH 2.5, .5 M)	.5-10.0	2-.2	2.0
Available chlorine NaOCl (ppm)	70-490	210-10	210
Ortho-tolidine, .1% in water (ml)	.5-6.0	3-.25	3.0
Time lapse between addition of ingredients 2 and 3 (min)	.5-30	2-.5	2.0
Time lapse between formation of haloquinone and addition of carrageenan (min)	.5-30	2-.5	2.0
Time lapse between addition of carrageenan and measurement of color (min)	.5-60	12-2 or 40-10	30

Source: Graham (1972 A).

TABLE 15.6

RECOVERY OF CARRAGEENAN FROM WATER, MILK AND OTHER DAIRY PRODUCTS[1]

Product	Carrageenan Recovered		
	Range (mg)	Average (mg)	(%)
Water	8.4-9.0	8.6	86.0
Milk (commercial, pasteurized homogenized)	8.4-8.7	8.5	85.0
Chocolate milk (1% commercial chocolate added to milk)	7.8-8.5	8.2	82.0
Ice cream (vanilla)	8.1-8.5	8.3	83.0
(chocolate)	7.5-8.1	7.7	77.0
Milk shake (strawberry, commercial product)	7.9-8.4	8.1	81.0
Evaporated milk	8.0-8.3	8.1	81.0
Process cheese	7.8-8.7	8.3	83.0
Baby milk formula A	7.9-8.5	8.3	83.0
Baby milk formula B	8.0-8.4	8.2	82.0

Source: Graham (1972A).
[1] 10 mg of carrageenan added to each product.

Phosphate buffer, 0.5 M, pH 1.4-3.0.—Mix 0.5 *M* H_3PO_4 and 0.5 *M* $NaH_2PO_4 \cdot H_2O$ and adjust to the desired pH with a pH meter.

Ortho-tolidine Hydrochloride (Fisher Scientific Co.)—0.1% in distilled water. Prepare 1 liter stock solution. Store in amber glass-stoppered bottle.

Sodium Hypochlorite, 500 Ppm.—Prepare from commercial stock (Fisher Scientific Co., 5% available chlorine) by dilution with distilled water. Determine available chlorine in diluted solution by arsenious oxide titration (Am. Public Health Assoc. 1960).

Enzymes.—Papain concentrate (72 milk clotting units per gram).

Diethanolamine.—Reagent grade (Fisher Scientific Co.). Prepare a 25% solution by dilution with deionized water.

Celite 535 (Johns Mansville Co., New York).—Before use, wash three times with running water, distilled water, once with 2% diethanolamine and finally with deionized water. Then dry at 110°C and store in a dry bottle.

Charcoal (Pfanstiehl Laboratories, Chicago, Ill.).—Wash, dry, and store as described for Celite 535.

Ion-exchange Resin.—Dowex 50W-X8, 100 mesh, H + form (Fisher Scientific Co.).

Buffer-EDTA-cysteine HCl Mixture.—Add 2 gm of disodium ethylenediametetraacetate (EDTA) and 1 gm of cysteine·HCl to 1 liter of 0.1 *M* phosphate buffer (0.1 *M* Na_2HPO_4 or NaH_2PO_4) of pH 6.5 Readjust pH of the mixture to 6.5 by the addition of 0.1 *M* NaH_2PO_4 or Na_2HPO_4, as necessary.

Papain-buffer-EDTA-cysteine HCl Mixture.—Add 20 gm of papain concentrate to 250 ml of the buffer-EDTA-cystein·HCl mixture. Immediately before use incubate the mixture for 30 min at 65°C.

Equipment.—Bausch and Lomb Spectronic 20 colorimeter-spectrophotometer, a Beckman Zeromatic pH meter, actinic flasks or bottles and incubators (Fisher Isotemo). Soak glassware in chromic acid, wash thoroughly with tap water, distilled water and finally deionized water.

Digestion of the Proteins in the Mixture.—Add 10 gm of milk or other food product containing 2.5-25 mg of carrageenan to 250-ml Erlenmeyer flasks. Add enough distilled water to bring the volume to 50 ml, if necessary, and 3 gm of NaCl. Adjust pH of mixture to 7.0 with 25% diethanolamine and heat it for 10 min in a boiling water bath. Cool the mixture to 70°C and add to it 25 ml of the papain-buffer-EDTA-cysteine·HCl mixture. Tape hard rubber stoppers onto the flasks and incubate the mixtures at 70°C for 16 hr.

Clarification of the Digest.—At end of incubation, remove the flasks, cool them to 30°C and add 10 ml of 25% diethanolamine to each one. Shake the mixture well and add 5 gm of Celite 535, 3 gm of charcoal, and 2 gm of Dowex 50W-X8. After 10-15 min filter the mixture through Reeve angel 202 paper. Wash the residue with five 25-ml portions of deionized water. Collect the filtrate and washings and adjust the pH to 2.5 with H_3PO_4 and, finally, make the volume up to 250 ml with deionized water.

Development of the Color.—Add to a 50-ml volumetric actinic flask 2 ml of 0.5 *M* phosphate buffer, pH 2.5, 5 ml of *O*-tolidine hydrochloride (0.1% in water, use a burette), 3 ml of sodium hypochlorite (210 ppm available chlorine) and 25 ml of the filtrate, pre-adjust to pH 2.5. Stopper the flask and invert it to mix the contents well. Allow 2 min between the addition of each of the last three ingredients. Thirty minutes after addition of filtrate to the flask, measure the absorbance of color developed at 570 nm.

TESTS FOR SODIUM CARBOXYMETHYL CELLULOSE

Apart from carrageenan, sodium carboxymethyl cellulose (CMC) is one of the most widely used stabilizers in food and food products. It finds use, generally, in combination with other food gums such as gum arabic, alginate, carrageenan or gum karaya, in such products as ice cream, ice bases, etc. General tests for this hydrocolloid have been described for paper pulp (Hercules Powder Co. 1963) and in pharmaceutical preparations (Szalkowski and Mader 1955). However, due to its strong interaction with other food components, when it is added to mixes and formulae, a preliminary separation is usually necessary, prior to its quantitative estimation. As in the case of carrageenan, complexing occurs most strongly with proteins, hence a preliminary digestion with papain is highly effective in releasing the bound hydrocolloid. For the colorimetric determination, 2,7-naphthalenediol and chromotropic acid (1,8-dihydroxynaphthalene, 3,6-disulfonic acid) proved to be the most sensitive and specific reagents. The latter reagent is less influenced by interfering substances in products which contain chocolate.

Hansen and Chang (1968) recovered carboxymethyl cellulose from milk quantitatively after digesting the protein with trypsin. Colorimetric determination was done with the phenol-H_2SO_4 reagent. Samples containing much fat must be defatted.

Determination of Carboxymethyl Cellulose in Food Products Using 2,7-Naphthalenediol as the Colorimetric Reagent

Principle.—Carboxymethyl cellulose in the food mixture is freed from any complexes with proteins by digestion of the protein in the sample with papain. Alginate and pectic acid in the digest are removed by precipitation as their insoluble calcium salts. Sulfated hydrocolloids are removed by precipitation as their cetylpyridinium complexes in the presence of 0.5 *M* NaCl. At this concentration of NaCl, any complexes formed by carboxymethyl cellulose and the cetylpyridinium chloride are dispersed.

After removing the previously mentioned interferences, the mixture is filtered and diluted to 0.2 *M* NaCl at which salt concentration, CMC-CP complex flocculates. The flocculate is treated with 30% H_2SO_4 and the ensuing mixture reacted with 2,7-naphthalenediol, to give a measure of the CMC present.

Materials.—The CMC was obtained from the Hercules Powder Company and was prepared and dialyzed as described for carrageenan. Cetylpyridinium chloride was obtained from the Fine Organic Chemicals Corp. $CaCl_2 \cdot 2\ H_2O$, obtained from the Fisher Scientific Co.

Buffer-EDTA-cysteine HCl-papain Mixture.—Add 8 gm of papain (72 milk clotting units, Schwartz-Mann Labs.) to 100 ml of 0.1 *M* sodium acetate-acetic acid buffer of pH 5.5. The buffer contains 2 gm of disodium ethylenediamine tetraacetate (EDTA) and 1 gm of cysteine hydrochloride per liter of solution.

2,7-Naphthalenediol Reagent.—Add 50 mg of 2,7-naphthalenediol in 100 ml of concentrated H_2SO_4 (Szalkowski and Mader 1955), place the mixture in an amber bottle, age it for 6 hr at 28 ± 1°C and store it under refrigeration, until needed. The 2,7-naphthalenediol is obtainable from the Fisher Scientific Co.

Equipment.—Spectronic 20 colorimeter; 25 × 300 ml borosilicate test tubes with metal caps.

Filtration Column.—Place a porcelain disc (15 mm diam) at the bottom of a 250-ml burette and put a 2-cm plug of glass wool on top of it. Add a slurry of Celite 535 or Hyflo Super Cel with the stopcock closed. After 5 min, open the stopcock and allow the water to drain through. Add enough Celite to provide a Celite bed of 1-2 cm. Prewash the Celite as follows. Wash the Celite consecutively with distilled water, 0.1 *M* acetic acid, and distilled water. Then wash the column with concentrated H_2SO_4, tap water, sodium bicarbonate and distilled water until a 2-ml aliquot of the washings gives no color with the phenol-H_2SO_4 reagent. Now add 20 ml of 5% CPC to the column, allow about 15 ml to pass through and close the stopcock.

Procedure.—*Digestion of Proteins in the Sample.*—Add 10 gm of the food product, dispersed in 10 ml of distilled water, or in the solid form, to a 250-ml Erlenmeyer flask. Add 10 mg of the hydrocolloid, dispersed in 10 ml of distilled water.[2] When necessary, the volume may be made up to 75 ml with distilled water. Add 6 gm of

[2] To assess recovery of the hydrocolloid from the mixture.

NaCl. Heat the mixture at 99 ± 1°C (boiling water bath) for 10 min, cool to 70°C and add 25 ml of the 0.1 *M* acetate buffer-papain-cysteine-HCl-EDTA mixture. Shake the flask gently to mix the contents well, stopper it with a hard rubber stopper and incubate it at 70°C for 16 hr.

Removal of Protein Degradation Products and Precipitation of Alginate and Pectic Acid.—At the end of the digestion period, place the flasks in tap water for about 10–15 min to cool the contents to about 40–37°C. Then add 10 ml of 1 *M* $CaCl_2$ to each flask and incubate for 15 min at 37°C. Add 2 gm of Celite 535 or Hyflo Super Cel to each flask, mix the contents thoroughly and after 5 min pour over Reeve angel No. 202 filter paper. Rinse the flask with five 20-ml portions of distilled water and pour the rinsings over the residue on the filter paper.

Removal of Sulfated Polysaccharides.—Collect the filtrate and washings in another 250-ml Erlenmeyer flask. Add 2 gm of CPC to each flask. Mix the contents well and, after 5 min, add 1 gm of Celite 535 or Hyflo Super Cel. Incubate the mixture at 45°C for 15 min, cool to 29 ± 1°C and filter over Reeve angel No. 202 paper.

Precipitation of CMC as the CMC-CP Complex.—Rinse the flasks with ten 25-ml portions of 0.01% CPC-0.01 *M* NaCl and pour the rinsings over the residue on the filter paper. Collect the filtrate and washings in a 1000-ml beaker and add distilled water to make the volume up to 600 ml.

Collecting the CMC-CP Precipitate.—Mix the contents thoroughly and incubate for 15 min at 35°C. Then filter the liquid through a glass column. Return the first 250 ml passing through to the column. Filtration is done at the rate of 15–20 ml per min. Wash the beakers with five 50-ml portions of 0.01% CPC-0.01 *M* NaCl and pour the washings over the column. When all the liquid has drained through, wash the column with 200 ml of 0.5 *M* Na_2SO_4 and then with the 0.01% CPC-0.01 *M* NaCl, dispensed from a glass bottle and making sure to wash the sides of the column. Continue washing in this manner until a 2-ml aliquot gives a negative test with the phenol-H_2SO_4 reagent.

Dissociation of the CMC-CP Complex and Elution of the CMC.—Wash the column with ten 20-ml portions of hot (80°C) 30% (v/v) H_2SO_4 and collect the eluate in a 200-ml Erlenmeyer flask. Cool the contents of the flask to 29 ± 1°C and, where necessary, make the volume up to mark with 30% H_2SO_4.

Determination of CMC in the Eluate.—Mix the eluate in the 200-ml flask thoroughly and place 1 ml in 25 × 300 borosilicate tubes (in duplicate). Add 9 ml of the 2,7-naphthalenediol reagent from a burette, allowing the reagent to run down the side of the tubes. Shake to mix the contents well and cover them with metal caps. Place them in a boiling water bath for 3 hr. At the end of this period remove them and cool them for 5 min in tap water and measure the color developed at 540 nm against a reagent blank.

Recovery data are shown in Tables 15.7 and 15.8. The mechanism of the reaction is shown in Fig. 15.3.

Determination of Carboxymethyl Cellulose with Chromotropic Acid

Preparatory steps in this procedure are essentially the same as those for the method employing 2,7-naphthalenediol. Although, 2,7-naphthalenediol is a highly sensitive reagent for the determination of carboxymethyl cellulose and offers stability as well as specificity, the heating time of 3 hr needed for maximum color development is somewhat lengthy. Moreover, when chocolate products are being analyzed, serious interference may occur. When chromotropic acid (1,8-dihydroxynaphthalene-3,6-disulfonic acid) is used for color de-

TABLE 15.7

RECOVERY OF CMC FROM MILK (10 gm) IN THE PRESENCE OF OTHER GUMS; 10 MG OF CMC ADDED IN ALL CASES (2,7-NAPHTHALENEDIOL METHOD)

Symbol	Gum Added[1]	CMC Recovered		Average Recovery	
		Range (mg)	Average (mg)	A[2] (%)	B[3] (%)
	None	8.3–9.0	8.4	84.0	97.7
(A)	Carrageenan (10)	8.5–8.8	8.4	84.0	97.7
(B)	Furcellaran (10)	8.1–8.6	8.3	83.0	96.5
(C)	Alginate (10)	8.0–8.7	8.4	84.0	97.7
(D)	Pectic acid (10)	8.2–8.5	8.3	83.0	96.5
(E)	Pectin (10)	8.3–8.7	8.5	85.0	98.8
(F)	Gum karaya (20)	8.2–8.7	8.3	83.0	96.5
(G)	Starch (20)	7.6–8.1	7.8	78.0	88.4
(H)	Locust bean gum (20)	8.1–8.4	8.3	83.0	96.5
(I)	Gum arabic (20)	8.3–9.0	8.5	85.0	98.8
(J)	Quince seed mucilage (10)	8.0–8.7	8.3	83.0	96.5
(K)	Gum tragacanth (10)	7.7–8.4	8.0	80.0	93.0
(L)	Gum ghatti (20)	8.0–8.5	8.3	83.0	96.5
(M)	Gum karaya (20)	8.2–8.6	8.3	83.0	96.5
(N)	Gum guar (20)	8.2–8.6	8.3	83.0	96.5
	A + C (10 of each)	7.8–8.3	8.0	80.0	93.0
	A + G (10 of each)	7.3–7.8	7.7	77.0	90.0
	A + G + I (10 of each)	7.2–8.0	7.8	78.0	90.7

Source: Graham (1971A).
[1] Numbers in parentheses indicate milligrams of gum added to the milk.
[2] Recovery based on amount of CMC added to the mixture and carried through the entire process.
[3] Recovery based on CMC recovered from water when 10 mg of the hydrocolloid was carried through the entire process. = Corresponding value in (column A ÷ 86.0) × 100.

velopment, interference from components in chocolate products is much less. Chromotropic acid is more specific but less sensitive than 2,7-naphthalenediol for the determination of CMC. Determination using it is more rapid than when 2,7-naphthalene is used since maximum color developments occurs in 90 min at 99 ± 1°C.

Mechanism of the Reaction.—Figure 15.4 outlines the steps in the reaction leading to color formation. Carboxymethyl cellulose (A) is degraded by hot concentrated H_2SO_4 to produce glycolic acid (B). Glycolic acid is further degraded by the hot acid to produce formaldehyde (C) which then condenses with chromotropic acid. The condensation product (D), in the hot acid, is transformed to its para-quinoidal form (E) which produces the typical purple color which has maximum absorption at 570 nm.

For maximum color development, concentrated H_2SO_4 must be used to prepare the chromotropic acid reagent and a heating time of 90 minutes at 99 ± 1°C must be employed.

As in the case of the 2,7-naphthalenediol reagent, color develop-

TABLE 15.8

RECOVERY OF CMC FROM WATER AND VARIOUS FOOD PRODUCTS (2,7-NAPHTHALENEDIOL METHOD)

Product	CMC Recovered		Average Recovery	
	Range (mg)	Average (mg)	A[1] (%)	B[2] (%)
Water	8.2-9.2	8.6	86.0	—
Milk	8.3-9.0	8.4	84.0	97.7
Chocolate milk (1% commercial cocoa powder added to commercial milk)	7.8-8.6	8.1	81.0	94.2
Ice cream (vanilla flavor)	8.2-9.0	8.4	84.0	97.7
Ice cream (chocolate flavor)	7.2-8.0	7.5	75.0	87.2
Dietetic food, liquid (strawberry flavor)	7.9-8.7	8.1	81.0	94.2
French salad dressing	7.8-8.6	8.1	81.0	94.2
Pancake syrup	7.7-8.4	7.9	79.0	92.2
Powdered beverage base (grape flavor)	7.8-8.6	8.2	82.0	97.7
Powdered beverage base (orange flavor)	8.0-8.8	8.3	83.0	96.5
Fruit slaw	7.8-8.2	7.9	79.0	92.2
Dietetic pancake cake	8.0-8.6	8.3	83.0	96.5
Canned sardines in oil	7.6-8.2	7.8	78.0	90.7
Cake mix	7.8-8.4	7.9	79.0	92.0

Source: Graham (1971A).
[1] Recovery based on amount of CMC added to the mixture and carried through the entire process.
[2] Recovery based on CMC recovered from water when 10 mg of the hydrocolloid was carried through the entire process. = Corresponding value in (column A ÷ 86.0) × 100.

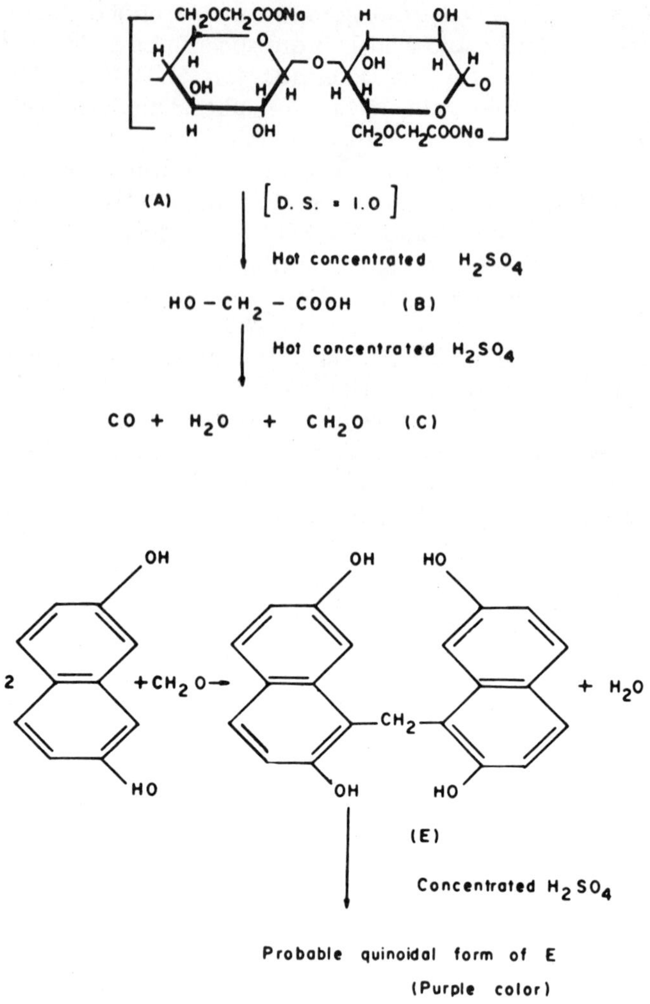

(A) [D. S. = 1.0]

Hot concentrated H_2SO_4

$HO-CH_2-COOH$ (B)

Hot concentrated H_2SO_4

$CO + H_2O + CH_2O$ (C)

2 (2,7-naphthalenediol) $+ CH_2O \rightarrow$ (E) $+ H_2O$

(E)

Concentrated H_2SO_4

Probable quinoidal form of E

(Purple color)

From Feigl (1966)

FIG. 15.3. MECHANISM OF REACTION OF CMC WITH 2,7-NAPHTHALENEDIOL IN HOT CONCENTRATED H_2SO_4

ment will vary with the degree of substitution of the carboxymethymethyl cellulose.

The Method.—Sodium carboxymethyl cellulose, obtained from Hercules Powder Company.

Acetate buffer, prepare as outlined for the 2,7-naphthalenediol procedure.

Chromotropic Acid Reagent.—Chromotropic acid (1,8-dihydroxynaphthalene-3-6-disulfonic acid) is obtainable from K and K Labora-

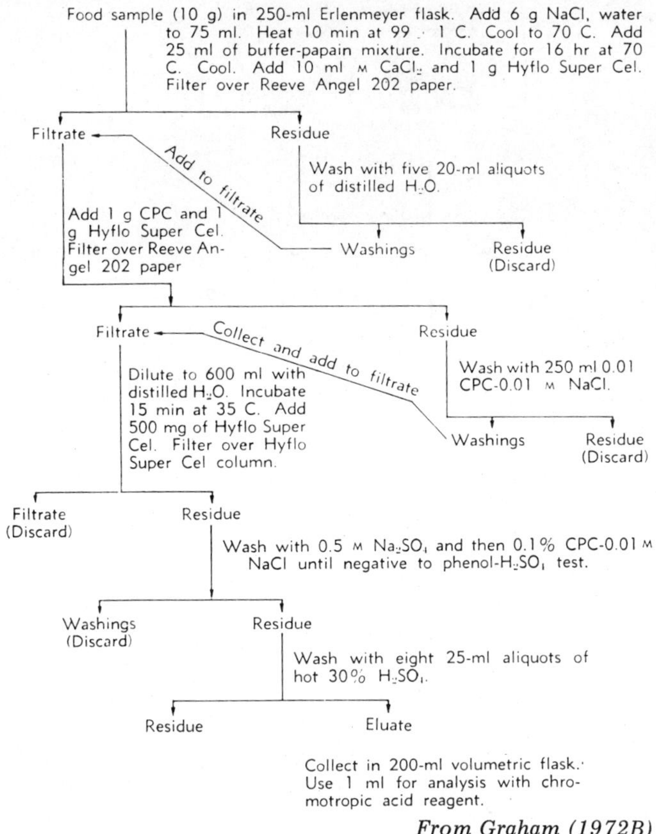

From Graham (1972B)

FIG. 15.4. FLOW SHEET FOR THE SEPARATION AND DETERMINATION OF CARBOXYMETHYL CELLULOSE BY CHROMOTROPIC ACID METHOD

tories, New York City. Dissolve 150 mg of the chemical per 100 ml of concentrated H_2SO_4.

Cetylpyridinium Chloride (*CPC*). Obtainable from Fine Organic Chemical Co., Philadelphia, or any other suitable source.

Equipment.—A Bausch and Lomb Spectronic 20 Colorimeter and 25 × 300 mm test tubes with metal caps, is used.

Procedure.—Digest samples of the food products (5–25 gm) with papain to degrade the proteins and to release any of the CMC bound to proteins. Precipitate alginate and pectic acid as insoluble calcium salts. Filter these off, along with the protein degradation products. Remove carrageenan and other sulfated polysaccharides as the insoluble cetylpyridinium complex in 0.5 *M* NaCl and filter to remove the precipitate. Then, precipitate the CMC by cetylpyridinium chloride (CPC) in 0.20 *M* NaCl. Collect the CMC-CP complex on a Hyflo

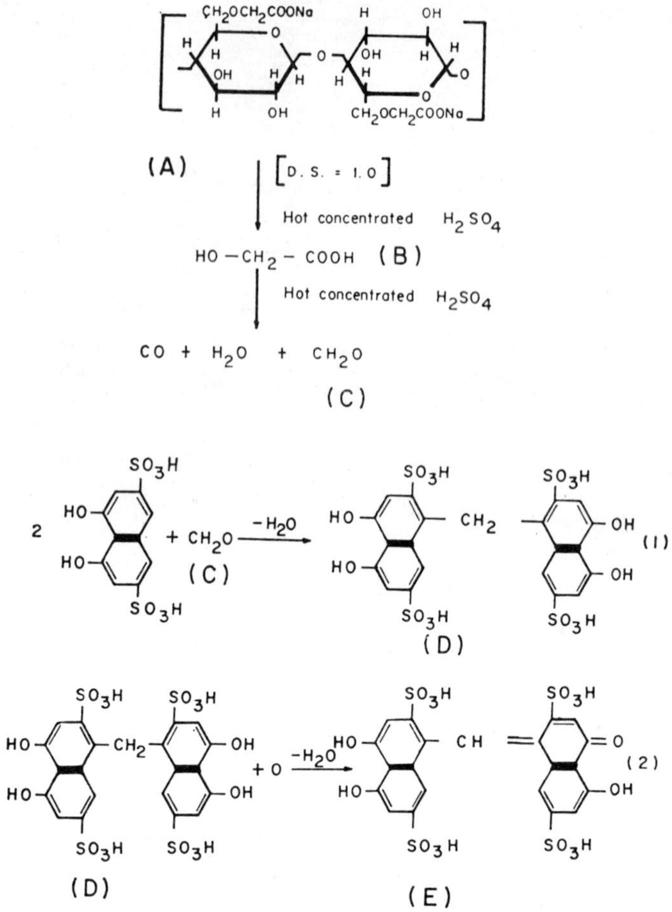

From Feigl (1968)

FIG. 15.5. MECHANISM OF COLOR PRODUCTION IN REACTION OF CARBOXYMETHYL CELLULOSE WITH CHROMOTROPIC ACID—H_2SO_4

Super Cel column, wash it with 0.5 *M* Na_2SO_4 and then with 0.1% CPC-0.01 *M* NaCl until the washings are negative to the phenol-H_2SO_4 test. Disperse the complex with hot 30% H_2SO_4 and collect 200 ml of the acidic eluate. Place 1 ml of this eluate in a 25 × 300 ml borosilicate test tube. Add 9 ml of the chromotropic acid reagent and heat the mixture for 90 min in a boiling water bath (99 ± 1°C). Cool mixture for 5 min in tap water and measure the color developed at 570 nm.

Recovery data are summarized in Tables 15.9 and 15.10. The steps in the procedure are summarized in Fig. 15.4 and the mechanism of the reaction in Fig. 15.5.

TABLE 15.9

RECOVERY OF CARBOXYMETHYL CELLULOSE FROM MILK (10 gm) IN THE PRESENCE OF OTHER GUMS (10 mg OF CMC ADDED IN ALL CASES) (CHROMOTROPIC ACID METHOD)

Gum Added	CMC Recovered		Recovery (Average)	
	Range (mg)	Average (mg)	A[2] (%)	B[3] (%)
Water	8.3-9.4	8.7	87.0	—
None	8.4-9.0	8.5	85.0	97.7
Carrageenan (10)[1]	8.5-8.9	8.6	86.0	98.8
Furcellaran (10)	8.3-8.7	8.5	85.0	97.7
Alginate (10)	8.2-8.7	8.5	85.0	97.7
Pectin (10)	8.5-8.9	8.6	86.0	98.8
Pectic acid (10)	8.2-8.7	8.4	84.0	96.6
Gum arabic (20)	8.3-9.0	8.5	85.0	97.7
Gum karaya (20)	8.1-8.7	8.5	85.0	97.7
Gum tragacanth (10)	7.6-8.2	8.0	80.0	92.0
Quince seed mucilage (10)	8.1-8.4	8.3	83.0	95.4
Gum guar (20)	8.2-8.6	8.3	83.0	95.4
Locust bean gum (20)	8.0-8.5	8.3	83.0	95.4
Gum ghatti (20)	8.0-8.5	8.3	83.0	95.4
Starch (20)	7.5-8.3	7.8	78.0	89.0
Carrageenan + alginate (10 of each)	7.7-8.5	8.0	80.0	92.0
Carrageenan + gum guar (10 of each)	7.4-8.0	7.7	77.0	90.0
Carrageenan + gum arabic + gum guar (10 of each)	7.3-8.0	7.7	77.0	90.0

Source: Graham (1972B).
[1] Numbers in parentheses indicate milligrams of gum added to the milk.
[2] Recovery based on amount of CMC added to the mixture and carried through the entire process.
[3] Recovery based on CMC recovered from water when 10 mg of the hydrocolloid was carried through the entire process. = Corresponding value in (column A ÷ 87.0) × 100.

QUANTITATIVE DETERMINATION OF ALGINATE

Sodium, potassium and propylene glycol alginates are used extensively in food products, because they have strong stabilizing properties. Several procedures for the determination of alginates in foods, especially in dairy products, have been proposed (Bundesen and Martinek 1954; Johnson 1956). However, drastic interference by other products which coprecipitated with alginate in the presence of the strong acid used as precipitant was noted. In an attempt to obviate this difficulty, a preliminary enzymatic digestion was done to degrade proteins, which constitute the major co-precipitants. After filtration, the pectic substances in the filtrate were degraded with pectin esterase and polygalacturonase. Then, the alginate was specifically precipitated as the insoluble calcium salt, the alginate released from the salt by the use of sodium hexametaphosphate and finally determined colorimetrically by the phenol-H_2SO_4 reagent or the Ferric-H_2SO_4 method. The details are outlined below.

TABLE 15.10

RECOVERY OF CARBOXYMETHYL CELLULOSE FROM WATER AND VARIOUS FOOD PRODUCTS (CHROMOTROPIC ACID METHOD)

Product	CMC Recovered		Recovery (Average)	
	Range (mg)	Average (mg)	A[1] (%)	B[2] (%)
Water	8.3-9.4	8.7	87.0	—
Milk	8.3-9.2	8.4	84.0	96.6
Chocolate milk (1% commercial chocolate added to commercial milk)	7.9-8.1	8.1	81.0	93.1
Ice cream (vanilla)	8.4-9.1	8.6	86.0	98.0
Ice cream (chocolate)	7.4-8.0	7.6	76.0	87.4
Powdered beverage base (grape)	7.7-8.5	8.2	82.0	94.2
Powdered beverage base (orange)	8.0-8.4	8.3	83.0	95.4
Dietetic food, liquid (strawberry)	7.8-8.4	7.9	79.0	90.8
Dietetic pancake syrup	7.8-8.7	8.3	83.0	95.4
Pancake syrup	7.5-8.6	7.8	78.0	90.0
Cake mix	7.6-8.2	7.9	79.0	90.8
Canned sardines in oil	7.7-8.2	7.9	79.0	90.8
French salad dressing	7.9-8.4	7.6	76.0	87.4
Fruit slaw	7.7-8.4	7.9	79.0	90.8

Source: Graham (1972B).

[1] Recovery based on amount of CMC added to the mixture and carried through the entire process.

[2] Recovery based on CMC recovered from water when 10 mg of the hydrocolloid was carried through the entire process. = Corresponding value in (column A ÷ 87.0) × 100.

Determination of Alginate Using Phenol-H_2SO_4 as the Colorimetric Agent

Materials.—*Sodium alginate.*—obtainable from the Kelco Company, San Diego, California.

Stock Solution.—Prepare a 0.5% stock dispersion as described for carrageenan. Dialyze it and store it in the refrigerator.

Calcium Chloride ($CaCl_2 \cdot 2H_2O$).—Reagent grade, Fisher Scientific Company. All the other salts used also of reagent grade.

Celite 535 and Charcoal.—These have been described under the tests for CMC.

Buffer-EDTA-cysteine Mixture.—Add 2 gm of sodium ethylenediaminetetraacetate (EDTA) and 1 gm of cysteine hydrochloride to one liter of *M* sodium acetate-acetic acid. Adjust the pH of the mixture to 6.5 with a pH meter and with addition of *M* NaOH or acetic acid as needed.

Papain-EDTA-cysteine HCl Mixture.—Disperse 50 grams of papain concentrate (Mann Research Laboratories, 72 milk clotting units per gram) in 500 ml of the buffer-EDTA-cysteine hydrochloride mixture. Just before use incubate the mixture for 30 min at 65° C.

Equipment.—Spectronic 20 colorimeter, Beckman Zeromatic pH meter and incubator (Fisher Isotemp). Soak all glassware in chromic acid, tap water, and finally wash thoroughly with distilled water.

Digestion of the Protein with Papain.—To a 500-ml Erlenmeyer flask containing 25 gm of the material dispersed in 200 ml of distilled water, add 6 gm of sodium chloride. Heat the flask for 10 min in a boiling water bath (99 ± 0.5° C) and cool to 65° C. Add 50 milliliters of the papain-EDTA-cysteine HCl mixture and mix the contents well. Incubate the tightly stoppered flask overnight (16 hr) at 65° C. Tape a rubber stopped onto the flask to prevent it from blowing off during incubation.

Clarification of the Digest.—At the end of incubation add 2 gm of Celite 535 and 2 gm of charcoal to each flask. Mix the contents well and hold the flasks at 65°C for 10 min. Filter the hot contents over Reeve Angel No. 202 filter paper. Collect the filtrate in a 1,000-ml Erlenmeyer flask. Wash the residue with ten 25 ml portions of boiling distilled water and collect the washings in the Erlenmeyer flask.

Precipitation of the Alginate.—If pectin, pectinic acid, or pectic acid are present, make the filtrate 0.1 *M* to NaCl and adjust to pH 4.8. Add pectin methylesterase and incubate the mixture at 30°C for 30 min. Then, add polygalacturonase and again incubate at 30°C for 30 min. Finally, adjust the mixture to pH 7.0. Add 10 gm of $CaCl_2$ to each flask, mix the contents well by shaking, and incubate at 30°C for 30 min with shaking every 5 min. Then add 1 gm of Celite 535 and allow the mixture to stand at 30°C for 10–15 min.

Separation of the Calcium Alginate.—Pour the contents of each flask over Reeve Angel No. 202, 11 cm filter paper, shaking well before each pouring. Wash the flask and the precipitate on the paper with 0.1% $CaCl_2$. Discard the filtrate and washings. Continue washing the precipitate until a 2 ml aliquot of the washings gives a negative test with the phenol-sulfuric acid reagent.

Dissolution of the Calcium Alginate.—Pour 25-ml portions of boiling 2% sodium hexametaphosphate over the precipitate on the filter allowing it to flow through at 10–15 ml per minute. Collect the filtrate in a 250-ml volumetric flask.

Colorimetric Determination of Alginate.—Determine the alginate content of the final filtrate by the phenol-sulfuric acid method. Figure 15.6 summarizes the entire process. The effect of several variables on the procedure is summarized in Table 15.11. Recovery data are shown in Table 15.12.

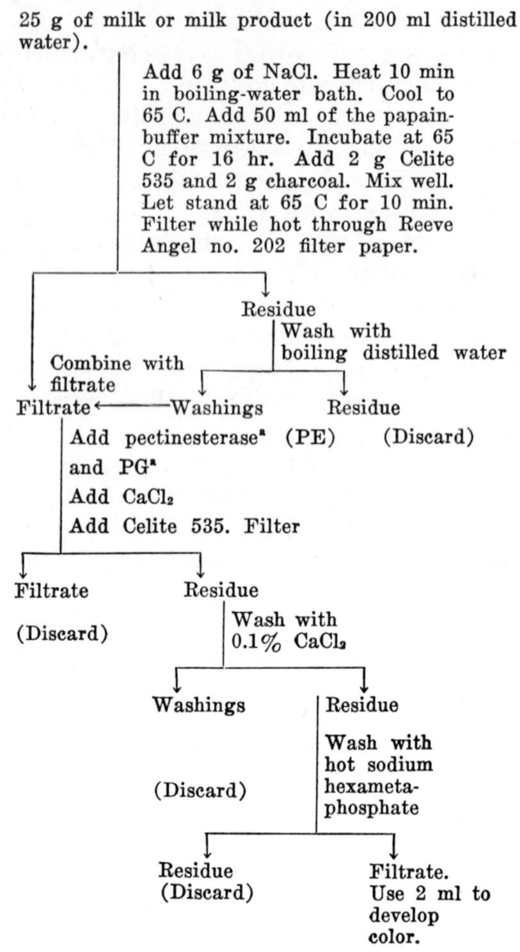

FIG. 15.6. FLOW SHEET FOR SEPARATION OF ALGINATES ($PHENOL-H_2SO_4$ METHOD)

Use of the Ferric-H_2SO_4 Reagent to Determine Alginate

This stable and rather simple reagent was first suggested by Bundesen and Martinek (1954), for the detection of alginate in dairy products. However, under the conditions outlined by them, many other compounds interfered. By isolating the alginate from a papain digest, as described previously for the phenol-H_2SO_4 test for alginate, a greater purity of the test material was obtained. Finally, in the de-

TABLE 15.11

EFFECT OF VARIABLES ON DETERMINATION OF ALGINATES (PHENOL-H_2SO_4 REAGENT)

	Levels or Conditions		
Variable	Investigated	Needed for Reproducibility	Selected
Removal of fat with dioxane	Used or not used	Not necessary	Not used
Amount of NaCl added (gm)	0–20	2–10	6
Preheating of sample (min)	0–30	5–15	10
Temperature of preheating (C)	50–100	95–100	99 ± 1
Buffer added (pH)	4–10	5–7.0	6.5
Strength of buffer added (N)	0.1–2	0.5–2.0	1.0
Volume of buffer added (ml)	5–50	25–50	50
Temperature of incubation (C)	40–70	60–70	65
Duration of incubation (hr)	0.5–24	4–24	16
Papain concentrate added per gram of protein (gm)	1–25	5–15	10
Celite 535 added to digest before filtration (gm)	0–10	2–6	2
Type of filter paper used	9 Types tried	Reeve Angel No. 202, Whatman No. 50 or glass fiber	Reeve Angel No. 202 or glass fiber
Temperature of digest during filtration (C)	25–99 ± 1	80–99	80–99
Temperature of water for washing of residue (C)	25–99 ± 0.5	95–100	Boiling (99 ± 1)
Number of washings necessary (25-ml portions) before filtration	0–20	5–15	10
Activated carbon added to digest before filtration (gm)	0–10	1–3	2
Concentration of $CaCl_2$ in filtrate (%)	0.5–5	1–3	2
Washing of Ca-alginate-Celite 535 mixture	H_2O or 0.1% $CaCl_2$	0.1% $CaCl_2$	0.1% $CaCl_2$
0.1% $CaCl_2$ needed for washington (ml)	0–2000	Until negative to phenol-H_2SO_4 test	Until negative to phenol-H_2SO_4 test
Concentration of sodium hexametaphosphate (%)	0.1–10	2 or more	2
Volume of 2% sodium hexametaphosphate needed (ml)	5–500	150–300	250

Source: Graham (1969).

velopment of the color, strict control of the time and temperature of heating the mixture with the ferric-H_2SO_4 reagent further enhanced the specificity of the test.

The preliminary steps in this procedure are identical to those described for the use of the phenol-H_2SO_4 test. The general outline is shown in Fig. 15.6. Other details are outlined as follows.

Preparation of the Ferric-H_2SO_4 Reagent.—Add 5 gm of ferric oxide to one liter of concentrated H_2SO_4 (Baker analyzed reagent grade, 97.2%, sp. gr. 1.84). Age the mixture for at least 7 days. When needed filter an aliquot through glass fiber paper into a glass-stoppered bottle.

Preparation of the Column.—Place a porcelain disc (15 mm diameter) in a Jones-Blair reductor (total length 300 mm). Place a 3–5 cm plug of glass wool on top of it and close the stopcock. Pour an aqueous slurry of Celite 535 into the column and after 5–10 min open the stopcock. Add enough slurry to provide a Celite 535 pack of 2–3 cm. Pour 20 ml of concentrated H_2SO_4 on to the column and allow it to drain through. Then wash the column consecutively with tap water, 1% sodium bicarbonate and distilled water until a 2 ml aliquot gives no color with the phenol-H_2SO_4 reagent. Close the stopcock and add 10 ml of 1% calcium chloride to the column.

Dispersion of the Ca-Alginate.—Pour boiling 1% sodium hexametaphosphate over the residue on the glass column at about 10 ml each time and collect the filtrate, passing at 5–10 ml per minute, in a 100-ml volumetric flask. Cool the flask to 30°C and, if necessary, make the volume to mark with 1% sodium hexametaphosphate.

Determination of Alginate with the Ferric-H_2SO_4 Reagent

Mix the contents of the flask well, place 1–5 ml (50–250 μg of sodium alginate) in 25 × 200 mm borosilicate tubes (in triplicate) and dry for 10–16 hr at 100°C. Include triplicate tubes containing 1–5 ml of 1% sodium hexametaphosphate (depending on aliquot of filtrate used) to serve as reagent blanks. Add 5 ml of the ferric-H_2SO_4 reagent to each tube and mix the contents well by gentle shaking. Heat the tubes and their contents for 12 min at 60°C. Begin timing at the moment the tubes are inserted into the bath. At the end of this period place the tubes in a 29 ± 1°C water bath for 90 min, then remove them and measure the color developed at 495 nm, against a reagent blank. Determine the amount of alginate present using a reference curve.

Preparation of Reference Curve.—Prepare stock dispersions containing 0, 50, 100, 150, 200, 250 and 300 μg of sodium alginate per milliliter dispersed in 1, 2 and 5% sodium hexametaphosphate. Put

1 ml from each dispersion in 25 × 200 mm borosilicate tubes (in triplicate). Dry the contents for 10–16 hr at 100° C. Add 5 ml of the ferric-H_2SO_4 reagent, heat the mixture for 12 min at 60° C, age it for min at 29 ± 1° C, measure the density of the color developed at 495 nm. A plot of the optical density versus the amount of added sodium alginate should produce a straight line. Recovery data appear in Table 15.13. Details of the procedure are summarized in Fig. 15.7.

QUANTITATIVE DETERMINATION OF XANTHAN GUM (KELTROL)

Xanthan gum (Keltrol, polysaccharide B-1459) is the only microbial polysaccharide which is used widely in food preparations as a

TABLE 15.12

RECOVERY OF ALGINATES FROM VARIOUS FOOD PRODUCTS USING PHENOL-H_2SO_4 METHOD[1]

	Protein (%)	Alginate Added (mg)	Recovered[2] (mg)	(%)
Milk, homogenized, pasteurized, commercial	3.3	10	9.4	94.0
		25	24.0	96.0
Chocolate milk, 1% chocolate added to commercial milk	3.4	5	4.6	92.0
		10	8.9	89.0
		25	22.6	90.4
Ice cream, vanilla flavor	4.2	5	4.5	90.0
		10	9.2	92.0
Cream cheese	7.9	5	4.8	90.0
		10	8.9	89.0
		25	23.1	88.4
Process cheese	10.6	5	4.5	90.0
		10	9.5	95.0
		25	23.8	92.8
Dietetic food, liquid, strawberry flavor	6.2	5	4.5	90.0
		10	9.1	91.0
		25	23.8	92.8
Salad dressing[3]	[4]	5	4.3	86.0
		10	8.9	89.0
		25	22.8	91.2

Source: Graham (1969).
[1] 25 mg of sodium alginate or propylene glycol alginate added.
[2] Average of three determinations.
Standard deviations:
5 mg alginate added:
Chocolate milk: ± 0.4 mg, n = 10
Ice cream : ± 0.3 mg, n = 10
All products : ± 0.4 mg, n = 41
25 mg alginate added:
Chocolate milk: ± 1.8 mg
Ice cream : ± 1.7 mg
All products : ± 1.9 mg
[3] Propylene glycol alginate added.
[4] Not determined.

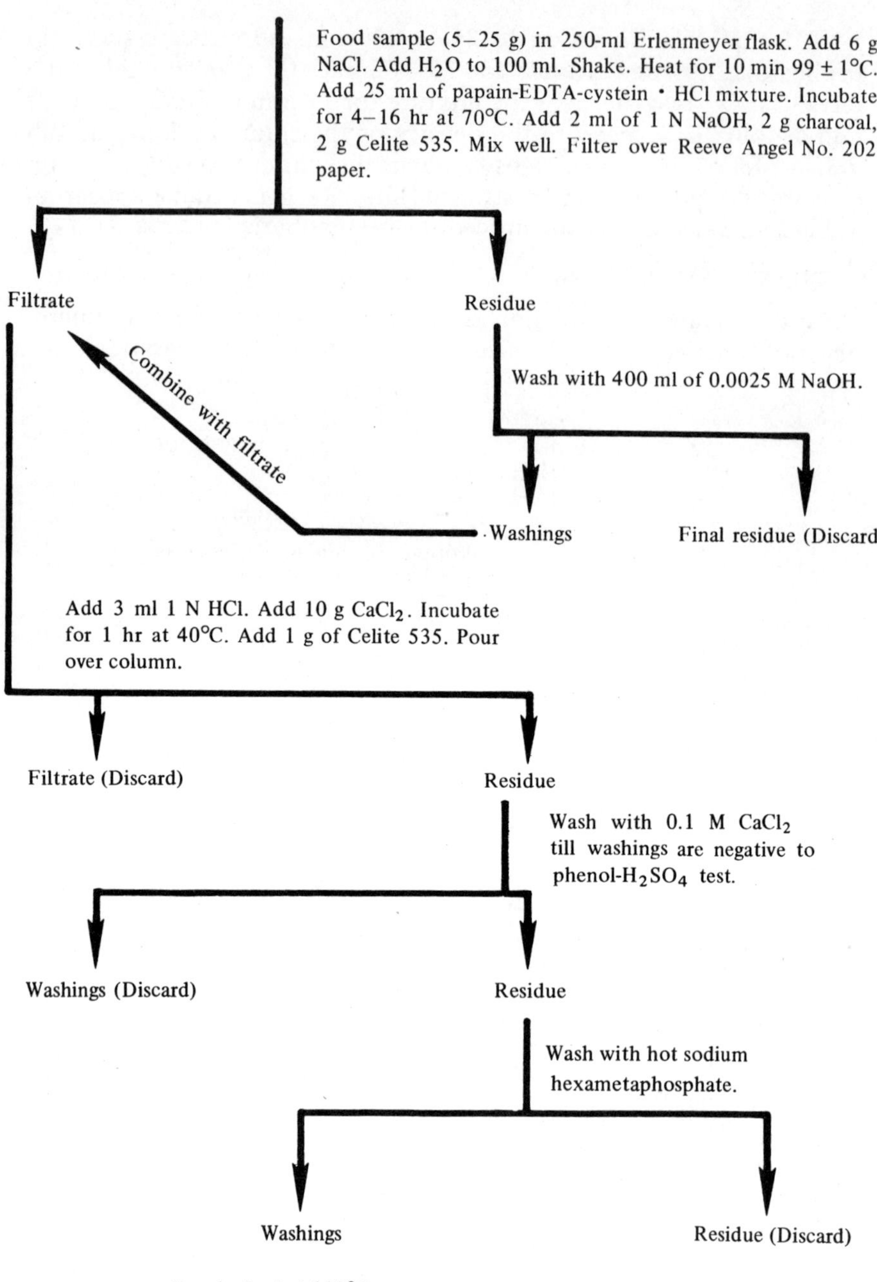

From Graham (1970)

FIG. 15.7. PROCEDURE FOR SEPARATION AND DETERMINATION OF ALGINATES BY FERRIC-H_2SO_4 METHOD

TABLE 15.13

RECOVERY OF ALGINATE FROM VARIOUS FOOD PRODUCTS (FERRIC H_2SO_4 METHOD)

Product	Alginate Added		Average Alginate Recovered[1] (mg)	Recovery (%)
	Amount (mg)	Range (mg)		
Water	5.0	4.6- 4.9	4.8	96.0
	10.0	9.7- 9.9	9.8	98.0
	25.0	24.6-24.9	24.8	99.2
Milk	5.0	4.4- 4.9	4.7	94.0
	10.0	9.2- 9.8	9.5	95.0
	25.0	23.9-24.8	24.2	96.8
Chocolate milk	5.0	4.2- 4.7	4.5	90.0
	10.0	8.9- 9.4	9.2	92.0
	25.0	23.6-23.5	23.1	92.4
Ice cream (vanilla flavor)	5.0	4.5- 4.8	4.7	94.0
	10.0	9.2- 9.6	9.4	94.0
	25.0	23.0-24.5	23.8	95.2
Ice cream (chocolate flavor)	5.0	4.2- 4.6	4.5	88.0
	10.0	8.6- 9.4	8.9	89.0
	25.0	21.8-23.7	22.5	90.0
Process cheese	5.0	4.2- 4.8	4.5	90.0
	10.0	8.7- 9.6	9.1	91.0
	25.0	21.9-23.7	22.5	90.0
Salad dressing	5.0	4.0- 4.7	4.3	86.0
	10.0	9.0- 9.4	9.2	92.0
	25.0	23.0-24.5	23.7	94.8

Source: Graham (1970).
[1] Average for 3 different samples. Duplicate colorimetric determinations done.

stabilizing agent. It is produced by *Xanthomonas campestris*. This high molecular weight, linear polysaccharide, is permitted for use in foods and finds applications, also, in pharmaceutical and cosmetic preparations. Having a molecular weight of more than 1 million, it has a β-linked backbone and contains D-glucose, D-mannose and D-glucuronic acid in the ratio of 2.8:3:2.0. It is partially acetylated and contains pyruvic acid attached to the glucose side chain residue (Smiley 1966).

As outlined in the Federal Register (Anon. 1969), xanthan gum can be detected and/or determined by two distinctly different methods. In the qualitative locust bean gum gel test, a gel forms when xanthan gum is mixed with locust bean gum under appropriate conditions. This test requires 1 gm of the hydrocolloid and special mixing conditions with respect to temperature and cooling. The pyruvic acid test quantitatively extracts phenylhydrazone in a sodium carbonate solution. This test requires 60 mg of the hydrocolloid, which is explicable by the small (3.0–3.5%) amount of pyruvic acid in the polymer.

Graham (1971) described a procedure for the microdetermination of xanthan gum in the presence or in the absence of locust bean gum. In principle, proteins are degraded with papain. Alginate and pectic acid in the mixture are removed as the insoluble calcium salts. Carrageenan or any other sulfated polysaccharide in the mixture is removed as the highly insoluble cetylpyridinium complex formed in the presence of 0.5 *M* or greater NaCl. Xanthan gum and CMC are then coprecipitated, in the presence of 0.2 *M* NaCl. The mixed coprecipitate is dispersed with hot (80° C) 30% H_2SO_4.

Carboxymethyl cellulose in the acidic dispersion was determined by the chromotropic acid method (Keltrol does not react). Carboxymethyl cellulose + Keltrol was determined by the phenol-H_2SO_4 and the anthrone methods. The amount of Keltrol present was determined by difference, using standard curves for the hydrocolloids when reacted with the aforementioned reagents. Recoveries of 90-96% were achieved from milk, chocolate milk, and salad dressing. The method can determine Keltrol in milk and other food products at 0.05% (w/w) or lower.

Microdetermination of Xanthan Gum[3]

Materials.—*Hydrocolloid.*—Prepare a 0.20% solution of Keltrol, dialyze it and store it in the refrigerator. The hydrocolloid is obtainable from the Kelco Company, San Diego, California.

Acetate Buffer.—Prepare a 0.1 *M* solution of acetate buffer (sodium acetate-acetic acid) of pH 5.5, and which contains 1 gm of cysteine dihydrochloride and 2 gm of ethylenediamine-tetracetic acid (EDTA) per liter of the solution.

Papain-EDTA-cysteine Hydrochloride-buffer Mixture.—Eight grams of papain concentrate (Schwartz-Mann Biochemicals, 72 milk clotting units per gm) per 100 ml. Preincubate the mixture at 70°C for 30 min before adding it to the food mixture which contains the hydrocolloid.

Other Solutions.—Solutions of *M* $CaCl_2$, 0.01 *M* $CaCl_2$ and a mixture of 0.01% cetylpyridinium chloride (CPC) and 0.001 *M* NaCl. Prepare these from commercial, reagent grade chemicals. Hereafter, the mixture of 0.01% CPC-0.001 *M* NaCl will be called solution A.

Preparation of the Celite Filtration Column.—Place a perforated disk or a 1-cm piece of glass tubing, diameter 3/8 mm, at the bottom of a 250-ml burette, and pack glass wool on top of this to a height of about 3 cm. Add 2 gm of Hyflo Super Cel and wash the column with 100 ml of tap water. Then add 25 ml of concentrated sulfuric acid

[3] Based on Graham (1971).

and allow to drain. Wash column consecutively with 200 ml of tap water, 100 ml of 10% Na_2CO_3, and finally, with 100 ml of 1% CPC. At this point a 2-ml aliquot of the washings should give a negative test with the phenol-H_2SO_4 reagent. Finally, close the stopcock of the burette and add 20 ml of 1% CPC. Leave CPC in the column until ready for filtration.

Isolation of Keltrol in the Absence of Locust Bean Gum (Procedure A)

To each of a pair of 250-ml Erlenmeyer flasks add 10 mg of Keltrol dispersed in distilled water. To a second pair of flasks, add 10 gm of the food product. To a third pair of flasks add 10 mg of Keltrol plus 20 gm of the food product.

Add distilled water to bring the volume to about 75 ml. Add 6 gm of NaCl. Heat the mixture for 10 min in a boiling water bath, cool it to about 70°C and add 25 ml of the papain-cysteine·HCl-buffer mixture. Incubate the mixture at 70°C for 16 hr. Close each flask with a hard rubber stopper and tape on the stopper to prevent it from blowing off during incubation.

At the end of the incubation period, add 10 ml of *M* $CaCl_2$, and 1 gm of Hyflo Super Cel. After 5 to 10 min, filter the mixture through Reeve angel 202 paper, rinse the flask with four 25-ml portions of 0.01 *M* $CaCl_2$ and pour the rinsings over the residue on the filter paper. Collect the filtrate and washings in a 500-ml Erlenmeyer flask, add 1 gm of CPC, 1 gm of Hyflo Super Cel and 5 ml of *M* $CaCl_2$. Stir the mixture well and, after 10 min, pour it through Reeve Angel 202 filter paper, rinse the flask with five 100-ml portions of Solution A and pour the rinsings over the residue on the filter paper. Collect the filtrate and washings (about 750 ml) in a 1000-ml beaker, stir it well and incubate it at 35°C for 15 min. Filter the contents of the beaker through the Hyflo Super Cel column at 15 to 20 ml per minute, and return the first 50 ml of filtrate to the column. After all liquid has passed through, wash the column with Solution A until a 2-ml aliquot of the filtrate gives no color with the phenol-H_2SO_4 reagent and then wash with 200 ml of 0.2 *M* Na_2SO_4, (v/v). Pour hot (80°C) 30% H_2SO_4 (v/v) over the column in about 25-ml aliquots and collect 200 ml of the acidic eluate in a 200-ml volumetric flask. Cool the flask and contents to 30°C and make the volume up to the mark with 30% H_2SO_4, if necessary.

Determination of Keltrol in the Presence of Locust Bean Gum (Procedure B)

Keltrol reacts with locust bean to form a characteristic gel. Such gel formation is used as one of the official tests for the detection of

Keltrol. Salts, especially potassium and calcium chloride, increase the viscosity of Keltrol considerably (Glicksman 1969). In Procedure A, when cetylpyridinium chloride is added to a mixture containing NaCl, $CaCl_2$, locust bean gum, and Celite, the gum cannot be recovered from such a mixture, by direct filtration over paper or by centrifugation and pouring off the supernatant. Apparently, the resulting highly viscous mixture is adsorbed onto the Celite and is lost in the preliminary filtration steps. It is necessary, therefore, to employ a special process for the isolation and determination of Keltrol in the presence of locust bean gum.

Details of the Procedure.—After digestion with papain, add 10 ml of *M* $CaCl_2$ to the flask. Mix the contents well and, after about 10 min pour the mixture over the sand-glass wool column (Figure 15.8) and collect the filtrate in a 300-ml Erlenmeyer flask. Wash the column four times with 25-ml portions of 0.01 *M* $CaCl_2$, collect the washings and combine them with the filtrate. Add 1 gm of cetylpyridinium chloride and 5 ml of *M* $CaCl_2$ to the filtrate and washings. Stir the mixture well. Incubate it for 15 min at 45°C and then pour it over the sand-glass wool column, return the filtrate to the column and filter it again. Finally, wash the column with 250 ml of 0.1% CPC-0.01 *M* NaCl, collect the filtrate and washings in a 1,000 ml beaker. Dilute the liquid to 750 ml with distilled water, incubate it for 15 min, at 35°C and filter it. Elute the precipitate and determine the Keltrol as in Procedure A. The procedures for this operation and for that in the absence of locust bean gum are summarized in Fig. 15.9 and 15.10, respectively.

Colorimetric Assay of the Acidic Eluates.—Aliquots of the eluates are tested with following reagents. (1) The anthrone reagent: 2 gm of anthrone in 100-ml of concentrated H_2SO_4; heat 3 ml of the eluate in a boiling water bath with 5 ml of the reagent for exactly 10 min. Cool to 30 ± 1°C and measure the absorbance of the color developed at 620 nm. (2) The phenol-H_2SO_4 reagent: add 1 ml of 5% phenol in water to 2 ml of the eluate. Add rapidly to the mixture 5 ml of concentrated H_2SO_4. Leave the mixture at room temperature for 10 min, cool it to 29 + 1° C, and measure the color developed at 490 nm. (3) The chromotropic acid reagent: dissolve 150 mg of chromotropic acid (1,8-dihydroxynaphthalene-3-6-disulfonic acid) in 100 ml of concentrated H_2SO_4. Heat 1 ml of the sample and 9 ml of the reagent in a boiling water bath for 90 min, cool it, and measure the color developed at 570 nm.

Prepare standard curves for each reagent after reaction with CMC and Keltrol (Fig. 15.11).

Calculations.—Based on the procedures outlined, the following formulations may be written.

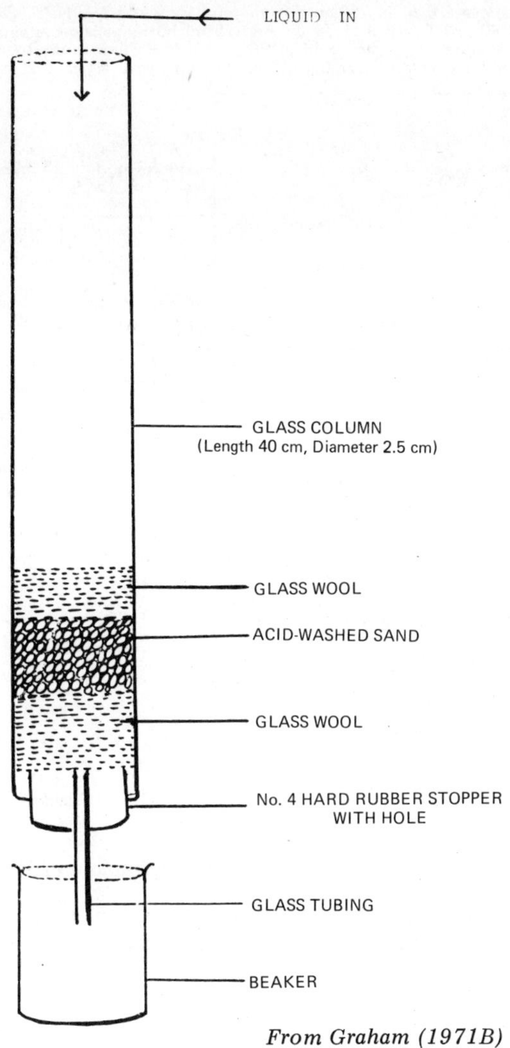

From Graham (1971B)

FIG. 15.8. SAND-GLASS WOOL COLUMN FOR FILTERING MIXTURE CONTAINING KELTROL AND LOCUST BEAN GUM

Let OD at 570 nm (chromotropic acid) = CMC in sample
Let OD at 490 nm (Phenol-H_2SO_4) = CMC + Keltrol in sample
Let OD at 620 nm (Anthrone) = CMC + Keltrol

The amount of CMC in the sample is determined with a standard curve prepared by reating varying amounts of CMC with the chromotropic acid reagent.

Since optical density is additive, and the amount of CMC in the

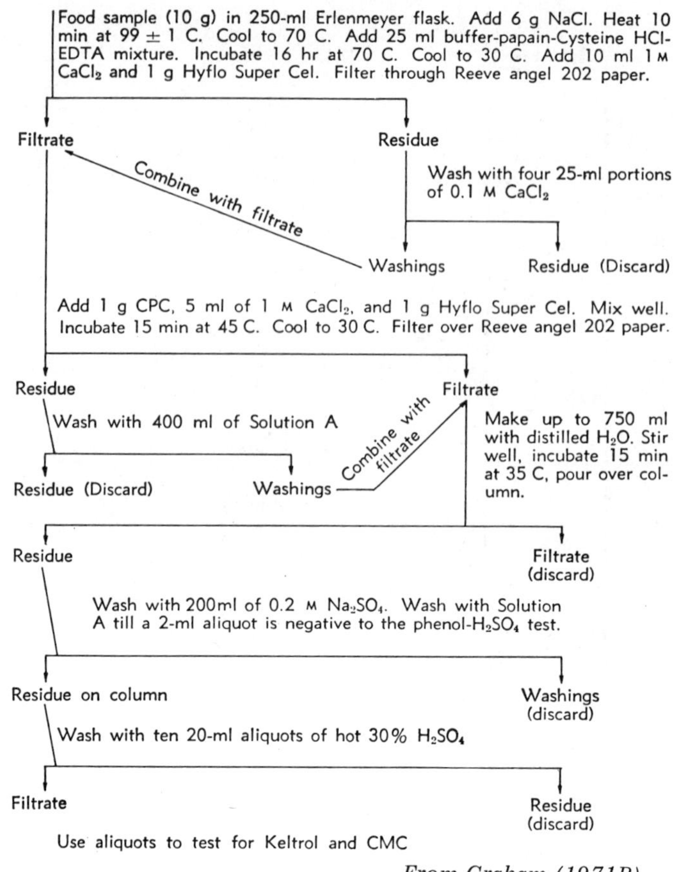

From Graham (1971B)

FIG. 15.9. FLOW SHEET FOR ISOLATION OF KELTROL IN ABSENCE OF LOCUST BEAN GUM (PROCEDURE A)

mixture is known from the chromotropic acid test, optical density contributed by CMC in the phenol-H_2SO_4 test or the anthrone test can be readily determined from standard curves.

In the phenol-H_2SO_4 test:

Let OD contributed by CMC = B

Let the total OD (OD CMC + OD Keltrol) = A

Then OD contributed by Keltrol only = A – B = C

From the phenol-H_2SO_4 standard curve for Keltrol, the amount of Keltrol corresponding to an optical density equal to C, can be readily determined. An example will suffice: 10 mg of CMC plus 10 mg of Keltrol were added to 10 gm of milk and the gums separated and determined by the suggested method. The 30% H_2SO_4 eluate = 200 ml.

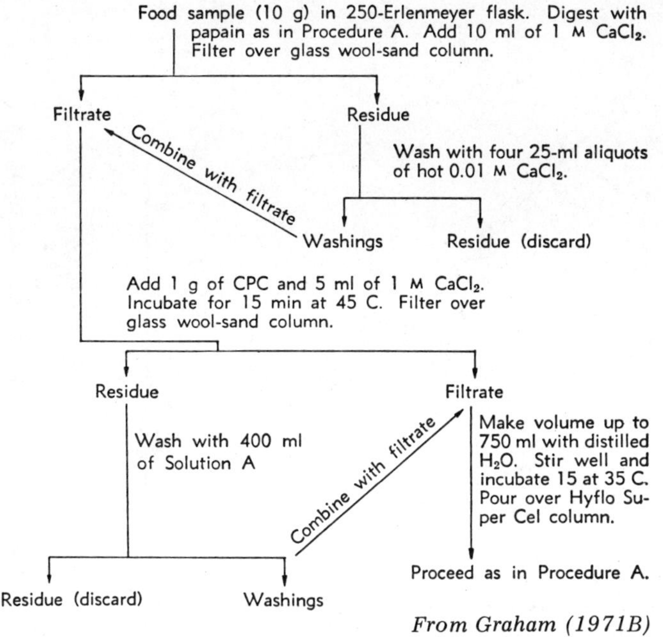

From Graham (1971B)

FIG. 15.10. FLOW SHEET FOR SEPARATION OF KELTROL IN PRESENCE OF LOCUST BEAN GUM (PROCEDURE B)

The optical density (of CMC) by the chromotropic acid test was found to be = 0.35 = 50 μg of CMC (1 ml used). The optical density (of CMC + Keltrol) by the phenol H_2SO_4 = 0.62 (A) (2 ml of filtrate used). Fifty micrograms of CMC will give an OD of 0.18 = (B) by the phenol-H_2SO_4. Therefore, since optical density is additive, the OD contributed by Keltrol = A - B = 0.62 - 0.18 = 0.44.

Now, from the standard curve for Keltrol by the phenol-H_2SO_4 test, OD of 0.44 = 82.5 μg of Keltrol.

The total amount of Keltrol present = 82.5 × 100 (dilution factor of 100),

= 8,250 μg

= 82.5 mg

= 82.5% recovery

Similarly, the amount of Keltrol present and the percent recovery could be calculated, if the anthrone test were used. Data on the recovery of Keltrol from milk are summarized in Table 15.14.

GENERAL COMMENTS

Accurate determination of any one of the hydrocolloids discussed depends on the use of a proper or appropriate standard for the prepa-

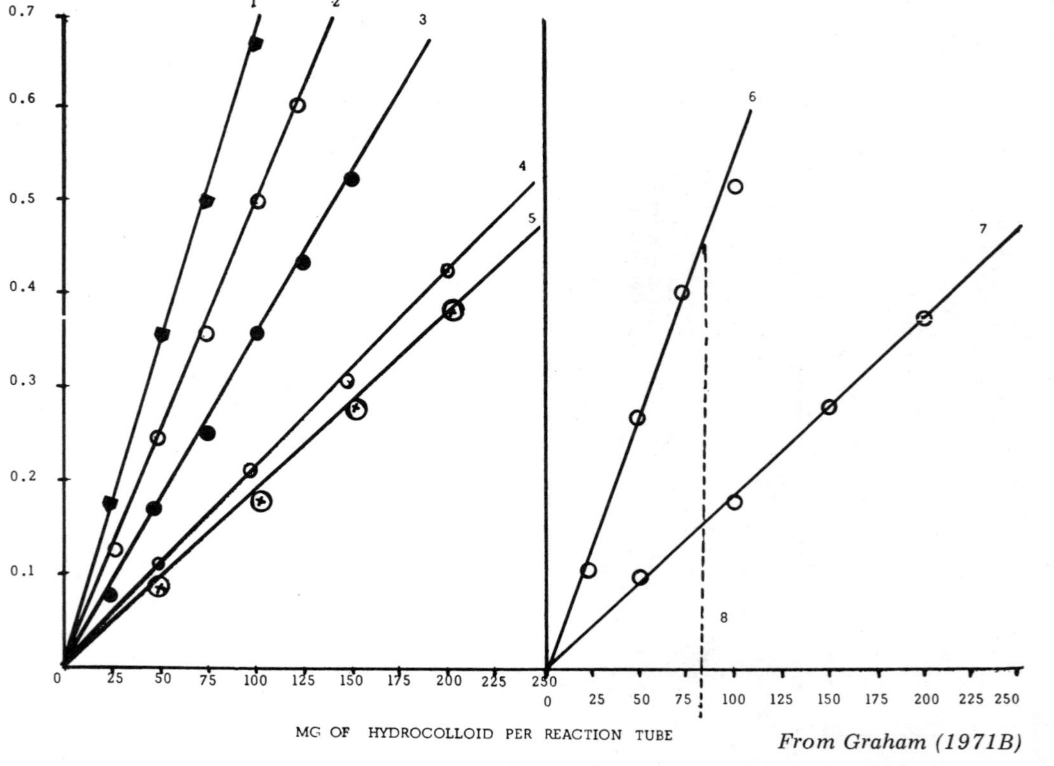

From Graham (1971B)

FIG. 15.11. STANDARD CURVES FOR DETERMINATION OF CARBOXYMETHYL CELLULOSE AND KELTROL

1 = CMC—Chromotropic acid (570 mμ); 2 = CMC + Keltrol (1:1 mixture) Phenol-H_2SO_4 (490 mμ); 3 = CMC – Phenol-H_2SO_4 (490 mμ); 4 = CMC – Anthrone (630 mμ); 5 = CMC + Keltrol (1:1 mixture) Anthrone (620 mμ); 6 = Keltrol – Phenol – H_2SO_4 (490 mμ); 7 = Keltrol – Anthrone (620 mμ); 8 = Intersecting line illustrating calculation of Keltrol in CMC – Keltrol mixture.

TABLE 15.14

RECOVERY OF KELTROL FROM MILK IN THE PRESENCE OF OTHER FOOD GUMS (10 MG OF KELTROL AND 10 MG OF EACH GUM ADDED TO 10 GM OF MILK)

	Keltrol Recovered[1]		Recovery	
Gum	Range (mg)	Average (mg)	A[2] (%)	B[3] (%)
None	8.7-9.5	8.9	89	94.5
Carrageenan	8.4-9.0	8.6	86	91.0
Furcellaran	8.5-9.2	8.7	87	93.0
Alginate	8.7-9.3	8.9	89	94.5
Pectic acid	8.4-9.2	8.7	87	93.0
Carboxymethyl cellulose	8.6-9.2	8.8	88	93.5
Quince seed mucilage	8.5-9.3	8.7	87	93.0
Locust bean gum	8.4-9.3	8.6	86	91.0
Starch	8.0-9.1	8.4	84	90.0
Pectin	8.6-9.5	8.8	88	93.5
Gum arabic	8.5-9.1	8.7	87	93.0
Gum tragacanth	8.7-9.3	8.9	89	94.5
Gum ghatti	8.6-9.0	8.5	85	90.4
Gum karaya	8.3-9.1	8.7	87	93.0
Gum guar	8.5-9.2	8.8	88	93.5
Agar	8.5-9.3	8.7	87	93.0
Gelatin	8.4-9.4	8.8	88	93.5
Phosphomannan Y-2154	8.6-9.2	8.9	89	94.5
Phosphomannon Y-2448	8.5-9.1	8.8	88	93.5

Source: Graham (1971B).
[1] Average of 5 different samples.
[2] Based on 10 mg added to milk.
[3] Based on recovery of 10 mg of Keltrol added to water and passed through the entire process.

ration of the "standard curve." However, due to the various modifications which have been made in some of these hydrocolloids to suit particular industrial uses, it might be rather difficult at times for a laboratory to have available a sample of the identical food gum which was used in the food sample being analyzed. In addition, during processing and storage of the products, and depending on the conditions obtaining during these periods, it is not improbable that changes might occur in the hydrocolloid. Therefore, even if an identical sample of the hydrocolloid was available for the preparation of the calibration curve, the use of such a curve for calculating quantities of the hydrocolloid isolated from a commercial product might involve some uncertainties. A recovery test should always be done, using the "standard" hydrocolloid.

Since the hydrocolloid dispersions are subject to microbial degradation and/or contamination they should be stored under refrigeration.

Specific methods for the quantitative determination of other plant hydrolloids such as gum arabic, gum karaya, locust bean gum and gum tragacanth are still lacking. Difficulties here are due partly to many common reactions given by these gums when quantitative separation is attempted.

Well-established methods are available for the determination of the pectic substances as mentioned in Chapter 10, which deals with this hydrocolloid.

BIBLIOGRAPHY

AM. PUBLIC HEALTH ASSOC. 1960. Standard Methods for the Examination of Dairy Products, 11th Edition. American Public Health Assoc., U.S. Govt. Printing Office, Washington, D.C.

ANON. 1969. Xanthan gum. Federal Register, Sect. 121.1224, 5376-5377. Mar. 19.

BUNDESEN, H. N., and MARTINEK, M. J. 1954. Procedures for the separation, detection and identification of the more common vegetable gums in dairy products. J. Milk Food Technol. *17*, 79-81.

CHANG, J. C., RENOLL, M. W., and HANSEN, P. M. T. 1974. Zone electrophoresis of food stabilizers in malonate buffer. J. Food Sci. *39*, 97-102.

DUBOIS, M., GILLIES, K. A., and HAMILTON, J. K. 1956. Colorimetric method for determination of sugars and related substances. Anal. Chem. *28*, 350.

FEIGL, F. 1966. Spot Tests in Organic Analysis, 7th Edition. Elsevier Publishing Co., New York.

GLICKSMAN, M. 1969. Gum Technology in the Food Industry. Academic Press, New York.

GRAHAM, H. D. 1966. Spectrophotometric determination of carrageenan ester sulfate in milk and milk products with barium chloranilate. J. Dairy Sci. *49*, 1102-1108.

GRAHAM, H. D. 1967. Reaction of tetrazolium salts with sulfated polysaccharides and other hydrocolloids. J. Food Sci. *32*, 489-495.

GRAHAM, H. D. 1968. Quantitative determination of carrageenan in milk and milk products using papain and cetyl pyridinium chloride. J. Food Sci. *33*, 390-394.

GRAHAM, H. D. 1969. Determination of alginate in dairy products. J. Dairy Sci. *52*, 443-448.

GRAHAM, H. D. 1970. Specificity of the Ferric-H_2SO_4 reagent for alginates. J. Food Sci. *35*, 494.

GRAHAM, H. D. 1971A. Determination of carboxymethylcellulose in food products. J. Food Sci. *36*, 1052-1055.

GRAHAM, H. D. 1971B. Microdetermination of Keltrol (xanthan gum). J. Dairy Sci. *54*, 1622-1628.

GRAHAM, H. D. 1972A. Ortho-tolidine and sodium hypochlorite for the determination of carrageenan and other ester sulfates. J. Dairy Sci. *55*, 1675-1682.

GRAHAM, H. D. 1972B. Determination of carboxymethyl cellulose with chromotropic acid. J. Dairy Sci. *55*, 42-50.

GRAHAM, H. D., and WILLIAMS, J. L. 1966. Quantitative aspects of the interaction of carrageenan and other hydrocolloids with polyvalent cobalt complexes. J. Food Sci. *31*, 362-372.

HANSEN, P. M. T. 1966. Distribution of carrageenan stabilizers in milk. J. Dairy Sci. *49*, 698.

HANSEN, P. M. T., and WHITNEY, R. McL. 1960. A quantitative test for carrageenan ester sulfate in milk products. J. Dairy Sci., *43*, 175-186.

HANSEN, P. M. T., and CHANG, J. C. 1968. Quantitative recovery of carboxymethyl cellulose from milk. J. Agr. Food Chem. *16*, 77.

HERCULES POWDER CO. 1963. Analytical Procedures for Assay of CMC and its Determination in Formulations. Cellulose and Protein Products Dept., Hercules Powder Co., Wilmington, Del.

JOHNSON, B. C. 1973. Private Communication.

JOHNSON, R. H. 1956. Report on gums in process cheese spreads. J. Assoc. Offic. Agr. Chem. *39*, 286-290.

SCOTT, J. E. 1956. The preparation and fractionation of acidic polysaccharides using long-chain quaternary ammonium compounds. Biochem. J. *62*, 31P.

SCOTT, J. E. 1960. Aliphatic ammonium salts in the assay of acidic polysaccharides from tissues. *In* Methods of Biochemical Analysis, D. Glick (Editor). Interscience Publishers, New York.

SLACK, G. B. 1958. Connective tissue growth stimulated by carrageenin. 3. The nature of and amount of polysaccharides produced in normal and scorbutic guinea pigs and the metabolism of a chondroitin sulphuric acid fraction. Biochem. J. *69*, 125-134.

SMILEY, K. L. 1966. Microbial polysaccharides—a review. Food Technol. *20*, 112-116.

SZALKOWSKI, C. R., and MADER, W. J. 1955. Determination of sodium carboxymethyl cellulose. J. Am. Pharm. Assoc. (Sci. Ed.) *46*, 533.

Index

LIBRARY
ATHROFA GOGLEDD DDWYRAIN CYMRU
...E NORTH EAST WALES INSTITUTE
...her education CONNAH'S QUAY
...E, CLWYD CH5 4BR

29 SEP 1977

30 MAR 1983

11 MAY 1988

13 DEC 1988